The Abel Prize 2008-2012

Niels Henrik Abel 1802–1829
The only contemporary portrait of Abel, painted by Johan Gørbitz in 1826

Helge Holden · Ragni Piene
Editors

The Abel Prize 2008-2012

ABEL PRISEN

DET NORSKE VIDENSKAPS-AKADEMI

Springer

Editors
Helge Holden
Department of Mathematical Sciences
Norwegian University of Science and Technology
Trondheim, Norway

Ragni Piene
Department of Mathematics
University of Oslo
Oslo, Norway

Photos on Part 1
Photos of Thompson and Tits (Frontpage) by Heiko Junge/Scanpix
Photo of Thompson (CV) by Renate Schmid/Oberwolfach Photo Collection
Photo of Tits (CV) by Jean-François Dars/CNRS Images
Photos on Part 2
Photo of Gromov (Frontpage) by Erlend Aas/Scanpix
Photos on Part 3
Photo of Tate (Frontpage) by Berit Roald/Scanpix
Photo of Tate (CV) by Charlie Fondville
Photos on Part 4
Photo of Milnor (Frontpage) by Scanpix
Photo of Milnor (CV) by Knut Falch/Scanpix
Photos on Part 5
Photo of Szemerédi (Frontpage) by Erlend Aas/Scanpix
Photo of Szemerédi (CV) by Harald Hanche-Olsen
Photos on the cover of the book:
Photos of Thompson, Tits and Szemerédi by Harald Hanche-Olsen
Photo of Milnor by Knut Falch/Scanpix
Photo of Tate by Scanpix
Photo of Gromov by Knut Falch/Scanpix

Videos to this book can be accessed at http://www.springerimages.com/videos/978-3-642-39449-2.

ISBN 978-3-642-39448-5 ISBN 978-3-642-39449-2 (eBook)
DOI 10.1007/978-3-642-39449-2
Springer Heidelberg New York Dordrecht London

Library of Congress Control Number: 2013955612

Mathematics Subject Classification: 00-XX

Printed on acid-free paper

Springer is part of Springer Science+Business Media (www.springer.com)

Au reste il me paraît que si l'on veut faire des progrès dans les mathématiques il faut étudier les maîtres et non pas les écoliers.

Niels Henrik Abel[1]

[1] "It seems to me that if one wants to make progress in mathematics, one should study the masters, not the pupils." In: "Mémoires de Mathématiques par N.H. Abel", Paris, August 9, 1826, in the margin of p. 79. Original (Ms.fol. 351 A) in The National Library of Norway. Reprinted with permission.

Preface

In 2002, the year marking the bicentennial of Abel's birth, the Norwegian Parliament established the Niels Henrik Abel Memorial Fund with the objective of creating an international prize for outstanding scientific work in the field of mathematics—the Abel Prize.

In this book we present the Abel Laureates for the period of 2008–2012. It is a sequel to our first book.[1]

The present volume starts with the history of the Abel Prize written by the historian Kim Helsvig. He had unrestricted access to all sources and interviewed the key people involved in the creation of the Abel Prize, and he presents his independent view on the prize.

There is one chapter for each of the years 2008–2012. Each chapter starts with an autobiographical piece by the laureate(s). Then follows a text on the laureate's work: R. Lyons and R.M. Guralnik write on the work of John G. Thompson while F. Buekenhout writes on the work of Jacques Tits. The work of Mikhail Gromov is described by M. Bestvina, D. Burago, F. Forstnerič, L. Guth, Y. Eliashberg, A. Nabutovsky, A. Phillips, J. Roe and A. Vershik, coordinated and edited by D. Burago and Y. Eliashberg. J. Milne writes about the work of John Tate. H. Bass, M. Lyubich, and L. Siebenmann treat different aspects of John Milnor's work. Finally, T. Gowers describes the work of Endre Szemerédi. Each chapter contains a complete bibliography and a curriculum vitae, as well as photos—old and new.

Each year Martin Raussen and Christian Skau interviewed the laureate in connection with the Prize ceremonies, and the interviews have been broadcast on Norwegian national television. The interviews can be streamed from the Abel Prize web site www.abelprize.no or the Springer web site http://www.springerimages.com/videos/978-3-642-39449-2. Transcripts of all interviews have been published in the *EMS Newsletter* and the *Notices of the AMS*.

We have included a facsimile of a letter dated September 25, 1828, from Niels Henrik Abel to his publisher August Leopold Crelle. The handwritten letter in Ger-

[1] H. Holden, R. Piene (eds.) *The Abel Prize 2003–2007. The First Five Years*, Springer, Heidelberg, 2010.

man is transcribed and also translated into English. More interestingly, the letter is set into a historical perspective by Christian Skau. In 2007 the original was procured from Institut Mittag-Leffler, Stockholm, Sweden, thanks to a generous donation by Alf Bjørseth and the company Scatec. The original letter resides at the National Library of Norway in Oslo. Finally, we have included an update to the first volume in this series.

We would like to express our gratitude to the laureates for collaborating with us on this project, especially for providing the autobiographical pieces and the photos. We would like to thank the mathematicians who agreed to write about the laureates, and thus are helping us in making the laureates' work known to a broader audience.

Thanks go to Martin Raussen and Christian Skau for letting us use the interviews and to Marius Thaule for his LaTeX expertise and the preparation of the bibliographies.

The technical preparation of the manuscript was financed by the Niels Henrik Abel Board.

Oslo, Norway
March 6, 2013

Helge Holden and Ragni Piene

Contents

The Abel Prize—The Missing Nobel in Mathematics?

Kim G. Helsvig

By the spring of 2001, the lobbying to establish a prize in mathematics in memory of the Norwegian mathematician Niels Henrik Abel (1802–1829) was well underway. On May 10, Professor Arnfinn Laudal from the Department of Mathematics at the University of Oslo sent an e-mail to the president of the International Mathematical Union, Jacob Palis. Laudal described a campaign that was recently launched by some activists in the department to establish a prize in mathematics. They hoped it would persuade the international mathematical community and Norwegian politicians "... to present the Abel Prize as the 'missing Nobel Prize' in mathematics".[1]

Until quite recently, no mathematician had suggested that the time was right to establish a scientific prize in mathematics comparable to the various Nobel Prizes. It is true that there existed several international prizes in mathematics, but the obvious absence of a prize that could be compared with the most prestigious prizes in physics, chemistry and medicine had been a source of frustration among mathematicians ever since the Nobel Prizes were first awarded in 1901. The mathematical community had apparently learned to live with this frustration, but when the initiative to establish a prize in mathematics on the same level as the Nobel was suddenly introduced from the mathematical periphery of Norway, it was a wakeup call that resonated with mathematicians. The initiative would in a short time gain support, not only among Norwegian mathematicians, but also in the international mathematical community as well as from influential sectors of Norwegian society. Only after a few months of the campaign being launched in the spring of 2001, the Norwegian government decided to set aside NOK 200 million (approximately USD 35 million)

[1]E-mail from Professor Arnfinn Laudal, Department of Mathematics, University of Oslo to the president of IMU Jacob Palis, 10.5.2001. Nils Voje Johansen's private archive, file "Abel-pris" (hereafter NVJ).

Also published in *Centaurus*, DOI 10.1111/1600-0498.12038.

K.G. Helsvig (✉)
Schønings gt. 14B, 0362 Oslo, Norway
e-mail: k.g.helsvig@gmail.com

H. Holden, R. Piene (eds.), *The Abel Prize 2008-2012*,
DOI 10.1007/978-3-642-39449-2_1, © Springer-Verlag Berlin Heidelberg 2014

of the rapidly increasing national petroleum fund to establish the Niels Henrik Abel Memorial Fund, and in June 2003 King Harald V presented the first Abel Prize of NOK 6 million to Jean-Pierre Serre—one of the most influential mathematicians of the twentieth century—during a solemn ceremony in the university assembly hall in an Abel-celebrating Oslo city center in full spring blossom.

The successful campaign was based on the mobilization of both the Norwegian history of mathematics and the history of the dissolution of the political union between Sweden and Norway in 1905. This mobilization was presented to the international mathematical community, which had waited for one hundred years for a scientific prize at the highest level, and to the Norwegian political community, who for several years had expressed increasing concerns at the lack of interest and competence in the natural science arena throughout the entire national schooling and higher education systems. The way in which this was done might explain why the campaign was well received by both the vast majority of the international mathematical community and Norway's politicians, but it might also explain why it seriously irritated the Swedish Nobel Foundation.

In this article, I hope to shed light on the dynamics of the history of the intensive campaign that led to the establishment of the Abel Prize in 2002. This investigation also leads to more tentative considerations regarding to what degree the prize has lived up to the high expectations placed on it during the first decade of its existence.

1 Science Prizes in Historical Perspective

Historians and sociologists of science have long noted the rich variety and changing nature of prizes in the world of science. Although national academies of science might occasionally award prizes to foreigners, the notion of a truly international prize first emerged with the establishment at the turn of the twentieth century of the Nobel prizes [3, 5]. Although Swedish and Norwegian national institutions would be responsible for awarding the various prizes, Alfred Nobel's testament underscored that they should be awarded based on merit whether or not the candidates were Scandinavian or not. This clearly stated goal, as well as the implementation a system of nomination of candidates that at the time was unprecedented in its international scope, reinforced the assumption of a prize that could serve as a so-called level playing field on which the so-called civilized nations of the world could compete peacefully to prove their cultural strength and vitality [5, 6, pp. 158–162]. Although the prize committees remained national, and although recent historical studies have demonstrated a range of biases that have entered into deliberations on awarding prizes, the will to believe in a truly impartial international prize, from the very start in 1901, tended to slow down criticism and skepticism. National scientific communities and competing elite institutions within a country generally understood the importance of international recognition as a resource for local prestige and benefits [6].

The emergence and rapid acceptance of international prizes by significant national scientific communities occurred at a time when internationalism blossomed in academic cultures. The rise during the 19th century of international organizations and international commissions for science not only fomented their supposed missions of the advancement of knowledge through facilitating communication and collaboration, they also enabled national scientific leaders to appeal locally for funds and authority based on their international activities and recognition. Appeals to the trans-national republic of letters to gain prestige in local cultural politics was nothing new, but seemingly such tendencies grew in intensity starting in the late nineteenth century and interrupted only occasionally during the First World War [6].

During the last quarter of the 20th century the number of international prizes in science really started to proliferate. Wealthy patrons of culture gladly donated fortunes to endow prizes that would carry their name in the name of advancing and stimulating science. Several such prizes attempted to outdo the Nobels in the amount of money to be awarded; some clearly aimed to bring prestige to the nation awarding the prizes. Universities and state ministries for research and higher education gladly turned to such alleged measures of excellence in their efforts to support national science [16].

The cult of excellence, as defined by prizes and awards, was largely a feature of American science and academia for much of the 20th century. Especially during the last two decades of the century, it spread to Europe and Asia. Although this phenomenon has been noted, its causes seem complex and not fully explored. New management thinking, an emerging global economy of higher education and research, higher costs for doing research and universities confronting post-Cold War reductions of state-funding of science seem to be some of the most obvious factors. Similarly the rise of global media coverage such as CNN, BBC World etc. helped transform the awarding of enormously huge prizes and award ceremonies into news events to an even greater extent than previously. Some of the prizes has thus became potentially much more powerful marketing tools for both institutions and nations alike in en ever more fierce competition for status, prestige and funds in the academic world.

The Abel prize plans were introduced in the context of this increasingly overheated culture and economy of international prizes in science. And the eventual birth of a major international prize for mathematics in a nation not normally considered a major player in international science was marked by multiple intentions and motivations, and what might be seen as pure luck. The story of how the Abel prize came about is a story of tight networks of contacts among academics, politicians, bureaucrats and industrial leaders in a small nation. It is at once both a piece of contemporary national history, offering a preliminary glimpse behind the scenes of how things might happen in a small and wealthy nation, as well as an account of how cultural heritage can be recruited as a resource for science policy: Here one of Norway's legendary highly-gifted mathematicians would be recruited for a wide spectrum of cultural-political purposes from differing constituencies with a variety of interests some 170 years after his death.

The article is based on a variety of sources: Private archives—to a large extent e-mail archives, project drafts and personal notes—from the central mathematicians involved in the process, interviews with relevant mathematicians, bureaucrats and politicians, Norwegian and international newspaper articles and magazines and primary sources from the Norwegian Ministry of Education and Research and the Norwegian Academy of Science and Letters. Most of the key actors in the Norwegian mathematical community seem to have been quite aware that the process might be of interest for historians in the future, and most of them have kept personal—and often overlapping—archives and notes almost in the wait of an historian knocking on their door. These archives have provided not only a very rich mosaic of primary sources, but also glimpses from different angles into the most important contacts with Norwegian politicians and bureaucrats as well as with international mathematical communities during the process. Seen together with the rest of the source material—especially documents and official records from the Ministry of Education and Research—these personal archives have proven most valuable for the study. Problems related to the study of contemporary science, including the challenges of source materials are further discussed in many of the essays in *The Historiography of Contemporary Science, Technology, and Medicine: Writing Recent Science* [4].

2 A National Icon

During the nineteenth century, Niels Henrik Abel became an important part of the Norwegian national narrative of the country's unique history and cultural identity, a narrative that became ever more important in the years leading up to the dissolution of the political union with Sweden in 1905. At the centennial of his birth in 1902, and three years before the break from Sweden, Abel was celebrated as a national hero in the Norwegian capital Christiania (renamed Oslo in 1925). In a letter to the Norwegian 1903 Nobel Prize in Literature laureate Bjørnstjerne Bjørnson, the scientist, polar hero and later Nobel Peace Prize laureate Fridtjof Nansen wrote as chairman of the Abel celebrations, "For me it stands as a duty to make the most of an event like the birth of Abel in our nation; by holding this up for the rest of the world, we prove our right to exist as an independent state" [10, p. 5].

The celebration was the perfect occasion to display the desire for national independence. When the Swedish and Norwegian king, Oscar II, invited 500 guests to supper at the Royal Palace, the students at the Royal Frederick University—later the University of Oslo—seized the opportunity and arranged the greatest torchlight parade ever seen in the city. The university invited mathematicians from all over the world and appointed numerous honorary doctors. A memorial publication was published in both Norwegian and French, and the National Theater put on a celebratory performance of the Norwegian play *Peer Gynt* by Henrik Ibsen (now the most well-known Norwegian play), followed by the national anthem sung by the entire auditorium [7]. A competition was also called to create an Abel memorial monument. Six years later—and three years after the dissolution of the political union with Sweden—a pedestaled monument by Gustav Vigeland nearly forty feet high was erected in the Royal Gardens (in what later came to be known as Abelhaugen) close to the university buildings.

It is in no doubt that Abel played an important role as a symbol of Norway's growth as an independent nation from the beginning of the twentieth century. He was perhaps most important as an academic ideal in a period when the natural sciences especially received both scientific and economic importance and prestige in Norway [12]. On the centenary of Abel's death in 1929, a series of Abel stamps were issued—an honor previously only given to royalty and Henrik Ibsen—and again international honorary doctors were appointed at the university. Abel's portrait was also printed on the Norwegian 500 kroner bill between 1948 and 1991, and many Norwegian cities and towns have streets and squares named after him. In 1966 the mathematicians at the University of Oslo, where Abel himself was a student from 1821 to 1825, moved into the new Niels Henrik Abel building at the modern Blindern campus on the western outskirts of the city center.

Abel's name was once again brought to the forefront in the late 1990s as part of plans for two major international mathematical events in Norway. The Abel bicentennial in 2002 was approaching while the International Mathematical Union (IMU) made preparations for the International Congress of Mathematicians to be held the same year. The IMU leadership was aware that China would apply to host the world congress, but the international union feared that such an event would be a highly controversial or even impossible because of China's human rights situation. In 1995, a professor of mathematics at the University of Oslo, Jens Erik Fenstad, shared a seat with the then IMU secretary and later IMU president Jacob Palis on the executive board of the International Council of Scientific Unions. Palis asked Fenstad if it was possible for Norway as both a politically and mathematically uncontroversial nation to prepare a back-up application to host the 2002 IMU International Congress of Mathematicians as a plan B if it proved politically unattainable to proceed with the original plan to hold the congress in Beijing.[2]

In a relatively short time, it became clear that the IMU plans were going ahead as scheduled, and that China would host the 2002 congress. The Norwegian preparations for a back-up plan, which had been started shortly after Palis' request, were soon transformed into plans for a major international mathematical conference in memory of Abel. A working committee of five mathematicians from the University of Oslo and the Norwegian University of Science and Technology in Trondheim was put in place and given the task to make appropriate preparations for a "*Niels Henrik Abel Bicentennial Conference*" in Oslo in the spring of 2002. This committee also made preparations for Norway's role in the World Mathematical Year 2000. Even though Abel's name and history was prominent in the preparations for both these events, it never occurred to anybody in the Norwegian mathematical community that these occasions could be used to advocate the creation of an international prize in mathematics in memory of Abel. One hundred years earlier, major efforts had been made to do so when Norwegian mathematicians had launched an extensive campaign to establish a prize as part of the Abel centennial celebrations in 1902.

---

[2]Jens Erik Fenstad, undated and unpublished, "*3.4 Abelåret og Abelprisen*" in "*Noen notater om matematikken etter annen verdenskrig*", in Jens Erik Fenstad's private archive (hereafter JEF).

3 The Initiative

The new initiative to create a prize in memory of Abel was presented from a different angle by the author and poet Arild Stubhaug. Stubhaug had published several poetry collections, but he had also studied mathematics at the University of Bergen. As a passionate stamp collector, Stubhaug was from an early age acquainted with and fascinated by Abel's life. Abel had died at the age of 26, only two days before the Frederick William University in Berlin—later known as Humboldt University—was to offer him the tenured position as professor he needed to be able to fully concentrate on his mathematical work. Abel also died convinced that his main work, the so-called Parisian manuscript, was lost forever. The same day as the letter was posted offering him the professorate in Berlin, the manuscript was found in Paris. More than anything this work laid the foundation for Abel's posthumous fame and acknowledgment both in Norway and in the international mathematical community [9, 14].

Arild Stubhaug had for a long time wanted to write about Abel. From the early 1980s he lived in Arendal, close to Gjerstad where Abel spent his childhood years, and not far from Froland where the mathematician died. In 1988 Stubhaug was encouraged by the publishing house Aschehoug to go forward with his plans. The research and writing took him eight years and resulted in the critically acclaimed biography *Et foranskutt lyn* (the English version, *Called Too Soon by Flames Afar: Niels Henrik Abel and His Times*, was published in 2000). The response of the Norwegian daily newspapers was close to panegyrical and wrote of a "biographical masterpiece" and a "first class cultural biography".[3] The Notices of the American Mathematical Society several years later, when the book had been translated into English, wrote of "a great contribution to our knowledge of Abel and his time".[4] The research and the positive reception of the book brought Stubhaug in close contact with the Norwegian mathematical community in general, particularly with mathematicians at the University of Oslo. Shortly after the release of the Abel biography, the mathematicians in Oslo encouraged him to write a biography about a further great nineteenth century Norwegian mathematician, Sophus Lie (1842–1899). In 1997 Stubhaug got his own office next to the Oslo mathematicians in the Niels Henrik Abel building on Blindern campus.

During his work on the two biographies, Stubhaug collected a great deal of material detailing how Sophus Lie dedicated a good part of the last years of his life to establishing an international mathematical prize in acknowledgment of Abel. Lie's work was clearly inspired by the grand Nobel prize plans presented in 1897. Those plans did not include any prize in mathematics because Nobel first and foremost wanted to encourage the practical outcomes of science and technology, and found

[3] *VG*—24.9.1996 "Biografi-bragd", *Aftenposten*—24.9.1996, "Storslått Abel-biografi", *Dagbladet*—30.9.1996, "Kultur-biografi av ypperste rang".

[4] Book review by Jesper Lützen, "Called Too Soon by Flames Afar: Niels Henrik Abel and His Times", in *Notices of the American Mathematical Society*, August 2002 Volume 49 Issue 7.

the mathematical world far too theoretical and abstract. Sophus Lie received much support for his plans from the leading mathematical circles in Europe.

International support for the award was closely connected to Lie personally, and it essentially dwindled away after his death in 1899. Efforts to create an Abel Prize were now carried along within a national framework. In close cooperation with some of Lie's colleagues, the Scientific Society of Christiania (now the Norwegian Academy of Science and Letters), represented by W.C. Brøgger—the future rector of the university—and Fridtjof Nansen, developed statutes and rules for an Abel Prize in mathematics. The prize of a gold medal worth 1000 kroner was to be awarded every fifth year for "outstanding work within pure mathematics". The Nobel medals were at this time worth 500 kroner and awarded annually [10, pp. 6–8].

The Brøgger/Nansen initiative was financially supported by the king of Sweden and Norway, Oscar II, who took a personal interest in mathematics, and it was applauded by leading Norwegian mathematicians and cultural personalities. However, after the dissolution of the political union between the two countries in 1905, the Norwegian Academy was not able to collect the necessary funds to finance the prize in the relatively poor, independent Norway. The dissolution of the union was eventually peaceful, even though some days in September 1905 saw tens of thousands of troops lined up on both sides of the border. It was nevertheless financial constraints and not resentment against the Swedes that made it impossible to go on with the prize plans after the separation. W.C. Brøgger was by far the most influential organizational entrepreneur in Norwegian science in the first decades of the 20th century, and he was above all a great admirer of Swedish science and Swedish scientific institutions. Almost all of Brøgger's influential institutional work within Norwegian science in the first three decades of the 20th century was in fact modeled after Swedish ideals on how to provide ever more substantial funding for science and scientific institutions [2, pp. 19–22]. It was a great disappointment for a resigned Norwegian mathematical and scientific community that the money for a prestigious Norwegian prize in mathematics disappeared with the dissolution of the union with Sweden. Fridtjof Nansen wrote: "the Abel Prize that we had been promised by good King Oscar went to heaven with the union" [10, pp. 6–8].

Arild Stubhaug gave many lectures in the years following the publication of his biography on Abel. Time and again he told of the plans to establish an Abel Prize around the 1902 centennial, and time and again he told of how the plans were aborted with the death of the mathematician Sophus Lie and the dissolution of the union with Sweden. Stubhaug frequently asked, would one hundred years later not be the perfect opportunity to revitalize these plans and finally create a prize in memory of Abel? He aired his thoughts among a great number of Norwegian mathematicians, but even though several found the history and the idea interesting, the general attitude in the mathematical community was that an Abel Prize was, in Stubhaug's words, "extremely unrealistic" [13, p. 5].

Stubhaug's efforts to bring the idea of an Abel Prize back to life seemed to amount to nothing when suddenly things brightened up in the summer of 2000, nearly four years after the publication of the biography. The book had now been translated into English, and on July 1, Stubhaug was invited to a bookstore in the small coastal town of Risør, close to where Abel both grew up and died, to sign books on the final day of a yearly chamber music festival. Every year this festival attracts many tourists and potential customers to town. Abel's mother grew up in Risør in a house close to the store, and the owner had arranged for the translated biography to be sent by courier in time for the festival. Thus, the bookstore in the small harbor town of Risør became the first to sell the English edition of the Abel biography.[5]

When Stubhaug turned up at the bookstore, a few minutes late, he met the CEO of Telenor (a Norwegian telecommunications company) Tormod Hermansen on his way out the door with a copy of the Abel biography under his arm. Towards the end of the 1990s, the prominent Labour Party politician and industrial leader Hermansen was characterized as both "the most powerful bureaucrat in Norway" and "Norway's most powerful man" [15]. He was primarily known as one of the most influential architects of the far-reaching and controversial public sector reforms of the past couple of decades, heavily inspired by the principles of New Public Management. Hermansen had held different key positions in the state bureaucracy since the early 1970s, and as Under-secretary of State he led a profound reorganization of the Ministry of Finance towards the end of the 1980s. As chairman of the Government Bank Insurance Fund he was in charge of the rescue plan for private banks during the Norwegian bank crisis in the early 1990s, and had been CEO of Telenor (the former state-owned telephone monopoly Televerket) since 1995 [15].

Stubhaug and Hermansen were not previously known to each other, but they sat down to talk in the bookstore. During the conversation Stubhaug told Hermansen about the failed attempts to create an Abel Prize one hundred years earlier. As on numerous occasions during his lectures about the Abel biography, he made sure again to mention that the forthcoming bicentennial celebrations provided a new opportunity for such an award. According to Stubhaug, Hermansen took special notice of how the dissolution of the union between Sweden and Norway in 1905 had halted the plans for an Abel Prize.[6] As CEO of Telenor, Hermansen had his own and very recent experiences on how relations with Norway's Swedish neighbors could put an end to grand-scale plans. At the time of the meeting with Stubhaug, Hermansen had just finished protracted and failed negotiations regarding a merger between Telenor and the Swedish telecommunication giant Telia. This had been one of the most contested and controversial political issues in both countries in the past two years. After fierce political, administrative and personal struggles for power, the two companies broke an already signed merger deal. Hermansen took much of the blame for the failure [15]. The relationship with the Swedes should not also end the plans for an Abel Prize, and the Telenor CEO promised Stubhaug that he would do everything he could to grab this new opportunity. The two separated with mutual assurances that they would promote the cause with a new strength, Hermansen through his po-

[5]Interview with Arild Stubhaug, 7.2.2012. Telephone interview with Torkjel Gudmund Johansen, owner of Risør Bokhandel, 9.2.2012.

[6]Interview with Arild Stubhaug, 7.2.2012.

litical contacts and Stubhaug through the mathematicians at the University of Oslo [13, p. 5].

4 Mobilization

Shortly after the meeting in Risør, Stubhaug made his colleagues in the Department of Mathematics at the university aware of the new development. The department immediately got in contact with Hermansen. Together with Stubhaug, the head of department and head of office "looked into what had to be done to create a powerful prize with international recognition". There "was no Nobel Prize in mathematics". Abel "was a world known name that enjoyed the highest scientific recognition", and the department suggested that the establishment of a prize in Abel's name was something that Telenor might take interest in as part of their internationalization strategies.[7]

However, Tormod Hermansen did not envision Telenor as a sponsor of an Abel Prize. This was a matter of national importance that had to be presented to the Labour Party government. In early September 2000, Hermansen called the Minister of Education and Research, Trond Giske, and some days later the Telenor CEO sent the Minister a letter regarding "the Abel Prize". In the letter Hermansen expressed his hopes that Giske would get people truly exited about the idea. The enclosed memorandum from the Department of Mathematics stated that Abel's name was among the very greatest in "the history of mathematical-technical science". An Abel Prize would "highlight a research area that is at the core of all scientific progress, and as such it would create both attention and respect for Norway all around the world".[8]

The Ministry did nothing about Hermansen's initiative during the fall and winter of 2000–2001. The pessimists seemed to be proven the most realistic; there was no obvious reason to believe that there would be a break-through to establish an Abel Prize one hundred years after it "went to heaven with the union" with Sweden. Stubhaug tried to call Hermansen several times, but was only able to speak to his secretary, who had no new information.[9]

Therefore, it came totally out of the blue when Hermansen called Stubhaug early in the morning on March 12, 2001, and said that they had to immediately set up a meeting with the mathematicians in Oslo. Hermansen had been involved in some promising talks with possible key political actors regarding establishing a mathematical prize. Not only had he spoken to the Minister of Education and Research,

---

[7]"Opprettelse av Abelpris", letter from the head of department Arne Bang Huseby and the head of office Yngvar Reichelt in the Department of Mathematics to Telenor CEO Tormod Hermansen, 22.8.2000, NVJ.

[8]Letter from Tormod Hermansen to the Minister of Education and Research, Trond Giske, 7.9.2000. Ministry of Education and Research archive (hereafter KD): Saksnr. 00/7309, FO./ESO, Archive code 757, "Abelprisen i matematikk, Niels Henrik Abels minnefond".

[9]Interview with Arild Stubhaug 7.2.2012.

he had also been in contact with the Prime Minister's office, and he had received positive signals from the experienced Under-secretaries of State in both the Ministry of Education and Research and in the Ministry of Finance.[10]

As noted above, a national committee had been set up in the late 1990s to prepare the program for the Abel bicentennial celebrations as well as the academic program for the international "Niels Henrik Abel Bicentennial Conference" in 2002. The Ministry of Education and Research contacted the leader of this committee, the mathematician Jens Erik Fenstad at the University of Oslo. In 1995 Fenstad was asked by the later president of IMU, Jacob Palis, to design a Norwegian alternative for the 2002 International Congress of Mathematicians, and in doing so Fenstad cooperated closely with the Ministry of Education and Research. The Ministry now apparently believed that it was the National Abel Bicentennial Committee that had come up with the initiative to create an Abel Prize as part of the 2002 celebrations. However, the national committee knew nothing of any such plans. The few involved at the Department of Mathematics at the University of Oslo had not deemed it necessary to inform the national committee in the early stages of the process, and as autumn passed and winter came without any new information or developments, the department members lost interest in the project and the issue was nearly forgotten.[11]

Up until March 2001, there were only a handful of people in the mathematical community at the University of Oslo who even knew of the plans for an Abel Prize; this was more than eight months after the biographer Stubhaug met the Telenor CEO Hermansen at the Risør bookstore during the chamber music festival. The future planning of the prize would be carried out by a small group of people at the Department of Mathematics in Oslo, where Stubhaug had his office and closest contacts with the Norwegian mathematical community.

5 The Abel Prize "Working Group"

On March 21, Tormod Hermansen and his Telenor colleague Kjell Stahl met with the Oslo mathematicians on one of the top floors of the Niels Henrik Abel building on Blindern campus. The head of department Arne Bang Huseby, head of office Yngvar Reichelt and university lecturer Nils Voje Johansen had all known of the plans from their inception nearly one year prior. The biographer Arild Stubhaug was of course present. From this date the University of Oslo mathematics professors Arnfinn Laudal, Jens Erik Fenstad and Ragni Piene became fully involved in the process. Laudal was head of the scientific committee for the forthcoming Abel Bicentennial Conference, and both Fenstad and Piene were members of the national committee for the Abel bicentennial celebrations.

[10]Jens Erik Fenstad in an e-mail to Trond Fevolden 3.4.2001, JEF. Telephone interview with Tore Eriksen 23.01.2012.

[11]Interview with Nils Voje Johansen, 16.12.2011.

During the meeting, Tormod Hermansen told of the contacts with the Office of the Prime Minister and the two core ministries, Education and Research and Finance. He made it clear that there was a good chance of success if the government could be persuaded to use the petroleum fund, where surplus profits from Norwegian petroleum income were deposited. By the end of 2000, the fund had reached NOK 386.4 billion (approximately USD 45 billion). He had reason to be optimistic. Two years earlier, the government had used the petroleum fund to establish a research fund of NOK 3 billion, and the profits were allocated to support Norwegian research. With such use of the funds, the Ministry of Finance had deviated from the normal practice to make all public spending visible in the national budget through explicit postings. A similar exception was possible for the Abel Prize. The Undersecretary of State of the Ministry of Finance said in a later interview that he made it explicitly clear to both Hermansen, as well as later contacts concerning the Abel Prize, that the Ministry would not normally encourage such a practice. Nevertheless he also stated that it was a widely held belief in the Ministry of Finance at the time that an Abel Prize could be important in future recruitment, within an academic field that many in the Ministry believed to be of the greatest importance. According to the Undersecretary of State, there existed nothing less than "a genuine enthusiasm for promoting mathematics" within the Ministry. Initially, there was a fear in the Ministry that an Abel Prize might become "a home-made Norwegian prize", but such concerns gradually diminished as the campaign for the prize proceeded into the spring of 2001.[12]

The meeting with Hermansen made it clear to all those present that it was possible that they would obtain more than anyone had dared to dream. Jens Erik Fenstad stated that the government "could just earmark a sum of money, and leave the profits from the interest to pay for both the award and the arrangement".[13] Hermansen's passion fuelled others; "Hermansen inspired us to aim high", Nils Voje Johansen voiced and "his enthusiasm rubbed off on us", wrote Jens Erik Fenstad.[14] Several of those present at this initial meeting tell that they immediately understood the importance of putting aside all potential conflicts and power struggles that often characterize the inner life of academic institutions. There were already more than enough potential time bombs in the room. Only a few years earlier it would have been unthinkable that Hermansen's colleague Kjell Stahl and Professor Arnfinn Laudal would sit together in the same room peacefully. Laudal had been a central force behind a strong university opposition in 1990 that ended Kjell Stahl's short career as director at the University of Oslo after the most fierce internal conflict and crisis within the university since WWII. As pro-rector at the time, Jens Erik also played a central role in this conflict [8, pp. 75–79]. Now, however, there were important matters of mutual interest that everybody focused on.

---

[12]Telephone interview with Tore Eriksen, 23.1.2012.

[13]*Aftenposten*, 28.5.2001, "Ber om 150 mill. til Abel-pris".

[14]"Noen notater ..." and "Abel-prisen—punkter fra historien", NVJ. Interview with Yngvar Reichelt, 13.2.2011.

From the very beginning of the work in this group, the Nobel Prizes were established as the "gold standard". Before the meeting, the Oslo mathematicians had prepared a presentation on the Nobel organization and its budget. During the meeting it was decided that the committee should seek advice from the director of the Norwegian Nobel Institute and the leader of the Norwegian Nobel Committee. It was suggested that a fund of approximately NOK 100–150 million was required to present an award of the same value and prestige as the Nobel Prizes.[15]

Before the meeting, Arnfinn Laudal had mentioned the plans to the General Secretary of the Norwegian Academy of Science and Letters. The Academy, which at this time was beginning to regain some strength after decades in decline, was more than willing to take on the task of administrating a scientific prize on a Nobel level [7, pp. 194–198]. As we have seen, the Academy would have also taken care of the Abel Prize if it had been created one hundred years earlier. There was also general agreement that the Academy would provide the prize with a greater national legitimacy than if it was handled by the University of Oslo. The latter option would have most probably started one of the very common and time-consuming Norwegian debates about localization, which might have jeopardized the whole project.

The meeting ended with the creation of an informal "working group" for the Abel Prize in the Department of Mathematics at the University of Oslo. It consisted of biographer Stubhaug, head of office Reichelt, lecturer Voje Johansen and the three professors Laudal, Fenstad and Piene. In the time that followed, the six people in this working group came to run an intense campaign directed towards the international mathematical community and the national public opinion to obtain a political green light for their plans.

6 Scientific Legitimization and Support

The members of the working group tell of an informal and level structure where each was given more or less a defined area of responsibility, and where the development of the work was coordinated via contact by e-mail and in frequent meetings. Partly depending on each member's personal networks in academic and political circles, and influential echelons of Norwegian society, the working group contacted people in the international mathematical community and those in Norwegian society that would benefit the campaign which was by now really getting underway. The work was proceeding without any major tensions and in an enthusiastic environment. The trustful atmosphere was also most likely strengthened by the fact that the initiative was being met with open arms from almost all arenas.

From the beginning, the working group was assured of full support from the Norwegian Academy of Science and Letters where all three professors of the working

[15]Minutes from the meeting with Hermansen 21.3.2001 noted in Ragni Piene's academic diary. Ragni Piene's private archive (hereafter RP). Minutes from the meeting with Hermansen 21.3.2001, JEF.

group were members. In just a few weeks after the first request, the working group received the Academy's formal acceptance that the institution would like to take responsibility for the prize.[16] By the end of April 2001, Oslo mathematicians Piene and Laudal contacted the European Mathematical Society (EMS) and IMU to obtain further support for the plans for an Abel Prize from the highest level in European and world mathematics.

Professor Piene wrote to the president and vice-president of EMS, Rolf Jeltsch and Bodil Branner, to tell them that they were now making plans in Norway for a prize in mathematics "comparable to the Nobel Prize in the other sciences". Piene asked EMS if they wanted to support such an idea, and if so, were they willing to write a letter to confirm "that the name of Niels Henrik Abel is indeed a good, internationally recognized name, worthy to be used for a prize of this magnitude". She added "in fact, it seems quite realistic, though of course it is yet too soon to say precisely how realistic it is". Over the duration of the day, many e-mails were sent between Piene, Jeltsch and Branner. Jeltsch got in contact with the EMS executive committee, and stated that he had received an important message from Norway about the establishment of a prestigious prize in mathematics. If he did not receive any objections from the committee, Jeltsch made clear that he was going to write a letter in support of the idea as soon as possible.[17]

Later the same day, EMS vice-president Bodil Branner could inform her Norwegian contact that the distinguished American mathematician John H. Hubbard from Cornell University was visiting her. Hubbard is especially known for his studies within complex dynamics, a core field in Abel's works. When Branner told Hubbard about the plans, he immediately wrote a text which—Branner added—"I think is what you need in Norway". In his letter Hubbard wrote that Abel's ideas—like those of "the very greatest scientists"—"have so permeated mathematics that one is no longer aware where they come from". According to Hubbard, Abel had "revolutionized the theory of equations, complex analysis, number theory and algebraic geometry". Abel was "responsible for the deepest work on algebraic integrals of the nineteenth century", and "the ideas he initiated are still, 170 years later, central to the best of mathematics". "Abelian groups" are part of the undergraduate vocabulary; "Abelian varieties", "Abelian integrals" are part of standard topics in graduate courses. The next day Hubbard's panegyric appreciation of Abel was enclosed in a letter of wholehearted support from the president Rolf Jeltsch on behalf of EMS: "It is with great pleasure that I hear that Norway plans to honor its famous Mathematician Niels Henrik Abel by introducing a prestigious prize. I can assure you that the European Mathematical Society is enthusiastically supporting this idea. For us, there is no doubt that his name is worthy to be used for this prize".[18]

---

[16]Ragni Piene to the Norwegian Academy of Science and Letters, 18.4.2001. Final confirmation in mail from the academy to the working group, 20.4.2001. RP and NVJ.

[17]Rolf Jeltsch to the EMS executive committee, 18.4, NVJ.

[18]President of the EMS Dr. Rolf Jeltsch to Ragni Piene, 19.4.2001. KD: Saksnr. 00/7309, FO./ESO, Archive code 757, "Abelprisen i matematikk, Niels Henrik Abels minnefond".

Professor Laudal had by now got in contact with the president of IMU, Jacob Palis, who liked the idea very much.[19] IMU already awarded the most prestigious international prize in mathematics, the Fields Medal. The Fields Medal—which history can be traced back to 1924—is awarded every four years during the International Congress of Mathematicians to recognize outstanding mathematical achievement for existing work and for the promise of future achievement to mathematicians under the age of 40. The prize amount was however modest. In 2000 it was well under NOK 100,000 (approximately USD 10,000). IMU did not want a new prize to compete with the Fields Medal, and Laudal assured Palis that the planned Abel Prize would not be comparable to any of the existing IMU prizes. As we saw in the introduction, the prize was therefore also presented as "the 'missing Nobel Prize' in mathematics".

This argument was very well received by IMU, which had for one hundred years been longing for a scientific prize comparable to the Nobel Prizes. This desire is evident in the strength of the stories retold in the international mathematical community, about how Nobel allegedly decided not to establish a prize in mathematics because of quarrels over a woman with the Swedish mathematician Gösta Mittag-Leffler. Even though this story has been proven false, the almost haunting issue of the "missing Nobel in mathematics" has been kept very much alive [11, 13, p. 17]. This might explain why the working group was soon able to present Norwegian politicians and state authorities with overwhelming support for the plans from the world's leading mathematical organization.

Through President Palis, IMU expressed its "full support to [the] initiative to create the Prize named after Abel, one of the greatest mathematicians of all time". The mathematical union considered the establishment of an Abel Prize no less than "the most important project in many years for the development of mathematics worldwide". A prize at this level was very much welcomed and would not interfere with the other IMU prizes such as the Fields Medal or the Nevanlinna Prize. According to IMU leadership, the absence of a prize "similar to the Nobel Prize for Mathematics is a century old one and an ever discussed missing feature of the scientific work of our community". The creation of the Abel Prize would therefore be most appropriate "to fill such a serious gap".[20]

7 Political Lobbying

During May 2001 the working group gathered support from the members of the parliamentary Standing Committee on Education, Research and Church Affairs, and key persons in Norwegian science, culture and public life. One main objective of

[19] Jacob Palis to Arnfinn Laudal 1.5.2001, NVJ.

[20] KD, Saksnr. 00/7309, FO./ESO, Archive code 757, "Abelprisen i matematikk, Niels Henrik Abels minnefond". (E-mail 19.7.2001. This message was sent in different versions, in both e-mail and letters through the spring and summer of 2001).

the prize was, according to the working group's four-page public presentation, "to create better understanding of the importance of mathematics and science in today's society". Mathematics, it was emphasized, is a broad discipline, ranging from "pure mathematics to a vast number of applications within science, society and technology". The prize would contribute to make these applications more visible to the general public, whether they dealt with understanding ecological systems, the study of financial markets, oil drilling, medical diagnostics or biotechnological development. The working group's presentation also highlighted the Nobel Prizes as the obvious ideal for the new prize, both in organizational and economical terms: "If handled properly, an annual Abel Prize might draw great attention and in time achieve the same status as the Nobel Prizes within the other sciences".[21]

The responses were welcomed and strongly supported the need to make the importance of mathematics much more visible to the general public. This, many of the respondents maintained, was essential to create a much needed interest in mathematics and science, and to improve the poor and declining recruitment numbers in those fields over the past decades. Soon the working group had a list of 29 prominent people supporting the plan. The list included the rectors of all four Norwegian universities, the presidents of the Norwegian Academy of Science and Letters in Oslo and the Royal Norwegian Society of Sciences and Letters in Trondheim, several other professors, a recent Minister of Education and Research and the most influential prime ministers of the past two decades, Kåre Willoch (Conservative Party) and Gro Harlem Brundtland (Labour Party).[22]

The then Minister of Education and Research, Trond Giske (Labour Party), was nevertheless somewhat reluctant. The initiative was presented to him and his office several months earlier and nothing had happened. Some days after the first meeting in late March, the working group invited the Minister to meet the group to be "informed and initiate a dialogue [...] about possible solutions".[23] But Minister Giske was still not very enthusiastic, and made it clear that he did not have millions of kroner just lying around. And if he did, he was not at all sure that he would spend them on a mathematics award.[24] In contrast, the working group had a "very encouraging meeting" with the Under-secretary of State in Giske's department.[25]

Towards the end of May, the group seemed somewhat uncertain about what their next move should be: was it strategically wise to proceed with talks with Giske's Ministry or would it be better to head straight for the Office of the Prime Minister?[26]

---

[21]Prospectus "Abel-prisen", spring 2001, JEF/NVJ. See also *Aftenposten* 27.6.2001, "Blant de største" by Jens Erik Fenstad.

[22]Prospectus "Abel-prisen", spring 2001, JEF/NVJ.

[23]Letter from Yngvar Reichelt on behalf of the Abel Prize working group to the Minister of Education and Research Trond Giske, "Abelpris i matematikk", 29.3.2001. Yngvar Reichelt's archive (hereafter YR).

[24]E-mail from Rolf Reikvam to Arnfinn Laudal, 18.5.2001, NVJ.

[25]E-mail from Jens Erik Fenstad to Kjell Stahl 27.4.2001, JEF.

[26]E-mail from Laudal to Reikvam, 16.5.2001, and e-mail from Reikvam to Laudal, 19.5.2001, NVJ.

They chose a broad approach. By the end of May the working group made three parallel moves: (1) they sent a letter to Prime Minister Stoltenberg and asked for a meeting; (2) Professor Laudal got the Socialist Left Party representative in the Standing Committee on Education, Research and Church Affairs—who was very enthusiastic about the plans—to present a formal question in Parliament (Stortinget) to Minister Giske; and (3) Professor Fenstad got full-page coverage in the large national newspaper Aftenposten, regarding the possible creation in the near future of a prestigious mathematics award in memory of Abel.[27]

Together with the letter to the Prime Minister and the news coverage, the formal question asked in Parliament raised the political temperature of the issue.[28] In his question, the Socialist Left Party representative first made clear that the prize had to be on level with Nobel Prizes to gain the necessary prestige. He summed up by asking "Will the Minister take action and establish a fund so that the prize can be awarded already next year?"[29] Based on the letter to the Prime Minister and the question in Parliament, a state secretary at the Office of the Prime Minister advised Minister Giske to meet with the Abel Prize working group. Even Giske's own Department of Research now urged the Minister to give Parliament a positive answer, and invite the working group to a follow-up meeting.[30]

In his answer to Parliament in the beginning of June, Giske stated that he was fully aware of the plans, and that he agreed that "the building up and strengthening of the national competence within mathematics and the natural sciences is needed". He had also noted that the plans were supported in "important academic circles". If the government was to establish a fund "that placed a prospective Abel Prize on level with the Nobel Prize", then the proposal had to "be evaluated together with other measures to strengthen recruitment and improve the scholarly quality of Norwegian mathematics and natural science". In accordance with the clear political signals from both the Office of the Prime Minister and his own Department of Research, Giske promised to look closer into the matter and invite the Abel working group to further talks.[31]

---

[27]KD, Saksnr. 00/7309, FO./ESO, Archive code 757, "Abelprisen i matematikk, Niels Henrik Abels minnefond" and Saksnr: 01/4385, FO./ESO, Archive code 001 "Spørsmål til skriftlig besvarelse fra representanten Rolf Reikvam om etablering av en Abelpris i matematikk". *Aftenposten* 28.5.2001. "Ber om 150 mill. til Abel-pris".

[28]E.g., *Aftenposten* 1.6.2001, "Økt press på Abel-prisen".

[29]"Spørsmål til skriftlig besvarelse" (spørsmål nr. 421), 30. May 2001. KD, Saksnr. 01/4385, FO./ESO, Archive code 001 "Spørsmål til skriftlig besvarelse fra representanten Rolf Reikvam om etablering av en Abelpris i matematikk".

[30]KD, Saksnr. 01/4385, FO./ESO, Archive code 001 "Spørsmål til skriftlig besvarelse fra representanten Rolf Reikvam om etablering av en Abelpris i matematikk". See also the letter from the secretary of state Tom Therkildsen at the Office of the Prime Minister to the Abel Prize working group, 18.6.2001, NVJ.

[31]Trond Giske's response 7.6.2001. KD, Saksnr. 01/4385, FO./ESO, Archive code 001 "Spørsmål til skriftlig besvarelse fra representanten Rolf Reikvam om etablering av en Abelpris i matematikk".

8 Breakthrough

On July 18, the working group received a call from a state secretary in the Ministry of Education and Research who asked if they could come to a meeting the following day. This was now “a matter of great importance”. The director general in the Ministry’s Department of Research and a state secretary from the Ministry of Finance would also take part in the meeting. The working group immediately got together to prepare their answer to the main question that would come up the next day: why an Abel Prize in mathematics?[32]

At the July 19 meeting, the State Secretaries of both the Ministry of Education and Research and the Ministry of Finance emphasized that for the Abel plans to come to fruition, it was absolutely crucial that the prize would contribute to strengthen the position and status of the natural sciences and mathematics in Norway, and increase public awareness of the societal importance of these fields of knowledge. The working group was asked to formulate some thoughts about this issue as soon as possible. Four days later the group presented a document on how the Abel Prize could meet those requirements with regard to both school-aged and university students, researchers and the general public. There already existed two Abel associated mathematics competitions in Norwegian schools: KappAbel at the final level of primary school and the Abel competition in upper-secondary school. These competitions could organizationally be placed under the umbrella of the Abel Prize, and this merger would increase media attention as well as the status of the competitions. Students at universities and university colleges could find motivation from the fact that the international mathematical stars would come to Norway on a regular basis, and Norwegian mathematicians would be given unique opportunities to create networks and work with the world’s top mathematicians and mathematical research centers. In this way—the working group argued—the Abel Prize would also contribute to enhance public understanding of mathematics and the natural sciences as important cornerstones of modern society. The Abel Prize working group ended with a quote from the IMU president Jacob Palis: “Abel’s prize would certainly make mathematics much more visible to society and perhaps this is more important than ever”.[33]

Then things really began to move. In early August the working group was again contacted by the Ministry, which wanted them to specify the organizational structures and budget. A couple of weeks later the Ministry wanted further biographical details on Abel and his scientific contributions and importance. The working group was now totally on home ground, and their paper “Facts about Niels Henrik Abel” was immediately—and without the knowledge of anyone in the working

[32]Jens Erik Fenstad in two e-mails to the working group, 18.7.2001. NVJ.

[33]Letter from Nils Voje Johansen on behalf of the working group to the secretary of state Randi Øverland in the Ministry of Education and Research, 23.7.2001, “Abel-prisen og rekruttering til realfagene” og notat til KUF og Finansdepartementet, “Virkninger av en Abel-pris i matematikk”. KD, Saksnr. 00/7309, FO./ESO, Archive code 757, “Abelprisen i matematikk, Niels Henrik Abels minnefond”. E-mail from Jacob Palis to Arnfinn Laudal, 19.7.2001, NVJ.

group—passed on from the Ministry of Education and Research to the Office of the Prime Minister.[34] Like many of the Ministry of Finance economists, Prime Minister Stoltenberg was educated at the University of Oslo's Department of Economics. And like many of the economists in the Ministry of Finance, Stoltenberg would soon prove to be very much in favor of creating an international mathematics prize that was believed to support the visibility of and interest in mathematics in Norway.[35]

The working group had no idea that Prime Minister Stoltenberg would announce the establishment of the Abel Prize during a speech to the local Labour Party youth organization Arbeidernes Ungdomsfylking (AUF) at the University of Oslo's Blindern campus on August 23, 2001. The working group knew that the Prime Minister was scheduled for a meeting with the university rector later in the day, and because of this they had provided the rector with further arguments in support of the prize.[36] Three of the six working group members nevertheless walked the few meters from the Niels Henrik Abel building to the auditorium in the neighboring Vilhelm Bjerknes building, where the Prime Minister was to meet with the AUF. Here the working group members hoped to have a quick word with Stoltenberg and deliver a letter they just had received from the IMU president. In the letter Jacob Palis gave an assurance that the IMU "in every possible way" would cooperate with the Norwegian government and the Norwegian Academy of Science and Letters "in establishing such a wonderful prize and in implementing it in the most dignified form".[37]

IMU would soon get the opportunity to do so. The Abel Prize working group members immediately understood what was coming when they saw that all of their bureaucratic and political contacts from the past month were present at the Prime Minister's meeting with the local branch of AUF at Blindern.

9 High Expectations

A press release from the Office of the Prime Minister later that same day stated that "the Government wanted to heavily increase the focus on mathematics and natural sciences". The Abel Prize should "serve as an encouragement for both students and researchers alike" at a time "when great parts of the Western world experienced a decline of interest in these fields of knowledge". The prize was intended "to make visible the importance of mathematics and the natural sciences". Prime Minister Stoltenberg hoped the prize would "improve the recruitment of young people to

[34] The paper "Fakta om Niels Henrik Abel" was sent from the Ministry of Education and Research to the Office of the Prime Minister til SMK on the 22.8.2001. KD, Saksnr. 00/7309, FO./ESO, Archive code 757, "Abelprisen i matematikk, Niels Henrik Abel minnefond". Interview with Nils Voje Johansen, 16.12.2011.

[35] Telephone interview with Prime Minister Stoltenberg 7.11.2011.

[36] E-mail from Jens Erik Fenstad to Helge Holden, 2.11.2001, NVJ.

[37] Telefax from Jacob Palis to the Abel Prize working group, 22.8.2001. KD, Saksnr. 00/7309, FO./ESO, Archive code 757, "Abelprisen i matematikk, Niels Henrik Abels minnefond".

mathematics and the natural sciences, the strengthening of Norwegian mathematical research, and the image of Norway as a knowledge society".[38]

The Labour Party government was not alone in wanting to strengthen the position and status of mathematics and the natural sciences. From around the turn of the century, this topic was considered to be one of the most important issues by all major Norwegian political parties. The first OECD Programme for International Student Assessment (PISA) survey in 2001 presented a gloomy picture of Norwegian students' knowledge about, and attitudes towards, mathematics and the natural sciences. This was a serious wake up call that seemed to confirm what had been an increasing suspicion recent years: the Norwegian educational system was in poor shape. High quality competence in mathematics, natural sciences and technology was now seen as the key factor to improve the future capacity for innovation and economic growth in the so-called global knowledge economy.

This shift was evident in all areas of Norwegian education and research policy. In research, the creation of the Norwegian Centres of Excellence system in 2002 represented a serious break with the traditional and essentially egalitarian national policy in that field. In higher education, the so-called Quality Reform in 2003 was inspired by similar concerns, as was the Knowledge Promotion reform in primary and secondary schools in 2006. Since 2002, both the center-right and the center-left governments have presented and updated their own strategic plans to support mathematics, natural science and technology studies and research. These fields of knowledge were all matters of increasing political attention, and simultaneously more elitist perspectives gained ground in the Norwegian debate regarding primary and secondary schools, higher education and research. The Abel Prize plans encompassed the very essence of this political wave as it was about to break over the Norwegian political landscape [8, pp. 115–144].

The need to improve within the fields of mathematics, natural science and technology in order to succeed in the developing so-called knowledge economy thus became an axiom of all major Norwegian political parties—from the Socialist Left Part on one side to the right-wing Progressive Party on the other—from about the turn of the century. All the members of the parliamentary Standing Committee on Education, Research and Church Affairs thus fully supported the prize when it was announced, and basically for the same reason: fear that the Norwegian educational system—from top to bottom—was unprepared to meet the challenges from the developing knowledge economy when it came to mathematics, natural science and technology. The Conservative Party representative stated that Norway "was facing a general natural sciences crisis", and that any measure taken to counteract this was "most appropriate". The Liberal Party committee member hoped the prize would help to turn the "decay of natural science in this country". The former minster of education and research and Christian Democratic Party representative said the prize might come to mean just as much for natural science in Norway "as the Nobel Prizes means for science in Sweden".[39] The right-wing Progressive Party had for a long

[38]Press release 155/2001, 23.8.2001. The Office of the Prime Minister.

[39]*Aftenposten*, 24.8.2001: "200 millioner kr til nyopprettet Abel-pris".

time warned against the economical consequences of a decline in the "hard subjects" within the national educational system, and the party's representative had early in the process encouraged the Abel Prize working group to ask for more money.[40] On the other side of the political spectrum, the Socialist Left Party representative became—as we have seen—the most important advocate of the committee when he brought the question about the prize to Parliament in the end of May.

A significant section of the international mathematical community now hoped that they would soon be enjoying a mathematical counterpart to the Nobel Prizes. Because of the frustration caused by—as IMU president Palis had put it—"a century old one and an ever discussed missing feature of the scientific work of our community", the history about the Abel Prize and how it "went to heaven" with the dissolution of the union between Sweden and Norway in 1905, did resonate extremely well in the international mathematical community. The story brought forward by the Abel Prize working group was nothing less than the story of how one hundred years earlier Norway had been on the verge of creating a prize in mathematics almost as prestigious as the Nobel Prizes, and at the same time and in the same kingdom! This was undoubtedly the single most important reason for the overwhelming and absolutely decisive support from the highest level in European and world mathematics via the EMS and IMU.

The establishment of the Abel Prize one hundred years later could therefore also be seen as the righting of old wrongs against mathematics. In influential parts of the international mathematical community, the Abel Prize was from the very beginning clearly seen as a very close relative to the Nobel Prizes, both historically and geographically. One week after the announcement of the prize, Science—the journal of the American Association for the Advancement of Science—stated it clear and simple: "For mathematics, Abel = Nobel" [1]. President of EMS Rolf Jeltsch wrote in his annual report that the creation of the Abel Prize would have enduring consequences for mathematics both in Europe and in the rest of the world. In a short time it also became common for mathematicians internationally to speak about the Abel Prize as the "Nobel Prize in Mathematics".[41]

However, some Canadian mathematicians soon became concerned about what the creation of the Abel Prize would mean for the status and prestige of the Fields Medal, and wrote an open letter to their prime minister: According to the Canadian mathematicians, few people were aware that John Charles Fields was a Canadian, that the prize was originally a Canadian idea and that the medal foundation was in Canada. They argued, largely in vain, that the newly established Abel Prize had to be met by the Canadian government with a strong manifestation of the Canadian identity of the Fields Medal and a marked increase in the modest prize value.[42] IMU—which awarded the Fields Medal—was nevertheless relatively clear in its

[40]Nils Voje Johansen's meeting with Ursula Evje (FrP) in Parliament (Stortinget) 23.5.2011, NVJ.

[41]President Rolf Jeltsch to EMS, December 2001, NVJ.

[42]"Open letter to the Prime Minister of Canada", 7.5.2002. From Nassif Ghoussoub, Arvind Gupta and Robert V. Moody. YR.

support for the new Abel Prize as a mathematics counterpart to the Nobel Prizes. And obviously the generous funding of the prize from the rich Norwegian state played an important role in gaining this support from large parts of the international mathematical community. The marketing of the Abel Prize during the first years thus came to show clear similarities to the establishment of the Nobel Prizes in the early twentieth century; both prize winners and their often very influential scholarly communities almost systematically praised their respective prizes very highly, thereby increasing the prestige of both the prizes and the academic elites that received them. In these processes, media attention has been of vital importance. Media fascination from the start seemed to reinforce a general will to believe that a prize that entails so much money as well as so much prestige simply must be just and important [5, especially pp. 268–272].

In summary, it is fair to say that when the initiative was finally brought out in the open in the spring of 2001, the entire spectrum of Norwegian political parties and significant and influential sections of the international mathematical community were fervent in their desire for the prize, but for entirely different reasons. For many years the campaign started by Abel's biographer Arild Stubhaug had garnered polite interest but little true support among Norwegian mathematicians. When the Telenor CEO and Labour Party veteran Tormod Hermansen came into the picture in the summer of 2000, the issue was put into a much larger cultural and political context, and soon the Abel Prize initiative proved to be much more in tune with the times than anybody had expected: Norwegian politicians across the whole political spectrum—from the Socialist Left Party to the Progressive Party—saw the prize as an opportunity to promote mathematics and science and counteract an alarming decline in these fields within the national educational system, as well as a way to promote Norway as a knowledge society. And the international mathematical community finally—after one hundred years—got a prize they both hoped and believed could equal the Nobel prizes. Well underway, the process was also characterized by close and informal contacts among academics, politicians, bureaucrats and industrial leaders, relationships that would possibly be much harder to manage in a larger political system.

The story of how the Abel Prize came about is a piece of contemporary national history about how politics could be made off the public scene through close networks in a small nation. The story would also almost repeat itself in 2005 when the Kavli Prizes in Astrophysics, Nanoscience and Neuroscience were established with an ambition to become as important for the development of these fields of knowledge in the 21st century as the Nobels had been for physics, chemistry and medicine in the 20th. As the Abel Prize, the Kavli prizes would also be awarded by the Norwegian Academy of Science and Letters. And again the establishment was a result of a close collaboration between an industrial leader, scientists, bureaucrats, and the Ministry of Education and Research. Only this time the sponsor providing the necessary funding to reach the level of the Nobel prizes was not the ever richer Norwegian state but the US-based Norwegian-born entrepreneur, billionaire and long time science benefactor Fred Kavli.[43] The story of the Abel prize is also an account of how cultural heritage can be recruited as a resource for science policy: how the prize could bring together politicians, bureaucrats and mathematicians for a spectrum of cultural-political purposes from differing constituencies working with a variety of interests.

10 Nobel Level?

It is too soon to accurately present an authoritative historical judgment of the first ten years of the Abel Prize. Nevertheless, it is possible to make some tentative observations regarding the extent that the prize has succeeded in establishing its reputation and status, both in the academic community and in the greater public, both in Norway and internationally.

It was a bold ambition to establish the Abel Prize as a mathematical counterpart to the Nobel Prizes, but as we have seen, the approach was warmly welcomed by the international mathematical community. This support was not only important but also absolutely necessary for the subsequent political acceptance of the prize. During the first ten years, the prize seems to have been firmly established, not the least because the winners are chosen by an international committee whose members are nominated by the EMS and IMU. Unlike the national Swedish Nobel committees and the national Norwegian Peace Prize committee, the Abel committee leans on the expertise of the highest European and world organizations within the field. So far few controversies have risen from the Abel awards. After just two years, and the awards to Jean-Pierre Serre (2003) and the shared prize to Michael F. Atiyah and Isadore M. Singer (2004), the mathematics section of the US National Academy of Sciences sent their most sincere congratulations:

> We extend our congratulations to the Abel Prize Committee, to the Norwegian Academy, and to the Norwegian Government, for their outstanding management of the Abel Prize these two first years. [...] In two unerringly placed awards, you have made the Abel Prize the leading international prize in Mathematics—the true "Nobel Prize" of Mathematics.[44]

In spite of the strong academic support, the prize has had problems in obtaining media attention both in Norway and internationally. To increase the general awareness and knowledge of the prize, key persons have tried to establish more formal links between the Abel Prize and the Nobel Prizes. Former IMU President David Mumford was a member of the first Abel committee, and he had very high ambitions for the prize. To ensure that the Abel Prize would reach "the general consciousness

[43] http://www.kavliprize.no.

[44] Letter from Richard V. Kadison, Chairman for the mathematics section in National Academy of Sciences, USA, to the Norwegian Academy of Science and Letters, 28.4.2004. The archives of the Norwegian Academy of Science and Letters (hereafter DNVA), file: Abelkomiteen.

of the world" in a way comparable to the Nobel Prizes, he advocated to use more actively the history behind the prize. This history—according to Mumford—showed "that the Abel Prize is logically the long missed mathematics component of the Nobel awards". Mumford thus suggested that the president of the Norwegian Academy of Science and Letters should get in contact with its Swedish sister academy to ask if they could officially present the creation of the Abel Prize during the 2002 Nobel Prize award ceremony in Stockholm.[45] The following year, the new IMU president John Ball followed up on Mumford's idea. If the Norwegian Academy of Science and Letters did not have any objections, he would very much like to try to persuade the Royal Swedish Academy of Sciences to, at least, "give a recognition of the Abel Prize".[46]

Both within the Norwegian Academy and the Niels Henrik Abel Memorial Fund Board it was generally known that such recognition from either the Swedish Academy or the Nobel Foundation was not likely given the traditional and strong exclusivity of the Nobel Prizes. At the same time, the Norwegians enjoyed the references in print and media stating that the Abel Prize was a Nobel in mathematics.[47] It took six years for the New York Times to include a full-page article on the Abel Prize; by then it was—according to the newspaper—"widely regarded as the Nobel of mathematics".[48] The Nobel Foundation was not at all pleased with the repeated presentation of the Abel Prize as a Nobel in mathematics, and after some years they took action to ensure that this practice at least was not officially promoted by the Abel Prize itself. In 2008 the foundation sent a stern message to the Norwegian Academy: "The Nobel Foundation has noticed that the 'Abel Prize' in some contacts with media, and in some e-mails, has been presented as 'the Nobel Prize of Mathematics'." Further, the letter stated that it was a registered trademark, and the Norwegian Academy should know that the Nobel Foundation always had and always would protect this trademark "from degeneration and watering down with great care and determined efforts". Those responsible for the Abel Prize were strongly requested to "carefully avoid" such practice in the future "in a, for us, matter of great importance".[49] Apart from this—and although the Abel Prize has not been the subject of much media attention in Sweden—it seems that the Abel Prize

[45]E-mail from David Mumford to the president of the Norwegian Academy of Science and Letters, Lars Walløe, 29.10.2002, NVJ.

[46]E-mail from IMU President John Ball to Chairman of the Abel Committee Erling Størmer, Chairman of the Abel Board Jens Erik Fenstad, and General Secretary of the Norwegian Academy of Science and Letters Reidun Sirevåg, 17.11.2003. JEF.

[47]See for example F. Thomas Bruss, "Homage to the Abel Prize. Homage to Norway" in EMS Newsletter June 2005, NVJ.

[48]New York Times, 31.5.2009: "Complex Math, Simple Sum: 3 Awards in 5 Years" http://www.nytimes.com/2009/06/01/nyregion/01nyu.html (visited 10.12.2011). The theme of the NYT article was that in the short history of the Abel Prize, already three professors at New York University had received the prize.

[49]Letter from the Nobel Foundation to the Norwegian Academy of Science and Letters, 21.4.2008, DNVA.

has been received in a rather welcoming way also there. Perhaps it helped that the Swedish mathematician Lennart Carleson was awarded the Abel prize in 2006, as three years later, the leading Swedish newspaper Dagens Nyheter presented both the Fields medal and the Abel Prize as the only prizes in mathematics on "Nobel level"?[50]

Even though only very few Norwegian mathematicians were involved in the planning for an Abel Prize, the whole Norwegian mathematical community did receive it with open arms. Initially it caused some frustration that the planning was carried out in secrecy as an exclusive project of a small group of Oslo mathematicians. For example, in 2001 Professor Helge Holden from the Norwegian University of Science and Technology in Trondheim was heading the national committee for the Abel bicentennial celebrations together with Jens Erik Fenstad from the University of Oslo. Holden was not very pleased that he, in this position, knew nothing about the prize plans until he read about them in the newspapers.[51]

When the Abel Prize plans became known to the public at the end of May 2001, both the Royal Norwegian Society of Sciences and Letters in Trondheim (in the middle of the country) and a planned university in Abel's home county Agder (in the south) wanted their share of the prize. The Academy in Trondheim—the oldest in the country—pointed to the fact that Abel had been an Academy member. In Agder, it was suggested that the planned university would be called the Niels Henrik Abel University. It was then, they argued, "a natural consequence that the prize would be awarded by the university which carries his name".[52] When the plans became known outside Oslo, they had nevertheless been developed to a point where there was little room for a traditional Norwegian regional or center vs. periphery debate.

Additional activities were soon put in place to secure support from Norway's mathematicians and universities. Parts of the official program took place in Trondheim, central Norway, Bergen in the west and Kristiansand in the south. The Norwegian Mathematical Society was thoroughly revitalized in numerous ways. Most importantly, the society was put in charge of the annual and week-long international Abel Symposia which since 2004 has attracted cutting edge mathematicians from all over the world, and every year different Norwegian universities have been given special responsibilities for the symposium. Abel stipends have been created to support recruitment at the universities, and there are numerous activities in Norwegian schools connected to the prize, most notably the primary school KappAbel competition and the upper secondary school Abel competition.

The Niels Henrik Abel Memorial Fund was from the start worth NOK 200 million (approximately USD 35 million), NOK 50 million more than the Abel Prize

[50] *Dagens Nyheter*, 11.10.2009, "Matematik i praktiken".

[51] E-mails between Jens Erik Fenstad and Helge Holden 2.11.2001, JEF.

[52] Letter from the President of the Royal Norwegian Society of Sciences and Letters in Trondheim, Karsten Jacobsen, to the Abel Prize Working Group, 12.6.2001, NVJ. Letter from the Agder County Council to Prime Minister Kjell Magne Bondevik (Christian Democratic Party), 2.4.2002, "Sørlandet som utdelingssted for matematikkpris og lokaliseringssted for fondstiftelse". KD, Saksnr. 00/7309, FO./ESO, Archive code 757, "Abelprisen i matematikk, Niels Henrik Abels minnefond". The new university was finally called the University of Agder.

working committee had asked for. This came close to an embarrassment for the mathematicians, as the symbolism of asking for 200 million at the bicentennial should be more than obvious. The NOK 200 million fund also provided almost exactly the income the group had wanted: a budget of NOK 12.4 million to cover the prize (NOK 6 million), administration and associated activities, such as the Abel Symposia and the school competitions.[53] These additional activities were also important from the view of Norwegian politicians. Because of this, the Abel Memorial Fund Board soon created a Child and Youth Committee to support and fund mathematical activities all over the country. A mathematics teachers' prize was also established in 2005 in honor of Abel's teacher, Bernt Michael Holmboe. An annual Holmboe Prize of NOK 50,000 (approximately USD 8,500) is awarded in Abel's old school, Oslo Cathedral School, by the Minister of Education and Research the day before the Abel Prize itself, and in the presence of the Abel Prize laureate.

11 Conclusion—And the Need for Future Adjustments?

In conclusion, it is fair to say that the Abel Prize has gained significant reputation and prestige in the international mathematical community, and has injected new life into Norwegian mathematics. This article has explained how the prize was established in light of multiple intentions and ambitions among a great variety of actors and stakeholders, both nationally and internationally, during a short period of time in 2000 and 2001.

From the outset, the plans for an Abel Prize in mathematics were presented by Abel's biographer, the poet and author Arild Stubhaug, without gaining much support from the Norwegian mathematical community. The plans were nevertheless eventually warmly embraced by the major Norwegian political parties, all increasingly concerned with educational performances in mathematics, natural science and technology and the future capacity for innovation and economic growth in the so-called global knowledge economy. In the process of creating political awareness of the prize plans, Stubhaug initially got substantial help from the prominent Labour Party politician and industrial leader Tormod Hermansen. But it was also of crucial importance that Norwegian mathematicians took real interest in the project in the spring of 2001. The small group of mathematicians at the University of Oslo was instrumental in collecting the decisive support from the highest level in European and world mathematics via the EMS and IMU. Without this strong international academic support, it is not likely that the plans would have seen a political breakthrough in Norway.

For the international mathematical community in large, and especially for EMS and IMU, the Abel Prize was an opportunity finally—after one hundred years—to establish "the missing Nobel in mathematics". For Norwegian mathematicians the Abel prize became a welcome opportunity to celebrate one of their own grand heroes and increase their international attention and cooperation. And, even though

[53]The fund was placed in government bonds with a ten-year fixed interest rate of 6.2 %.

the prize was unanimously supported by all major Norwegian political parties, the Labour party government could proudly present a prize that—as previously noted—aimed to "improve the recruitment of young people to mathematics and the natural sciences, the strengthening of Norwegian mathematical research, and the image of Norway as a knowledge society".[54] In this way the story of the creation of the Abel Prize stands as no exception when it comes to how it promised pride and prestige to important stakeholders. But the story to some extent stands out in the way both political authorities and national and international academic communities could read quite different desires and ambitions into the prize in a given place at a given time: the wealthy oil-nation Norway concerned with educational performances at the time of the Niels Henrik Abel bicentennial. In a matter of months, this nexus of desires and ambitions gave rapid and strong momentum to the establishment of a science prize that just a handful of people could only imagine one year earlier.

There is nothing to suggest that the broad support obtained through the process described in this article has diminished in any way during the first ten years. However, the finance model changed from 2012 because of the steep fall in interest rates during the international financial crisis. The fund was liquidated, and the Abel Prize was now listed clearly as a separate post on the national budget. This of course means greater insecurity for the future, but in the first year after the change the Abel budget was increased by a few hundred thousand NOK. As we have seen, the prize has also had problems with low levels of media attention. In addition, it has been especially difficult to ensure that the laureates' work is understandable to the general public. As the need to increase the popularity and visibility of mathematics was a major argument in obtaining political support for the creation of the prize, these problems are challenging. The Abel Prize working group stated in the spring of 2001, the prize would "demonstrate the numerous mathematical applications in science, society and technology"; the prize has only lived up to this promise to a small extent.

It would appear that first and foremost the Norwegian Academy of Science and Letters and the Abel committees wanted to firmly establish the prize via uncontroversial awards to the grand old men of "pure" mathematics. In the short run, this seems to have been a successful strategy. In the longer run however, this strategy may prove somewhat risky, when much of what is of most—at least obvious—importance for the rest of the world, goes on in more applied parts of the broad mathematical field. A systematic opening of the prize towards more applied areas of mathematics could instigate controversies within the mathematical community and lead to more controversial awards. It might also increase the level of outside and media attention. Such an adjustment would be a bold one after just ten years, but then again, the history of the Abel Prize—both the long and the short one—is paved with bold ideas.

Acknowledgements This article was written at the request of and mainly financed by the Niels Henrik Abel Board at the Norwegian Academy of Science and Letters. I want to thank my colleagues at the University of Oslo, Edgeir Benum, John Peter Collett and Robert Marc Friedman

[54]Press release 155/2001, 23.8.2001. The Office of the Prime Minister.

for their encouragement and constructive comments, but of course, the responsibility for the final result is mine alone.

References

[1] Cipra, B.: For Mathematics, Abel = Nobel. Science **293**(5536) (2001). http://news.sciencemag.org/sciencenow/2001/08/30-03.html. Accessed on 12 February 2012

[2] Collett, J.P.: Videnskap og politikk. Samarbeid og konflikt om forskning for industriformål, 1917–1930 (1983). University of Oslo

[3] Crawford, E.: The Beginnings of the Nobel Institution: the Science Prizes, 1901–1915. Cambridge University Press, Cambridge (1984)

[4] Doel, R.E., Søderqvist, T. (eds.): The Historiography of Contemporary Science, Technology, and Medicine: Writing Recent Science. Routledge, London (2006)

[5] Friedman, R.M.: The Politics of Excellence. Behind the Nobel Prize in Science. Times Books, New York (2001)

[6] Friedman, R.M.: Human frailty etched in gold: demystifying the Nobel Prize in science. In: Collet, J.P., Myhre, J.E., Skeie, J. (eds.) Kunnskapens betingelser. Festskrift til Edgeir Benum, pp. 146–167 (2009). Vidarforlaget

[7] Helsvig, K.G.: Elitisme på norsk. Det Norske Videnskaps-Akademi 1945–2007. Novus Forlag, Oslo (2007)

[8] Helsvig, K.G.: Universitetet i Oslo 1975–2011. Mot en ny samfunnskontrakt? (2011), Unipub.

[9] Kragh Sørensen, H.: Niels Henrik Abel's political and professional legacy in Norway. In: Siegmund-Schultze, R., Kragh Sørensen, H. (eds.) Perspectives on Scandinavian Science in the Early Twentieth Century. The Norwegian Academy of Science and Letters, I. Mat.-Nat. Klasse. Skrifter og avhandlinger 1, pp. 197–219. Novus Forlag, Oslo (2006)

[10] Reichelt, Y.: Medaljen som for til himmels med unionen, pp. 5–13 in NNF-nytt, 1/2007 (2007)

[11] Ross, P.: Math. Horiz. **Nov**, 9 (1995). http://mathforum.org/social/articles/ross.html. Accessed 10 February 2012

[12] Sem Fure, J.: Universitetet i Oslo 1911–1940. Inn i forskningsalderen (2011), Unipub

[13] Stubhaug, A.: The history of the Abel prize. In: The Abel Prize 2003–2007: The First Five Years, pp. 1–6. Springer, Berlin (2010)

[14] Stubhaug, A.: Niels Henrik Abel. In: Store Norske Leksikon (2011). http://snl.no/.nbl_biografi/Niels_Henrik_Abel/utdypning. Accessed 15 December 2011

[15] Thue, L.: Tormod Hermansen. In: Store Norske Leksikon (2011). http://snl.no/.nbl_biografi/Tormod_Hermansen/utdypning. Accessed 10 December 2011

[16] Zuckerman, H.: The proliferation of prizes: Nobel complements and Nobel surrogates in the reward system of science. Theor. Med. **13**, 217–231 (1992)

2008

John G. Thompson
and
Jacques Tits

"for their profound achievements in algebra and in particular for shaping modern group theory"

ABEL
PRISEN

Some Reflections

John G. Thompson

> "My enthusiasm for mathematics had perhaps had as its principal basis my horror of hypocrisy; hypocrisy, in my eyes, meant my Aunt Séraphie, Mme Vignon, and their priests. In my opinion, hypocrisy was impossible in mathematics, and in my youthful simplicity, I thought it was also the case in all sciences to which I had heard they were applied. What were not my feelings when I perceived that nobody could explain to me how it came about that minus multiplied by minus gives plus ($- \times - = +$)? (This is one of the fundamental bases of the science called *algebra*.)"[1]

The entire chapter, indeed the entire work is vibrant and gossipy. My own difficulties with minus multiplied by minus show that math had the power to engage my full attention. I asked my father why this rule was true, and when he could not explain it, I started crying, tears of frustration welling up. Such a passionate response to incomprehension was an indicator of possible future devotion to math, which my life has borne out.

My older brother showed me how to cast out nines as a check on the accuracy of computed sums. It was not until I learned of modular arithmetic in the first few pages of the *Disquisitiones* that I understood why it works.

Having read Stendhal's autobiography, I sense his genuine interest in math, which was superceded by his commitment to literature. In my case, there were no viable alternate career paths to impede my choice.

It was my good fortune to team up with Walter Feit. Our work generated interest and results in the theory of finite groups. The program of Gorenstein to classify

[1] Stendhal, *The Life of Henry Brulard*, Chap. 34.

Electronic supplementary material Supplementary material is available in the online version of this chapter at http://dx.doi.org/10.1007/978-3-642-39449-2_2. Videos can also be accessed at http://www.springerimages.com/videos/978-3-642-39449-2.

J.G. Thompson (✉)
16 Millington Road, Cambridge CB3 9HP, UK
e-mail: johngriggst@aol.com

H. Holden, R. Piene (eds.), *The Abel Prize 2008-2012*,
DOI 10.1007/978-3-642-39449-2_2,

At Oberwolfach (Archives of the Mathematisches Forschungsinstitut Oberwolfach)

Lecturing at Oberwolfach (Archives of the Mathematisches Forschungsinstitut Oberwolfach)

all the finite simple groups was begun energetically by him, carried forward by Lyons and Solomon, and fundamentally aided by Aschbacher, and by Aschbacher and Smith. The simple groups discovered by Janko, Conway, McLaughlin, Sims, D. Higman, Lyons, O'Nan, M. Hall, G. Higman, Held, Norton, Harada, Rudvalis, Fisher, Suzuki, Ree, and Steinberg, and the deep work of Steinberg and Tits contributed mightily to the completion of the classification of all finite simple groups. The combined efforts of these mathematicians and others too numerous to mention here have created a body of results which I hope and expect will serve diverse areas of scientific research.

John G. Thompson in Oslo, 2008 (Photo Harald Hanche-Olsen)

John G. Thompson (2012)

A Biography of Jacques Tits

Francis Buekenhout

1 1930–1944

1.1 A Belgian Mathematician

Jacques Léon Tits was born on August 12, 1930, in Uccle (Ukkel in Dutch) in Belgium, a southern township of Brussels. All of his publications, except one, are signed Jacques Tits and this is how I [Francis Buekenhout] will refer to him from now. Tits, as well as his ancestors over several generations, are Belgians. However, he became a French citizen in 1975. This was required by French law in order to become a Professor at the Collège de France. Tits always remained faithful to Belgium and proud of his roots despite some bad experiences.

This article is based on a conversation between Jacques and Marie-Jeanne Tits, and Francis and Monique Buekenhout, which took place in the apartment of Tits in Paris on June 30, 2011.

Electronic supplementary material Supplementary material is available in the online version of this chapter at http://dx.doi.org/10.1007/978-3-642-39449-2_3. Videos can also be accessed at http://www.springerimages.com/videos/978-3-642-39449-2.

F. Buekenhout (✉)
Département de Mathématique, Université libre de Bruxelles, CP216, Boulevard du Triomphe, 1050 Brussels, Belgium
e-mail: fbueken@ulb.ac.be

F. Buekenhout
Académie Royale des Sciences, des Lettres et des Beaux-Arts de Belgique, Brussels, Belgium

H. Holden, R. Piene (eds.), *The Abel Prize 2008-2012*,
DOI 10.1007/978-3-642-39449-2_3,

1.2 Ancestors

I have had the privilege to have in my possession a family document[1] centered on Yvonne Tits, the youngest sister of Jacques. This document contains an accurate list of the ancestors of Jacques Tits going back to 1719. The text also includes a chapter, which we will return to, written by Gertrude Tits (1905–2006) who was known to Jacques's family as Aunt Gertrude.

Another most valuable source has been the "Interview with Jacques Tits"[2] by Francis Buekenhout, Bernhard Mühlherr, Jean-Pierre Tignol and Hendrik Van Maldeghem.

1.3 Parents

Jacques's father Léon Tits (1880–1943) was a mathematician and received his education from the Université Catholique de Louvain. His mother, Maria Louisa André (1889–1957) was a piano teacher. She used to be called Louisa, and I will do so hereafter. Léon and Louisa got married in 1917. Their first child lived only 5 weeks. However, they had four other children, all of whom received a degree in science at the Université libre de Bruxelles: Jean (1923–1993) was awarded a degree in civil engineering; he married Violet de la Ruwière (1921–). Ghislaine (1926–2003) was awarded a degree in chemistry (1948). She married Jean-Claude Piret (1926–2003), who was the best friend of Jacques and a mathematician. Yvonne (1929–) was awarded a degree in physics (1951). She married Roald Bingen (1929–1963), an engineer. Jacques, born on August 12, 1930 in Uccle, was awarded his first degree in mathematics (1948). He married Marie-Jeanne Dieuaide (1932–), a historian educated at the Université libre de Bruxelles. They met in 1953 at the Academia Belgica in Rome where both were living as researchers of the Belgian F.N.R.S. ("Fonds National de la Recherche Scientifique"). They married in Brussels in 1956, and had no children. From 1956 on, they have been inseparable: For example, when Tits entered a lecture room to teach, Marie-Jeanne would leave, and come back when, according to her, time was over. The year after Jacques got married, Louisa died feeling secure that someone was going to take care of her youngest son.

There are more mathematicians in Tits's family, namely the two children of Ghislaine and Jean-Claude Piret: Claude Piret (1954–) was awarded a degree in mathematics from the Université libre de Bruxelles. She married the mathematician Didier Misercque (1957–). Their son Corentin (1991–) is an engineer. Jacques Piret (1959–) was awarded a degree in mathematics from the Université libre de Bruxelles.

[1] Johan D'Hondt, *Livre de famille préparé à l'occasion du quatre-vingtième anniversaire d'Yvonne Bingen-Tits* (130 pages, February 2009). Warm thanks are due to my friend Franz Bingen who provided me with this text.

[2] The interview was made in Paris, September 21, 2006, and will appear in "*The Collected Works of Jacques Tits*", The Publishing House of the European Mathematical Society, Zürich.

1.4 Grandparents

The parents of Léon were Auguste Tits (1850–1901) and Anne-Marie Rigaux (1853–1928). The parents of Louisa were Louis André (1851–1927) and Cornelia Becquaert (1852–1912). Auguste became an orphan at the age of three. He was welcomed and raised by his aunt Adèle and her brother Coët, who lived in Louvain. He was a very good pupil at school. He got a job at the Volksbank (later called Kredietbank) and he was a director there. His aunt and uncle were good friends with their neighbors, the Rigaux family who were merchants and whose daughter Anne-Marie would become the spouse of Auguste. They had been friends since childhood. When a young man visited Mr. Rigaux in order to ask if he could marry Anne-Marie, the father consulted his daughter, and she again consulted her friend Auguste. The latter ran immediately to the father and asked the same. Their marriage was celebrated soon thereafter. Gertrude explained that these people were all bilingual. French was used in official matters and for studies. The Flemish language of Leuven (Louvain) was used in daily life. Not long after, French became the exclusive language of the Tits family.

Auguste Tits and Anne-Marie had eight children and Léon was the third. He was reputed to have been the quietest child in the world in contrast to his elder brother. Léon was extremely docile, and he cried a lot. He was very clever, gifted in mathematics and music, and he was a very good pupil. Apparently, his mother always thought that he would become a priest, which he eventually became. He made brilliant studies at the Université Catholique de Louvain, where he later became an assistant. Furthermore, he won the prestigious "Concours Interuniversitaire", and he became a Professor at the Institut Saint-Louis, Bruxelles, which was and remains a top level catholic secondary school in Brussels.

In the interview with Tits (*loc. sit.*), Jacques stated "... he [Léon] was a very timid man. In fact, he was about to become a full professor at the University of Louvain when he left religion, and then he was thrown out. At that time it was very difficult to find a job in any official position, and in order to earn a living ... he had to resort to private tutoring, and it took a lot of his time and he was ... really overworked. Furthermore he smoked, and that is probably why he died rather early ... at the age of 63."

At some point in 1914, Léon visited his superior, Cardinal Mercier, and explained that he had lost his faith and that he felt it to be dishonest to remain in the Church.[3] I continue with the testimony of Aunt Gertrude. In those times, the Church was dominated by the crisis of modernism to which Rome was violently opposed. This resulted in many conscience problems. In 1914, Léon left the priesthood. It was a major drama in his mother's life. Léon faced the vindication of all those who were right-believers. However, Cardinal Mercier did all he could do in order to help him. He found a position for him in the United States as a university professor. Léon was too afraid to cross the ocean in times of war and he refused. He was also reputed

[3] This statement is due to Jacques.

to be fearful, refusing, in particular, to visit a dentist despite his awful sufferings. He got a position from the official "schools network" as a professor at the Athénée d'Arlon. The mayor of the city established such a cabal against him that he was forced to leave. From then on he lived from very badly paid courses he could give at private institutes or private lessons. All those who benefited from his courses remembered his remarkable pedagogical talent. In 1917, he married Louisa André, an excellent pianist. He had known her at home in his youth. She was the daughter of a cousin of his father. She was a warm, cordial and deeply honest woman. Later in life, Léon suffered from Parkinson's disease.

1.5 Child Prodigy—Always Ahead of His Age and of His Time

The Tits family was living in their house at 21 Avenue Victor-Emmanuel III in Uccle. The age difference between Yvonne and Jacques is one and a half year, and in their early childhood they were together most of the time. When Louisa explained something to Yvonne, Jacques got it as well, as by osmosis. It soon became clear that he was learning with an exceptional facility. He could read and write before the age of five. At the École Decroly, where he was from 1936 to 1940, he immediately entered the second primary class, where he was from 1936 to 1940. In 1940, he should have entered the sixth and final year. However, Yvonne entered the first year of secondary school at the Athénée d'Uccle. Their mother insisted that the school accept Jacques for an entrance examination. Thus, he skipped the sixth and last primary class and entered secondary school at the age of ten. He studied there from 1940 to 1944. In the interview with Tits (*loc. sit.*), Jacques stated "When I was very young, say one and a half years (laughs) I got very quickly interested in mathematics, and my father explained things to me; for instance, very early he told me how to use complex numbers to solve algebraic equations, and so on. All that was fantastic for me, and I learned very quickly a lot of things especially in analysis." At the family table, his elder brother asked his father mathematical questions. Jacques captured pieces of the conversations, and he reconstructed the facts. He mastered second degree equations when he was only nine and in the fifth primary class. In 1942 his elder brother entered the Faculty of Engineering, called the École Polytechnique. Jacques was now hearing of differential and integral calculus. As always, he wanted to understand. A great reference on the matter was a treatise in two volumes by Charles-Jean de La Vallée Poussin (1866–1962), the most reputed Belgian mathematician in the first half of the 20th century. These volumes were on his father's bookshelves. His father forbade Jacques to look at these books because he feared that the boy was going too fast and that he might neglect his courses in school. At the time, Léon was severely ill and remained in bed. Profiting from this situation, Jacques got the two volumes and read them. Soon after, he went to his father saying: "Now I know." In April 1943, his father passed away. Jacques understood the economic pressure put on his mother and his family. Some months later, he was a pupil in the fourth class of secondary school. He gave lessons to his brother

Charles Nootens (*right*) and Jacques Tits (*left*)

and friends, explaining topics in their course on differential and integral calculus in the second year at the École Polytechnique. At that time, the Université libre de Bruxelles was closed. The students had to work on their own in little groups. They had to pass examinations before a State Jury or at another university. Jacques was helping a friend from the final year to prepare for the entrance examination at the École Polytechnique.

1.6 Charles Nootens and Petit Jacques

Tits entered the fifth year at the Athénée d'Uccle in September 1944 at the age of 14. School started with a delay of several days. Indeed, since May 1940, Belgium had been occupied by the German troops and ruled by the Nazi regime. Now, after the fierce battle of Normandy the Allies were progressing through Normandy and the North of France. The liberation of Brussels took place on September 3 and 4. It was a time of popular celebration and reorganization in all respects.

Tits was in the class taught by the mathematician Charles Nootens, who received his degree in mathematics at the Université libre de Bruxelles in the thirties. He passed away in 1999. We have a photo of Tits and Nootens from the nineties.

Nootens always called him "Petit Jacques," as he was indeed younger than the other pupils by two years and he was not tall for his age. Nootens had a strong grip. Soon after "Petit Jacques" became his pupil and at the beginning of some lesson, he sent Tits to the blackboard and asked him to state the theorem he had to study. "Petit Jacques" started a proof. He was interrupted by the teacher who said he wanted the statement of the theorem first. "Petit Jacques" replied that he would reconstruct the statement at the end of the proof. Nootens said: "You did not learn your lesson, you get a zero. Go back to your seat." Many mathematicians who had the privilege of listening to Tits later on, will recognize this style of his: reconstruction from the memorized basics. Over the months, Nootens realized that "Petit

Jacques" was a mathematical genius. He realized that the boy already knew calculus, and Nootens realized that it would be advantageous for Tits to skip the last year of high school and actually to interrupt the ongoing fifth year as well. Nootens found out how to proceed. In Uccle, one is within walking distance of the Université libre de Bruxelles. In November 1941, when the Nazi Regime increased its pressure, the Administration Council decided to close the university. In September 1944, with the liberation, the Université libre de Bruxelles started to reorganize itself rapidly with the goal of starting a short academic year on January 1, 1945, which was late by three months with respect to the traditional beginning. In September 1944, Nootens knew and told Jacques that he might perhaps enter the university right away. However, how could it be done without a high school diploma? Again, Nootens knew of a solution. In order to enter the École Polytechnique, candidates had to pass an entrance examination extending over a week in October. If "Petit Jacques" passed the examination, he would be accepted as a student, and he would even be admitted to the study of mathematical sciences. Tits decided to do so. Nootens knew that there was a subject included in the examination that was not studied in high school, namely spherical trigonometry. He lent a book of about 60 pages on that subject to "Petit Jacques." The book was returned the next day, to the teacher's surprise. Was "Petit Jacques" unwilling to study it? Tits said that he had studied the essentials and that he could reconstruct the rest from it. In the reputed entrance examination of the École Polytechnique, Tits scored the highest! He once told me the secret: "When reading a math book, start at the end. If you understand it, you do not need to read the rest." About the examinations, I was given the testimony of Henri Levarlet.[4] He was in charge of the section on solid geometry. Tits was asked to establish the volume of a sphere and to solve a numerical example. The boy of fourteen calmly asked whether he was allowed to use integral calculus. Levarlet was surprised for an instant because this approach was not part of the entrance curriculum. He nevertheless agreed. In the audience were teachers and prospective candidates ... they reacted to Levarlet's acceptance. Levarlet had great fun when colleagues reproached him for having accepted the proposal of Tits.

2 1945–1949

2.1 At the Age of Fourteen, Tits Entered University

At the beginning of 1945, Jacques Tits entered the Department of Mathematics at the Universitté libre de Bruxelles. He was already renowned as a little genius by those who had heard of him. He followed courses given by Paul Libois (Géométrie analytique), Théophile Lepage (Calcul différentiel et intégral, Algèbre), Frans van den Dungen (Mécanique analytique), Robert Debever (Géométrie descriptive), etc. Every course had a component of "project work" supervised by an assistant. The

[4] A letter dated July 17, 1987.

course by Libois was called "Geometry" by everybody. Its leading theme was an introduction to spaces in dimensions two and three with an eye to other dimensions, especially dimension one which was going to matter for Tits's first research. Euclidean, affine, projective, inversive, Lobachevski, de Sitter and Minkowski spaces all came to the forefront, together with groups and their inclusions. Unity of the various geometries came from projective spaces and projective quadrics over the real and the complex fields.

2.2 *Jean-Claude Piret, a Friend for Life*

Jacques was not lost at the university. His sister Ghislaine undertook the study of chemistry in January 1945, at the same time as he started with mathematics. Their older brother Jean was already in the fourth year at the École Polytechnique.

The first session of the "project work" in geometry was held after three weeks. Libois organized this in an original way, inspired by the great pedagogue Ovide Decroly with whom he had a lasting friendship. The students were distributed in teams of four or five on a spontaneous basis. Each team included a second year student. Older students circulated from team to team. The idea was to discuss geometry questions within the team. Every person was supposed to participate without any restrictions. Libois had a pile of note cards. On every card he had written a problem. In principle, every problem could keep a team busy for several two hours sessions. The pile circulated from team to team and every team made its choice. In the team of Tits there was a tall freshman, Jean-Claude Piret, whom Tits met for the first time. When the pile of cards was in their hands, Tits looked at the cards. When he had read a card, he made a comment like "Yes, of course," meaning that it was not really a problem. Then he put the card in the bottom of the pile. Piret who was a well-prepared student, objected as he wanted more explanations, but Tits continued. There was a unique problem that he found worthy of study, which remained famous for many generations of students. It reads as follows: Candles of lengths $l_1, l_2, \ldots, l_n$ stand on a horizontal plane. They burn simultaneously. What is the locus of their center of mass? Tits and Piret became inseparable friends. Later on, Piret married Ghislaine, the sister of Jacques. The close friendship of Tits and Piret lasted for dozens of years, until the death of Piret in 2003. In May–June 1945, Jacques's friends were working hard in order to prepare for their exams. Jacques had nothing to do; he knew it all and much more. Ghislaine was dealing with the huge chemistry course in the first year. Tits decided to help his sister and in this way he learned a great deal of basic chemistry. At Tits's own exam in geometry, Libois surprisingly asked him whether he could compare the projective quadrics that he had studied to the cubic surfaces of which he had never thought. When Tits left the examination room he said to his friends that he had failed. In fact, Libois had given him the highest mark, 20 out of 20, after a moment of hesitation. One day in 1964, Libois did it again for Pierre Deligne.

2.3 Lectures of Libois in 1945–1946

Tits was attracted by Analysis, the preferred name given to the course of Lepage that was more precisely devoted to Real Variables. A renowned "Treatise of Analysis" by Goursat, comprised three thick volumes. It went well beyond the material that could be covered in the three first years at the Université libre de Bruxelles. Tits got to know it completely in 1945–1946, at age fifteen. He was a little disappointed in Lepage. Indeed, after six weeks in the first year, the boy had decided to try to obtain his first degree in two years rather than four. Lepage was rather surprised. However, the laws did not allow this. The relationship was taking another turn with Libois who was teaching "Géométrie projective" for second year students. It is relevant here to mention that Libois was an important communist leader. When the university closed in November 1941, he started to live clandestinely, in various tiny hiding places provided by friends. At some point, the Nazi authority put a prize on his head due to his direct connection with the Kremlin and the transmission of orders to the Resistance in some part of Belgium. Hours before the closure of the university, he went to the library where he borrowed two major geometric works: "Conics"[5] by Apollonius (around 262 BC–190 BC), and "Geometrie der Lage" (Geometry of position) from 1847 by Karl von Staudt (1798–1867). These works became part of his small luggage. They were intensively studied for two and a half years. Before the war, Libois had realized from analytical geometry that there was a gap in the foundations of projective geometry. Indeed, homogeneous coordinates apply perfectly in order to describe a projective line as a projective space of dimension one. However, the theory of projective spaces as presented in Veblen and Young's "Projective Geometry" (1910), does not allow for a non-trivial space of dimension one. Thanks to von Staudt, Libois came up with a new idea. In a projective plane over a field, central mappings from line to line in all possible ways provides every line with an intrinsic group acting 3-transitively on its points and the action is sharp: the stabilizer of three points is reduced to the identity. This is a key result achieved by von Staudt in the real plane. It is a combination of mobility and rigidity that characterizes the work of Tits and its developments till now. The idea of Libois was that progress in one-dimensional projective geometry required the study of a set equipped with a sharply 3-transitive permutation group. Also, with respect to affine geometry, he wanted the study of sharply 2-transitive permutation groups. He had no other references. These ideas were in his course on Projective Geometry in 1945–1946 for the class of Tits. Libois raised one of the right questions: does there exist a sharply 2-transitive permutation group on six points? Why six? Because there is no field of order six! He did not have the answer. Tits did not yet study it. Tits had two more years before he could obtain his first degree. Every year he was passing the examinations with "la plus grande distinction." For the years 1946–1948 he became a student-assistant, supervising students in mathematics and physics during one afternoon every week.

[5] In the famous translation to French by the Belgian engineer and historian of mathematics, Paul Ver Eecke (1867–1959).

2.4 Research

In 1946–1947, Libois further developed the subject in his course "Géométrie Supérieure" for the third-year students including Tits. This time, Tits responded forcefully. For six weeks in a row, he devoted most of his time to research about the subject. His family and friends feared for his health. Libois had a conversation with the young man of 16, and discovered with stupefaction the results accumulated by his student. Tits produced a complete classification of the finite sharply n-transitive groups for all $n \geq 2$. His theory did not limit itself to the finite case. In addition to this, he had started a theory which applied to the affine and projective groups in any dimension. Libois calmed the volcano by a proposal: the first part of Tits's results could constitute the "Mémoire de Licence" required as the student's main work in 1948. The rest of his work could be the subject of his "Thèse de doctorat" to be completed in 1949 (actually defended early in 1950). In those days, it usually took five or six years in order to get a PhD after the first degree.

2.5 First Degree in 1948

The Dissertation was conceived along the proposal made by Libois. Writing a text was a hard task for the young man. He needed to write the Dissertation five times on his typewriter in order to obtain Libois's approval. This was his first mathematical writing. Actually, it was also his first writing in French. Indeed, he had skipped the final year and most of the fifth year at the Athénée. Libois was training him on this matter. There was no time for research.

2.6 Paris and Emil Artin

Tits became a "Boursier" (1948–1956) of the Belgian F.N.R.S. This gave him total freedom to travel, study and do research. The results were immense. In 1949, Libois took his student to a conference in Paris. It was the "Colloque d'Algèbre et Théorie des Nombres." At first glance, Tits did not observe any significant differences between Brussels and Paris. He was going to change his mind completely. He gave a lecture and provided a paper for the proceedings entitled "Groupes triplement transitifs et généralisations." Among those present were Claude Chevalley (1909–1984) and Emil Artin (1898–1962). The latter was a Professor at Princeton University from 1946 to 1958. He was impressed by Tits. He explained to Tits that some of his results had been obtained already in 1936 by Hans Zassenhaus, who was one of his students. Tits included a footnote in his paper to mention the work of Zassenhaus.

3 1950–1963

3.1 Docteur ès Sciences Mathématiques

His thesis concluded his work in 1947. A gem among several others was: a sharply 4-transitive permutation group cannot be infinite! Fifty years later, Tits was still a bit angry about the fact that Marshall Hall Jr. had overlooked this result in his famous "Theory of Groups" of 1959. Indeed, Hall mentioned the work of Camille Jordan (1872) classifying the finite sharply 4-transitive groups and he proved a stronger result of his own superseding Tits's result.

3.2 To Heinz Hopf in Zürich in 1950, 1951, and 1953

The visits of three months to Zürich were made upon the initial advice of Paul Libois in order to go and listen to Heinz Hopf and to study Lie groups. This was highly fruitful! Hopf told Tits about the famous Helmholtz–Lie problem, which roughly asked for all "spaces" equipped with a "convenient" group of motions. Kolmogorov had provided a major contribution in 1930. He gave a system of axioms and provided the classification. He did not provide the proof. Hopf suggested that Tits might work it out. He definitely did! One of Kolmogorov's axioms was asking a property M_n indexed by a natural number n. Tits showed that M_2 sufficed to get a classification, and he wrote down the proof in May 1954. It appeared in his Thèse d'Agrégation (1955). This work may be considered as a final solution to the Riemann–Helmholtz–Lie space problem, namely the search for "all possible" non-Euclidean spaces, for example. It was also under the influence of Hopf that Tits got interested in the projective plane over the octonions, and he started to work with the exceptional groups G_2, E_6 and F_4 in papers of 1953 and 1954. He got to know the work of Freudenthal (1951) on the geometry of the exceptional groups. Freudenthal was active in Utrecht which was only some 150 kilometers from Brussels, and there was a direct train connection. During the next years, Tits went regularly to visit Freudenthal. The latter was one of the few who understood the work of Tits. They worked independently on the same subjects. Freudenthal did much to publicize the mathematics of Tits. This is how Tits got the famous "Magic Square" concept in 1954, without calling it that. This is a symmetric 4×4 matrix whose rows and columns are indexed by the division algebras of finite dimension over the reals, namely the real field, the complex field, the quaternions and the octonions. As to the entries, there are different versions: Lie algebras, Lie groups, geometries. The exceptional Lie groups E_6, E_7, E_8 occupy some of the entries.

3.3 *Institute for Advanced Study, Princeton and H.C. Wang (1951–1952)*

After the "Colloque" in Paris (see Sect. 2.6), Artin had written to Herman Weyl in order to recommend that Tits be invited by the Institute for Advanced Study. Heinz Hopf wrote to Weyl with the same purpose: "a star is born." Tits got invited for a year and received a scholarship from the Institute for Advanced Study. In Princeton, Tits shared an office with H.C. Wang (1918–1978), and they became good friends. Wang taught Tits a lot on Lie groups, but nothing specific on the exceptional groups as Tits recalls in the Interview with Tits (*loc. sit.*).

3.4 *The Cremona Plane*

Paul Libois had been working in the context of Italian algebraic geometry. He spent a year in Rome in the early thirties. He worked under the daily supervision of Federigo Enriques, and he became friends with Guido Castelnuovo and his daughter Emma. Furthermore, he met Oscar Zariski. When Libois defended his Thèse d'Agrégation (1934) in Brussels, Enriques was a member of the jury. Libois became a full professor at the Université libre de Bruxelles in 1937. Libois advised several students in algebraic geometry over the next 30 years. Among other subjects, he pursued the following problem: Consider the complex Cremona plane equipped with the group of birational transformations in three homogeneous coordinates. Such a transformation maps a point to a point with exceptions related to a "blow up." In other words, a point is not a Cremonian invariant. The problem was to look for a Cremonian invariant and to build a Cremonian foundation for the plane. Progress was made by Pierre Defrise who wrote his "Thèse d'Agrégation" on the subject (1949). Jacques Tits closely followed this work. In 1950, he found a stunning solution for Libois' problem. He wrote down a system of geometric axioms for the Cremona plane over any algebraically closed field. However, his manuscript remained unpublished. He came back to the subject in 1999, for a Collège de France course taught at the Université libre de Bruxelles. Notes were taken, and the resulting paper will be part of the Collected Works of Jacques Tits.[6]

3.5 *The Thèse d'Agrégation (1955)*

In May 1954, Tits completed the manuscript of his monumental Thèse d'Agrégation: "Sur certaines classes d'espaces homogènes de groupes de Lie" (268 pages). The purpose of the jury was to ask for a publication in the Memoirs of the Académie

[6]"The Collected Works of Jacques Tits" (*loc.sit.*).

Royale de Belgique and to solicit advice on the work from reputed mathematicians. The chairman of the jury would take the manuscript to the Academy. After some months without news, some anxiety developed. It turned out that nobody knew where to locate the manuscript. Some day in 1955, Tits visited the Perpetual Secretary of the Academy, the astronomer Jacques Cox, for some other purpose. At some point he saw his manuscript there on the shelves in some pile among other piles of papers. Eventually, the Academy decided to publish the work in May 1955. The Examination Jury included Heinz Hopf (Zürich) and Deane Montgomery (Princeton). The report written by Hopf to Lepage on July 27, 1955 stated[7]

> "Dear colleague, It is already close to one year since you asked me to give my advice on Mister Tits and his work, and more particularly on his thesis "Sur certaines classes d'espaces homogènes de groupes de Lie." I do not want to excuse my long silence, but I insistently beg you not to interpret my silence as indifference with respect to Mister Tits and his work. All to the contrary, among the young mathematicians known to me, there are few indeed who I do appreciate as much as Mister Tits and on whom I have great expectations. His spirit, acuteness, his scientific training, his sane and natural manner to choose and attack problems, his working capacity still keep me under a strong impression. The conversations I remember to have had with him on mathematical subjects were always interesting and stimulating. His works rest on a deep synthesis of algebraic and geometrical methods and they bring testimony of an acute sense of essentials. His results range from elementary geometry to the difficult domain of present mathematics as the theory of the exceptional Lie groups in which he has made an important contribution and—to take but one example, the results of Tits (Thèse, 4è partie, E) are completing and ending in a most satisfactory way the set of theorems of Kolmogorov and Wang. The thesis is imposing, not only by the abundance of its original ideas, but also because it offers a masterly broad view on a great deal of modern geometry and I believe that its author may be counted among the best geometers of our time."

In the report by Montgomery from August 19, 1954, on his way to the International Congress in Amsterdam, he stated

> "...I am writing to say that I have very high regard for this paper by Tits as well as a great respect for all the work he has produced. He seems to me to be an unusually brilliant and well-informed young man who has made a great many valuable contributions especially for one so young. The present paper on homogeneous spaces is certainly the best work he has done so far."

A year later Montgomery confirmed: "...it is a first-rate piece of work. My opinion of Tits is very high in all ways." The Thesis was defended on December 19, 1955. The event included a public lecture entitled: "Mathematical Foundations of General Relativity." He was 25 years old.

[7] Translated from German by Ms. Charlotte Bouckaert.

3.6 Memoir for the Prix Louis Empain (1955)

Tits received the "Prix scientifique interfacultaire Louis Empain." He submitted the original memoir "Les espaces doublement homogènes et les espaces homogènes et isotropes" (124 pages, completed on December 31, 1954): The memoir remained unpublished,[8] yet it contained brand new geometry, and constituted the best genuine realization of Klein's Erlangen Program. Here, there are difficult questions dealt with for every simple Lie group, in particular, the exceptional group E_8. All available resources on Lie groups were used.

3.7 Prehistory of Buildings (1955–1961)

In his Thèse, Tits had a uniform construction of a geometry attached to every simple Lie group G. Eventually, these would be buildings. The geometry was constructed from subgroups of G now called parabolic subgroups. The geometry was closely related to the Dynkin diagram underlying G. Tits studied these geometries systematically. They were called "R-espaces" in different papers from 1954 on. The Chevalley groups were readily integrated in Tits's theories and observations. They provided "R-espaces" right away. In 1961, the birth of buildings, apartments, Tits systems and abstract regular polytopes was announced publicly during a conference in Florence. Tits had the idea in the Metro of Paris on his way to the Séminaire Chevalley. Tits wanted to grasp the Bruhat BNB property from a geometric viewpoint. His terminology did not use the words building, apartment, etc., rather he used "generalized polyhedron," "skeleton," group with a BN-pair, and "regular polyhedron."

3.8 Birth of the General Theory of Coxeter Groups (1961)

In his unpublished paper "Groupes et géométries de Coxeter,"[9] Tits founded a great theory, which he named after Coxeter in respectful generosity to the great pioneer. These notes were written to serve Bourbaki, via Tits's friend François Bruhat, for the volume "Groupes et Algèbres de Lie." In my opinion, this was Bourbaki's best volume (1968) ever, but I did not read all of them. By the way, a person who has read every volume and solved all the exercises is Pierre Deligne. In this volume, Bourbaki made plain use of Tits's work, acknowledging his contributions in a footnote: "Pour la rédaction de ces trois chapitres, de nombreuses conversations avec J. Tits nous ont apporté une aide précieuse. Nous l'en remercions très amicalement." The paper

[8]It will be included in "The Collected Works of Jacques Tits" (*loc.sit.*).

[9]To appear in "The Collected Works of Jacques Tits" (*loc.sit.*).

V. Georges, M. Lazard, J. and M. Tits (from left to right) in 1957 (Archives of the Mathematisches Forschungsinstitut Oberwolfach)

by Tits was not mentioned by Bourbaki, and it remained unpublished until 2001, when it appeared in a volume devoted to the Wolf Prizes.[10] The editors quote

> "Part II is the reproduction *ne varietur* of mimeographed notes [attributed to Tits] entitled *Groupes et Géométries de Coxeter* ... which never appeared in print. This is the first paper ever written on *arbitrary* Coxeter groups (the terminology was coined there); it played a rather important, though somewhat hidden role in the early history of those groups. Indeed, the most commonly used reference concerning them is Chap. 4 of Bourbaki, *Groupes et Algèbres de Lie*, and the mimeographed notes in question were precisely written to serve Bourbaki's work on that volume. But while Bourbaki's book, and in particular its "Note historique," fully acknowledges Tits's contributions to the subject, it does not explicitly mention the preprint reproduced here, which is the original source for those contributions."

In this book, buildings were revealed in the exercises. Here, the word "immeuble" appeared for the first time. The concept had been made public by Tits in 1961 at a conference in Florence and in his paper for the proceedings.

3.9 Denied Access to the US from 1953 to 1963

Every year from 1953 on, Tits was invited to visit Princeton or other important places in the US. His visa was refused every year. In 1963 he went to the US Em-

[10]S.S. Chern and F. Hirzebruch (editors), "*Wolf Prize in Mathematics*", vol. 2, 2001, World Scientific, Singapore.

bassy in Brussels and inquired about the reasons why he was denied visa. Eventually he heard that he was suspected of being a communist, which he forcefully denied. Then, after some activity at the embassy, he heard that he was suspected of being a member of the "Amitiés Belgo–Soviétiques." Again, he forcefully denied the allegations. His sister may have attended a conference organized by that association. However, at this time, the witch hunt period had come to an end, and he obtained his visa in a matter of days. During that year 1963, he was invited for long stays in Chicago and Berkeley. In 1964, he was back at the Institute for Advanced Study.

3.10 International Congress of Mathematicians (1954–1994)

In 1962 Tits was an invited plenary speaker at the International Congress of Mathematicians (ICM) in Stockholm. He was also an invited speaker at the ICM in 1970 in Nice, and again a plenary speaker at the ICM in 1974 in Vancouver. He was a member of the jury awarding the Fields Medals in 1978 (Helsinki) and in 1994 (Zürich). In 1978, he read the laudatio for the Fields Medalist Grigori Margulis. The Soviet authorities had not allowed Margulis to attend the Congress. A similar situation had occurred already in Nice in 1970 and in Vancouver in 1974.

Tits's speech on Margulis included a courageous political conclusion:

> "I wish to conclude this report by a nonmathematical comment. This is probably neither the time nor the place to start a polemic. However, I cannot but express my deep disappointment—no doubt shared by many people here—in the absence of Margulis from this ceremony. In view of the symbolic meaning of this city of Helsinki, I had indeed grounds to hope that I would have a chance at last to meet a mathematician whom I know only through his work and for whom I have the greatest respect and admiration."

A footnote recalled that "The address was delivered in Finlandia Hall, where the 1975 Helsinki Agreements were concluded." The Helsinki Agreements dealt with security and cooperation in Europe. One of its priorities was the respect of human rights and of fundamental liberties. The absence of invited Soviet mathematicians at the International Congresses was explained repeatedly by Lev Semenovich Pontryagin who expressed the profound dissatisfaction of the National Committee of Mathematics of the USSR concerning the choice of Soviet mathematicians who were invited to deliver a talk and to receive a Fields Medal. The first visit of Margulis to the West was in 1979 when he stayed at the University of Bonn for three months. A little ceremony was organized and Tits could give his medal to Margulis. Tits was already present at the ICM in 1954 in Amsterdam with a short (contributed) lecture and abstract. Hans Freudenthal devoted his invited lecture to the mathematics of Tits.

4 1964–1975

4.1 *Professor at the Universität Bonn (1964–1974)*

Jacques and Marie-Jeanne Tits moved from Brussels to Bonn. Here Tits was recognized in all respects at last, as a great mathematician and great professor.

4.2 *Buildings Coming of Age (1974)*

The fantastic theory of buildings appeared in the Springer Lecture Notes.[11] It dealt with a totally new geometry already foreseen in 1961. There had been a preliminary preprint on Chap. 7 devoted to polar spaces, circulated in 1968. The book also dealt with his theory of finite groups with a BN-pair (or Tits's systems).

4.3 *Collège de France (1975)*

Jacques and Marie-Jeanne Tits moved to an apartment near Place d'Italie in Paris. Jacques became an Associate Professor at the Collège de France (1973–1975). He acquired French citizenship in order to conform to law when he applied for a full professorship. He was a professor at the Collège de France (1975–2000) in the Chaire de Théorie des Groupes. He delivered his inaugural lecture in 1975, from which we quote: "It has to be expected that, in my lectures, geometry will often take its revenge from the Erlangen program, the theory of groups serving as a pretext this time."

5 1976–2000

5.1 *Professor at the Collège de France (1973–2000)*

A course was taught every year on an entirely new subject or new developments for a former subject. The course was followed by a most interesting text called "Résumé de cours." These texts were published by Collège de France.

[11] "Buildings of spherical type and finite BN-pairs." Springer Lecture Notes, vol. 386, 1974.

Jacques Tits at Oberwolfach in 1980 (Peter van Emde Boas, Archives of the Mathematisches Forschungsinstitut Oberwolfach)

5.2 No Pension in Belgium (1994)

Every person who has worked in Belgium is entitled to a pension. Jacques Tits had been working for the Université libre de Bruxelles from 1957 to 1962. However, he was unable to get recognition for his rights to pension. The question was not a material one. He got a pension from Germany and from France. The question was a matter of justice. Tits did all he could. Friends and colleagues such as Franz Bingen and Joseph Thas tried to help Tits. However, they remained unsuccessful.

A letter from Tits to the Minister of Pensions, Frank Vandenbroucke in June 2002, reveals interesting details on his career. After an explanation about his position at the Université libre de Bruxelles where he was full professor in 1962, he continues:

> "Les enseignements qui m'étaient confiés par cette université ...étaient d'un niveau très élémentaire. Des discussions que j'ai eues à plusieurs reprises avec les autorités académiques, m'ont fait comprendre qu'il n'était pas possible, même à moyen terme, d'améliorer cette situation, à la longue très défavorable à mon développement scientifique. Or, à partir de 1957, j'avais reçu diverses offres d'universités étrangères: allemandes, américaines, suisses. Cela m'a finalement conduit à accepter, en 1964, une chaire de mathématique à l'Université de Bonn. En 1974, j'ai été élu au Collège de France, à une chaire de Théorie des Groupes créée à mon intention. Accepter cette chaire impliquait que je me fasse naturaliser français, ce que j'ai fait (je me suis, à l'époque, renseigné auprès de l'Ambassade de Belgique à Paris sur la possibilité de conserver également la nationalité belge, mais il s'est avéré qu'une telle possibilité était exclue)."

Marie-Jeanne and Jacques Tits in 2007 (Archives of the Mathematisches Forschungsinstitut Oberwolfach)

6 2001–2012

6.1 *The Book with Weiss*

A major mathematical feat is the book "Moufang polytopes" (2002) by Tits with Richard Weiss in 2002. It is a masterpiece of 535 pages! Richard Weiss had an immense courage and love for the master. He came over from Boston to Paris a great number of times to record their conversations and write from them. I note in passing that Tits never did use email, but he did use a fax.

6.2 *Editor of Mathematical Journals*

Tits served as an editor for several journals. We can mention here that he served as Chief Editor of the premier journal 'Publications Mathématiques de l'Institut des Hautes Études Scientifiques (IHES)" from 1980 to 1999. Furthermore, he was a founder of "Inventiones Mathematicae" and served as an editor during the period 1966–1975.

7 Postscript

As a preparation for writing this biographical sketch, my wife Monique and I visited Jacques and Marie-Jeanne Tits in their apartment in Paris near Place d'Italie on June 30, 2011. During our visit, Jacques was stuck in his armchair. He kept smiling and joking as he always did.

Jacques Tits in Oslo in 2008 (Photo Harald Hanche-Olsen)

The Work of John Griggs Thompson: A Survey

Richard Lyons and Robert M. Guralnick

Up to the early 1960's, really nothing of real interest was known about general simple groups of finite order ... Since he [John Thompson] first appeared at the International Congress in Stockholm eight years ago, finite group theory simply is not the same any more.

Richard Brauer, 1970
ICM Nice

1 Thompson's Thesis, and Local Analysis

In 1954, Richard Brauer, the leading finite group theorist of the time, wrote on the occasion of the International Congress of Mathematicians in Amsterdam [18]

> The theory of finite groups has been rather in a state of stagnation in recent years. This has certainly not been due to a lack of unsolved problems.

In rather short order, however, waves of exciting new results washed away such pessimism. By the end of the decade Claude Chevalley, Robert Steinberg, Michio Suzuki and Rimhak Ree had discovered and constructed all the finite simple groups of Lie type; Philip Hall and Graham Higman had published an influential study of the structure of solvable and p-solvable groups; and Suzuki had proved the solvabil-

---

Electronic supplementary material Supplementary material is available in the online version of this chapter at http://dx.doi.org/10.1007/978-3-642-39449-2_4. Videos can also be accessed at http://www.springerimages.com/videos/978-3-642-39449-2.

R. Lyons (✉)
Department of Mathematics, Rutgers University, 110 Frelinghuysen Road, Piscataway, NJ 08854, USA
e-mail: lyons@math.rutgers.edu

R.M. Guralnick
Department of Mathematics, University of Southern California, Los Angeles, CA 90089-2532, USA
e-mail: guralnic@usc.edu

H. Holden, R. Piene (eds.), *The Abel Prize 2008-2012*,
DOI 10.1007/978-3-642-39449-2_4, © Springer-Verlag Berlin Heidelberg 2014

ity of groups of odd order of a certain kind (so-called CA-groups, in which commuting is assumed to be a transitive relation on the set of all nonidentity elements). In 1959, John Griggs Thompson received his doctorate, having been supervised by Saunders Mac Lane at the University of Chicago.

Thompson's thesis [93] was the first of three seminal papers of ever-widening scope. Next came the Feit-Thompson Odd Order Paper [27], which appeared in 1963, and the third was the so-called N-Group Paper [110, 115, 117, 120], on which Thompson lectured during the 1960's and which appeared in installments between 1968 and 1974. These three articles led finite group theory into a new world, rich in new structures, new problems, new vocabulary, and new tools. Certain sections of these papers—much of the first eight sections of the Odd Order Paper, and Sect. 5 of the N-Group Paper—were the new essential text for students of the subject. One could not find much of this material anywhere else.

In his thesis, Thompson proved a celebrated conjecture about Frobenius groups. A *Frobenius group* G is a subgroup of a finite symmetric group $Sym(n)$, that is, a permutation group on n letters, which is transitive and has the properties that each non-identity element fixes at most one letter, while some non-identity element fixes some letter. Familiar examples are the dihedral group acting on the vertex set of a regular n-gon, n odd; the alternating group on 4 letters; and for any finite field, the group of all affine transformations of the affine line. Now in a transitive permutation group, an elementary count shows that on average, each group element fixes one letter; it then follows that there must be precisely $n-1$ elements of G fixing no letter. Frobenius proved that these elements, together with the identity, form a normal transitive subgroup K of G, now called the "Frobenius kernel" of G. The main difficulty in Frobenius' theorem is to show that K is actually a subgroup of G; this is overcome by using the theory of group characters.

If we let H be the stabilizer of some arbitrarily chosen letter, then $G = HK$ and $H \cap K = 1$, so that the structure of G is determined by the structures of H and K and the action of H by conjugation on K. This action turns out to be necessarily "fixed-point-free," which is to say that the centralizer $C_K(h) = 1$ for every $h \in H$. Consequences of this for the structure of H (which with essentially one exceptional isomorphism type must be solvable) were understood by the middle of the twentieth century—see for instance [54]—but in the 1950's it was an open question whether the kernel K must always be nilpotent. (A finite group is nilpotent if and only if any two elements whose orders are relatively prime must commute, or equivalently if the group is the direct product of its Sylow subgroups. In the examples of Frobenius groups cited above, the kernel is actually more than nilpotent in each case—it is Abelian.) In his thesis [93] Thompson settled the matter.

Theorem 1 [93, 94] *The Frobenius kernel of any Frobenius group is nilpotent.*

At least as significant as the theorem itself were the revolutionary tools Thompson created—the beginning of a series of robust inventions and insights to analyze the architectural structure of a finite group. Let us digress briefly to examine the goals of such an analysis.

By Sylow's Theorem a group G of order

$$|G| = \prod_{i=1}^{n} p_i^{a_i},$$

where the p_i are distinct primes, possesses Sylow subgroups—subgroups of order $p_i^{a_i}$—for each i. "Local analysis," which had its beginnings around the turn of the twentieth century in the work of Frobenius and William Burnside, aims to understand how these subgroups fit together inside G. For example, a theorem of Burnside and Philip Hall asserts that G is solvable if and only if $n = 1$ or G possesses subgroups P_i of order $p_i^{a_i}$, $i = 1, \ldots, n$, such that for any $i \neq j$, the subgroup $\langle P_i, P_j \rangle$ of G generated by P_i and P_j has order $p_i^{a_i} p_j^{a_j}$. Burnside's part in this was to deal with the rock-bottom case $|G| = p_1^{a_1} p_2^{a_2}$. In honor of Hall's contribution, $\langle P_i, P_j \rangle$ is now called a "Hall $\{p_i, p_j\}$-subgroup of G."

Nowadays, with the classification of finite simple groups (CFSG) in hand, we know by inspection of all examples that the "local" structure of a nonabelian finite simple group is highly restricted. For one thing, its Sylow subgroups have one of a handful of structures, related to wreath products, complex reflection groups, or crystallographic root systems. Most striking of all, the simple group must contain elements of order 2—this is the Odd Order Theorem published by Feit and Thompson a few years after Thompson's thesis.

To what extent can one show *from the axioms of group theory* that such restrictions are necessary consequences of simplicity? Even without knowledge of the full list of finite simple groups, and indeed with some not even yet discovered, this is the question that fascinated Thompson and drove his work for nearly two decades.

Proving theorems about a simple group G, one has at first glance a negative hypothesis to use: there are no normal subgroups other than 1 or G, i.e., no *proper* normal subgroups. Indeed, until the 1950's, theorems about simple groups were typically proved indirectly, with the eventual contradiction arising from the existence of a representation or other homomorphism with a proper kernel. With Thompson's work, and others' before him in the 1950's, notably Richard Brauer and Michio Suzuki [18, 68], a positive spin on simplicity took hold as well: for every proper subgroup H of a simple group G, the normalizer $N_G(H)$ is again a proper subgroup. So perhaps one can find a rich enough inductive setting in which to analyze G.

Suppose, for example, that one wishes to prove that if in a finite group G every two elements generate a solvable group, then G is solvable. This hypothesis is inherited by subgroups and quotients, so one immediately is reduced by induction to the case in which G is simple and nonabelian, and all its proper subgroups are solvable. In particular $N_G(H)$ is solvable for every proper subgroup H of G, and from this one must prove that G is generated by 2 elements, hence solvable. In fact Thompson does this and more in his N-Group Paper (q.v.): he deduces the possible isomorphism types of G and then observes in each case that G is generated by 2 elements.

Most important among the various normalizers $N_G(H)$ are those for which H is a p-group, that is, a group whose order is an integral power of a prime p. Such normalizers are called *local*, or *p-local*, subgroups of G. The goal of local analysis is then to deduce "global" properties—properties of G itself—from properties of the local subgroups of G. More specifically, when G is simple, one then hopes to deduce properties of the local subgroups from the simplicity of G.

The technical result behind the proof of Theorem 1 is just such a local-global theorem. It is a criterion for a Sylow p-subgroup P of a finite group G to have a normal complement, i.e., for G to have a normal subgroup of order $|G|/|P|$. Equivalently it is a criterion for a Sylow p-subgroup P of G to be a homomorphic image of G. What was new was that the criterion only mentions normalizers of those subgroups of P that are normal in P and invariant under automorphisms of G leaving P invariant.

Theorem 2 [94, Theorem A] *Let K be a finite group with a p-Sylow subgroup P, p an odd prime, and let A be a group of automorphisms of K which leaves P invariant. Suppose that for every A-invariant normal subgroup Q of P, all elements of order prime to p which normalize Q also centralize Q. Then K possesses a normal p-complement.*

This generalized—at least for odd primes p—a theorem of Frobenius: if for *all* subgroups $Q \leq P$, all elements of order prime to p normalizing Q actually centralize Q, then K possesses a normal p-complement. The restriction of the hypothesis to A-invariant normal subgroups of P enabled Thompson to prove Theorem 1 by induction, taking K to be the Frobenius kernel K above, and A to be the point stabilizer H—or some conjugate of H. Theorem 2 and induction imply that K has an H-invariant normal subgroup N other than K and the trivial subgroup. Then with an appeal to induction, K/N and N must be solvable, so K is solvable; and routine arguments then imply that K is nilpotent. The key all along was to prove the solvability of K.

2 The Thompson J-Subgroup and Weak Closure Arguments

In Thompson's proof of Theorem 2, and in the Feit-Thompson Odd Order Paper [27], a technique was introduced and developed that over the years acquired the tag "weak closure arguments". This technique, heavily reliant on the representation theory of subquotients of G in finite characteristic, was refined and generalized by Thompson and others, in many steps, all the way through the final general theorems of the CFSG. For the "quasi-thin" books of Michael Aschbacher and Stephen D. Smith [10, 11], and the "uniqueness case theorem" of Aschbacher [7–9], for example, the authors needed to develop elaborate "weak closure machinery".

Thompson was aware of ground-breaking work of Philip Hall and Graham Higman dating to 1956 [39], in which they studied the representations of solvable groups in characteristic $p > 0$ and applied their results to the Burnside problem. This famous problem aside, their striking Theorem B was of wider interest. It asserts that for an irreducible solvable subgroup $G \leq GL(n, p)$, any element $x \in G$ of order p^n must have minimal polynomial equal to $(t-1)^{p^n}$ or $(t-1)^{p^n-1}$, and even the latter case can only occur if p is a Fermat prime or 2.

The Hall-Higman Theorem B plays a critical role in Thompson's arguments in his thesis, and later in his refinements and extension of those arguments.

An early refinement was the introduction by Thompson of what is now called the **Thompson subgroup** of a p-group P. It is defined [100] as

$$J(P) := \langle A \leq P \mid A \text{ is Abelian of maximal order among all subgroups of } P \rangle.$$

(Walter Feit tried calling this $\mathbf{T}(P)$ in his *Characters of Finite Groups* [22], but the notation did not stick.) It is obvious that $J(P)$ is invariant under all automorphisms of P. In [100] Thompson strengthened Theorem 2 as follows. Here $Z(P)$ is the center of P.

Theorem 3 *Suppose that p is an odd prime and P is a Sylow p-subgroup of the finite group G. If the subgroups $N_G(J(P))$ and $C_G(Z(P))$ have normal p-complements, then G has a normal p-complement.*

Thompson saw that the Hall–Higman theorem would provide the punch line to the proof. Through an original kind of reduction involving a new ordering on the set of p-local subgroups of G, he reduced for all practical purposes to the following (admittedly oversimplified) case: $G = SQR$, $S \lhd G$, S is the direct product of copies of Z_p, and $S = C_G(S)$. Moreover R has order p, R acts nontrivially on Q, Q acts nontrivially on S, and $|Q|$ is a power of a prime $q \neq p$. The subgroup $P = SR$ is then a Sylow p-subgroup of G. Moreover, S can be regarded as a vector space over the field of p elements, and by the conjugation action we can regard $QR \cong G/S \leq \mathrm{Aut}(S) \cong GL(n, p)$.

Here we have a kind of configuration that one sees frequently in local analysis: if we have information about the action of R on Q and the action of Q on S, what can we deduce about the action of R on S? In this case we know that $|Q|$ is relatively prime to both $|R| = p$ and $|S|$. Then Theorem B provides an answer: the fixed subspace of S under the action of R has codimension at least $p-2$. With a little extra argument when $p = 3$, S is then seen to be the unique Abelian subgroup of its order in the Sylow p-subgroup SR of G. So $S = J(SR)$. But then $G = N_G(J(SR))$ has a normal p-complement by assumption, which finishes the proof.

The proof of Theorem 3 just squeaks through for $p = 3$, and the punch line utterly fails for $p = 2$. This is for good reason. Indeed $G = SQR$ could be the symmetric group $Sym(4)$ on four letters, which is a counterexample to the assertion of the theorem for $p = 2$. Looking ahead to the Odd Order Theorem and the N-Group Theorem, Thompson's next grand projects, we can say that the normal p-complement theorem boded well for studying groups of odd order, but not necessarily for studying simple N-groups, which of course have even order. However, in both cases the J-subgroup, and variations of it, would play a critical part.

3 Groups of Odd Order Are Solvable

William Burnside, in his classic treatise *The Theory of Groups of Finite Order*, remarked [20, Note M] that "the contrast ... between groups of odd and of even order suggests inevitably that [nonabelian] simple groups of odd order do not exist." Sixty-odd years later, in his *Laudatio* for Thompson at the 1970 Fields Medal ceremonies, Richard Brauer commented of the years leading up to the 1950's that "while it [Burnside's remark] was usually mentioned in courses on algebra, it is only fair to say that nobody ever did anything about it, simply because nobody had any idea how to get even started." One could say of course that a minimal counterexample to Burnside's suggestion would be a minimal simple group of odd order—i.e, all its proper subgroups would be solvable—but one stalled almost immediately.

In 1963 Walter Feit and Thompson published their answer to this question [27], the proof occupying an entire issue of the Pacific Journal of Mathematics, dubbed the "Odd Order Paper."

Theorem 4 *Every finite group of odd order is solvable.*

Equivalently, the only simple groups of odd order are the cyclic groups of prime order. The proof is carefully laid out in six chapters, of which the first three are introductory in nature. Although these contain much original material, we focus on Chap. IV of the Odd Order Paper for most of our remarks.

In this chapter the proof hits its stride, facing and analyzing a minimal simple group G of odd order. The eventual aim, achieved in some 98 pages, is a taxonomy of the maximal subgroups of G, together with an analysis of their possible intersections. This information then feeds into character-theoretic analysis in Chap. V. The taxonomy is too technical to include here, but a key intermediate result is the following Maximal Subgroup Theorem:

Theorem 5 (Maximal Subgroup Theorem) *Let G be a minimal simple group of odd order and p a prime. Then every elementary Abelian p-subgroup of G of rank at least* 3 (*i.e. every direct product of at least three copies of Z_p*) *lies in a unique maximal subgroup of G.*

Thus, the configuration of maximal subgroups of G, which *a priori* is an arbitrary tangle of solvable groups, is much closer to a partition than one might have expected: the intersection of distinct maximal subgroups contains no $Z_p \times Z_p \times Z_p$-subgroup for any prime p.

(It should be remarked that it was known that if for every prime p, G has no $Z_p \times Z_p \times Z_p$ subgroup, then a contradiction to the existence of G could be reached using 1-dimensional representations and the transfer map—methods known to Frobenius and Burnside.)

A maximal subgroup theorem similar to Theorem 5 had been established in similar contexts in two previous attacks on groups of odd order, both with very strong extra hypotheses. Suzuki [68] had proved the nonexistence of simple CA-groups of odd order, in which all centralizer subgroups $C_G(x)$, $x \in G$, $x \neq 1$, were assumed to be Abelian; and Feit and Thompson, together with Marshall Hall, Jr. [23] had generalized from "abelian" to "nilpotent." (In both cases, once the maximal subgroup theorem had been established in a minimal counterexample, character theory provided a final contradiction.) The passage from [23] to the maximal subgroup theorem of the Odd Order Paper was the long jump from "nilpotent" to "solvable."

Feit and Thompson introduce a dichotomy with implications far beyond groups of odd order, as it turned out. The set $\pi_0 = \pi_0(G)$ of prime divisors p of $|G|$ for which G possesses a subgroup isomorphic to $Z_p \times Z_p \times Z_p$ is introduced and then partitioned into subsets $\pi_3 = \pi_3(G)$ and $\pi_4 = \pi_4(G)$. A prime p lies in π_3 if and only if some Sylow p-subgroup lies in the normalizer of some nonidentity q-subgroup for some prime $q \neq p$. Otherwise $p \in \pi_4$. (This is not actually their definition but a consequence of their definition.) Such a partition makes sense in any finite group containing $Z_p \times Z_p \times Z_p$, and it approximates the distinction between unipotent and semisimple subgroups in a connected linear algebraic group. In fact, in any finite simple group G descended from a linear algebraic group in characteristic p, such as the projective special linear groups $PSL(n, p^a)$, it is the case that $p \in \pi_4(G)$, as long as G is not tiny—if, say, $a(n-1) \geq 3$. On the other hand it is rather unusual for $\pi_4(G)$ to contain any other prime, and it seems only to happen in connection with number-theoretic coincidences.

The case $p \in \pi_4$ Feit and Thompson's analysis of subgroups containing a Sylow p-subgroup P of G for some $p \in \pi_4$ is an archetype of what Thompson called "the successful translation of the theory of solvable groups to the theory of simple groups" [110, p. 383]. He was talking about the N-Group Paper but his characterization fits Chap. IV of the Odd Order Paper as well. Among other things they want to prove that P lies in only one maximal subgroup of G. Here is the relevant theory of solvable groups of odd order revealed in their paper, slightly updated to include later simplifications. The subgroup $J_1(P)$ in the theorem is only a slight variant of $J(P)$, including more Abelian subgroups; we omit the precise definition (cf. [27, Lemma 24.4], [101]).

Theorem 6 ("3 Against 2 Theorem") *Let M be a solvable group of odd order. Let P be a Sylow p-subgroup of M for some prime $p \geq 5$, and assume that P normalizes no nonidentity subgroup of M of order relatively prime to p.*

Define $M_1 = N_M(J(P))$, $M_2 = C_M(Z(P))$ and $M_3 = N_M(Z(J_1(P)))$. Then

$$M = M_1 M_2 = M_1 M_3 = M_2 M_3.$$

In particular if we define $G_1 = N_G(J(P))$, $G_2 = C_G(Z(P))$ and $G_3 = N_G(Z(J_1(P)))$, these three G_i's are solvable groups of odd order and so by the theorem each of the three is the product of its intersections with the other two. From this comes the elegant conclusion that the product of any two of the G_i is a proper *subgroup* of G:

$$\begin{aligned} G_1G_2 &= (G_1 \cap G_2)(G_1 \cap G_3)(G_2 \cap G_3)(G_2 \cap G_1) \\ &\subseteq G_2G_3G_1 = G_2(G_3 \cap G_2)(G_3 \cap G_1)G_1 = G_2G_1, \end{aligned}$$

whence $G_1G_2 = G_2G_1$ by symmetry, and so G_1G_2 is a subgroup of G. Another application of the theorem, to any maximal subgroup M of G containing P, yields $M = G_1G_2 = G_1G_3 = G_2G_3$. In particular M is unique.

In the original, the J and J_1 subgroups were not yet introduced, but their place was taken by similar objects, namely weak closures of certain Abelian subgroups of P, whence the term "weak closure arguments." (The weak closure of a subgroup A of P in P with respect to G is the subgroup of P generated by all those G-conjugates of A that happen to be subgroups of P.)

Theorem 6 was later strengthened by Glauberman [30] to conclude that $Z(J(P)) \lhd M$, i.e., $M = N_M(Z(J(P)))$, enabling the case $p \in \pi_4$ to be streamlined.

The case $p \in \pi_3$ As we noted earlier, solvability would follow from the existence of a "Sylow system", which is a set of Sylow subgroups $P_1, \ldots, P_n$, one for each prime dividing $|G|$, such that the group P_{ij} generated by any P_i and P_j, $i \neq j$, has order precisely $|P_i||P_j|$—no extraneous primes entering into its group order. One possible point of entry here was the question of when a Sylow p-subgroup P can normalize a Sylow q-subgroup Q, as Thompson described in his Colloquium Lectures at the American Mathematical Society annual meeting in 1983. More generally, since the hoped-for group P_{ij} has order $p^a q^b$ for some primes $p \neq q$, it would have a nontrivial normal subgroup N of prime power order, an easy consequence of solvability. If N were, say, a q-group, then a Sylow p-subgroup of G would normalize N, and so $p \in \pi_3$ as long as G contains a $Z_p \times Z_p \times Z_p$ subgroup. Moreover, as this heuristic applies for any pair (P_i, P_j), the set π_3 should be a substantial subset of π_0.

Looking for ways to establish that primes are in π_3, Feit and Thompson turned the traditional point of view of local analysis upside-down. Instead of asking, given a proper subgroup X of G, what can be said of its normalizer $N_G(X)$, they asked: given P, a Sylow p-subgroup of G, what are all the P-invariant q-subgroups of G? The set of all such subgroups is obviously permuted by the normalizer $N_G(P)$ under conjugation.

There is advantage as well in asking the question for p-subgroups A of G other than Sylow p-subgroups. Feit and Thompson write

$$\mathsf{И}_G(A; q)$$

for the set of all A-invariant q-subgroups of G, and

$$\mathsf{И}^*_G(A; q)$$

for the set of all maximal members of $\mathsf{И}_G(A; q)$ with respect to inclusion. $N_G(A)$ permutes both $\mathsf{И}_G(A; q)$ and $\mathsf{И}^*_G(A; q)$ by conjugation, but what is the nature of this action? The relevance and accessibility of the question depends on the choice of A. If A is chosen, for example, to be normal in some Sylow p-subgroup P, then $P \leq N_G(A)$ and the answer to the question may reveal useful information about the set of P-invariant q-subgroups of G. If A is chosen to be Abelian, then the subgroups $C_G(B)$, $B \leq A$, are all P-invariant, and one may be able to compare pieces of two elements of $\mathcal{M}_G(A; q)$ inside the solvable subgroup $C_G(B)$. Indeed if A is Abelian of exponent p then for any $W \in \mathcal{M}_G(A; q)$, the groups $C_W(B)$, for B maximal in A (of "codimension 1"), generate W. If, moreover, $A \cong Z_p \times Z_p \times Z_p$ then any two such B's have a common element $b \neq 1$. Thus for any two nontrivial subgroups W, $W' \in \mathcal{M}_G(P; q)$, there is some b such that $C_W(b) \neq 1 \neq C_{W'}(b)$, by which some relationship between W and W' might be deduced.

The rank $m(A)$ of an Abelian group A is the minimum number of factors in the decompositions of A into a direct product of cyclic groups. The Odd Order Paper brought to the fore the condition $m(A) \geq 3$ for Abelian p-subgroups A, in the analysis of a simple group. The condition became the standard dividing point between "large" and "small" simple groups in the classification of finite simple groups.

The ideas above culminated in what is commonly called now a "transitivity" theorem, and a powerful corollary:

Theorem 7 (14.1, [27]) *Let G be a minimal simple group of odd order. Suppose that p and q are distinct odd primes, P is a Sylow p-subgroup of G, and A is a normal Abelian subgroup of P of rank at least* 3 *and such that $C_P(A) = A$. Then $C_G(A)$ permutes $\mathcal{M}^*_G(A; q)$ transitively by conjugation.*

Corollary 3.1 *Suppose that p, q, P and A are as in the above theorem. Then P normalizes some element of $\mathcal{M}^*_G(A; q)$.*

And so if $\mathcal{M}_G(A; q) \neq 1$, then $p \in \pi_3$ and a good start has been made to finding a Hall $\{p, q\}$-subgroup of G.

The Odd Order Paper is famous for its length. Part of the reason for its length, and the length of the N-Group Paper as well, is that their early sections contain what amounts to textbooks for the new methods that they introduce. And of course it was a first proof, presumably to be condensed and streamlined with time—as it has been. The basic conception, however, has not been improved; rather, in deep studies by Helmut Bender, George Glauberman, Thomas Peterfalvi and others [12, 55, 56], the tools have continued to be sharpened.

Feit and Thompson lead the reader on a long journey studying objects—minimal simple groups G of odd order—which in the end turn out not to exist. After the maximal subgroups of G are brought into focus, finely detailed character-theoretic arguments are used in Chap. V to reduce the local structure of G to a single strange but resistant configuration. With character theory apparently exhausted, they resort to manipulating generators and relations for G in the final Chap. VI until eventually a contradiction is found.

With the object of study thus further and further removed from the "reality" of the structure of familiar simple groups, one might expect Feit and Thompson's arguments to be of limited value in characterizations of the extant simple groups. But in fact, an imprint of the overall structure of their proof can be seen in much of the subsequent work toward the CFSG.

4 N-Groups and Minimal Simple Groups

Thompson's N-Group Paper [110, 115, 117, 120] classifies the minimal simple groups as well as a somewhat larger class of groups which he calls N-groups.

A group G is a *minimal simple group* if and only if

- G is simple but not cyclic of prime order, and
- For any proper subgroup H of G (i.e., $H \neq G$) and any simple quotient $Q = H/K$ of H, Q is cyclic of prime order.

The second condition in this definition may be replaced by the equivalent condition that

- Any proper subgroup of G is solvable.

Theorem 8 (Minimal Simple Group Theorem) *Let G be a minimal simple finite group. Then G is isomorphic to one of the following minimal simple groups*:

1. $PSL_2(2^p)$ *or* $PSL_2(3^p)$, *p any prime*;
2. $PSL_2(p)$, *p any prime exceeding* 3 *such that* $p \equiv \pm 2 \pmod 5$;
3. $Sz(2^p)$, *p any odd prime*; *or*
4. $PSL_3(3)$.

Here $PSL_n(q)$, for q a prime power and $n \geq 2$, is the projective special linear group corresponding to a vector space of dimension n over the finite field with q elements. The groups $Sz(2^n)$, n odd, form a family of noncyclic simple groups named for its discoverer, Michio Suzuki. ($Sz(2^n)$ is a subgroup of the four-dimensional symplectic group $Sp_4(2^n)$, and is the group of fixed points of a special involutory automorphism of this symplectic group.)

A finite group G is an *N-group* if and only if every *local* subgroup of G is solvable, i.e., for every nonidentity subgroup $R \leq G$ of prime power order, its normalizer $N_G(R)$ is solvable. It is an elementary observation that every minimal simple group is a simple N-group; and Thompson classifies the latter:

Theorem 9 (N-Group Theorem) *Let G be a noncyclic simple N-group. Then G is isomorphic to one of the following simple N-groups*:

- $PSL_2(q)$, *q any prime power exceeding* 3;
- $Sz(2^n)$, *n odd and at least* 3;
- $PSL_3(3)$;

- A_7, *the alternating group of degree* 7;
- M_{11}, *the Mathieu group of degree* 11 *and order* 11.10.9.8;
- $PSU_3(3)$, *the projective special unitary group corresponding to a nondegenerate sesquilinear form on a* 3*-dimensional vector space over the field of* 3^2 *elements*; *or*
- *The Tits simple group* $[^2F_4(2), {}^2F_4(2)]$ *of order* $2^{11}.3^3.5^2.13$.

Besides the Minimal Simple Group Theorem, the N-Group Theorem had other corollaries that had seemed completely untouchable at the time. Here are two of them:

Corollary 4.1 *Let G be a finite group. If $\langle g, h\rangle$ is solvable for every $g, h \in G$, then G is solvable.*

Corollary 4.2 *Let G be a finite group. Then G is nonsolvable if and only if there exist nonidentity elements $a, b, c \in G$ whose orders are pairwise relatively prime and such that $abc = 1$.*

In the case of both corollaries, the proof reduces to checking the assertion in the case that G is a minimal simple group, so that G is one of the handful of groups in Theorem 8.

The N-Group Paper established a model of a broad framework for solving general classification problems. The same framework has stood since then as the chief strategy for attacking such problems; it is clearly visible in the actual CFSG proof, including the in-progress second and third generation efforts. At the same time the paper built its own tools, some quite well-developed, others in a form which was later improved and strengthened, both by Thompson himself and by others.

One sees again the three-stage structure, as in the Odd Order Paper, consisting of (a) analysis of the structures and mutual relationships among the maximal local subgroups of a minimal counterexample; (b) a refining stage for the structure of these maximal subgroups, whether by representation theory or other means; and (c) identification of the target simple groups of the theorems (or a contradiction, as in the Odd Order Paper, when the set of target groups is empty), by geometric means or by generator-and-relation arguments.

In the N-Group Paper one sees this pattern in more than one context, corresponding to the overarching subdivision of the proof into cases.

Thompson begins as usual with a minimal counterexample to the theorem, a simple group G in which all local subgroups are solvable, but such that G is not isomorphic to one of the target groups of the N-Group Theorem.

As was done in the Odd Order Paper, Thompson introduces a partition of the set of all prime divisors of $|G|$, into sets $\pi_1, \pi_2, \pi_3, \pi_4$; the main case division is according to which of these sets contains the prime 2. One of the four cases is easy. If we let T be any Sylow 2-subgroup of G, then the condition $2 \in \pi_1$ means that T is cyclic. A theorem going back to Frobenius and Burnside implies that G cannot be simple unless it is cyclic of order 2.

Thompson's analysis of each of the other three cases turned out to be a prototype, in whole or in part, for the analysis of an analogous case in the classification of *all* finite simple groups. For a remarkable number of key ideas in the classification of finite simple groups, one can point to their origins in some section or other of the N-Group Paper.

For example, the condition $2 \in \pi_2$ means that T, a Sylow 2-subgroup of G, is noncyclic but has no normal subgroup isomorphic to $Z_2 \times Z_2 \times Z_2$; again the main divide is drawn between rank at most 2 and rank at least 3. Later Daniel Gorenstein and Koichiro Harada [33] were able to classify *all* finite simple groups satisfying this condition.

The condition $2 \in \pi_3$ means that a Sylow 2-subgroup T of G has a normal Abelian subgroup of rank at least 3, and furthermore that the set $\mathcal{V}_G(T; 2')$ of subgroups of G of odd order normalized by T is nontrivial, i.e., does not consist only of the trivial subgroup 1. Thompson's idea in Sect. 12 was to prove that the group $\langle \mathcal{V}_G(T; 2') \rangle$ generated by all such "2-signalizers" again had odd order. This foreshadowed the later development of signalizer functor theory by Gorenstein and John Walter and others [32, 34], a vital tool for understanding the array of subgroups of odd order in a simple group.

The condition $2 \in \pi_4$, whose analysis occupies the last 8 sections and 200-odd pages of the N-Group Paper, is a precursor of the important notion [31] of a simple group's having "characteristic 2-type." (G is of characteristic 2-type if and only if for every nonidentity 2-subgroup $U \leq G$, the normalizer $N_G(U)$ possesses a normal 2-subgroup W such that the centralizer order $|C_G(W)|$ is a power of 2. A theorem of Borel and Tits implies that $PSL_n(2^m)$ is of characteristic 2-type, as is more generally any simple group of Lie type over a field of characteristic 2.)

To sort out this rather unwieldy case, Thompson defines a new invariant $e(G)$, measuring the complexity of the 2-local subgroups of G. The subdivisions of the final sections correspond neatly to major steps of the classification of all finite simple groups in the characteristic 2-type case. Section 13 argues the case $e(G) \geq 3$, the wide case, to a contradiction, corresponding to the "uniqueness case" of the general classification treated by Aschbacher [8, 9]; Sect. 14 manages the case $e(G) = 2$, the "quasi-thin case", corresponding to the quasi-thin case of the general classification, treated by Aschbacher and Smith [10, 11], and the remaining sections deal with the case $e(G) = 1$, the "thin" case, corresponding to the thin case of the general classification, treated by Aschbacher [6].

5 The B-Conjecture and the Grand Conjecture

From early on, Thompson called attention to the strong influence of 2-signalizers on the structure of a finite simple group. For example in an early note [97], he conjectures that in a finite simple group, any subgroup of odd order invariant under a Sylow 2-subgroup is Abelian, and he calls such subgroups "2-signalizers." In the late 1960's, Daniel Gorenstein and John Walter began a systematic study of the embedding of 2-signalizers (in a broader sense of the word) in finite groups. A linchpin of their analysis is the following theorem of Thompson:

Theorem 10 (Thompson $A \times B$ Lemma) *Suppose that the group $A \times B$ acts on the group P, with B and P being p-groups for the same prime p, and A having order relatively prime to p. If A acts trivially on $C_P(B)$, then A acts trivially on P.*

In 1973, Thompson defined the "bad" subgroup $B(G)$ of a finite group G and formulated the crucial B-Conjecture. This conjecture is roughly analogous to the theorem about linear algebraic groups that in a simply connected reductive group, the centralizer of any semisimple element x is connected and reductive. The conjecture was established in the mid-to-late 1970's for finite groups and involutions x by the combined work of a number of authors, providing a critical step in the classification of finite simple groups.

To state the B-Conjecture we need some notation. First, $O(X)$ is defined as the largest normal subgroup of X of odd order. (The notation $O(X)$ is due to Richard Brauer, with "O" standing for "odd.") Next, as defined by Helmut Bender, $E(X)$ is the largest normal subgroup of X which is the commuting product of quasisimple groups (with "E" for "einfach," perhaps). Of those quasisimple factors in $E(X/O(X))$, certain ones—the "good" ones—are images of the quasisimple factors of $E(X)$, modulo $O(X)$; the others are the "bad" ones, and generate $B(X)O(X)/O(X)$. In particular if $O(X) = 1$ then $B(X) = 1$. The subgroup $B(X)$ is the natural obstruction to the condition that all quasisimple factors of $E(X/O(X))$ be good, i.e. to the condition

$$E\big(X/O(X)\big) = E(X)O(X)/O(X),$$

and $B(X)$ is characterized by the four conditions

$$B(X) \lhd X, \qquad \big[B(X), E(X)\big] = 1, \qquad B(X) = \big[B(X), B(X)\big], \quad \text{and}$$
$$E\big(X/O(X)\big) = B(X)E(X)O(X)/O(X).$$

The B-Conjecture *For any involution z in any finite group G, $B(C_G(z)) \leq B(G)$.*

For simple groups G, we have $G = E(G)$ so $B(G) = 1$, and the B-Conjecture thus becomes $B(C_G(z)) = 1$ in this special case, that is,

$$E\big(C_G(z)/O\big(C_G(z)\big)\big) = E\big(C_G(z)\big)O\big(C_G(z)\big)/O\big(C_G(z)\big).$$

In the early 1970's, Aschbacher proceeded from this starting point to develop deeper consequences of the B-conjecture [5]. The first of these was that in a simple group G in which $E(C_G(z)) \neq 1$ for some involution $z \in G$, there exists an involution z' such that $E(C_G(z'))$ has a "large" normal subgroup L which is quasisimple (or—a small exception—which is isomorphic to a 4-dimensional orthogonal group), and which is normal in $C_G(t)$ for all involutions $t \in C_G(L)$. Aschbacher would go on, partly in collaboration with Gary Seitz and using work of John Walter, to finish the classification of such groups when L is a group of Lie type in odd characteristic, except for one of the smallest such groups, that is, $PSL_2(q)$ for some odd q. The result was a characterization of all the groups of Lie type in odd characteristic except for the "rank 1" groups—$PSL_2(q)$ and a rather bizarre family of groups, the Ree groups ${}^2G_2(3^{2n+1})$, $n \geq 1$, in characteristic 3.

A proof of the B-conjecture would therefore lead to a local and natural characterization of this large family of groups. Although Thompson could essentially reduce the proof of the B-conjecture to considering a simple counterexample G in which $O(B(C_G(z)))$ is a p-group for some odd prime p, he and others stalled there.

On the other hand, it was well-known that the B-Conjecture held for all groups G whose composition factors were simple groups known to exist at that time.

In 1975 Thompson announced on a visit to Rutgers that the B-conjecture appeared to have been "busted"—a favorite expression of his—by his proof, with critical contributions from Nick Burgoyne, of the most general case of a stronger conjecture, which Thompson called "The Grand Conjecture." (It also came to be known as the "U-Conjecture".)

Conjecture 5.1 (Grand Conjecture) *Let G be a group with a normal simple subgroup N such that $C_G(N) = 1$. Let z be any involution of G. Then either $O(C_G(z)) = 1$ or N is isomorphic to a known simple group, specifically to one of the following*:

1. *An alternating group A_n*;
2. *$PSL_2(q)$ for some odd q*;
3. *$PSL_3(4)$ or the sporadic Held group He*; *or*
4. *A group of Lie type over a field of odd characteristic, but not $PSL_2(q)$ for any q.*

By that time the theory of signalizer functors had been sufficiently developed that one knew that in a minimal counterexample G to the Grand Conjecture, there would exist an involution z and a subnormal subgroup L of $C_G(z)$ such that $L/O(L)$ would be isomorphic to a central extension of one of the groups listed in the conjecture. In a paper of Burgoyne [19] the fourth and most general of these possibilities is shown to lead to a contradiction. Eventually the other three possibilities were likewise shown by various authors to lead to a contradiction, establishing both the Grand Conjecture and the B-Conjecture.

6 Factorizations, Quadratic Action, and Quadratic Pairs

Factorization theorems such as Theorem 6, or a partial analogue for $p = 2$ at the heart of the N-group paper [110, 5.53, 5.54], were so powerful in the analysis of simple groups that an industry developed to study the most general conditions under which such theorems could be proved. More precisely, the industry aimed for a complete list of obstructions. In a couple of very short papers

[101, 112] Thompson clarified the terms of this study, for the most basic factorization $G = N_G(J(P))C_G(Z(P))$ of a finite group G with Sylow p-subgroup P. The first paper essentially showed that obstructions arise from (and only from) a quotient H of G with no nontrivial normal p-subgroup and a faithful H-module M over the field of p elements such that for some elementary Abelian p-subgroup A of H,

$$|M| \leq |A|\big|C_M(A)\big|,$$

where $C_M(A)$ denotes the set of fixed points of A on M. Such a module has come to be known as an "F-module," and A is called an "offender." The second paper showed, in this terminology, that for any F-module M there is an offender A that is "quadratic" in the sense that

$$M(a-1)(b-1) = 0 \quad \text{for all } a, b \in A.$$

In particular $M(a-1)^2 = 0$ for all $a \in A$.

Thompson followed up with a lovely theorem determining the possible groups H in the above terminology, when $p \geq 5$. He defined a *quadratic pair* to consist of a finite group $G \neq 1$ and a faithful G-module M over the field of p-elements for some prime p, such that G is generated by the set Q of all its elements x such that $M(x-1)^2 = 0$.

The elements of Q are said to be "quadratic on M." Any element x of order p satisfies

$$M(x-1)^p = M\big(x^p - 1\big) = 0,$$

so the minimal polynomial of x, considered as a linear transformation of M, has the form $(x-1)^n$, $n \leq p$; x is quadratic on M if and only if $n = 2$. On the other hand any element x that is quadratic on M satisfies the above equations, so x has order p by the faithfulness of the module M.

Thus for example, if $p = 2$ then every involution of G is quadratic on M, and the hypothesis is only that G is generated by its involutions. Even for $p = 3$ the hypothesis is considerably stronger, but stability does not occur until $p \geq 5$. Thompson classified quadratic pairs in this stable case [69]:

Theorem 11 *If (G, M) is a quadratic pair for p and $p \geq 5$, then*

1. *For some $n \geq 1$ there exist quadratic pairs (G_i, M_i) for p, for each $i = 1, 2, \ldots, n$, such that $G = G_1 G_2 \cdots G_n$, $[G_i, G_j] = 1$ for all $i \neq j$, and $M \cong M_1 \otimes \cdots \otimes M_n$ as $\mathbb{F}_p G$-modules.*
2. *For each i, $G_i = [G_i, G_i]$ and $G_i/Z(G_i)$ is isomorphic to one of the following groups of Lie type:*

$$A_n(q_i),\ B_n(q_i),\ C_n(q_i),\ D_n(q_i),\ G_2(q_i),\ F_4(q_i),\ E_6(q_i),\ E_7(q_i),$$
$$^2A_n(q_i),\ ^2D_n(q_i),\ ^3D_4(q_i),\ ^2E_6(q_i).$$

Here q_i is the order of the center of a Sylow p-subgroup of G_i.

Substantial progress on a companion result for $p = 3$ was made by Ho [40]. There is a longer list of examples, including among other groups the Conway group $Co_0 = \mathrm{Aut}(\Lambda)$, where Λ is the Leech lattice, as well as the unitary groups $PSU_n(2)$ over the field of 2 elements and the double covers $2A_n$ of the alternating groups.

7 The Ree Groups

During the 1950's the finite groups of Lie type, those long known to exist as well as some newcomers, were constructed systematically by Chevalley, Steinberg, Suzuki and Ree. Chevalley constructed the "untwisted" groups $\mathscr{L}(q)$ with $\mathscr{L}$ a simple Lie algebra over $\mathbb{C}$ and q a prime power; Steinberg constructed twisted variations of those Chevalley groups whose Dynkin diagram has only single bonds and possesses a nontrivial symmetry; Suzuki discovered the Suzuki groups, rediscovered by Ree as ${}^2B_2(2^{2n+1})$; and Ree constructed the new groups ${}^2F_4(2^{2n+1})$ and ${}^2G_2(3^{2n+1})$, $n \geq 0$. It is to the centralizer-of-involution characterization of these last groups of type ${}^2G_2$ that Thompson devoted considerable effort over a ten-year period [43, 107, 118, 123].

In 1954 Richard Brauer had suggested centralizers of involutions as providing an effective avenue for characterizing finite simple groups. By the mid-1960's his suggestion had been vindicated in practice by a number of successful characterizations, as well as in principle by the fundamental Brauer-Fowler theorem that for a simple group G, the order $|G|$ is bounded by a function of $|C_G(z)|$ for any involution $z \in G$. (The Brauer-Fowler bound, however, is astronomically larger than the actual bounds coming from specific theorems in which the isomorphism type of $C_G(z)$ was also assumed.)

In $G = {}^2G_2(q)$, q an odd power of 3, the centralizer of every involution is isomorphic to $Z_2 \times PSL_2(q)$, and G has Sylow 2-subgroups isomorphic to $Z_2 \times Z_2 \times Z_2$.

Janko independently was investigating the same question and in the process discovered his first sporadic group J_1, fitting the centralizer of involution $Z_2 \times PSL_2(5)$. In the end they jointly published an article [43] whose main thrust is the following theorem.

Theorem 12 (Janko and Thompson) *Let G be a finite simple group and z an involution in G. Assume that Sylow 2-subgroups of G are Abelian and $C_G(z) = \langle z \rangle \times K$ with $K \cong PSL_2(q)$ for some prime power $q > 5$. Then q is an odd power of* 3.

This is only a first step toward the desired conclusion that $G \cong G^*$, where we define $G^* := {}^2G_2(q)$. The work of H.N. Ward, who also addressed this problem, shows among other things that G is a doubly transitive group on the cosets of $B := N_G(P)$, where P is a Sylow 3-subgroup of G, and P has order q^3. These properties of course also hold in G^* with respect to the corresponding subgroups P^* and B^*. Thompson begins by partially settling the isomorphism type of B. He defines for each odd power q of 3, and for each automorphism σ of the field of q elements, a certain group $N(q,\sigma)$. It is clear from his definition that

$$B^* \cong N(q,\sigma_0),$$

where σ_0 is the (unique!) automorphism of the field of q elements such that

$$\sigma_0^2 = 3,$$

the Frobenius. Thompson proves that

$$B \cong N(q,\sigma) \quad \text{for some } \sigma.$$

The definition of $N(q,\sigma)$ gives some idea of the complexity of the problem: $N(q,\sigma)$ consists of all quadruples $(\kappa,\alpha,\beta,\gamma)$ of elements of the field of q elements and with $\kappa \neq 0$. The multiplication is

$$\begin{aligned}(\kappa,\alpha,\beta,\gamma)(\lambda,\xi,\eta,\zeta) = \big(&\kappa\lambda, \alpha\lambda+\xi, \beta\lambda^{1+\sigma}+\eta+\alpha\lambda\xi^{\sigma}-\alpha^{\sigma}\lambda^{\sigma}\xi,\\ &\zeta+\gamma\lambda^{2+\sigma}+\xi\beta\lambda^{1+\sigma}+\alpha^{\sigma}\lambda^{\sigma}\xi^2+\alpha\lambda\xi^{1+\sigma}-\alpha^3\lambda^2\xi^{\sigma}\big)\end{aligned}$$

The subgroup $H := \{(\kappa,0,0,0) \mid 0 \neq \kappa \in \mathbb{F}_q\}$ of B is the stabilizer of two points, and its normalizer is $N := N_G(H) = H\langle\tau\rangle$, with $\tau^2 = 1$. Also $B = HU$ where U is the subgroup of B defined by $\kappa = 1$. The "structure functions" $f : U - \{1\} \to B$ and $g : U - \{1\} \to U$ are defined by the equation

$$\tau u \tau = f(u)\tau g(u), \quad 1 \neq u \in U,$$

and they determine the isomorphism type of G. In [118] Thompson studies the structure functions, and he proves eventually in [123] that

$$\text{if } \sigma = \sigma_0, \quad \text{then } G \cong G^*.$$

In addition this last paper derives some restrictions on σ, but falls short of proving that $\sigma = \sigma_0$. Later Bombieri, Hunt, and Odlyzko [14] were able to use Thompson's properties to prove that $\sigma = \sigma_0$. With this combined work of Thompson, Ward, Bombieri, Hunt and Odlyzko, and Janko's work on the case $q = 5$ [42], this chapter of the classification of finite simple groups was closed.

Theorem 13 *Let G be a finite simple group and z an involution in G. Assume that Sylow 2-subgroups of G are Abelian and $C_G(z) = \langle z\rangle \times K$ with $K \cong PSL_2(q)$ for some prime power q. Then either $G \cong J_1$, Janko's first sporadic simple group, with $q = 5$, or $G \cong {}^2G_2(q)$ for some odd power $q > 3$ of 3.*

8 The Finite Sporadic Simple Thompson Group *Th*, also Known as F_3

In a lovely but not widely circulated article in the proceedings of a 1974 conference in Sapporo [122], Thompson narrates his discovery of the sporadic simple group that bears his name, of order

$$|Th| = 2^{15}.3^{10}.5^3.7^2.13.19.31.$$

Because this paper is not widely available—in particular it was not reviewed in *Mathematical Reviews*—we describe it in some detail.

At about the same time as this discovery, which took "several months" according to Thompson, parallel investigations of the Monster simple group by Thompson's student Robert Griess, Bernd Fischer, and John Conway were proceeding apace. It was known that the Monster M, if it existed, would possess an element x of order 3 such that $C_M(x) = \langle x\rangle \times Th$, where Th is simple of the above order. Griess would not construct M for another several years, but Thompson was able to construct Th as a subgroup of $GL(248, \mathbb{Z})$, and prove that modulo 3 this representation gives an embedding $Th \leq E_8(3)$. Thompson's hypotheses were taken from properties of Th that would have to hold if the Monster existed.

Theorem 14 *Up to isomorphism there exists a unique finite simple group Th with the properties*

- *For some involution* $z \in Th$, $C_{Th}(z)$ *is an extension of an extraspecial group of order* 2^9 *by the alternating group* A_9;
- *All involutions of Th are conjugate.*

Furthermore, the following conditions hold.

- *The order of Th is as above.*
- *Th acts irreducibly on the complex Lie algebra* E_8, *preserving the Killing form and a lattice* Λ, *but not preserving the Lie multiplication in* E_8.
- *Let* $M = \mathbb{Z}[\frac{1}{2}]\Lambda$. *Then* $[MM] \subseteq M$, *and Th preserves the Lie multiplication on* $M/3M$.

Thompson describes the "clear" strategy: assuming that Th is a finite simple group satisfying the two conditions above, determine the fusion in a Sylow 2-subgroup, and then all the conjugacy classes of Th meeting $C := C_{Th}(z)$. Then use exceptional characters and counting to determine $|Th|$. Finally, build the character table and "hope for the best" in proving existence and uniqueness. And the strategy produces a unique irreducible complex character θ of Th of degree 248. Moreover, the analysis of fusion of involutions shows that Th must contain a subgroup D which is a nonsplit extension of $GL(5, \mathbb{F}_2)$ by its natural module, a subgroup of order 2^5 and exponent 2 containing z.

The existence and uniqueness proofs both hinge on an analysis of the subgroups D, $C := C_{Th}(z)$, and the intersection $C_0 := C \cap D$. They also depend on machine calculations, for which Thompson collaborated with Peter Smith. First, D is shown to have a unique nonprincipal irreducible complex character θ_0 of minimal degree, this degree being 248. Using this character, he shows that D exists as a subgroup of the Chevalley group $E_8(\mathbb{C})$. (A sidelight is that D is the stabilizer in $E_8(\mathbb{C})$ of a "Dempwolff decomposition" of E_8, a decomposition of the Lie algebra E_8 as the direct sum of a set of 31 Cartan subalgebras, the bracket of any two of which again lies in the set.) Then there are exactly two subgroups C_i, $i = 1, 2$ of $GL(248, \mathbb{C})$ such that $C_i \cong C$ and $C_i \cap D = C_0$. One checks that subscripts can be chosen so that $\langle D, C_1\rangle$ is an infinite subgroup of $E_8(\mathbb{C})$, while $\langle D, C_2\rangle$ is finite. This proves existence and uniqueness. Finally, C_1 and C_2 stabilize $\mathbb{Z}[\frac{1}{2}]\Lambda$ for some lattice Λ, and C_1 and C_2 have the same character modulo 3, whence $C_2 \leq E_8(3)$.

9 "Elementary" Group-Theoretic Results

Among the tools that Thompson has created are several gems, such as the $A \times B$-Lemma mentioned earlier, whose proofs are short but which have had significant consequences for the classification of finite simple groups. We mention four here.

Theorem 15 (Thompson Order Formula) *Let G be a group of even order and let $t, u \in G$ be involutions that are not G-conjugate. For every involution $z \in G$ let n_z be the number of ordered pairs (t', u') of elements of $C_G(z)$ such that t' is G-conjugate to t, u' is G-conjugate to u, and z is a power of $t'u'$. Let $\{z_1, \dots, z_m\}$ be a set of representatives for the conjugacy classes of involutions in G. Then*

$$|G| = |C_G(t)||C_G(u)| \sum_{i=1}^{m} \frac{n_{z_i}}{|C_G(z_i)|}.$$

When $m = 2$ the formula is prettiest:

$$|G| = n_t|C_G(u)| + n_u|C_G(t)|.$$

The group order is thus determined exactly by information all of which is observable in local subgroups. Indeed, n_z depends only on information contained in $C_G(z)$, together with a determination of the sets $t^G \cap C_G(z)$ and $u^G \cap C_G(z)$, where $t^G := \{t^g \mid g \in G\}$. The latter sets, in turn, are determined by conjugations taking place in certain 2-local subgroups of G, by Alperin's Fusion Theorem.

The Thompson Order Formula has been used in the discovery and characterization of a number of the sporadic simple groups.

Theorem 16 (Thompson Transfer Lemma [110, 5.38]) *Let S be a Sylow 2-subgroup of the finite group G, and let T be a subgroup of S of index 2. Let $\tau \in S$ be an involution. If $\tau^g \notin T$ for all $g \in G$, then G has a subgroup G_0 of index 2 such that $G = G_0\langle\tau\rangle$.*

When applied to a nonabelian simple group G, this result implies that $\tau^g \in T$ for some $g \in G$. In other words τ has a fixed point in the action of G on the coset space G/T. Since the cardinality of this coset space is twice an odd number, the only alternative would be that τ acts as the product of an odd number of transpositions, i.e., as an odd permutation. The set of all elements of G inducing even permutations would serve as the subgroup G_0 of the lemma. Though very elementary, and well-known in particular cases, this fact may have been first observed in its full generality by Thompson. In any case it has been sharpened by various authors and has played a crucial role in the analysis of 2-fusion throughout the classification of finite simple groups.

The last two lemmas are of frequent use in the classification of finite simple groups of characteristic 2-type.

Theorem 17 (Thompson Dihedral Lemma [110, 5.34]) *Let the elementary Abelian 2-group $E \cong (Z_2)^n$ act faithfully on the p-group P of odd order. Then the semidirect product EP contains the direct product of n copies of the dihedral group of order $2p$ (the symmetry group of the regular p-gon).*

The final lemma was discovered independently by Bender.

Theorem 18 (Bender-Thompson Signalizer Lemma [99]) *Let p be an odd prime and let X be a p-constrained group. Let E be an elementary Abelian p-subgroup of X, that is, $E \cong (Z_p)^n$ for some n. Suppose that every element of $C_X(E)$ of order p already lies in E. Then any E-invariant p'-subgroup of X lies in $O_{p'}(X)$, the largest normal subgroup of X whose order is not divisible by p.*

10 The Inverse Galois Problem

The major conjecture in Galois theory has been the inverse Galois problem: is every finite group G the Galois group of $K/\mathbb{Q}$ for some number field K? It was proved using more number-theoretic methods that Abelian groups, then nilpotent groups and finally solvable groups are all Galois groups of number fields. Hilbert observed that symmetric (and alternating) groups occur as well.

More recently, the approach has been to consider what is called the Regular Inverse Galois Problem (RIGP): is every finite group G the Galois group of some Galois extension $K/\mathbb{Q}(t)$ with $\mathbb{Q}$ algebraically closed in K? Thus, K will be the function of field of some curve over $\mathbb{Q}$. Using the fact that $\mathbb{Q}$ is Hilbertian, an affirmative answer will imply that there are infinitely many number fields over $\mathbb{Q}$ with Galois group isomorphic to G.

It is well known that this can be done with $\mathbb{Q}(t)$ replaced by $\mathbb{C}(t)$. This is because $\mathbb{C}(t)$ is the function field of the Riemann sphere and the fundamental group of the Riemann sphere with r points removed is a free group of rank $r-1$. Thus, the whole problem is to consider such covers over $\mathbb{C}$ and try to descend to $\mathbb{Q}$.

Using the description of the fundamental group of the Riemann sphere with r points removed, one is led to consider a finite group G with conjugacy classes $C_1, \ldots, C_r$ such that the following set is nonempty.

Let

$$\mathrm{Ni}(G; C_1, \ldots, C_r) := \left\{ (g_1, \ldots, g_r) \in C_1 \times \cdots \times C_r \,\middle|\, G = \langle g_1, \ldots, g_r \rangle, \prod_{i=1}^{r} g_i = 1 \right\}.$$

A necessary condition that one can descend to $\mathbb{Q}(t)$ is that the set of conjugacy classes C_i is rational—i.e. as a multiset $\{C_1, \ldots, C_r\} = \{C_1^e, \ldots, C_r^e\}$ for all e prime to $|G|$. This does not seem to be close to being sufficient though. There is a moduli space (called the Hurwitz space) parametrizing such G-covers where the inertia groups of points over the branch points of the cover are generated by elements in $C_1, \ldots, C_r$. The rationality condition above guarantees that the variety is defined over $\mathbb{Q}$. There has been a lot of progress in producing examples where the variety is irreducible. However, it seems quite difficult to decide when the variety has a rational point (which gives a cover over $\mathbb{Q}(t)$). This method has led to showing that groups occur as Galois groups over "big" Hilbertian fields.

One criterion that has proved extraordinarily productive is the notion of rigidity. This was a concept that seemed to be independently discovered by many authors including Belyi, Fried, Matzat and Thompson. We say that $C_1, \ldots, C_r$ is rigid if G acts transitively on $\mathrm{Ni}(G; C_1, \ldots, C_r)$. The rigidity criterion says that if $C_1, \ldots, C_r$ is rational and rigid and $Z(G) = 1$, then G occurs as a Galois group of a regular Galois extension of $\mathbb{Q}(t)$. In his initial paper on rigidity, Thompson [125] showed that the Monster (the largest sporadic group) has a rationally rigid triple of conjugacy classes (consisting of elements of orders 2, 3 and 71) and so the Monster is a Galois group over $\mathbb{Q}$.

Numerous authors have used this to prove that many simple groups (and related groups) are Galois groups over $\mathbb{Q}$. This method seems especially fruitful for the finite groups of Lie type defined over prime fields. There are typically two parts to proving that a family of conjugacy classes is rigid. The first part is to count the number of r-tuples in $C_1 \times \cdots \times C_r$ with product 1. This can be done by a basic formula involving the irreducible characters of the group—although this can often be quite difficult. In particular, for the case of the finite groups of Lie type, this involves using the Deligne-Lusztig theory of characters. It can also involve working with the corresponding algebraic group and showing the system is rigid for the algebraic group. The second part is to show that there are such r-tuples which generate the group. This often requires detailed knowledge about maximal subgroups. Indeed, one wants to count the r-tuples that do not generate and then show that what is left over is a single G-orbit.

In joint work with Völklein [92], Thompson introduced what are now called Thompson tuples. These were large numbers of tuples of conjugacy classes in classical groups that were shown to be rationally rigid for finite fields, including nonprime order fields. In [92], Thompson and Völklein considered another family that was not rigid. Still they could show that the corresponding variety parametrizing covers corresponding to that family was unirational, irreducible and defined over $\mathbb{Q}$ whence it has $\mathbb{Q}$ points. The results are still not at all comprehensive over $\mathbb{Q}$. Most of the finite simple groups have been shown to be Galois groups over some cyclotomic field. This approach has interesting connections with algebraic geometry and other areas (for example, see the book of Katz [45]). We refer the reader to the books [50, 141] for more details and for many examples of how this idea has been used.

11 The Genus of a Permutation Group

Because of his interest in Galois groups, Thompson asked the following natural question:

Let $f : X_g \to \mathbb{P}^1$ be a nonconstant rational map from a curve X_g of genus g. Let G be the monodromy group of f—i.e. the Galois group of the Galois closure of $\mathbb{C}(X_g)/\mathbb{C}(\mathbb{P}^1)$. Given g, what are the possible G that can occur? Thompson asked some algebraic geometers if he thought that fixing g imposed any restrictions on the possibilities for G. The consensus was that the geometry seemed to impose no such restrictions. However, considering this as a problem about transitive permutation groups (which one can), it is clear that there are serious restrictions. In particular, Guralnick and Thompson [38] conjectured that the set of nonabelian composition factors of monodromy groups of maps from genus g curves to the Riemann sphere was finite aside from the alternating groups.

It is easy to see that if $g = 0$ and f is a generic map of degree n, then the monodromy group will be S_n. The map $x \to x^p$ shows that cyclic groups occurs as monodromy groups of maps from the Riemann sphere to itself. In [38], a reduction theorem was proved (to the case of minimal covers and so primitive permutation groups) and the conjecture was proved in many cases. Moreover, an outline for a possible complete proof was laid out. Indeed, this is a template for various kinds of problems using permutation groups. Reduce to the case of primitive permutation groups and then further reduce to properties of simple groups.

In a series of papers by various authors (including Aschbacher, Guralnick, Neubauer, Liebeck, Saxl, Shalev), this problem was attacked. The final step in the proof of the conjecture was provided by Frohardt and Magaard [28]. Indeed, it now seems possible that one can classify all indecomposable covers of degree n of the Riemann sphere by curves of small genus (say at most 2) where the monodromy group is not the alternating group or the symmetric group.

This conjecture also led to quite a lot of interesting related work. For example, this led to a vast generalization of a result of Zariski's thesis (which answered a question of Enrique). Zariski proved that there is no solvable map from the generic Riemann surface of genus $g > 6$ to the Riemann sphere (note that every curve of genus 6 admits a degree 4 cover to the Riemann sphere and in particular a solvable map). Interestingly, Zariski's proof was essentially done by translating the problem to group theory and using facts about solvable primitive permutation groups. In a series of papers (see [37]), it is shown that for $g \geq 4$, if $f : X \to \mathbb{P}^1$ is an indecomposable map of degree n from the generic Riemann surface of genus g, then the monodromy group of f is either S_n, $n > (g+2)/2$ or A_n, $n > 2g$. It was well known that S_n could occur but it was only recently shown [49] that A_n could occur.

The Guralnick–Thompson conjecture also led to a resurgence into the study of the number of fixed points of permutations in primitive permutation groups. The idea is that aside from some special cases, most elements in primitive permutation groups have very few fixed points. This had been studied extensively as early as the late 1800's. Using the classification of finite simple groups, all primitive permutation groups containing an element fixing at least $1/2$ the points were classified [36].

The Riemann–Hurwitz formula implies that in most cases the genus grows linearly with the degree of the map. Kleidman [46] used these ideas to prove a conjecture of Kegel and Wielandt characterizing subnormal subgroups of finite groups: a subgroup H of a finite group G is subnormal if and only if for every prime p and every Sylow p-subgroup P of G, $P \cap H$ is a Sylow p-subgroup of H.

12 Representation Theory

Thompson did quite a lot of outstanding work in representation and character theory of finite groups. In particular, the Odd Order Paper developed a substantial amount of character theory. Thompson also write a number of relatively short and incredibly clever papers that solved various interesting questions and are still of considerable interest today.

For example, in [126], Thompson proved a result about Frobenius–Schur indicators for modular representations. Recall that if k is algebraically closed field of characteristic $p \neq 2$, G is a group and V is an irreducible finite dimensional kG-module, then the Frobenius–Schur indicator is $0, \pm 1$. It is $+1$ if G leaves invariant a symmetric bilinear form on V, -1 if G leaves invariant an alternating bilinear form and 0 otherwise (i.e. if V is not self-dual). If G is finite, then an indicator of $+1$ is equivalent to the representation being defined over $\mathbb{R}$. If $p = 0$, there is a very nice character formula to compute the indicator. This is not available in positive characteristic. Thompson showed that if $p > 0$ and V has indicator ± 1, then there is an irreducible $\mathbb{C}G$-module W such that V occurs with odd multiplicity in the reduction of W modulo p and that W and V must have the same indicator. In particular, this furnishes a quick proof that all irreducible representations of S_n in characteristic $p \neq 2$ admit an invariant quadratic form. Thompson's result has very recently been extended to certain families of Hopf algebras [51].

The paper [108] makes an important contribution to block theory. Brauer's theory of blocks relative to a fixed prime p partitions the set of irreducible complex characters of a finite group G in a way reflecting how the corresponding representations reduce modulo p. (Any complex representation of G is similar to one in which the entries of the representing matrices all lie in a ring R that is a finite extension of the p-adic integers, enabling reduction modulo p. However, in the case of interest, when p is a divisor of $|G|$, the reduction of an irreducible complex representation modulo

p is in general not irreducible.) Each subset in the partition—i.e., each "block"—has an associated "defect group" according to Brauer's theory. The defect group is a p-subgroup of G defined up to conjugacy in G, and it is a basic theme of block theory that the defect group governs much about the structure of the corresponding block. For instance if the defect group is trivial, then the block consists of a single representation that stays irreducible and projective modulo p. Brauer [17] obtained thorough results about blocks whose defect group has order p, for example showing that the reduction mod p of a complex irreducible representation in the block is multiplicity-free, i.e., has no repeated composition factors. An evident question was whether the multiplicity-free conclusion would still hold for representations in blocks with cyclic defect group. Thompson writes that he undertook the work in [108] to try to understand Brauer's results. He succeeds in establishing this generalization under the assumption that a Sylow p-subgroup P of G is cyclic, equal to its own centralizer in G, and disjoint from its distinct conjugates. The methods involve the "Green Correspondence" between indecomposable modules for G and indecomposable modules for the normalizer $N_G(P)$, in both cases over an algebraically closed field of characteristic p. (The paper includes an elegant proof of part of the famous Hall-Higman Theorem B, using the same techniques.)

With the help of the methods of this paper, E.C. Dade [21] was able to analyze decisively the structure of blocks with arbitrary cyclic defect groups.

In [24] with Feit, it is shown that if p is a prime and G has a faithful irreducible complex representation of dimension n with $p > 2n + 1$, then the Sylow p-subgroup of G is normal and Abelian. This elegant paper generalized some results of Blichfeldt and Brauer. Note that $SL_2(p)$ has irreducible representations of dimension $(p - 1)/2$ (Weil representations) and so the result is best possible.

In [79], Thompson proves that if G is a finite subgroup of $GL_n(\mathbb{C}) = GL(V)$, then for some $m \leq 4n^2$, G has an invariant 1-dimensional subspace on the mth symmetric power. The proof is quite clever. First note that $GL_n(\mathbb{F})$ has a 1-dimensional composition factor in the $(p - 1)n$th symmetric power (for $\mathbb{F}$ of characteristic $p > 0$). Thus if p does not divide $|G|$, by reducing modulo p and lifting the invariant subspace, G will have a 1-dimensional invariant subspace in the $(p - 1)n$th symmetric power of V. He then picks a prime $p > 2n + 1$ and uses the result just discussed to show that the Sylow p-subgroup of G is central (after reducing to the case that G is irreducible) and so one may assume that p does not divide $|G|$. He conjectures that there might be a bound of the form cn for some constant c. This is still an open problem and of considerable interest to algebraic geometers.

In [139], Thompson studies the problem of the possible composition factors of rational finite groups. A finite group G is called rational if all characters on G are rational valued (equivalently g and g^e are conjugate in G for any e prime to $|G|$). Feit and Seitz, using the classification of finite simple groups, determined all possible nonabelian composition factors (aside from alternating groups, only four others are possible). The question is, what cyclic groups of prime order p can be composition factors of such groups? It is conjectured that $p \leq 5$. In an elegant proof, Thompson shows that $p \leq 11$. The proof uses the Feit-Seitz result and reduces to the case where G has a minimal normal elementary Abelian p-subgroup and analyzes the possible module structure of this subgroup.

In [78], Thompson studies finite dimensional representations of amalgamated free products of cyclic groups and uses this in the study of the Monster and the sporadic group J_4.

Brauer made an important conjecture about blocks which was only recently proved, called the $k(B)$ conjecture. This conjecture is about the number $k(B)$ of irreducible complex characters of a finite group in a block B with defect group D. The conjecture is that $k(B) \leq |D|$.

The key idea to prove it was to show that if G is a finite group and V an irreducible module over $\mathbb{F}_p$ with p prime to $|G|$, then the number of conjugacy classes in the semidirect product VG is at most $|V|$. In joint work with Robinson [61], this inequality is proved under the assumption that the stabilizer H of a nonzero vector v has the property that V has a faithful self-dual H-submodule—in particular, this holds in the case G has a regular orbit. This is used to show the inequality for p large enough. It was a crucial ingredient in the eventual solution of the problem.

13 Projective Planes

Thompson wrote several interesting papers on projective planes (the main questions in the area are still wide open).

In [48], MacWilliams, Sloane and Thompson proved various results about a projective plane of order 10 (assuming one exists).

In [131], Thompson observed that if c is any characteristic root of the incidence matrix of a projective plane of order n, then either $c = 1 + n$ or c has all its conjugates on the circle $|z| = n$.

In [137], the minimal polynomial of an incidence matrix was studied and some interesting results reported. Indeed, this led Thompson to an interesting question in linear algebra.

14 Cosets

In [106], Thompson answered a 1954 question of Paige proving that there exist finite groups G such that some coset of a Sylow 2-subgroup consists of elements of even order. It is easy to see that this cannot occur for solvable groups. The example Thompson gives is the group $G = PSL_2(53)$. In fact, with some extra work, one can show that there are infinitely many such groups G (namely any $PSL_2(q)$ such that 16 does not divide $q^2 - 1$ and $q \geq 53$ and odd). Until recently, this was thought to be a curiosity. Indeed, in the math review written by Paige, there was no mention of any consequences. It turns out that Thompson's method can be used to prove a similar result for all primes p and this has been used to answer a question of Richard Taylor regarding Galois representations. The idea of Thompson's proof was to define a certain variety and show that for certain values of q if the variety had points over $\mathbb{F}_q$, then the coset with the desired property exists (and indeed there should be many such cosets). Thompson observed that for $q = 53$, this variety did have points. Using various estimates for the number of points on varieties over finite fields, the result holds for any sufficiently large q (satisfying the appropriate congruence conditions). For odd p, there is a different variety but the idea is quite similar.

Interestingly, Ho and Völklein proved a result which leads to a related problem: do there exist nontrivial cosets of a Sylow p-subgroup consisting of only p-elements? There are also some intriguing connections with Gowers' theory of pseudorandom properties.

15 Divisor Matrix

Let A be the infinite integral matrix where the rows and columns are indexed by positive integers and the (i, j) entry is 1 if i divides j and 0 otherwise. This is called the divisor matrix. In [65], Sin and Thompson produce a representation of $SL_2(\mathbb{Z})$ where A corresponds to one of the natural unipotent generators. Moreover, this representation preserves, in a natural way, the ring of Dirichlet series that converge at some nonzero complex number.

In particular, if $\zeta(s)$ is the Riemann zeta function, they show that there exists a Dirichlet series $\phi(s)$ such that the orbit of $\zeta(s)$ is contained in the $\mathbb{C}$-span of $\zeta^m(s)$ and $\phi(s)\zeta^m(s)$ where m is a natural number. While there is no explicit formula given for $\phi(s)$, it does satisfy

$$\bigl(\zeta(s) - 1\bigr)\phi^2(s) + \zeta(s)\phi(s) + \zeta(s)\bigl(\zeta(s) - 1\bigr) = 0.$$

While this work is still in its preliminary stages, the connection between the Riemann zeta function and the divisor matrix seems mysterious and may lead to unexpected results.

16 Other Work

Thompson also studied a myriad of other areas including problems in number theory, modular forms, Kleinian groups, coding theory and combinatorics. He also was involved in some of the early aspects of monstrous moonshine and the connections of the largest sporadic group (the Monster) and modular forms. See [74, 75].

References

1. *Carleson and Thompson receive Wolf prize*, Notices Amer. Math. Soc. **39** (1992), no. 4, 321–322. 1153175 (92m:01084)
2. A. A. Albert and John Thompson, *Two element generation of the projective unimodular group*, Bull. Am. Math. Soc. **64** (1958), 92–93. MR 0096728 (20 #3211)

3. ———, *Two-element generation of the projective unimodular group*, Ill. J. Math. **3** (1959), 421–439. MR 0106951 (21 #5681)
4. Richard P. Anstee, Marshall Hall Jr., and John G. Thompson, *Planes of order 10 do not have a collineation of order 5*, J. Comb. Theory, Ser. A **29** (1980), no. 1, 39–58. MR 577542 (82c:51012)
5. Michael Aschbacher, *On finite groups of component type*, Ill. J. Math. **19** (1975), 87–115. MR 0376843 (51 #13018)
6. ———, *Thin finite simple groups*, J. Algebra **54** (1978), no. 1, 50–152. MR 511458 (82j:20032)
7. ———, *Weak closure in finite groups of even characteristic*, J. Algebra **70** (1981), no. 2, 561–627. MR 623826 (82j:20034)
8. ———, *The uniqueness case for finite groups. I*, Ann. of Math. (2) **117** (1983), no. 2, 383–454. MR 690850 (84g:20021a)
9. ———, *The uniqueness case for finite groups. II*, Ann. of Math. (2) **117** (1983), no. 3, 455–551. MR 701255 (84g:20021b)
10. Michael Aschbacher and Stephen D. Smith, *The classification of quasithin groups. I*, Mathematical Surveys and Monographs, vol. 111, Am. Math. Soc., Providence, 2004, Structure of strongly quasithin K-groups. MR 2097623 (2005m:20038a)
11. ———, *The classification of quasithin groups. II*, Mathematical Surveys and Monographs, vol. 112, Am. Math. Soc., Providence, 2004, Main theorems: the classification of simple QTKE-groups. MR 2097624 (2005m:20038b)
12. Helmut Bender and George Glauberman, *Local analysis for the odd order theorem*, London Mathematical Society Lecture Note Series, vol. 188, Cambridge University Press, Cambridge, 1994, With the assistance of Walter Carlip. MR 1311244 (96h:20036)
13. J. Boen, O. Rothaus, and J. Thompson, *Further results on p-automorphic p-groups*, Pac. J. Math. **12** (1962), 817–821. MR 0152576 (27 #2553b)
14. E. Bombieri, A. Odlyzko, and D. Hunt, *Thompson's problem* ($\sigma^2 = 3$), Invent. Math. **58** (1980), no. 1, 77–100, Appendices by A. Odlyzko and D. Hunt. MR MR570875 (81f:20019)
15. G. W. Bond, *Eulogy: John G. Thompson*, Bull. Lond. Math. Soc. **19** (1987), no. 6, 635–636, Dual Latin-English text. MR 915435 (88i:01088)
16. R. Brauer, *On the work of John Thompson*, Actes du Congrès International des Mathématiciens (Nice, 1970), Tome 1, Gauthier-Villars, Paris, 1971, pp. 15–16. MR 0414281 (54 #2384)
17. Richard Brauer, *Investigations on group characters*, Ann. of Math. (2) **42** (1941), 936–958. MR 0005731 (3,196b)
18. ———, *On the structure of groups of finite order*, Proceedings of the International Congress of Mathematicians, Amsterdam, 1954, Vol. 1, Noordhoff, Groningen, 1957, pp. 209–217. MR 0095203 (20 #1709)
19. N. Burgoyne, *Finite groups with Chevalley-type components*, Pac. J. Math. **72** (1977), no. 2, 341–350. MR 0457550 (56 #15755)
20. W. Burnside, *Theory of groups of finite order*, Dover, New York, 1955, 2d ed. MR 0069818 (16,1086c)
21. E. C. Dade, *Blocks with cyclic defect groups*, Ann. of Math. (2) **84** (1966), 20–48. MR 0200355 (34 #251)
22. Walter Feit, *Characters of finite groups*, Benjamin, New York, 1967. MR 0219636 (36 #2715)
23. Walter Feit, Marshall Hall Jr., and John G. Thompson, *Finite groups in which the centralizer of any non-identity element is nilpotent*, Math. Z. **74** (1960), 1–17. MR 0114856 (22 #5674)
24. Walter Feit and John G. Thompson, *Groups which have a faithful representation of degree less than* $(p - 1/2)$, Pac. J. Math. **11** (1961), 1257–1262. MR 0133373 (24 #A3207)
25. ———, *Finite groups which contain a self-centralizing subgroup of order 3*, Nagoya Math. J. **21** (1962), 185–197. MR 0142623 (26 #192)
26. ———, *A solvability criterion for finite groups and some consequences*, Proc. Natl. Acad. Sci. USA **48** (1962), 968–970. MR 0143802 (26 #1352)

27. ———, *Solvability of groups of odd order*, Pac. J. Math. **13** (1963), 775–1029. MR 0166261 (29 #3538)
28. Daniel Frohardt and Kay Magaard, *Composition factors of monodromy groups*, Ann. Math. **154** (2001), 327–345. MR 1865973
29. G. Glauberman and J. G. Thompson, *Weakly closed direct factors of Sylow subgroups*, Pac. J. Math. **26** (1968), 73–83. MR 0235025 (38 #3337)
30. George Glauberman, *A characteristic subgroup of a p-stable group*, Can. J. Math. **20** (1968), 1101–1135. MR 0230807 (37 #6365)
31. Daniel Gorenstein, *On finite simple groups of characteristic 2 type*, Inst. Hautes Études Sci. Publ. Math. (1969), no. 36, 5–13. MR 0260864 (41 #5484)
32. ———, *On the centralizers of involutions in finite groups*, J. Algebra **11** (1969), 243–277. MR 0240188 (39 #1540)
33. Daniel Gorenstein and Koichiro Harada, *Finite groups whose 2-subgroups are generated by at most 4 elements*, Am. Math. Soc., Providence, 1974, Memoirs of the American Mathematical Society, vol. 147. MR 0367048 (51 #3290)
34. Daniel Gorenstein and John H. Walter, *Centralizers of involutions in balanced groups*, J. Algebra **20** (1972), 284–319. MR 0292927 (45 #2008)
35. Otto Grün, *Beiträge zur Gruppentheorie I*, J. Reine Angew. Math. **174** (1935), 1–14.
36. Robert Guralnick and Kay Magaard, *On the minimal degree of a primitive permutation group*, J. Algebra **207** (1998), 127–145. MR 164307
37. Robert Guralnick and John Shareshian, *Symmetric and alternating groups as monodromy groups of Riemann surfaces. I. Generic covers and covers with many branch points*, Amer. Math. Soc., Providence, 2007, Memoirs of the American Mathematical Society, vol. 886. MR 2343794
38. Robert M. Guralnick and John G. Thompson, *Finite groups of genus zero*, J. Algebra **131** (1990), no. 1, 303–341. MR 1055011 (91e:20006)
39. P. Hall and Graham Higman, *On the p-length of p-soluble groups and reduction theorems for Burnside's problem*, Proc. London Math. Soc. (3) **6** (1956), 1–42. MR 0072872 (17,344b)
40. C. Y. Ho, *Quadratic pairs for odd primes*, Bull. Am. Math. Soc. **82** (1976), no. 6, 941–943. MR 0417273 (54 #5330)
41. D. R. Hughes and J. G. Thompson, *The H-problem and the structure of H-groups*, Pac. J. Math. **9** (1959), 1097–1101. MR 0108532 (21 #7248)
42. Z. Janko, *A new finite simple group with Abelian Sylow* 2*-subgroups and its characterization*, J. Algebra **3** (1966), 147–186. MR 0193138 (33 #1359)
43. Zvonimir Janko and John G. Thompson, *On a class of finite simple groups of Ree*, J. Algebra **4** (1966), 274–292. MR 0201504 (34 #1386)
44. ———, *On finite simple groups whose Sylow* 2*-subgroups have no normal elementary subgroups of order* 8, Math. Z. **113** (1970), 385–397. MR 0283076 (44 #309)
45. Nicholas M. Katz, *Rigid local systems*, Princeton University Press, Princeton, 1996, Annals of Mathematical Studies, vol. 139. MR 1366651
46. Peter B. Kleidman, *A proof of the Kegel-Wielandt conjecture on subnormal subgroups*, Ann. of Math. **133** (1991), 369–428. MR 1097243
47. F. J. MacWilliams, N. J. A. Sloane, and J. G. Thompson, *Good self dual codes exist*, Discrete Math. **3** (1972), 153–162. MR 0307799 (46 #6919)
48. ———, *On the existence of a projective plane of order* 10, J. Comb. Theory, Ser. A **14** (1973), 66–78. MR 0313089 (47 #1644)
49. Kay Magaard and Helmut Völklein, *The monodromy group of a function on a general curve*, Israel J. **141** (2004), 355–368. MR 2063042
50. Gunter Malle and B. Heinrich Matzat, *Inverse Galois theory*, Springer, Berlin, 1999, Springer Monographs in Mathematics. MR 1711577
51. Susan Montgomery and A. Jedwab, *Modular representations and indicators for bismash products*, preprint.
52. Alexander Moretó, *An answer to a question of J. G. Thompson on: "Some generalized characters" [J. Algebra **179** (1996), no. 3, 889–893]*, J. Algebra **275** (2004), no. 1, 75–76. MR

2047441 (2005a:20014)

53. Gabriel Navarro, *John G. Thompson and finite groups*, Gac. R. Soc. Mat. Esp. **9** (2006), no. 1, 183–189. MR 2236305
54. Donald Passman, *Permutation groups*, Benjamin, New York, 1968. MR 0237627 (38 #5908)
55. Thomas Peterfalvi, *Simplification du chapitre VI de l'article de Feit et Thompson sur les groupes d'ordre impair*, C. R. Acad. Sci. Paris Sér. I Math. **299** (1984), no. 12, 531–534. MR 770439 (86d:20020)
56. ———, *Character theory for the odd order theorem*, London Mathematical Society Lecture Note Series, vol. 272, Cambridge University Press, Cambridge, 2000, Translated from the 1986 French original by Robert Sandling and revised by the author. MR 1747393 (2001a:20016)
57. Vera Pless and John G. Thompson, *17 Does not divide the order of the group of a (72, 36, 16) doubly even code*, IEEE Trans. Inf. Theory **28** (1982), no. 3, 537–541. MR 672889 (83j:94022)
58. Martin Raussen and Christian Skau, *Mathematics develops with a natural speed, and that is quickly*, Nieuw Arch. Wiskd. (5) **9** (2008), no. 4, 248–254, Interview with John G. Thompson and Jacques Tits. MR 2561950
59. ———, *Interview with John G. Thompson and Jacques Tits*, Not. Am. Math. Soc. **56** (2009), no. 4, 471–478. MR 2493482
60. Geoffrey R. Robinson and John G. Thompson, *Sums of squares and the fields* $\mathbf{Q}_{A_n}$, J. Algebra **174** (1995), no. 1, 225–228. MR 1332869 (96b:12004)
61. ———, *On Brauer's* $k(B)$*-problem*, J. Algebra **184** (1996), no. 3, 1143–1160. MR 1407890 (97m:20014)
62. J. E. Roseblade, *Obituary: Philip Hall*, Bull. Lond. Math. Soc. **16** (1984), no. 6, 603–626, With contributions by J. G. Thompson and J. A. Green. MR 758133 (86c:01056)
63. Oscar Rothaus and John G. Thompson, *A combinatorial problem in the symmetric group*, Pac. J. Math. **18** (1966), 175–178. MR 0195934 (33 #4130)
64. Leonard Scott, Ronald Solomon, John Thompson, John Walter, and Efim Zelmanov, *Walter Feit (1930–2004)*, Not. Am. Math. Soc. **52** (2005), no. 7, 728–735. MR 2159686
65. Peter Sin and John G. Thompson, *The divisor matrix, Dirichlet series, and* SL(2, **Z**), The legacy of Alladi Ramakrishnan in the mathematical sciences, Springer, New York, 2010, pp. 299–327. MR 2744269
66. N. J. A. Sloane and J. G. Thompson, *The nonexistence of a certain Steiner system* $S(3, 12, 112)$, J. Comb. Theory, Ser. A **30** (1981), no. 3, 209–236. MR 618528 (82m:05023)
67. Gernot Stroth and B. Heinrich Matzat, *John Griggs Thompson und die Theorie der endlichen Gruppen*, Mitt. Dtsch. Math.-Ver. **16** (2008), no. 3, 173–175. MR 2522905
68. Michio Suzuki, *The nonexistence of a certain type of simple groups of odd order*, Proc. Am. Math. Soc. **8** (1957), 686–695. MR 0086818 (19,248f)
69. J. G. Thompson, *Quadratic pairs*, Actes du Congrès International des Mathématiciens (Nice, 1970), Tome 1, Gauthier-Villars, Paris, 1971, pp. 375–376. MR 0430043 (55 #3051)
70. ———, *Simple* $3'$*-groups*, Symposia Mathematica, Vol. XIII (Convegno di Gruppi e loro Rappresentazioni, INDAM, Rome, 1972), Academic Press, London, 1974, pp. 517–530. MR 0360802 (50 #13249)
71. ———, *A conjugacy theorem for* E_8, J. Algebra **38** (1976), no. 2, 525–530. MR 0399193 (53 #3044)
72. ———, *Finite groups and even lattices*, J. Algebra **38** (1976), no. 2, 523–524. MR 0399257 (53 #3108)
73. ———, *Remarks on finite groups*, Proceedings of the 5th School of Algebra (Rio de Janeiro, 1978) (Rio de Janeiro), Soc. Brasil. Mat., 1978, pp. 75–77. MR 572056 (81j:20006)
74. ———, *Finite groups and modular functions*, Bull. Lond. Math. Soc. **11** (1979), no. 3, 347–351. MR 554401 (81j:20029)
75. ———, *Some numerology between the Fischer-Griess Monster and the elliptic modular function*, Bull. Lond. Math. Soc. **11** (1979), no. 3, 352–353. MR 554402 (81j:20030)

76. ———, *Uniqueness of the Fischer-Griess monster*, Bull. Lond. Math. Soc. **11** (1979), no. 3, 340–346. MR 554400 (81e:20024)
77. ———, *A finiteness theorem for subgroups of* PSL(2, **R**) *which are commensurable with* PSL(2, **Z**), The Santa Cruz Conference on Finite Groups (Univ. California, Santa Cruz, CA, 1979), Proc. Sympos. Pure Math., vol. 37, Amer. Math. Soc., Providence, 1980, pp. 533–555. MR 604632 (82b:20067)
78. ———, *Finite-dimensional representations of free products with an amalgamated subgroup*, J. Algebra **69** (1981), no. 1, 146–149. MR 613864 (82e:20007)
79. ———, *Invariants of finite groups*, J. Algebra **69** (1981), no. 1, 143–145. MR 613863 (82d:20042)
80. ———, *Rational functions associated to presentations of finite groups*, J. Algebra **71** (1981), no. 2, 481–489. MR 630609 (83a:20048)
81. ———, *Primitive roots and rigidity*, Proceedings of the Rutgers group theory year, 1983–1984 (New Brunswick, N.J., 1983–1984) Cambridge, Cambridge Univ. Press, 1985, pp. 327–350. MR 817267 (87g:12001)
82. ———, *Rational rigidity of* $G_2(5)$, Proceedings of the Rutgers group theory year, 1983–1984 (New Brunswick, N.J., 1983–1984) Cambridge, Cambridge Univ. Press, 1985, pp. 321–322. MR 817265 (87e:12006a)
83. ———, PSL_3 *and Galois groups over* **Q**, Proceedings of the Rutgers group theory year, 1983–1984 (New Brunswick, N.J., 1983–1984) Cambridge, Cambridge Univ. Press, 1985, pp. 309–319. MR 817264 (87e:12005)
84. ———, *Algebraic numbers associated to certain punctured spheres*, J. Algebra **104** (1986), no. 1, 61–73. MR 865887 (88e:11103)
85. ———, *Fricke, free groups and* SL_2, Group theory (Singapore, 1987), Gruyter, Berlin, 1989, pp. 207–214. MR 981843 (90f:20074)
86. ———, *Hecke operators and noncongruence subgroups*, Group theory (Singapore, 1987), Gruyter, Berlin, 1989, Including a letter from J.-P. Serre, pp. 215–224. MR 981844 (90a:20105)
87. ———, *Unipotent elements, standard involutions, and the divisor matrix*, Commun. Algebra **36** (2008), no. 9, 3363–3371. MR 2441119 (2009j:15016)
88. J. G. Thompson and H. Völklein, *Symplectic groups as Galois groups*, J. Group Theory **1** (1998), no. 1, 1–58. MR 1490157 (99c:12007)
89. John Thompson, *A method for finding primes*, Am. Math. Mon. **60** (1953), 175. MR 0052448 (14,621b)
90. ———, *Finite groups with fixed-point-free automorphisms of prime order*, Proc. Natl. Acad. Sci. USA **45** (1959), 578–581. MR 0104731 (21 #3484)
91. ———, *Finite groups with normal* p*-complements*, Proc. Sympos. Pure Math., Vol. 1, Am. Math. Soc., Providence, 1959, pp. 1–3. MR 0117290 (22 #8071)
92. John Thompson and Helmut Völklein, *Braid-Abelian tuples in* $\mathrm{Sp}_n(K)$, Aspects of Galois theory (Gainesville, FL, 1996), London Math. Soc. Lecture Note Ser., vol. 256, Cambridge Univ. Press, Cambridge, 1999, pp. 218–238. MR 1708608 (2000f:12004)
93. John G. Thompson, *Thesis*, University of Chicago (1959)
94. ———, *Normal* p*-complements for finite groups*, Math. Z. **72** (1959/1960), 332–354. MR 0117289 (22 #8070)
95. ———, *A special class of non-solvable groups*, Math. Z. **72** (1959/1960), 458–462. MR 0117288 (22 #8069)
96. ———, *Two results about finite groups*, Proc. Internat. Congr. Mathematicians (Stockholm, 1962), Inst. Mittag-Leffler, Djursholm, 1963, pp. 296–300. MR 0175972 (31 #248)
97. ———, *2-signalizers of finite groups*, Pac. J. Math. **14** (1964), 363–364. MR 0159874 (28 #3090)
98. ———, *Automorphisms of solvable groups*, J. Algebra **1** (1964), 259–267. MR 0173710 (30 #3920)
99. ———, *Fixed points of* p*-groups acting on* p*-groups*, Math. Z. **86** (1964), 12–13. MR 0168653 (29 #5911)

100. ———, *Normal p-complements for finite groups*, J. Algebra **1** (1964), 43–46. MR 0167521 (29 #4793)
101. ———, *Factorizations of p-solvable groups*, Pac. J. Math. **16** (1966), 371–372. MR 0188296 (32 #5735)
102. ———, *Hall subgroups of the symmetric groups*, J. Comb. Theory **1** (1966), 271–279. MR 0197546 (33 #5711)
103. ———, *Centralizers of elements in p-groups*, Math. Z. **96** (1967), 292–293. MR 0209357 (35 #255)
104. ———, *Defect groups are Sylow intersections*, Math. Z. **100** (1967), 146. MR 0213432 (35 #4296)
105. ———, *An example of core-free quasinormal subgroups of p-groups*, Math. Z. **96** (1967), 226–227. MR 0207844 (34 #7658)
106. ———, *On a question of L. J. Paige*, Math. Z. **99** (1967), 26–27. MR 0212093 (35 #2968)
107. ———, *Toward a characterization of $E_2^*(q)$*, J. Algebra **7** (1967), 406–414. MR 0223448 (36 #6496)
108. ———, *Vertices and sources*, J. Algebra **6** (1967), 1–6. MR 0207863 (34 #7677)
109. ———, *Characterization of finite simple groups*, Proc. Internat. Congr. Math. (Moscow, 1966), Mir, Moscow, 1968, pp. 158–162. MR 0236258 (38 #4555)
110. ———, *Nonsolvable finite groups all of whose local subgroups are solvable*, Bull. Am. Math. Soc. **74** (1968), 383–437. MR 0230809 (37 #6367)
111. ———, *Envelopes and p-signalizers of finite groups*, Ill. J. Math. **13** (1969), 87–90. MR 0237633 (38 #5914)
112. ———, *A replacement theorem for p-groups and a conjecture*, J. Algebra **13** (1969), 149–151. MR 0245683 (39 #6989)
113. ———, *Bounds for orders of maximal subgroups*, J. Algebra **14** (1970), 135–138. MR 0252500 (40 #5720)
114. ———, *A non-duality theorem for finite groups*, J. Algebra **14** (1970), 1–4. MR 0251125 (40 #4356)
115. ———, *Nonsolvable finite groups all of whose local subgroups are solvable. II*, Pac. J. Math. **33** (1970), 451–536. MR 0276325 (43 #2072)
116. ———, *Normal p-complements and irreducible characters*, J. Algebra **14** (1970), 129–134. MR 0252499 (40 #5719)
117. ———, *Nonsolvable finite groups all of whose local subgroups are solvable. III*, Pac. J. Math. **39** (1971), 483–534. MR 0313378 (47 #1933)
118. ———, *Toward a characterization of $E_2^*(q)$. II*, J. Algebra **20** (1972), 610–621. MR 0313377 (47 #1932)
119. ———, *Isomorphisms induced by automorphisms*, J. Aust. Math. Soc. **16** (1973), 16–17, Collection of articles dedicated to the memory of Hanna Neumann, I. MR 0330293 (48 #8630)
120. ———, *Nonsolvable finite groups all of whose local subgroups are solvable. IV, V, VI*, Pac. J. Math. **48** (1973), 511–592, ibid. 50 (1974), 215–297; ibid. 51(1974), 573–630. MR 0369512 (51 #5745)
121. ———, *Weighted averages associated to some codes*, Scr. Math. **29** (1973), no. 3–4, 449–452, Collection of articles dedicated to the memory of Abraham Adrian Albert. MR 0403994 (53 #7803)
122. ———, *A simple subgroup of $E_8(3)$*, Finite Groups (Sapporo and Kyoto, 1974) (Nagayoshi Iwahori, ed.), Japan Society for the Promotion of Science, Tokyo, 1976, pp. 113–116
123. ———, *Toward a characterization of $F_2^*(q)$. III*, J. Algebra **49** (1977), no. 1, 162–166. MR 0453858 (56 #12111)
124. ———, *Finite nonsolvable groups*, Group theory, Academic Press, London, 1984, pp. 1–12. MR 780566 (86h:20017)
125. ———, *Some finite groups which appear as* Gal L/K, *where* $K \subseteq \mathbf{Q}(\mu_n)$, J. Algebra **89** (1984), no. 2, 437–499. MR 751155 (87f:12012)

126. ———, *Some finite groups which appear as* Gal L/K*, where* $K \subseteq \mathbf{Q}(\mu_n)$, Group theory, Beijing 1984, Lecture Notes in Math., vol. 1185, Springer, Berlin, 1986, pp. 210–230. MR 842445 (87i:11160)
127. ———, *Archimedes and continued fractions*, Math. Medley **15** (1987), no. 2, 67–75. MR 931712 (89d:01008)
128. ———, *Groups of genus zero and certain rational functions*, Groups—Canberra 1989, Lecture Notes in Math., vol. 1456, Springer, Berlin, 1990, pp. 185–190. MR 1092232 (92h:12005)
129. ———, *Rigidity,* GL(n, q)*, and the braid group*, Bull. Soc. Math. Belg. Sér. A **42** (1990), no. 3, 723–733, Algebra, groups and geometry. MR 1316220 (96d:20018)
130. ———, *Galois groups*, Groups—St. Andrews 1989, Vol. 2, London Math. Soc. Lecture Note Ser., vol. 160, Cambridge Univ. Press, Cambridge, 1991, pp. 455–462. MR 1123999 (93a:12007)
131. ———, *Algebraic integers all of whose algebraic conjugates have the same absolute value*, Coding theory, design theory, group theory (Burlington, VT, 1990), Wiley-Intersci., New York, 1993, pp. 107–110. MR 1227123 (94d:11082)
132. ———, *Note on* $H(4)$, Commun. Algebra **22** (1994), no. 14, 5683–5687. MR 1298742 (95h:20048)
133. ———, *Note on realizable sequences of partitions*, Commun. Algebra **22** (1994), no. 14, 5679–5682. MR 1298741 (96d:05008)
134. ———, *4-punctured spheres*, J. Algebra **171** (1995), no. 2, 587–605. MR 1315914 (96f:20072)
135. ———, *Sylow 2-subgroups of simple groups*, Séminaire Bourbaki, Vol. 10, Soc. Math. France, Paris, 1995, pp. Exp. No. 345, 543–545. MR 1610481
136. ———, *Some generalized characters*, J. Algebra **179** (1996), no. 3, 889–893. MR 1371748 (96k:20014)
137. ———, *Incidence matrices of finite projective planes and their eigenvalues*, J. Algebra **191** (1997), no. 1, 265–278. MR 1444501 (98e:51008)
138. ———, *Power maps and completions of free groups and of the modular group*, J. Algebra **191** (1997), no. 1, 252–264. MR 1444500 (98a:20026)
139. ———, *Composition factors of rational finite groups*, J. Algebra **319** (2008), no. 2, 558–594. MR 2381796 (2008k:20015)
140. Helmut Voelklein and Tanush Shaska (eds.), *Progress in Galois theory*, Developments in Mathematics, vol. 12, New York, Springer, 2005. MR 2150438 (2006a:00014)
141. Helmut Völklein, *Groups as Galois groups, an introduction*, Cambridge University Press, Cambridge, 1996, Cambridge Studies in Advanced Mathematics, vol. 53. MR 1405612
142. Helmut Völklein, David Harbater, Peter Müller, and J. G. Thompson (eds.), *Aspects of Galois theory*, London Mathematical Society Lecture Note Series, vol. 256, Cambridge University Press, Cambridge, 1999, Papers from the Conference on Galois Theory held at the University of Florida, Gainesville, FL, October 14–18, 1996. MR 1708599 (2000c:14004)
143. P. Sin and J. G. Thompson, The divisor matrix, Dirichlet series, and SL(2;Z), in The legacy of Alladi Ramakrishnan in the mathematical sciences, pp. 299–327, New York: Springer, 2010

A Report on the Scientific Contributions of Jacques Tits

Francis Buekenhout

1 Introduction

In 1975, the Collège de France in Paris, created a Chair in Group Theory for Jacques Tits. He gave an inaugural lecture [T102][1] and, in his application for the Chair he wrote a detailed survey of his work: "Titres et travaux scientifiques de Jacques Tits" (32 pages, 1972). In both of these texts he gave his definition of group theory:

> "...theory resulting from the mathematical development of the neighbouring ideas of *homogeneity*, of *symmetry* and of *indistinguishability* or, in more philosophical terms, of the synthesis of same and of different".

The contributions of Jacques Tits belong to a broad and difficult range of mathematics whose leitmotiv is groups, homogeneity, geometry, topology, and algebra. He belongs to the tradition of Felix Klein, Federigo Enriques, Sophus Lie, Elie Cartan, Herman Weyl, among other great men. In a deep way, his work often represents a revenge of geometry on group theory, with respect to Klein's Erlanger Program, as he said in 1975 during his inaugural lecture at the Collège de France.

According to MathSciNet the number of publications of Tits is 153. Some of his papers have escaped this remarkable undertaking of the American Mathematical Society. One of these is:

- "Géométries polyédriques et groupes simples." Atti della Seconda Riunione del Groupement des Mathématiciens d'Expression Latine 1963 [T52]. This is no

Electronic supplementary material Supplementary material is available in the online version of this chapter at http://dx.doi.org/10.1007/978-3-642-39449-2_5. Videos can also be accessed at http://www.springerimages.com/videos/978-3-642-39449-2.

F. Buekenhout (✉)
Département de Mathématique, Université libre de Bruxelles, CP216, Boulevard du Triomphe, 1050 Brussels, Belgium
e-mail: fbueken@ulb.ac.be

[1] The citation "[T*n*]" refers to the *n*th publication in the list of publications of Jacques Tits in the present volume.

H. Holden, R. Piene (eds.), *The Abel Prize 2008-2012*,
DOI 10.1007/978-3-642-39449-2_5, © Springer-Verlag Berlin Heidelberg 2014

less than the founding paper of the theory of buildings and of Tits systems. See Sect. 11.

Quite a number of great works remained unpublished. Here are some of them.

- "Travaux sur le plan Crémonien." 1950–1952. Typewritten manuscript. Not quite recovered yet. An important paper. Completed by "The Cremona plane" 1999 [T198]. See Sect. 3.
- Empain Prize Memoir 1954. Typewritten manuscript entitled "Espaces Homogènes et Isotropes et Espaces Doublement Homogènes" (retyped, 124 pages) [T22]. Again an important paper. See Sect. 5.
- "Remarks on Griess' construction of the Griess–Fischer sporadic group (I), (II), (III), (IV)." 35 pages, 1983 [T143]. See Sect. 7.

The European Mathematical Society has taken on a project in order to publish the "Oeuvres de Jaques Tits" in four volumes (about 3800 pages) in 2013 [T209]. The editorial team comprises Hendrik Van Maldeghem, Jean-Pierre Tignol, Bernhard Mühlherr, and Francis Buekenhout. With the help of Jacques Tits, they have established an official list of publications: it includes 203 documents. Here we shall discuss only a sample of these, leaving true gems untouched.

2 The Projective Line

Tits started research at age 16 in 1946 in the context of a course taught by Paul Libois on projective geometry the year before and pursued in his course on "Géométrie Supérieure". Tits was a third year undergraduate student. He got impressive insights on sharply multiply transitive permutation groups, inspired by affine and projective geometry. On the basis of this work, he achieved at once international recognition, in particular from Emil Artin, as early as 1949, at the Colloque d'Algèbre et Théorie des Nombres in Paris. His numerous results cover the publications [T1] to [T4] and [T10]. The latter, his PhD thesis, includes a novel geometry of projective and affine spaces (over a field) in terms of a transformation group from which the geometry is extracted.[2]

Let us state one of Tits's results. Let G be a permutation group which is triply transitive on a set S. Let p, q be distinct elements of S. Assume that the G-stabilizer of p and q is Abelian. Then G is isomorphic to some $\mathrm{PGL}(2, K)$, K a commutative field, in its natural action on the projective line $P(1, K)$.

This work by a youngster should not be underrated. The projective lines belong to the rank one buildings whose invention came 15 years later when Tits was 31 years old. The projective lines are a class of Moufang sets whose classification started with

[2]Note also the paper [12] by Libois. In the introduction he says: "Cette communication, élaborée en étroite collaboration avec MM. P. Defrise et J. Tits, a pour but d'indiquer une conception de cette synthèse, de souligner les éléments déjà obtenus et qui nous paraissent les plus importants, d'apporter des résultats nouveaux."

Tits and remains a major open problem. Another related open problem is to decide whether a classification of infinite sharply 2-transitive groups can be accomplished. The open problems in the thesis have remained untouched: this should be a potential for development.

3 The Cremona Plane Made Invariant Under the Cremona Group

Tits's advisor, Paul Libois, drew his attention to a gap in birational geometry of the complex plane. Here, the concept of a projective point is not invariant under the Cremona group, which ought to be a group of automorphisms for some structure. Actually, in view of the phenomenon of blow-up, a projective point is not an invariant. Tits solved the problem.

In his words for one of the papers [T23] he presented in 1955: "The introduction of an adequate notion of point allows the elaboration of a purely geometric axiomatic of the Cremona plane".

In 1950–1951, Tits found a system of axioms for the Cremona plane over an algebraically closed field. His primitive concepts were invariant under the Cremona group. This remarkable piece of work remains unpublished. About four copies of the typewritten manuscript were circulated. One copy was saved at the library of the Université libre de Bruxelles—I do not know where the other are. I have my handwritten copy of most of it made in the early 1960's. Tits told me (1999) that Libois was not entirely happy with the paper. He found it too formal and not philosophical enough. This is why the paper was not published.

In 1999, Tits gave eight hours of lectures on the subject. It was a course of the Collège de France given at the Université libre de Bruxelles—an innovation. The notes were written by Hendrik Van Maldeghem and myself. Tits wrote a summary which can be found in [T198].

The basic idea in 1950 was to start with a partially ordered group whose positive elements are called figures. Additional structure consists of figures that are called points. These are the only primitive elements. In his axioms, Tits made the requirement that for every point p, the stabilizer of p acts as a sharply triply transitive group on the set of points covered by p. This was his way of getting his hands on a ground field.

The fact that a point may contain another point, expresses the old view of infinitely near points that gave rise to so much polemical discussions in former times. This is the philosophical component advocated by Libois. In Tits's view, it collapsed without words.

4 Lie Groups and the Riemann–Helmholtz–Lie Problem

The years 1951–1954 were devoted to the study of Lie groups, a subject in which Tits became rapidly a leading expert. A great achievement was an impressive solution of the Helmholtz–Lie problem that actually goes back to Riemann. The question is to provide a common characterization of Euclidean and non-Euclidean geometries on the basis of a "free mobility" property of their motion group. In other words, describe the set of all spaces in which non-Euclidean geometry may be developed.

In 1930, Kolmogorov gave an accurate answer, in terms of a system of axioms based on a group of homeomorphisms acting on a metrizable, connected, locally compact topological space. He had a recurrent set of axioms (M_n), n a natural number, and he stated that his spaces were either Euclidean or non-Euclidean. He did not publish the proof. Very surprisingly Tits classified all spaces satisfying (M_1). From his list, he checked that $(M_1) + (M_2)$ implies (M_n) for every natural number n, with a few exceptions. Moreover, he provided the missing proof. His methods gave a series of other classifications. They gave, in particular, all models of a universe of general relativity satisfying a mild homogeneity condition. Tits also obtained a new way to classify the simply connected homogeneous compact complex manifolds, a result obtained earlier by H.C. Wang. We are dealing here with [T23], which is the Thèse d'agrégation or Habilitation (1955), and the remarkable survey [T29]. The latter was written for a conference on 14 October 1954, at the Riemann Tagung in Berlin, which was held to commemorate the lecture given by Riemann on June 10 1854, "Über die Hypothesen welche die Geometrie zu Grunde liegen".[3] Riemann's memorable memoir which founded Riemannian geometry, was published by R. Dedekind in 1867. We follow [T29] in a loose translation:

> Speaking of the spaces of constant curvature, Riemann indicates that "the common characteristic of these varieties may also be expressed by saying that the figures may be translated and rotated arbitrarily".

Later on, dealing again with constant curvature, he says that "bodies have an existence independent of their position". Tits continues:

> "Later on, the characterization of spaces of constant curvature by an axiom of free mobility, was pushed further by Helmholtz (1868), then Lie (1893), who considers the group of motions. At this stage, Riemann's free mobility of bounded solid bodies is replaced by a new idea which is the free mobility of spaces."

Let us express the results of Tits, based on purely topological axioms due to A. Kolmogorov (1930).

Consider a pair (E, G) where:

(1) E is a connected locally compact topological space and G is a group of homeomorphisms acting transitively on E.
(2) For any two points p, q and a neighborhood U of q, there exists a neighborhood V of p and a neighborhood W of q such that every element of G mapping p into W, maps V into W.

[3] On the hypotheses which underlie geometry.

(3) For any distinct points p, q, r such that q and r have distinct G_p[4]-orbits Q and R, either Q separates p from R or R separates p from Q.

This is the axiom (M_1) mentioned at the beginning of this section.

As we said, Kolmogorov has an axiom (M_n), n a natural number, namely: For all points $p_1, \ldots, p_n$ such that for every $i \leq n$, p_i is contained in the orbit of p_{i-1} under the G-stabilizer of $p_1, \ldots, p_{i-2}$, the condition (3) holds if E is replaced by the orbit of p_n under the G-stabilizer of $p_1, \ldots, p_{n-1}$, and G is replaced by the G-stabilizer of $p_1, \ldots, p_{n-1}$.

Theorem *If the conditions* (1), (2), (3) *and the axiom* (M_2) *hold, then E may be identified with a Riemann space of constant curvature by a homeomorphism which is unique up to isometry, or up to similarity in the Euclidean case, in such a way that G is a group of isometries of that space.*

This conclusion means that (M_n) for every $n \geq 1$, follows from $(M_1) + (M_2)$.

Tits's way of proof is to establish a much stronger result, namely to classify the pairs (E, G) satisfying (1), (2), (3), and a mild condition (4). He obtains a list including affine spaces over $\mathbb{R}$, $\mathbb{C}$, $\mathbb{H}$, the affine plane over $\mathbb{O}$ (octonions), a spinorial space, a semi-spinorial space, a conformal space and a few more ... In each case, G is given as well. Condition (4) is used only in the final step of the theory. On page 221 of [T23], Tits states a conjecture which would provide the classification also when (4) does not hold. I must confess that I don't know the status of this conjecture.

A detailed, deep and broad study of the views explained here can be found in Freudenthal's article [8].

Freudenthal received frequent visits by Tits in Utrecht. He understood Tits immediately, and he worked with him. He contributed himself to the final solution of the Riemann–Helmholtz–Lie Problem. He writes:

> "The question was finally settled by Tits in 1952". See [T13, T16].

5 Doubly Homogeneous Spaces, and Homogeneous and Isotropic Spaces

Let me quote a report on this work written by Tits in 1954 (there exists a typewritten manuscript from 31 December 1954).

> "The spaces that are considered are homogeneous Klein spaces, namely ordered pairs (E, G) where E is a set (possibly equipped with a topological, differentiable, or analytic structure) and G is a transitive group of automorphisms of E. The space (E, G) is called *doubly homogeneous* if G is doubly

[4] G_p denotes the stabilizer of p in G.

transitive on E; it is called *homogeneous and isotropic* if G is transitive on the tangent directions (elements of contact) of E."

The main result of the memoir is the determination of

(i) all locally compact doubly homogeneous spaces
(ii) all homogeneous and isotropic spaces in the case where G is a Lie group, and where $E = G/H$ is a connected homogeneous space for that group.

Tits did not provide a full proof, but he took care of the most difficult parts, like homogeneous spaces of the exceptional Lie groups in order to convince the readers.

6 Geometric Interpretation of the Five Exceptional Simple Lie Groups and the Magic Square

All simple Lie groups of classical type A_n, B_n, C_n, D_n have a geometric interpretation in terms of projective geometry, projective quadrics and symplectic polarities. A geometric interpretation for the exceptional groups G_2, F_4, E_6, E_7 and E_8 represented an open question that had been pending since the time of S. Lie, W. Killing and E. Cartan. In 1954, Tits obtained an epoch making solution, and he made a major step in the direction of buildings. The geometry is constructed in group theoretical terms from the group and the maximal subgroups containing a Borel subgroup, namely a maximal connected solvable subgroup. The construction fits perfectly with projective spaces, projective quadrics, and symplectic polarities of the classical groups, and it provides a new view on them. Very soon, the construction applied to all Chevalley groups revealed in 1955, also to all semi-simple isotropic algebraic groups, and more would come until the late 1980's (Kac–Moody groups). A major paper switching from algebraic constructions to geometric axioms is [T30].

7 A World of Incidence Geometries

Tits had created a "world out of nothing" in the phrasing of Jeremy Gray [10]. It was a world of incidence geometries. A geometry can be seen as a multipartite graph in which the elements of a given component are called points (resp. lines, planes, quads, etc.). These geometries ought to be called Tits geometries.

Tits was axiomatizing his fantastic universe of geometries at various levels of generality. While doing this he obtained geometric constructions and existence proofs of various simple groups. His construction for a group of type E_6 over any field was superseded by Chevalley's uniform algebraic construction including the classical groups as well as G_2, F_4, E_6, E_7 and E_8 over any field. The exceptional groups were becoming classical too.

Tits's many different methods applied further through the years to provide several new classes of finite simple groups of "twisted" Lie–Chevalley type. He found the surprising "Tits group", namely the simple group of index 2 in ${}^2F_4(2)$. This group is sometimes considered as the 27th sporadic group, which is probably what it will be called in the future.

Tits greatly contributed to the sporadic groups. He provided the first computer free existence proof of the Hall–Janko group in a most elegant geometric way from an extension of the generalized hexagon of order 2.

In 1982–1984 he got a deeper understanding and nontrivial simplifications of Griess's construction of the Monster group. The reference is given in Sect. 1. It will appear in the "Oeuvres" [T209].

8 Generalized Polygons

Paper [T34], published in 1959, is a great one indeed. It provides a complete classification for the trialities of the Tits diagram geometries of type D_4. This parallels the classification of polarities in projective spaces which is, algebraically speaking, the theory of sesquilinear forms.

Here Tits discovers the class of twisted Chevalley groups ${}^3D_4(F)$, F a field, independently of Steinberg. His approach is similar to the derivation of classical groups from forms. Provided with a triality T in the D_4-geometry, and a point p, the latter is called isotropic, if it is incident to $T(p)$. In that case, $T(p)$ is incident with $T^2(p)$, and the latter is incident to p. A line L is called isotropic if $T(L) = L$. Then, every point of L is isotropic. Provided there exists an isotropic line, Tits obtains a geometry G of isotropic points and lines that he starts comparing to a projective plane and to a quadric of Witt index 2. He gets a totally new concept that he calls a generalized hexagon.

By further abstraction he gets a fantastic new concept called generalized n-gon, n a natural number ≥ 1, or $n = \infty$. The case $n = 3$ consists of the projective planes, the case $n = 4$ is modeled after the quadrics of Witt index 2, and its general member is called a generalized quadrangle, the case $n = 6$ is modeled after G and called a generalized hexagon. The case $n = \infty$ consists of all trees without elements of valency 1. A geometric world sprang out of this concept. It is still growing. Two years later, the generalized n-gons became the buildings of rank two, those relying on points and lines. A superb self-contained and deep survey is the book by Hendrik Van Maldeghem [15]

9 Moufang Polygons

In 1964, W. Feit and G. Higman showed that a finite thick generalized n-gon can exist only for $n = 2, 3, 4, 6, 8$ [7]. This was based on eigenvalue techniques for the incidence graph of a generalized polygon. Such a result was hopeless in the infinite case in view of the free constructions due to Tits (see Sect. 16). However, Tits detected an analogy with the projective Moufang planes that are related to the Graves–Cayley–Dickson octonions. Indeed, all models of generalized n-gons derived from simple groups, gave Tits a property he called the Moufang condition (1973) in honor of Ruth Moufang who had studied the projective planes over an octonion division algebra in the early 1930's. The property is expressed by the existence of particular automorphisms called elations. Assuming his Moufang condition, Tits proved that $n = 2, 3, 4, 6$, or 8. Moreover, Tits was expecting a complete classification of those Moufang polygons. That classification existed already for $n = 3$ (due to R.H. Bruck in 1951 and E. Kleinfeld in 1951). The case $n = 2$ is uninteresting from a purely geometric viewpoint. A generalized 2-gon is any complete bipartite graph. Tits obtained a complete classification for $n = 8$ and $n = 6$. He started on the case $n = 4$. The latter case would stimulate efforts towards completion over 30 years. (See Sect. 19.)

This may be a good opportunity to mention, that unlike most other mathematicians, Tits was constructing all of his mathematics in his head, without using paper and pen, including the proofs of difficult results. Years later, he could reconstruct them. The task became really hard for Moufang quadrangles.

10 General Theory of Coxeter Groups

In 1961, Tits had developed the general theory of Coxeter groups for which he is not well enough recognized. He wrote it down in a document that was fairly well distributed, but not published at the time, "Groupes et Géométries de Coxeter". This deep theory generalized fundamental results of Coxeter obtained in the 1930's. Tits paid a tribute to the great man by the creation of a "Coxeter language" to which many researchers in science adhere nowadays.

This task was necessary for Tits himself in view of his theory of buildings, which was born the very same year. It was also intended for Bourbaki, who used it in his book [4], published in 1968. Bourbaki writes

> "Pour la rédaction de ces trois chapitres, de nombreuses conversations avec J. Tits nous ont apporté une aide précieuse. Nous l'en remercions trés amicalement".

However, there is no trace in the book of the fact that Tits wrote down the general theory of Coxeter groups published by Bourbaki. Much later, in 2001, the paper was finally published by S.S. Chern and F. Hirzebruch in the Wolf Prize Volume [T204]. A lot of confusion prevails about the general theory of Coxeter groups. Some persons claim that it is due to Coxeter. Some others claim that it is due to Bourbaki. It is due to Tits!

11 Theory of Buildings: Birth 1961

The birth of buildings as well as Tits systems occurred in 1961, and Tits was lecturing on it at a conference held in Firenze and Bologna during the autumn. His paper

[T52] appeared in the proceedings of that conference in 1963. He did not use the words "building", "apartment", "Tits system", but rather "generalized polyhedron", "skeleton", "group with a BN-pair".

Tits gave a plenary lecture on the subject at the International Congress of Mathematicians in Stockholm 1962 (see [T53]).

Bourbaki, whose philosophy was to write only on theories whose permanent importance is well established and whose presentation has been reaching an optimal and concise shape, took up immediately a great deal of Tits's theories. The Tits systems are there, in the title of Chap. IV. The buildings (immeubles) are in the exercises! [4].

In 1954, Bruhat had discovered that for a classical simple group, there exists a decomposition $G = WBW$ where B is a Borel subgroup and W the Weyl group of the related root system. A moment of thought is required in order to explain WB, since W is not a subgroup of G. The result was extended to all real and complex semi-simple Lie groups by Harish Chandra (1956). Tits wanted to understand this in geometric terms. It was one of the roads to buildings. The concept of a Tits system is particularly handy and simple. It requires only the group structure. Providing a group G with a Tits system and a few additional group theoretical conditions, Tits proved the simplicity of G. The proof supersedes former proofs of simplicity for particular classes of groups. This is a major result presented in Bourbaki.

Every building comes with a Coxeter group and the corresponding Weyl group of a root system. It also has a Coxeter diagram. The buildings whose Weyl group is finite, are called spherical. This amounts to saying that all apartments are finite.

Tits started to classify all spherical buildings of rank $r > 2$, following the model of the classification of projective spaces of rank 3, whose principle is due to Veblen and Young (1910). They introduced the concept of a division ring (they called it a "number system") and showed that every projective space is isomorphic to a space constructed over a division ring.

In many respects, the most difficult part in the theory is to classify the buildings of type B_n that are also called "polar spaces". In order to get through this, Tits had to develop, among other things, a theory of pseudo-quadratic forms, pseudo-quadratic groups, and pseudo-quadratic polar spaces.

His book, actually a complete research paper, appeared in the Springer Lecture Notes series in 1974 [T98]. It allows the reader to profit from the author's imagination and insight. It is another testimony of Tits's total honesty with respect to references. He knows that every definition requires a reference if it has been borrowed! The book is written in a clear style, and it provides inspiration for a long time.

Nowadays buildings can be defined and studied in quite different ways. Sections 15 and 19 provide such views. A particularly attractive theory in terms of graphs appears in the book by Weiss [16]. There is a book which provides a new conceptual approach to spherical buildings [6].

12 Applications of Buildings

The theory of buildings has many applications, far beyond the initial purpose which was a geometric interpretation of the exceptional groups. Already in 1974, Tits mentioned applications to the study of p-adic simple groups, the cohomology of arithmetic groups, the representation theory of and the harmonic analysis on finite and p-adic simple groups, etc. [T98].

Around 1990, it had become an impossible task to keep track of all applications. Let me mention some. In his book of 1974 [T98] Tits used his classification of spherical buildings in order to derive a classification of the finite Tits systems. This became a major tool in the classification programme of all finite simple groups (see [9]). The role of Tits systems is crucial in the recognition of groups of Lie–Chevalley type, and as such it contributed to the sudden acceleration of Gorenstein's programme in the work of Aschbacher after 1976.

Let me also mention outstanding results of M. Gromov, in particular about the rigidity of locally symmetric spaces in the class of all manifolds of nonpositive curvature. Here too, a crucial role is played by the underlying buildings. Gromov emphasizes the use of a Tits metric as a promising tool for further investigations (see [2]). Another reference is the chapter "Applications of Buildings" by J. Rohlfs and T.A. Springer in [5] which discusses works by A. Borel, P. Cartier, J.P. Serre, G.A. Margulis, V.P. Platonov to mention but a few names.

The theory of buildings has gradually been brought to the attention of most mathematicians. This is due to the simplicity of its axioms that allows for a rather handy use in applications. It is also due to its fundamental depth, as well as its relationship to as many domains as the theory of Lie groups. There has been a growing interest in this theory. Textbooks have appeared. See Sect. 11. Let us add some more references: [1, 3, 13, 14].

13 Affine Buildings

Consider a Chevalley group over a local field. It carries a spherical building as does every Chevalley group. However, it carries a second Tits system whose Weyl group is no longer finite, but of affine type, or just affine in the present terminology.

François Bruhat and Jacques Tits developed a general theory of affine Tits systems leading in particular to the classification of the maximal compact subgroups of the p-adic groups and to a p-adic analog of Cartan's symmetric spaces (see [T155, T162, T163]). Tits obtained a remarkable theory and classification of the affine buildings of rank $r \geq 4$ (1985) together with developments on the rank 3 case (1990).

A remarkable property: every affine building has a spherical building at infinity. A path to the classification of affine buildings is to start with the classification of spherical buildings.

A beautiful book on this subject is one by Weiss, [17].

14 Diagram Geometries and Sporadic Groups

Another application of the incidence geometry developed by Tits from 1954 on, occurs in extensions of buildings, namely incidence geometries that contain buildings among their "residual" geometries. Every sporadic group acts on such geometries. See [T126], also a survey by Buekenhout–Pasini in [5] (see Sect. 12) and major work with MAGMA by Leemans [11].

The combinatorial, geometric and group theoretical work of Tits from 1956 to 1974, lead several people from 1975 on to consider more general diagrams and geometries inspired by and applicable to the sporadic groups. This point of view was strongly developed by various authors, in particular Ronan–Smith (1980), Ronan–Stroth (1984), etc. Tits paid much interest to this as we see from his paper [T126]. See also [6].

15 The Local Approach to Buildings

The developments of diagram geometry lead Tits to totally new and deep views on geometries and buildings. This became his "local approach", presented in a lecture at the Santa Cruz Conference on Finite Groups organized in 1979 by the American Mathematical Society and published in the Coxeter Festschrift in 1981 [T133]. Here he introduced chamber systems. Briefly, a chamber system consists of a graph in which every edge bears an index (a name) in such a way that for every index, all edges with that index partition the set of vertices. This had an immediate influence on various works, including the revision of the classification of finite simple groups.

Here, the "amalgam" method aims at the control over an unknown group G generated by two known subgroups whose intersection is also known. A concept of universal cover G is available.

16 Free Constructions

On the model of free projective planes which had been discovered by Marshall Hall (1943) with inspiration from free groups, Tits made various observations in the direction of free constructions for buildings and for Tits systems. Together with Mark Ronan (1987) he gave a short and simple construction procedure of a purely combinatorial nature, for many classes of buildings and groups along the ideas expressed in the local approach [T133]. They used the local approach (see Sect. 15). This included for instance, the buildings and groups of type E_6, E_7 and E_8. It provided a satisfactory and tempting alternative to the algebraic construction starting with Chevalley's construction of Lie algebras, then groups. The construction by Tits starts with a Chevalley group G and a Borel subgroup B of G. The subgroups of G containing B constitute a boolean lattice providing a building whose elements are the left cosets of those subgroups. However, the new approach did not apply to all spherical buildings.

17 Algebraic Groups

Tits was a major expert of algebraic groups together with Armand Borel [T62] (1965) when they extended the theory over any base field, for reductive groups, in particular simple groups. See also [T83, T86, T93] for work by Borel–Tits on algebraic groups.

Tits extended likewise the theory of linear representations of complex semi-simple groups due to Elie Cartan and Herman Weyl in the complex case and to Claude Chevalley for groups over an algebraically closed field.

A gigantic theory of reductive groups over local fields was gradually built by François Bruhat and Jacques Tits from 1966 to 1987 (see [T71, T72, T73, T74, T75, T77, T92, T141, T146, T162, T163]).

In 1987, William M. Kantor, Robert Liebler, and Tits classified the discrete chamber-transitive automorphism groups of affine buildings. They provided a characterization of the discrete subgroups acting on the affine building of a simple adjoint algebraic group of relative rank $r \geq 2$ over a locally compact field [T160].

18 Kac–Moody Groups and Twin Buildings

During the 1980's, Tits was involved to a large extent in a profound unification and generalization of earlier work of himself and others, concerning the Kac–Moody groups over fields. His paper [T161] in 1987 is an important landmark, as can be seen in particular from the unusual length of the report on it written for Mathematical Reviews by James Hurley. The Kac–Moody algebras have many applications in mathematics and in physics. The Kac–Moody groups appear as natural companions for the algebras. However, the "right" definition was not immediately clear to Tits. As it should, a geometric interpretation was looked for. This was started in joint work of Ronan and Tits around 1986. Tits published [T175] in 1988 and [T182] in 1992. The main idea is a new concept of twin building. It consists of a pair of spherical buildings together with a relation of "opposition" on their chambers. It extends the potential of applications of spherical buildings.

That work also provided joint papers with Ronan [T185, T197] devoted to twin trees.

19 Moufang Polygons: Thirty Years Later

Finally, to end this survey, let us quote from the introduction to the great book by Tits and Weiss [T205]:

> "Spherical buildings are certain combinatorial simplicial complexes introduced, at first in the language of "incidence geometries", to provide a systematic geometric interpretation of the exceptional complex Lie groups ... Via

the notion of a BN-pair [Tits system], the theory turned out to apply to simple algebraic groups over an arbitrary field. More precisely, to any absolutely simple algebraic group of positive relative rank k is associated a thick irreducible spherical building of the same rank ... and the main result of Buildings of Spherical type [T98] is that the converse, for $l \gg 3$, is almost true:

Theorem *Every thick irreducible spherical building of rank at least three is classical, exceptional or mixed.*

Classical buildings are those defined in terms of the geometry of a classical group (e.g. unitary, orthogonal, etc. of finite Witt index or linear of finite dimension) over an arbitrary field or skew-field ... Mixed buildings are more exotic; they are related to groups which are in some sense algebraic groups defined over a pair of fields k and K of characteristic p, where $K_p \subset k \subset K$ and p is two or (in one case) three.

Irreducible spherical buildings of rank two are called generalized polygons. Generalized polygons themselves are too numerous to classify ... but in the addenda of [T98], the Moufang condition for spherical buildings was introduced, and it was observed that ... every thick irreducible spherical building of rank at least three as well as every irreducible residue of such a building satisfies the Moufang condition. In particular, *all generalized polygons which are the irreducible rank two residues of thick irreducible spherical buildings of higher rank are Moufang*. As a consequence, every thick irreducible spherical building of higher rank is a kind of amalgam of Moufang (generalized) polygons.

In this book, we classify Moufang polygons and use this classification ... to simplify the classification of thick irreducible spherical buildings of higher rank. These are the two main projects proposed in the addenda of [T98]. It is probably more appropriate, however, to regard this book as a "prequel" to [T98] rather than a subsequent volume. Moufang polygons are a class of graphs which can be studied without any reference to the theory of buildings (or as an introduction to the theory of buildings) and, in fact, the classification of Moufang polygons which we give in ... this book is entirely elementary and self-contained".

References

1. Abramenko, P., Brown, K.S.: Buildings, Theory and Applications. Springer, Berlin (2008)
2. Ballmann, W., Gromov, M., Schroeder, V.: Manifolds of Nonpositive Curvature. Birkhäuser, Basel (1985)
3. Brown, K.S.: Buildings. Springer, Berlin (2009)
4. Bourbaki, N.: Groupes et Algèbres de Lie. Hermann, Paris (1968). Chapitres 4, 5, 6
5. Buekenhout, F. (ed.): Handbook of Incidence Geometry. North Holland, Amsterdam (1995)
6. Buekenhout, F., Cohen, A.M.: Diagram Geometry, Classical Groups and Buildings. Ergebnisse der Mathematik und ihrer Grenzgebiete. Springer, Berlin (2013)

7. Feit, W., Higman, G.: The nonexistence of certain generalized polygons. J. Algebra **1**, 114–131 (1964)
8. Freudenthal, H.: Lie groups in the foundations of geometry. Adv. Math. **1**, 145–190 (1964)
9. Gorenstein, D., Lyons, R., Solomon, R.: Classification of the Finite Simple Groups. Am. Math. Soc., Providence (1996–2004), six volumes
10. Gray, J.: Worlds out of Nothing, a Course in the History of Geometry in the 19th Century. Springer, London (2007)
11. Leemans, D.: Residually weakly primitive and locally two-transitive geometries for sporadic groups. Académie Royale de Belgique, Classe des Sciences (2007)
12. Libois, P.: La synthèse de la géométrie et de l'algèbre. Colloque de Géométrie Algébrique. Tenu à Liège les 19–21 décembre 1949. CBRM, Georges Thone, Liège, Masson & Cie, Paris, pp. 143–153 (1950)
13. Ronan, M.A.: Lectures on Buildings. Academic Press, New York (1989)
14. Timmesfeld, F.G.: Abstract Root Subgroups and Simple Groups of Lie-Type. Birkhäuser, Basel (2001)
15. Van Maldeghem, H.: Generalized Polygons. Birkhäuser, Basel (1998)
16. Weiss, R.M.: The Structure of Spherical Buildings. Princeton University Press, Princeton (2003)
17. Weiss, R.M.: The Structure of Affine Buildings. Princeton University Press, Princeton (2009)

List of Publications for John Griggs Thompson

1953

[1] A method for finding primes. *Amer. Math. Monthly*, 60:175.

1958

[2] (with A.A. Albert). Two element generation of the projective unimodular group. *Bull. Amer. Math. Soc.*, 64:92–93.

1959

[3] Finite groups with fixed-point-free automorphisms of prime order. *Proc. Nat. Acad. Sci. USA*, 45:578–581.

[4] (with A.A. Albert). Two-element generation of the projective unimodular group. *Illinois J. Math.*, 3:421–439.

[5] (with D.R. Hughes). The H-problem and the structure of H-groups. *Pacific J. Math.*, 9:1097–1101.

[6] Finite groups with normal p-complements. In *Proc. Sympos. Pure Math., Vol. 1*, pages 1–3. American Mathematical Society, Providence, RI.

[7] Normal p-complements for finite groups. *Math. Z.*, 72:332–354.

[8] A special class of non-solvable groups. *Math. Z.*, 72:458–462.

1960

[9] (with W. Feit and M. Hall, Jr.). Finite groups in which the centralizer of any non-identity element is nilpotent. *Math. Z.*, 74:1–17.

1961

[10] (with W. Feit). Groups which have a faithful representation of degree less than $(p - 1/2)$. *Pacific J. Math.*, 11:1257–1262.

1962

[11] (with W. Feit). Finite groups which contain a self-centralizing subgroup of order 3. *Nagoya Math. J.*, 21:185–197.

[12] (with W. Feit). A solvability criterion for finite groups and some consequences. *Proc. Nat. Acad. Sci. USA*, 48:968–970.

H. Holden, R. Piene (eds.), *The Abel Prize 2008-2012*,
DOI 10.1007/978-3-642-39449-2_6,

[13] (with J. Boen and O. Rothaus). Further results on p-automorphic p-groups. *Pacific J. Math.*, 12:817–821.

1963

[14] (with W. Feit). Solvability of groups of odd order. *Pacific J. Math.*, 13:775–1029.

[15] Two results about finite groups. In *Proc. Internat. Congr. Mathematicians (Stockholm, 1962)*, pages 296–300. Inst. Mittag-Leffler, Djursholm.

1964

[16] 2-Signalizers of finite groups. *Pacific J. Math.*, 14:363–364.

[17] Normal p-complements for finite groups. *J. Algebra*, 1:43–46.

[18] Fixed points of p-groups acting on p-groups. *Math. Z.*, 86:12–13.

[19] Automorphisms of solvable groups. *J. Algebra*, 1:259–267.

1966

[20] Factorizations of p-solvable groups. *Pacific J. Math.*, 16:371–372.

[21] (with O. Rothaus). A combinatorial problem in the symmetric group. *Pacific J. Math.*, 18:175–178.

[22] Hall subgroups of the symmetric groups. *J. Combinatorial Theory*, 1:271–279.

[23] (with Z. Janko). On a class of finite simple groups of Ree. *J. Algebra*, 4:274–292.

1967

[24] An example of core-free quasinormal subgroups of p-groups. *Math. Z.*, 96:226–227.

[25] Vertices and sources. *J. Algebra*, 6:1–6.

[26] Centralizers of elements in p-groups. *Math. Z.*, 96:292–293.

[27] On a question of L.J. Paige. *Math. Z.*, 99:26–27.

[28] Defect groups are Sylow intersections. *Math. Z.*, 100:146.

[29] Toward a characterization of $E_2^*(q)$. *J. Algebra*, 7:406–414.

1968

[30] Nonsolvable finite groups all of whose local subgroups are solvable. *Bull. Amer. Math. Soc.*, 74:383–437.

[31] (with G. Glauberman). Weakly closed direct factors of Sylow subgroups. *Pacific J. Math.*, 26:73–83.

[32] Characterization of finite simple groups. In *Proc. Internat. Congr. Math. (Moscow, 1966)*, pages 158–162. Izdat. "Mir", Moscow.

1969

[33] Envelopes and p-signalizers of finite groups. *Illinois J. Math.*, 13:87–90.

[34] A replacement theorem for p-groups and a conjecture. *J. Algebra*, 13:149–151.

1970

[35] A non-duality theorem for finite groups. *J. Algebra*, 14:1–4.
[36] Normal p-complements and irreducible characters. *J. Algebra*, 14:129–134.
[37] Bounds for orders of maximal subgroups. *J. Algebra*, 14:135–138.
[38] Nonsolvable finite groups all of whose local subgroups are solvable. II. *Pacific J. Math.*, 33:451–536.
[39] (with Z. Janko). On finite simple groups whose Sylow 2-subgroups have no normal elementary subgroups of order 8. *Math. Z.*, 113:385–397.

1971

[40] Nonsolvable finite groups all of whose local subgroups are solvable. III. *Pacific J. Math.*, 39:483–534.
[41] Quadratic pairs. In *Actes du Congrès International des Mathématiciens (Nice, 1970), Tome 1*, pages 375–376. Gauthier-Villars, Paris.

1972

[42] (with F.J. MacWilliams and N.J.A. Sloane). Good self dual codes exist. *Discrete Math.*, 3:153–162.
[43] Toward a characterization of $E_2^*(q)$. II. *J. Algebra*, 20:610–621.

1973

[44] (with F.J. MacWilliams and N.J.A. Sloane). On the existence of a projective plane of order 10. *J. Combinatorial Theory Ser. A*, 14:66–78.
[45] Isomorphisms induced by automorphisms. *J. Austral. Math. Soc.*, 16:16–17. Collection of articles dedicated to the memory of Hanna Neumann, I.
[46] Nonsolvable finite groups all of whose local subgroups are solvable. IV, V, VI. *Pacific J. Math.*, 48:511–592, ibid. 50:215–297, 1974; ibid. 51:573–630, 1974.
[47] Weighted averages associated to some codes. *Scripta Math.*, 29(3–4):449–452. Collection of articles dedicated to the memory of Abraham Adrian Albert.

1974

[48] Simple $3'$-groups. In *Symposia Mathematica, Vol. XIII (Convegno di Gruppi e loro Rappresentazioni, INDAM, Rome, 1972)*, pages 517–530. Academic Press, London.

1976

[49] A simple subgroup of $E_8(3)$. In Nagayoshi Iwahori, editor, *Finite Groups (Sapporo and Kyoto, 1974)*, pages 113–116. Japan Society for the Promotion of Science, Tokyo.
[50] Finite groups and even lattices. *J. Algebra*, 38(2):523–524.
[51] A conjugacy theorem for E_8. *J. Algebra*, 38(2):525–530.

1977

[52] Toward a characterization of $F_2^*(q)$. III. *J. Algebra*, 49(1):162–166.

1978

[53] Remarks on finite groups. In *Proceedings of the 5th School of Algebra (Rio de Janeiro, 1978)*, pages 75–77, Rio de Janeiro. Soc. Brasil. Mat.

1979

[54] Uniqueness of the Fischer–Griess monster. *Bull. London Math. Soc.*, 11(3): 340–346.

[55] Finite groups and modular functions. *Bull. London Math. Soc.*, 11(3):347–351.

[56] Some numerology between the Fischer–Griess Monster and the elliptic modular function. *Bull. London Math. Soc.*, 11(3):352–353.

1980

[57] A finiteness theorem for subgroups of PSL(2, $\mathbf{R}$) which are commensurable with PSL(2, $\mathbf{Z}$). In *The Santa Cruz Conference on Finite Groups (Univ. California, Santa Cruz, Calif., 1979)*, volume 37 of *Proc. Sympos. Pure Math.*, pages 533–555. Amer. Math. Soc., Providence, RI.

[58] (with R.P. Anstee and M. Hall, Jr.). Planes of order 10 do not have a collineation of order 5. *J. Combin. Theory Ser. A*, 29(1):39–58.

[59] Finite simple groups. II. In Michael J. Collins, editor, *Proceedings of the Symposium held at the University of Durham, Durham, July 31–August 10, 1978*, pages 321–337, London. Academic Press.

1981

[60] Invariants of finite groups. *J. Algebra*, 69(1):143–145.

[61] Finite-dimensional representations of free products with an amalgamated subgroup. *J. Algebra*, 69(1):146–149.

[62] (with N.J.A. Sloane). The nonexistence of a certain Steiner system $S(3, 12, 112)$. *J. Combin. Theory Ser. A*, 30(3):209–236.

[63] Rational functions associated to presentations of finite groups. *J. Algebra*, 71(2):481–489.

[64] Ovals in a projective plane of order 10. In *Combinatorics (Swansea, 1981)*, volume 52 of *London Math. Soc. Lecture Note Ser.*, pages 187–190. Cambridge Univ. Press, Cambridge.

1982

[65] (with V. Pless). 17 does not divide the order of the group of a $(72, 36, 16)$ doubly even code. *IEEE Trans. Inform. Theory*, 28(3):537–541.

1983

[66] (with N.J.A. Sloane). Cyclic self-dual codes. *IEEE Trans. Inform. Theory*, 29(3):364–366.

1984

[67] Some finite groups which appear as Gal L/K, where $K \subseteq \mathbf{Q}(\mu_n)$. *J. Algebra*, 89(2):437–499.

[68] Finite nonsolvable groups. In *Group theory: essays for Philip Hall*, pages 1–12. Academic Press, London.
[69] Obituary: Philip Hall. *Bull. London Math. Soc.*, 16:612–616.

1985

[70] PSL_3 and Galois groups over $\mathbf{Q}$. In *Proceedings of the Rutgers group theory year, 1983–1984 (New Brunswick, N.J., 1983–1984)*, pages 309–319, Cambridge. Cambridge Univ. Press.
[71] Rational rigidity of $G_2(5)$. In *Proceedings of the Rutgers group theory year, 1983–1984 (New Brunswick, NJ, 1983–1984)*, pages 321–322, Cambridge. Cambridge Univ. Press.
[72] Primitive roots and rigidity. In *Proceedings of the Rutgers group theory year, 1983–1984 (New Brunswick, NJ, 1983–1984)*, pages 327–350, Cambridge. Cambridge Univ. Press.

1986

[73] Some finite groups which appear as $\mathrm{Gal}\, L/K$, where $K \subseteq \mathbf{Q}(\mu_n)$. In *Group theory, Beijing 1984*, volume 1185 of *Lecture Notes in Math.*, pages 210–230. Springer, Berlin.
[74] Algebraic numbers associated to certain punctured spheres. *J. Algebra*, 104(1):61–73.

1987

[75] Archimedes and continued fractions. *Math. Medley*, 15(2):67–75.

1989

[76] Fricke, free groups and SL_2. In *Group theory (Singapore, 1987)*, pages 207–214. de Gruyter, Berlin.
[77] Hecke operators and noncongruence subgroups. In *Group theory (Singapore, 1987)*, pages 215–224. de Gruyter, Berlin. Including a letter from J.-P. Serre.

1990

[78] (with R.M. Guralnick). Finite groups of genus zero. *J. Algebra*, 131(1):303–341.
[79] Groups of genus zero and certain rational functions. In *Groups—Canberra 1989*, volume 1456 of *Lecture Notes in Math.*, pages 185–190. Springer, Berlin.
[80] Rigidity, $\mathrm{GL}(n, q)$, and the braid group. *Bull. Soc. Math. Belg. Sér. A*, 42(3):723–733. Algebra, groups and geometry.

1991

[81] Galois groups. In *Groups—St. Andrews 1989, Vol. 2*, volume 160 of *London Math. Soc. Lecture Note Ser.*, pages 455–462. Cambridge Univ. Press, Cambridge.

1992

[82] Discrete groups and Galois theory. In *Groups, combinatorics & geometry (Durham, 1990)*, volume 165 of *London Math. Soc. Lecture Note Ser.*, pages 476–479. Cambridge Univ. Press, Cambridge.

1993

[83] Algebraic integers all of whose algebraic conjugates have the same absolute value. In *Coding theory, design theory, group theory (Burlington, VT, 1990)*, Wiley-Intersci. Publ., pages 107–110. Wiley, New York.

1994

[84] Note on realizable sequences of partitions. *Comm. Algebra*, 22(14):5679–5682.

[85] Note on $H(4)$. *Comm. Algebra*, 22(14):5683–5687.

1995

[86] 4-Punctured spheres. *J. Algebra*, 171(2):587–605.

[87] (with G.R. Robinson). Sums of squares and the fields $\mathbf{Q}_{A_n}$. *J. Algebra*, 174(1):225–228.

[88] Sylow 2-subgroups of simple groups. In *Séminaire Bourbaki, Vol. 10*, pages Exp. No. 345, 543–545. Soc. Math. France, Paris.

1996

[89] Some generalized characters. *J. Algebra*, 179(3):889–893.

[90] (with G.R. Robinson). On Brauer's $k(B)$-problem. *J. Algebra*, 184(3):1143–1160.

1997

[91] Power maps and completions of free groups and of the modular group. *J. Algebra*, 191(1):252–264.

[92] Incidence matrices of finite projective planes and their eigenvalues. *J. Algebra*, 191(1):265–278.

1998

[93] (with H. Völklein). Symplectic groups as Galois groups. *J. Group Theory*, 1(1):1–58.

1999

[94] (with H. Völklein). Braid-abelian tuples in $\mathrm{Sp}_n(K)$. In *Aspects of Galois theory (Gainesville, FL, 1996)*, volume 256 of *London Math. Soc. Lecture Note Ser.*, pages 218–238. Cambridge Univ. Press, Cambridge.

2005

[95] (with L. Scott, R. Solomon, J. Walter, and E. Zelmanov). Walter Feit (1930–2004). *Notices Amer. Math. Soc.*, 52(7):728–735.

2008

[96] Composition factors of rational finite groups. *J. Algebra*, 319(2):558–594.
[97] Unipotent elements, standard involutions, and the divisor matrix. *Comm. Algebra*, 36(9):3363–3371.

2010

[98] (with P. Sin) The divisor matrix, Dirichlet series, and SL(2; Z), in The legacy of Alladi Ramakrishnan in the mathematical sciences, pp. 299–327, New York: Springer, 2010.

List of Publications for Jacques Tits

1949

[1] Généralisations des groupes projectifs. *Acad. Roy. Belgique. Bull. Cl. Sci. (5)*, 35:197–208.

[2] Généralisations des groupes projectifs. II. *Acad. Roy. Belgique. Bull. Cl. Sci. (5)*, 35:224–233.

[3] Généralisations des groupes projectifs. III. *Acad. Roy. Belgique. Bull. Cl. Sci. (5)*, 35:568–589.

[4] Généralisations des groupes projectifs. IV. *Acad. Roy. Belgique. Bull. Cl. Sci. (5)*, 35:756–773.

1950

[5] Groupes triplement transitifs et généralisations. In *Algèbre et Théorie des Nombres*, Colloques Internationaux du Centre National de la Recherche Scientifique, no. 24, pages 207–208. Centre National de la Recherche Scientifique, Paris.

[6] Généralisation d'un théorème de Kerékjártó. In IIIe *Congrès National des Sciences, Bruxelles, 1950, Vol. 2*, pages 64–65. Fédération Belge des Sociétés Scientifiques, Bruxelles.

[7] Collinéations et transitivité. In IIIe *Congrès National des Sciences, Bruxelles, 1950, Vol. 2*, pages 66–67. Fédération Belge des Sociétés Scientifiques, Bruxelles.

1951

[8] Les groupes projectifs: évolution et généralisations. *Bull. Soc. Math. Belgique*, 3:1–10.

[9] Sur les groupes triplement transitifs continus; généralisation d'un théorème de Kerékjártó. *Compositio Math.*, 9:85–96.

1952

[10] Généralisations des groupes projectifs basées sur leurs propriétés de transitivité. *Acad. Roy. Belgique. Cl. Sci. Mém. Coll. in* 8°. 27, no. 2, 115 pp.

H. Holden, R. Piene (eds.), *The Abel Prize 2008-2012*,
DOI 10.1007/978-3-642-39449-2_7,

[11] Sur les groupes doublement transitifs continus. *Comment. Math. Helv.*, 26:203–224.

[12] Caractérisation topologique de certains espaces métriques. In *Nachr. Österr. Math. Ges. 22/23 (Bericht III. Österr. Mathematikerkongr., Salzburg, sept. 1952)*. p. 51.

[13] Étude de certains espaces métriques. *Bull. Soc. Math. Belgique*, 5:44–52.

[14] La notion d'homogénéité en géométrie. Sém. de synthèse scient., Université Libre de Bruxelles, 1 p.

1953

[15] Le plan projectif des octaves et les groupes de Lie exceptionnels. *Acad. Roy. Belgique. Bull. Cl. Sci. (5)*, 39:309–329.

[16] Sur un article précédent: "Étude de certains espaces métriques". *Bull. Soc. Math. Belgique*, 6:126–127.

1954

[17] Le plan projectif des octaves et les groupes exceptionnels E_6 et E_7. *Acad. Roy. Belgique. Bull. Cl. Sci. (5)*, 40:29–40.

[18] Espaces homogènes et groupes de Lie exceptionnels. In *Proc. Internat. Congr. Mathematicians 1954*, volume 1, Erven P. Noordhoff N.V., Groningen, pages 495–496, North-Holland Publishing Co, Amsterdam.

[19] Étude géométrique d'une classe d'espaces homogènes. *C. R. Acad. Sci. Paris*, 239:466–468.

[20] Espaces homogènes et isotropes, et espaces doublement homogènes. *C. R. Acad. Sci. Paris*, 239:526–527.

[21] Sur les R-espaces. *C. R. Acad. Sci. Paris*, 239:850–852.

[22] Espaces homogènes et isotropes et espaces doublement homogènes. Mémoire présenté au concours scient. interfacultaire L. Empain, VII. To appear in [209].

1955

[23] Sur certaines classes d'espaces homogènes de groupes de Lie. *Acad. Roy. Belg. Cl. Sci. Mém. Coll.* 8°(2), 29(3). 268 pp.

[24] Groupes semi-simples complexes et géométrie projective. In *Séminaire Bourbaki 1954/1955*. Exposé 112, 11 pp.

[25] Sous-algèbres des algèbres de Lie semi-simples (d'après V. Morozov, A. Malčev, E. Dynkin et F. Karpelevitch). In *Séminaire Bourbaki 1954/1955*. Exposé 119, 18 pp.

1956

[26] Espaces homogènes et isotropes de la relativité. In *Fünfzig Jahre Relativitätstheorie, Cinquantenaire de la théorie de la relativité, Jubilee of relativity theory*. Verhandlungen – Actes – Proceedings, Bern, 11.–16. Juli 1955, Helv. Phys. Acta, Suppl. IV, pages 46–47, Bern, Birkhäuser Verlag.

[27] Sur les groupes doublement transitifs continus: corrections et compléments. *Comment. Math. Helv.*, 30:234–240.

[28] Les groupes de Lie exceptionnels et leur interprétation géométrique. *Bull. Soc. Math. Belg.*, 8:48–81.

1957

[29] Transitivité des groupes de mouvements. *Schr. Forschungsinst. Math.*, 1:98–111.

[30] Sur la géométrie des R-espaces. *J. Math. Pures Appl. (9)*, 36:17–38.

[31] Sur les analogues algébriques des groupes semi-simples complexes. In *Colloque d'algèbre supérieure, tenu à Bruxelles du 19 au 22 décembre 1956*, Centre Belge de Recherches Mathématiques, pages 261–289. Établissements Ceuterick, Louvain.

1958

[32] Les "formes réelles" des groupes de type E_6. In *Séminaire Bourbaki 1957/1958*. Exposé 162, 15 pp.

[33] Sur la trialité et les algèbres d'octaves. *Acad. Roy. Belg. Bull. Cl. Sci. (5)*, 44:332–350.

1959

[34] Sur la trialité et certains groupes qui s'en déduisent. *Inst. Hautes Études Sci. Publ. Math.*, 2:13–60.

[35] Isotropie des espaces de Klein. In *Colloque Géom. Diff. Globale (Bruxelles, 1958)*, pages 153–161. Centre Belge Rech. Math., Louvain.

[36] Sur la classification des groupes algébriques semi-simples. *C. R. Acad. Sci. Paris*, 249:1438–1440.

[37] Sur une classe de groupes de Lie résolubles. *Bull. Soc. Math. Belg.*, 11:100–115.

1960

[38] Les espaces isotropes de la relativité. In *Colloque sur la théorie de la relativité 1959*, pages 107–119. Centre Belge Rech. Math.

[39] Une remarque sur la structure des algèbres de Lie semi-simples complexes. *Nederl. Akad. Wetensch. Proc. Ser. A 63 = Indag. Math.*, 22:48–53.

[40] Sur les groupes algébriques affins. Théorèmes fondamentaux de structure. Classification des groupes semisimples et géométries associées. In *Gruppi, anelli di Lie e teoria della coomologia*. Istituto Matematico dell'Università, Roma, 74 pp. Also in C.I.M.E. Summer Sch., 20, 185–264. Springer, Heidelberg.

[41] Sur les groupes d'affinités sans point fixe. *Acad. Roy. Belg. Bull. Cl. Sci. (5)*, 46:954–956.

1961

[42] Les groupes simples de Suzuki et de Ree. In *Séminaire Bourbaki 1960/1961*. Exposé 210, 18 pp.

[43] Groupes et géométries de Coxeter. Notes polycopiées, I.H.E.S., Bures-sur-Yvette, 26 pp.

1962

[44] Groupes algébriques semi-simples et géométries associées. In *Algebraical and Topological Foundations of Geometry (Proc. Colloq., Utrecht, 1959)*, pages 175–192. Pergamon, Oxford.

[45] "Sur une classe de groupes de Lie résolubles", corrections et additions. *Bull. Soc. Math. Belg.*, 14:196–209.

[46] Ovoïdes à translations. *Rend. Mat. e Appl. (5)*, 21:37–59.

[47] Ovoïdes et groupes de Suzuki. *Arch. Math.*, 13:187–198.

[48] Espaces homogènes complexes compacts. *Comment. Math. Helv.*, 37:111–120.

[49] Théorème de Bruhat et sous-groupes paraboliques. *C. R. Acad. Sci. Paris*, 254:2910–2912.

[50] Une classe d'algèbres de Lie en relation avec les algèbres de Jordan. *Nederl. Akad. Wetensch. Proc. Ser. A 65 = Indag. Math.*, 24:530–535.

[51] Groupes semi-simples isotropes. In *Colloq. Théorie des Groupes Algébriques (Bruxelles, 1962)*, pages 137–147. Librairie Universitaire, Louvain.

1963

[52] Géométries polyédriques et groupes simples. In *Atti della Seconda Riunione del Groupement de Mathématiciens d'Expression Latine*, pages 66–88. Edizioni Cremonese, Roma.

[53] Groupes simples et géométries associées. In *Proc. Internat. Congr. Mathematicians (Stockholm, 1962)*, pages 197–221. Inst. Mittag-Leffler, Djursholm.

[54] (with F. Bingen and L. Waelbroeck). *Séminaire sur les Algèbres de Banach, 1962–1963*. Centre Belge d'Algèbre et de Topologie. Université Libre de Bruxelles, Faculté des Sciences, Institut de Mathématique, Brussels.

1964

[55] A theorem on generic norms of strictly power associative algebras. *Proc. Amer. Math. Soc.*, 15:35–36.

[56] Automorphismes à déplacement borné des groupes de Lie. *Topology*, 3(suppl. 1):97–107.

[57] Algebraic and abstract simple groups. *Ann. of Math. (2)*, 80:313–329.

[58] Géométries polyédriques finies. *Rend. Mat. e Appl. (5)*, 23:156–165.

[59] Sur les systèmes de Steiner associés aux trois "grands" groupes de Mathieu. *Rend. Mat. e Appl. (5)*, 23:166–184.

1965

[60] Algèbres alternatives, algèbres de Jordan et algèbres de Lie exceptionelles. Notes polycopiées, IAS, Princeton, 5 pp.

[61] Sur une conjecture de L. Solomon. *C. R. Acad. Sci. Paris*, 260:6247–6249.

[62] (with A. Borel). Groupes réductifs. *Inst. Hautes Études Sci. Publ. Math.*, (27):55–150.

[63] Simple groups over local fields. Summer Institute on Algebraic Groups, Boulder, July 1965. Notes polycopiées, I–G, 6 pp.

[64] Structures et groupes de Weyl. In *Séminaire Bourbaki 1964/1965*. Exposé 288, 15 pp.

1966

[65] Une propriété caractéristique des ovoïdes associés aux groupes de Suzuki. *Arch. Math. (Basel)*, 17:136–153.

[66] (with M. Lazard). Domaines d'injectivité de l'application exponentielle. *Topology*, 4:315–322.

[67] Algèbres alternatives, algèbres de Jordan et algèbres de Lie exceptionnelles. I. Construction. *Nederl. Akad. Wetensch. Proc. Ser. A 69 = Indag. Math.*, 28:223–237.

[68] Normalisateurs de tores. I. Groupes de Coxeter étendus. *J. Algebra*, 4:96–116.

[69] Sur les constantes de structure et le théorème d'existence des algèbres de Lie semi-simples. *Inst. Hautes Études Sci. Publ. Math.*, (31):21–58.

[70] Classification of algebraic semisimple groups. In *Algebraic Groups and Discontinuous Subgroups (Proc. Sympos. Pure Math., Boulder, Colo., 1965)*, pages 33–62. Amer. Math. Soc., Providence, R.I., 1966.

[71] (with F. Bruhat). Un théorème de point fixe. Notes polycopiées, I.H.E.S., Bures-sur-Yvette, 34 pp.

[72] (with F. Bruhat). BN-paires de type affine et données radicielles. *C. R. Acad. Sci. Paris Sér. A-B*, 263:A598–A601.

[73] (with F. Bruhat). Groupes simples résiduellement déployés sur un corps local. *C. R. Acad. Sci. Paris Sér. A-B*, 263:A766–A768.

[74] (with F. Bruhat). Groupes algébriques simples sur un corps local. *C. R. Acad. Sci. Paris Sér. A-B*, 263:A822–A825.

[75] (with F. Bruhat). Groupes algébriques simples sur un corps local: Cohomologie galoisienne, décompositions d'Iwasawa et de Cartan. *C. R. Acad. Sci. Paris Sér. A-B*, 263:A867–A869.

1967

[76] *Tabellen zu den einfachen Lie Gruppen und ihren Darstellungen*. Springer-Verlag, Berlin.

[77] (with F. Bruhat). Groupes algébriques simples sur un corps local. In *Proc. Conf. Local Fields (Driebergen, 1966)*, pages 23–36. Springer, Berlin.

[78] Lectures on algebraic groups. Notes rédigées par P. André et D. Winter, Yale University, 72 pp.

1968

[79] Formes quadratiques, groupes orthogonaux et algèbres de Clifford. *Invent. Math.*, 5:19–41.

[80] (with L. Waelbroeck). The integration of a Lie algebra representation. *Pacific J. Math.*, 26:595–600.

1969

[81] Le problème des mots dans les groupes de Coxeter. In *Symposia Mathematica (INDAM, Rome, 1967/68), Vol. 1*, pages 175–185. Academic Press, London.
[82] Le groupe de Janko d'ordre 604.800. In *Theory of finite groups*, pages 91–95. W. A. Benjamin, Inc., New York.
[83] (with A. Borel). On "abstract" homomorphisms of simple algebraic groups. In *Algebraic geometry: papers presented at the Bombay Colloquium 1968*, Tata Inst. Fund. Res. Stud. Math. 4, pages 75–82. Oxford University Press, Oxford.

1970

[84] Sur le groupe des automorphismes d'un arbre. In *Essays on topology and related topics (Mémoires dédiés à Georges de Rham)*, pages 188–211. Springer, New York.

1971

[85] Groupes finis simples sporadiques. In *Séminaire Bourbaki 1969/1970*. Exposé 375. Also available as Lecture Notes in Math. 180, pages 187–211, Springer-Verlag, Berlin, 1971.
[86] (with A. Borel). Éléments unipotents et sous-groupes paraboliques de groupes réductifs. I. *Invent. Math.*, 12:95–104.
[87] Représentations linéaires irréductibles d'un groupe réductif sur un corps quelconque. *J. Reine Angew. Math.*, 247:196–220.
[88] Homomorphismes et automorphismes "abstraits" de groupes algébriques et arithmétiques. In *Actes du Congrès International des Mathématiciens (Nice, 1970), Tome 2*, pages 349–355. Gauthier-Villars, Paris.
[89] Non-existence de certaines extensions transitives. I. Groupes projectifs à une dimension. *Bull. Soc. Math. Belg.*, 23:481–492. Collection of articles honoring P. Burniat, Th. Lepage and P. Libois.
[90] Non-existence de certaines extensions transitives. II. To appear in [209].

1972

[91] Free subgroups in linear groups. *J. Algebra*, 20:250–270. Also available in *Matematika: Periodicheskii sbornik perevodov inostrannykh statei*. Tom 16:47–66 (1972), vyp. 2, Izdat. "Mir", Moscow.
[92] (with F. Bruhat). Groupes réductifs sur un corps local. *Inst. Hautes Études Sci. Publ. Math.*, (41):5–251.
[93] (with A. Borel). Compléments à l'article: "Groupes réductifs". *Inst. Hautes Études Sci. Publ. Math.*, (41):253–276.
[94] Une propriété des systèmes de racines (appendice à l'article de J. Dixmier: "Sur les homomorphismes d'Harish-Chandra"). *Invent. Math.*, 17:174–176.

1973

[95] (with A. Borel). Homomorphismes "abstraits" de groupes algébriques simples. *Ann. of Math. (2)*, 97:499–571.

1974

[96] (with S. Koppelberg). Une propriété des produits directs infinis de groupes finis isomorphes. *C. R. Acad. Sci. Paris Sér. A*, 279:583–585.

[97] Homorphismes "abstraits" de groupes de Lie. In *Symposia Mathematica, Vol. XIII (Convegno di Gruppi e loro Rappresentazioni, INDAM, Rome, 1972)*, pages 479–499. Academic Press, London.

[98] *Buildings of spherical type and finite BN-pairs*. Lecture Notes in Mathematics, Vol. 386. Springer-Verlag, Berlin.

[99] Immeubles: classification et automorphismes. *Ann. Collège France*, 74:631–637.

1975

[100] On buildings and their applications. In *Proceedings of the International Congress of Mathematicians (Vancouver, BC, 1974), Vol. 1*, pages 209–220. Canad. Math. Congress, Montreal, Que.

[101] Sur certains groupes dont l'ordre est divisible par 23. *Bull. Soc. Math. Belg.*, 27(4):325–332.

[102] Leçon inaugurale de la chaire de théorie des groupes. Collège de France, 18 pp.

[103] Groupes réductifs sur un corps local. *Ann. Collège France*, 75:49–56.

1976

[104] Two properties of Coxeter complexes. *J. Algebra*, 41(2):265–268. Appendix to "A Mackey formula in the group ring of a Coxeter group" (J. Algebra **41** (1976), no. 2, 255–264) by Louis Solomon.

[105] Non-existence de certains polygones généralisés. I. *Invent. Math.*, 36:275–284.

[106] Systèmes générateurs de groupes de congruence. *C. R. Acad. Sci. Paris Sér. A-B*, 283(9):A693–A695.

[107] Classification of buildings of spherical type and Moufang polygons: a survey. In *Colloquio Internazionale sulle Teorie Combinatorie (Roma, 1973), Tomo I*, pages 229–246. Atti dei Convegni Lincei, No. 17. Accad. Naz. Lincei, Rome.

[108] Travaux de Margulis sur les sous-groupes discrets de groupes de Lie. In *Séminaire Bourbaki, 28ème année (1975/76), Exposé 482*, pages 174–190. Lecture Notes in Math., Vol. 567. Springer, Berlin.

[109] Quadrangles de Moufang, I. Prépublication, Paris, 16 pp.

[110] Rigidité et arithméticité des sous-groupes discrets de groupes de Lie. *Ann. Collège France*, 76:51–55.

1977

[111] A "theorem of Lie–Kolchin" for trees. In *Contributions to algebra (collection of papers dedicated to Ellis Kolchin)*, pages 377–388. Academic Press, New York.

[112] Endliche Spiegelungsgruppen, die als Weylgruppen auftreten. *Invent. Math.*, 43(3):283–295.

[113] Groupes de Whitehead de groupes algébriques simples sur un corps (d'après V.P. Platonov et al.). In *Séminaire Bourbaki, 29e année (1976/77) Exposé 505*, pages 218–236. Lecture Notes in Math., Vol. 677. Springer, Berlin.

[114] Groupes finis simples sporadiques. *Ann. Collège France*, 77:57–66.

1978

[115] (with W. Feit). Projective representations of minimum degree of group extensions. *Canad. J. Math.*, 30(5):1092–1102.

[116] (with A. Borel). Théorèmes de structure et de conjugaison pour les groupes algébriques linéaires. *C. R. Acad. Sci. Paris Sér. A-B*, 287(2):A55–A57.

[117] Polygones de Moufang et groupes de rang 2. *Ann. Collège France*, 78:73–80.

1979

[118] Reductive groups over local fields. In *Automorphic forms, representations and L-functions (Proc. Sympos. Pure Math., Oregon State Univ., Corvallis, Ore., 1977), Part 1*, Proc. Sympos. Pure Math., XXXIII, pages 29–69. Amer. Math. Soc., Providence, R.I.

[119] Non-existence de certains polygones généralisés. II. *Invent. Math.*, 51(3): 267–269.

[120] Le principe d'inertie en relativité générale. *Bull. Soc. Math. Belg. Sér. A*, 31(2):171–197.

[121] Problèmes de théorie des groupes en relativité Einsteinienne et chronogéométrie. *Ann. Collège France*, 79:65–69.

1980

[122] The work of Gregori Aleksandrovitch Margulis. In *Proceedings of the International Congress of Mathematicians (Helsinki, 1978)*, pages 57–63, Helsinki. Acad. Sci. Fennica.

[123] (with C. W. Curtis and G. I. Lehrer). Spherical buildings and the character of the Steinberg representation. *Invent. Math.*, 58(3):201–210.

[124] Quaternions over $\mathbf{Q}(\sqrt{5})$, Leech's lattice and the sporadic group of Hall–Janko. *J. Algebra*, 63(1):56–75.

[125] Four presentations of Leech's lattice. In Michael J. Collins, editor, *Finite Simple Groups II, Proceedings of the Symposium held at the University of Durham, Durham, July 31–August 10, 1978*, pages 303–307, London. Academic Press Inc.

[126] Buildings and Buekenhout geometries. In Michael J. Collins, editor, *Finite Simple Groups II, Proceedings of the Symposium held at the University of Durham, Durham, July 31–August 10, 1978*, pages 309–320, London. Academic Press Inc.

[127] Schéma en groupes sur un anneau de valuation: immeubles affines. *Ann. Collège France*, 80:75–79.

1981

[128] Exposé sur les mathématiques, fait à l'occasion du 450e anniversaire du Collège de France. C. R. de la réunion extraordinaire de l'Assemblée des professeurs, Collège de France, 3 pp.

[129] Appendix to: "Groups of polynomial growth and expanding maps" [Inst. Hautes Études Sci. Publ. Math. No. 53 (1981), 53–73] by M. Gromov. *Inst. Hautes Études Sci. Publ. Math.*, (53):74–78.

[130] Définition par générateurs et relations de groupes avec BN-paires. *C. R. Acad. Sci. Paris Sér. I Math.*, 293(6):317–322.

[131] Algèbres enveloppantes et groupes de Chevalley généralisés. Actes des Journées 'Groupes et langages', Amiens, 5 pp.

[132] Groupes à croissance polynomiale (d'après M. Gromov et al.). In *Séminaire Bourbaki (1980/81), Exposé 572*, pages 176–188. Lecture Notes in Math., Vol. 901. Springer, Berlin.

[133] A local approach to buildings. In *The geometric vein*, pages 519–547. Springer, New York.

[134] Géométrie de l'espace, du temps et de la causalité: la voie axiomatique (extraits d'une conférence faite à Halle en 1980). Mélanges Paul Libois, Bruxelles, 1981, 9–11.

[135] Algèbres et groupes de Kac–Moody. *Ann. Collège France*, 81:75–86.

1982

[136] Évariste Galois: son oeuvre, sa vie, ses rapports avec l'Académie. *C. R. Acad. Sci. Paris Vie Académique*, 295(suppl. 12):171–180.

[137] Fields Medals for 1978. *Fiz.-Mat. Spis. Bulgar. Akad. Nauk.*, 24(57)(3):203–207.

[138] Algèbres de Kac–Moody et groupes associés (suite). *Ann. Collège France*, 82:91–105.

1983

[139] Moufang octagons and the Ree groups of type ${}^2F_4$. *Amer. J. Math.*, 105(2):539–594.

[140] On the distance between opposite vertices in buildings of spherical type (appendix to the article of A.E. Brouwer and A.M. Cohen: "Some remarks on Tits geometries"). *Nederl. Akad. Wetensch. A86 = Indag. Math.*, 45(4):400–402.

[141] (with F. Bruhat). Groupes réductifs sur un corps local. II. Schémas en groupes. Existence d'une donnée radicielle valuée. *Inst. Hautes Études Sci. Publ. Math.*, (60):197–376.

[142] *Liesche Gruppen und Algebren.* Hochschultext. Springer-Verlag, Berlin. With the collaboration of Manfred Krämer and Hans Scheerer.

[143] Remarks on Griess' construction of the Griess–Fischer sporadic group. I–IV. To appear in [209].

[144] Le Monstre (d'après R. Griess, B. Fischer et al.) Séminaire Bourbaki (1983/84), Exposé 620. *Astérisque*, (121–122):105–122.

[145] Le groupe sporadique de Griess–Fischer. *Ann. Collège France*, 83:89–102.

1984

[146] (with F. Bruhat). Schémas en groupes et immeubles des groupes classiques sur un corps local. *Bull. Soc. Math. France*, 112(2):259–301.
[147] On R. Griess' "Friendly giant". *Invent. Math.*, 78(3):491–499.
[148] Affine buildings, arithmetic groups and finite geometries. Notes by R. Scaramuzzi and D. White, Yale University, 117 pp.
[149] Immeubles de type affine, classification et application aux groupes finis. *Ann. Collège France*, 84:85–96.

1985

[150] Symétries. *C. R. Acad. Sci. Sér. Gén. Vie Sci.*, 2(1):13–25.
[151] Groups and group functors attached to Kac–Moody data. In *Workshop Bonn 1984 (Bonn, 1984)*, volume 1111 of *Lecture Notes in Math.*, pages 193–223. Springer, Berlin.
[152] (with A. M. Cohen). On generalized hexagons and a near octagon whose lines have three points. *European J. Combin.*, 6(1):13–27.
[153] Avatars des grands théorèmes de classification d'Élie Cartan. In *Élie Cartan et les mathématiques d'aujourd'hui*, pages 439–440. Société Mathématique de France, Paris. Papers presented at the seminar held at the University of Lyon, Lyon, June 25–29, 1984.
[154] Immeubles affines, groupes arithmétiques et géométries finies. *Ann. Collège France*, 85:93–110.

1986

[155] Immeubles de type affine. In *Buildings and the geometry of diagrams (Como, 1984)*, volume 1181 of *Lecture Notes in Math.*, pages 159–190. Springer, Berlin.
[156] Ensembles ordonnés, immeubles et sommes amalgamées. *Bull. Soc. Math. Belg. Sér. A*, 38:367–387.
[157] (with J. Dieudonné). La vie et l'œuvre de Claude Chevalley. *C. R. Acad. Sci. Sér. Gén. Vie Sci.*, 3(6):559–565.
[158] Buildings and group amalgamations. In *Proceedings of Groups—St. Andrews 1985*, volume 121 of *London Math. Soc. Lecture Note Ser.*, pages 110–127, Cambridge. Cambridge Univ. Press.
[159] Le groupe de Griess–Fischer: construction et "Moonshine". *Ann. Collège France*, 86:101–112.

1987

[160] (with W.M. Kantor and R.A. Liebler). On discrete chamber-transitive automorphism groups of affine buildings. *Bull. Amer. Math. Soc. (N.S.)*, 16(1):129–133.
[161] Uniqueness and presentation of Kac–Moody groups over fields. *J. Algebra*, 105(2):542–573.

[162] (with F. Bruhat). Schémas en groupes et immeubles des groupes classiques sur un corps local. II. Groupes unitaires. *Bull. Soc. Math. France*, 115(2):141–195.

[163] (with F. Bruhat). Groupes algébriques sur un corps local. Chapitre III. Compléments et applications à la cohomologie galoisienne. *J. Fac. Sci. Univ. Tokyo Sect. IA Math.*, 34(3):671–698.

[164] Le module du "moonshine" [d'après I. Frenkel, J. Lepowsky et A. Meurman]. Séminaire Bourbaki (1986/87), Exposé 684. *Astérisque*, (152–153):285–303 (1988).

[165] (with P. Lentoudis). Sur le groupe des automorphismes de certains produits en couronne. *C. R. Acad. Sci. Paris Sér. I Math.*, 305(20):847–852.

[166] (with M.A. Ronan). Building buildings. *Math. Ann.*, 278(1–4):291–306.

[167] (with J. Dieudonné). Claude Chevalley (1909–1984). *Bull. Amer. Math. Soc. (N.S.)*, 17(1):1–7.

[168] Geometrie von Raum, Zeit und Kausalität: Ein axiomatischer Zugang. *Nova Acta Leopoldina (N.F.)*, 54(244):101–111.

[169] Unipotent elements and parabolic subgroups of reductive groups. II. In *Algebraic groups Utrecht 1986*, volume 1271 of *Lecture Notes in Math.*, pages 265–284. Springer, Berlin.

[170] Suite du précédent. *Ann. Collège France*, 87:90–97.

1988

[171] Groupes de type E sur les corps globaux. Collège de France, Paris, 4 pp.

[172] Sur le groupe des automorphismes de certains groupes de Coxeter. *J. Algebra*, 113(2):346–357.

[173] Groupes associés aux algèbres de Kac–Moody. Séminaire Bourbaki (1988/89), Exposé 700. *Astérisque*, (177–178):7–31.

[174] Formes et sous-groupes des groupes algébriques simples sur les corps et les corps locaux. *Ann. Collège France*, 88:85–100.

1989

[175] Immeubles jumelés. *Ann. Collège France*, 89:81–95.

1990

[176] Strongly inner anisotropic forms of simple algebraic groups. *J. Algebra*, 131(2):648–677.

[177] Spheres of radius 2 in triangle buildings. I. In *Finite geometries, buildings, and related topics (Pingree Park, CO, 1988)*, Oxford Sci. Publ., pages 17–28. Oxford Univ. Press, New York.

[178] Suite du précédent. *Ann. Collège France*, 90:87–103.

1991

[179] Symmetrie. In *Miscellanea mathematica*, pages 293–304. Springer, Berlin. Also in *Amer. Math. Monthly* 107(5):454–461, 2000 (in English).

[180] Cohomologie galoisienne des groupes semi-simples sur les corps de nombres. *Ann. Collège France*, 91:125–137.

1992

[181] (with G. Lusztig). The inverse of a Cartan matrix. *An. Univ. Timişoara Ser. Ştiinţ. Mat.*, 30(1):17–23.

[182] Twin buildings and groups of Kac-Moody type. In *Groups, combinatorics & geometry (Durham, 1990)*, volume 165 of *London Math. Soc. Lecture Note Ser.*, pages 249–286. Cambridge Univ. Press, Cambridge.

[183] Sur les degrés des extensions de corps déployant les groupes algébriques simples. *C. R. Acad. Sci. Paris Sér. I Math.*, 315(11):1131–1138.

[184] Groupes algébriques sur les corps non parfaits. *Ann. Collège France*, 92:115–132.

1993

[185] (with M.A. Ronan). Twin trees. I. *Invent. Math.*, 116(1–3):463–479.

[186] Sur les produits tensoriels de deux algèbres de quaternions. *Bull. Soc. Math. Belg. Sér. B*, 45(3):329–331.

[187] Groupes algébriques linéaires sur les corps séparablement clos. *Ann. Collège France*, 93:113–130.

1994

[188] Moufang polygons. I. Root data. A tribute to J.A. Thas (Gent, 1994). *Bull. Belg. Math. Soc. Simon Stevin*, 1(3):455–468.

[189] Groupes algébriques simples de rang 2 et algèbres de Clifford de petites dimensions (classification des polygones de Moufang). *Ann. Collège France*, 94:101–114.

1995

[190] Polygones de Moufang (suite), description des quadrangles de Moufang connus. *Ann. Collège France*, 95:101–114.

1996

[191] Ein Fixpunktsatz für Gebäude und Anwendungen (Zusammenfassung eines Vortages). Jahrestagung der DMV, Jena, 15.–21. Sept. 1996, Programm, p. 97.

[192] Arbres jumelés. *Ann. Collège France*, 96:79–101.

1997

[193] The work of Gregori Aleksandrovitch Margulis. In *Fields Medallists' lectures*, volume 5 of *World Sci. Ser. 20th Century Math.*, pages 263–269. World Sci. Publ., River Edge, NJ.

[194] Homomorphismes "abstraits" de groupes algébriques. *Ann. Collège France*, 97:89–102.

1998

[195] M.-A. Knus, A. Merkurjev, M. Rost, and J.-P. Tignol. *The book of involutions*, volume 44 of *American Mathematical Society Colloquium Publica-*

tions. American Mathematical Society, Providence, RI, 1998. With a preface in French by J. Tits.
[196] Immeubles jumelés: théorèmes d'existence. *Ann. Collège France*, 98:97–112.

1999

[197] (with M.A. Ronan). Twin trees. II. Local structure and a universal construction. *Israel J. Math.*, 109:349–377.
[198] Travaux sur le "Plan Crémonien". (1950–1952) To appear in [209].
[199] Éléments unipotents et sous-groupes paraboliques de groupes algébriques simples. *Ann. Collège France*, 99:95–107.
[200] Le plan de Cremona. *Ann. Collège France*, 99:108–114.

2000

[201] Groupes de rang 1 et ensembles de Moufang. *Ann. Collège France*, 100:93–109.

2001

[202] (with H. Bass). Discreteness criteria for tree automorphism groups. Appendix to: *Tree lattices* by H. Bass and A. Lubotzky, Progr. Math. 176, Birkhäuser, Boston, pages 185–212.
[203] Résumé des travaux antérieurs à 1972. In S.S. Chern and F. Hirzebruch, editors, *Wolf Prize in Mathematics*, volume 2, pages 706–729. World Scientific.
[204] Groupes et Géométrie de Coxeter. In S.S. Chern and F. Hirzebruch, editors, *Wolf Prize in Mathematics*, volume 2, pages 740–754. World Scientific.

2002

[205] (with R.M. Weiss). *Moufang polygons*. Springer Monographs in Mathematics. Springer-Verlag, Berlin.

2004

[206] Armand Borel as I knew him. *Notices Amer. Math. Soc.*, 51(5):515–518.
[207] Quarante-cinq ans et cent numéros des Publications Mathématiques de l'I.H.É.S. *Publ. Math. Inst. Hautes Études Sci.*, (100):1–4.

2006

[208] (with B. Mühlherr). The center conjecture for non-exceptional buildings. *J. Algebra*, 300(2):687–706.

2013

[209] *Jacques Tits, Œuvres—Collected Works*. F. Buekenhout, B. Mühlherr, J.-P. Tignol, H. van Maldeghem, editors. European Mathematical Society Publ. House, Zürich.

Curriculum Vitae for John Griggs Thompson

Born:	October 13, 1932 in Ottawa, USA
Degrees/education:	Bachelor, Yale University, 1955
	PhD, University of Chicago, 1959
Positions:	Assistant Professor, Harvard University, 1959–1962
	Professor, University of Chicago, 1962–1970
	Rouse Ball Professor of Mathematics, University of Cambridge, 1970–1993
	Graduate Research Professor, University of Florida, 1993–
Memberships:	National Academy of Sciences, USA, 1967
	Royal Society of London, 1979
	American Academy of Arts and Sciences, 1998
	Accademia Nazionale dei Lincei,
	Norwegian Academy of Science and Letters, 2008
Awards and prizes:	Cole Prize, 1965
	Fields Medal, 1970
	Senior Berwick Prize, 1982
	Sylvester Medal, 1985
	Poincaré Medal, 1992
	Wolf Prize, 1992
	National Medal of Science, US, 2000
	Abel Prize, 2008
Honorary degrees:	University of Illinois, 1968
	Yale University, 1985
	University of Oxford, 1987
	Ohio State University, 2009

H. Holden, R. Piene (eds.), *The Abel Prize 2008-2012*,
DOI 10.1007/978-3-642-39449-2_8,

Curriculum Vitae for Jacques Tits

Born: August 12, 1930 in Uccle, Belgium

Degrees/education: Licencié en Sciences Mathématiques, Université de Bruxelles, 1948
Docteur en Sciences Mathématiques, Université de Bruxelles, 1950
Agrégé de l'Enseignement Supérieur, Université de Bruxelles, 1955

Positions: Boursier du Fonds National de la Recherche Scientifique, 1948–1956
Assistant, Université de Bruxelles, 1956–1957
Chargé de Cours, Professeur extraordinaire, Professeur ordinaire, Université de Bruxelles, 1957–1963
Professor, Universität Bonn, 1964–1974
Professeur associé, Collège de France, 1973–1974
Professeur, Collège de France, Chaire de Théorie des Groupes, 1974–2000

Visiting positions: ETH, Zürich, 1950, 1951, 1953
Università di Roma, 1955, 1956
Paris, 1956
Institute for Advanced Studies, Princeton, 1951–1952, 1963, 1969, 1971–1972
Institut des Hautes Études Scientifique, Paris, 1960, 1961, 1962, 1964, 1965, 1966, 1968, 1970, 1973
University of Chicago, 1963
University of California, Berkeley, 1963
Summer School of the AMS; Boulder, Colorado, 1965
Yale University, 1966–1967, 1976, 1980, 1984, 1990, 1995

H. Holden, R. Piene (eds.), *The Abel Prize 2008-2012*,
DOI 10.1007/978-3-642-39449-2_9,

	Universities of Tokyo and Kyoto, 1971
	Universities of Tianjin, Shangai and Beijing, 1987
	University of Ottawa, 1998
	Chaire de la Vallée-Poussin, Université catholique de Louvain, 2001
Memberships:	Deutsche Akademie der Naturforscher Leopoldina, 1977
	Académie des sciences, France (correspondant: 1977, titulaire: 1979)
	International Jury, Balzan Prize, 1985–
	Royal Netherlands Academy of Arts and Sciences, 1988
	Academia Europaea (Founding Member), 1988
	Académie royale de Belgique, 1991
	American Academy of Arts and Sciences, 1992
	National Academy of Sciences (USA), 1992
	London Mathematical Society (Honorary Member), 1993
	Norwegian Academy of Science and Letters, 2008
Awards and prizes:	Lauréat du Concours des bourses de voyage organisé par le gouvernement belge, 1949
	Prix scientifique interfacultaire Louis Empain, Belgium, 1955
	Prix Adolphe Wettrems, Académie royale des Sciences de Belgique, 1958
	Prix décennal de mathématiques du gouvernement belge, Belgium, 1965
	Médaille Poincaré, Grand Prix des Sciences Mathématiques et Physiques de l'Académie des Sciences de Paris, 1976
	Commandeur des Palmes Académiques, France, 1993
	Wolf Prize in Mathematics, 1993
	Pour le mérite für Wissenschaften und Künste, Germany, 1995
	Chevalier de la Légion d'Honneur, France, 1995
	Georg Cantor Medal, Deutsche Mathematiker-Vereinigung, Germany, 1996
	Officier de l'Ordre national du Mérite, France, 2001
	Abel Prize, 2008
Honorary degrees:	Utrecht University, 1970
	Universiteit Gent, 1979
	University of Bonn, 1988
	Université catholique de Louvain, 1992

2009

Mikhail Gromov

"for his revolutionary contributions to geometry"

ABEL
PRISEN

A Few Recollections

Mikhail Gromov

After a few frustratingly unsuccessful attempts to write my biography, I have arrived at the inevitable conclusion that this is a *logically impossible* task.

Mind you, there are many counterexamples to this "non-existence conjecture". I enjoyed a lot reading the autobiographies in the first Abel's volume. Yet, I think the conjecture is true in a narrow sense, if you separate "mathematician—a human being" from "mathematician—mathematician".

Our non-mathematical lives are, mathematically speaking, not that interesting, unless somebody had a misfortune to live through interesting times or undergo "interesting" personal experiences.

The life of a mathematician is reflected in the ideas we expound in our papers, what else can we add to this? Is there any *non-trivial* "else" to our lives?

Being trivial is our most dreaded pitfall: you say stupid things, not original things, outrageously wrong things—all will be forgotten when the dust settles down. But if you pompously call $a+b=c$ "Theorem" in your paper, you will be forever remembered as "this $a+b$ guy", no matter you prove bloody good theorems afterwards. (Caution: $2+2=_3 4$ is something quite non-trivial, or at least, not quite trivial.)

I was introduced to the idea on September 1st 1960 at the then Leningrad University when our analysis professor Boris Mikhailovich Makarov said to me after our first calculus class—he expressed this in somewhat metaphorical terms—that I should've kept my mouth shut unless I had something non-trivial to say.

Electronic supplementary material Supplementary material is available in the online version of this chapter at http://dx.doi.org/10.1007/978-3-642-39449-2_10. Videos can also be accessed at http://www.springerimages.com/videos/978-3-642-39449-2.

M. Gromov (✉)
Institut des Hautes Études Scientifiques, Le Bois-Marie 35, route de Chartres, 91440 Bures-sur-Yvette, France
e-mail: gromov@ihes.fr

M. Gromov
Courant Institute of Mathematical Sciences, New York University, 251 Mercer Street, New York, NY 10012-1185, USA

H. Holden, R. Piene (eds.), *The Abel Prize 2008-2012*,
DOI 10.1007/978-3-642-39449-2_10, © Springer-Verlag Berlin Heidelberg 2014

Further encouraged by my teachers and fellow students, I tried to follow his advice and, apparently, have succeeded—I hear nothing disrespectful about my mouth for the last 10–20 years. Strangely, this does not make me feel a lot happier.

"Trivial" is relative. Anything grasped as long as two minutes ago seems trivial to a working mathematician. But it may be amusing, looking from afar, to recall personal eureka transition points. (According to Terry Pratchett's Revised Ancient Greek Dictionary, "eureka" translates as "give me a towel".)

Another concept you learn at some point is that of "unsolved problem". David Ruelle has once put it that he sees a problem when he feels annoyed by non-understanding something. Children, like scientists, are good at non-understanding, except that the annoyed ones are their parents bombarded with endless What and Why and When And How and Where and Who.

As your adult personality properly matures without being sidetracked by your scientific or artistic inclinations, you resolve these WWW problems with a single: "This is the stupidest question I have ever heard"—said to a child. (Lipman Bers once boasted to me that he had received this response when he had asked his high school mathematics teacher if there could be two different infinities.)

My parents, were medical doctors rather than mathematicians, actually pathologists; they often discussed the problems they were encountering during autopsies, with their friends—also pathologists.

One story, I recall, was very funny, at least everybody laughed. My father spent several hours carefully checking and rechecking everything *inside* a body on the dissecting table but was unable to find the cause of death. When he was ready to surrender and shamefully write it off to "the heart failure", the man responsible for moving and cleaning the bodies, said: "Hey, doctor, isn't it funny, the man did not wash his left foot, look at the black marks over there". At a glance my father realized that the course of death was electrocution, apparently, the poor stepped on a high-voltage wire.

A few comments are in order. By the book, one starts an autopsy with a careful *external* examination of the body before performing dissection. My father, experienced as he was, probably was absent-minded at the moment: neglecting external

(Archives of the Mathematisches Forschungsinstitut Oberwolfach)

inspection strikes as funny to a pathologist as $\mathrm{d}x/\mathrm{d}y = x/y$ to a mathematician. (Maynard Smith—a great theoretical biologist, complained that editors of biology journals had sometimes "simplified" $\mathrm{d}x/\mathrm{d}y \to x/y$ in his papers.)

Autopsies have been routinely performed in Russian hospitals. Treating physicians were in a constant dread of the final word by a pathologist like students waiting for results of an examination. Eventually, physicians revolted—the autopsy in most countries if performed, then only rarely and usually on decade-old exhumed corpses—deaths of patients can be safely attributed to "heart failures".

There is an obvious moral to this story for pathologists and mathematicians alike. But the "stupid question" might have escaped you: "What, How and Why is the heart failure when you die?"

It is *not* that the heart just stops—this is what the so called "common sense" would tell you. Actually, stopping and "resetting" the heart is what defibrillators are for—you see them at the airports—they *save* lives, if used promptly—the brain survives a couple of minutes after the oxygen delivery by the blood stops.

What happens to the heart at the critical transition moment before it goes to the final rest is a change in the dynamics of electric/chemical currents in the heart muscular tissue—a switch to a non-quasiperiodic "chaotic" regime. (A high external voltage can provoke this, but it may also "disperse the chaos".)

Isn't it a "New application of the chaos theory to living systems?!" a bright mathematician may exclaim. Indeed, this is not a bad idea. I bet there are several articles in Nature with this title. The catch is that biological chaotic systems do not live long, their life spans are even shorter than the half-life of such articles—there is no accepted theory of arrhythmia in general and of the ventricular fibrillation (this is what we all will end up with) in particular. The physiology of the heart and mathematics at the bottom of it are not that trivial. And, probably, the true "stupid child question" has not been asked yet.

Biology in general and medicine in particular are full of annoying nearly mathematical puzzles. At 5 you ask:

Can four elephants beat a whale in a fight?

Twenty years later you come up with:

How, in principle, a humble bacterium, a tiny virus, e.g. HIV whose all "knowledge of the world" is written down in four letters on a 9749-long string of RNA, outsmart all of humanity with terabytes (10^{12}) of "information" stored in our individual synaptic memories and as much in our libraries?

What is it that the virus know that we don't? How many bits do we have to add to (to erase from?) our knowledge banks to beat 9749?

My second story needs a little preamble. There are several innocuous reactions turning water-like solutions into red ones looking like blood.

Something more amusing you get of a mixture of potassium permanganate with concentrated sulfuric acid,

$$6KMnO_4 + 9H_2SO_4 \rightarrow 6MnSO_4 + 3K_2SO_4 + 9H_2O + 5O_3.$$

The O_3 (ozone) vapor will ignite paper soaked with alcohol; with some luck, an explosion throws sulfuric acid into your eyes.

According to the basic chemistry safety rule, you *first* produce artificial blood, place a large bowl B with it in front of you and *only then* proceed with mixing $KMnO_4$ and H_2SO_4 in a test-tube T making sure that B is *strictly on the line* between T and your eyes.

When T explodes, the bowl in between protects your eyes from the sulfuric acid, while the bloody contents of B picturesquely splash all over your face.

This happened to me at a demonstration of "miracles of chemistry" at our high school when I was about 13. The audience was duly impressed, especially our chemistry professor. But myself, I missed the best of the show as I could not see my face all in "blood" with no mirror near at hand.

I had no idea, of course, why the damn thing had exploded (some readers might have already guessed what was wrong in the above protocol), but afterwards, our chemistry teacher—Ivan Ivanovich Taranenko—said to me that it was he who had made the mistake: accidentally, when I started, I was about to mix $KMnO_4$ and H_2SO_4 in a flat dish, but Ivan Ivanovich suggested to use a test-tube instead. The heat escape as well as the escape of the gaseous product were limited in the relatively narrow test-tube and the explosion followed.

At the time, I was not much impressed by my teacher's honesty, I assumed this was an ordinary human behaviour.

Then I found out how psychologically difficult it was to emulate even a minor version of this, e.g. by *properly* acknowledging an influence of somebody's remark on your own theorem. For example, writing my early paper on the Banach conjecture, I convincingly persuaded myself, that the advice by Dima Fuks to look at the homotopy groups of the classical groups for evaluating dimensions of their k-classifying spaces, was too trivial to deserve being mentioned.

I am afraid I accumulated a score of such "unmentionable" remarks and many of my colleagues told me of similar painful fights with themselves they have had while resolving the "acknowledgement problem". But others could not see there any problem at all. Probably, honesty comes naturally to certain people and some see no difficulty because they have never tried to be honest.

When I lived in Russia, the main output of the Soviet radio transmitters was the white (it always felt grayish to me) noise. (2–7 years in prison was an alternative to the official point of view that no such thing as "white noise" existed. But undaunted Soviet admirers in the West admitted its existence and suggested plausible explanations for it, where the most convincing one was preventing flying sauces from landing on Soviet agricultural fields with little green men hungry for the tasty green crops.)

This "white noise" did not cover the FM (40–50 MHz) and television (around 70 MHz) frequencies being unneeded for an obvious reason. But one evening TV-jamming began. People in the apartment house where we lived were opening doors and worriedly looking at each other. They did not dare to ask aloud what they thought was happening but "yes, it is" was transparent in everybody's eyes.

Of course, there were no secrets in the family and my mother hurried to tell me the news. I was triumphant: the first (and the last) time in my life something made by my hands worked! This "something"—a small radio transmitter I assembled—was supposed to generate 42 MHz. But who cares for 40 % error, the very fact it functioned made me bubble with pride.

My involvement into make-it-yourself-radio-something was influenced by my close friend, Lev Slutsman, with whom we went through the high school and the math department at the University together. The mathematics of the electricity laws was for him something real, something he felt with his fingers, devices made by him worked. His was a quite special and rare facet of mathematical gift—mathematics in the bones as much as in the head. (Lev now works in US and authors a multitude of patents on algorithms for testing large communication networks and something else of this kind.)

There was another boy, Dima Smirnov, in our high school class with a similar, albeit not with apparently mathematically colored, ability. Dima was the worst, the laziest student in the class, he hardly managed to graduate.

Once, we were supposed to do something at home and to bring it to the class. Many boys, myself included, brought up models of gliders, which we had assembled from standard pieces bought in a store along with an instruction of how to make it.

The teachers evaluated our creations according to how pretty they looked. Mine was the second dirtiest, Dima's was four times as dirty and fully asymmetrical. Obviously, he was too lazy even to read the instruction. His was the only glider that glided.

Neither the teachers nor the fellow students were impressed by Dima's glider. We felt embarrassed. It looked unjust, incongruous, completely absurd, that this ugly thing soaked with oil and covered by smudges of dirty glue could so gracefully glide a dozen of meters in the air, while all beautifully assembled clean ones were heading straight down to the ground despite all efforts to propel them horizontally. (After graduation, Dima entered the physics department at the University and became a very successful experimental physicist.)

I met later on two experimental physicists in US and in France (whose names I forgot since it was not so long ago). One of them was working on quantum computers and the second one was making nano-devices, if I recall correctly, with the

(Photo by Jean-François Dars)

atomic force microscope—a device for "touching atoms" rather than for "looking" at them. All mathematics of quantum mechanics, at least all I have ever heard of, including representation of C^*-algebras, for instance, was at their fingertips.

What level of mathematics do you need to sustain in a scientific community that so much would percolate to somebody's fingers?!

Learning and understanding mathematics is difficult, both by reading articles and/or by talking to people. (Actually, not so much by talking but by listening—"You can not learn much with your mouth open"—Dennis Sullivan used to say to me.)

Rarely, something you read will inspire you right on the spot, but I remember an exception—Tony Phillips' 1966-paper in Topology on the existence of submersions.

We studied earlier, at the students' seminar run by our professor Vladimir Abramovich Rokhlin, the immersion theory of Smale and Hirsch. I thought I had a fair idea of what was going on.

The fact that submersions, something quite opposite to immersions, had, however, followed in steps of Smale–Hirsch was a revelation to me. It took me about a year to understand what was on the bottom of this similarity.

Something else written by Tony, a private letter to me, also kept me puzzled for quite awhile. This letter contained a couple of pages of incomprehensible mathematics, starting with something like:

…an involutive gromomorphism $G \to SU \to US$ of admissible type … T transforms $MG \to SB$ …

I could not understand a single sentence in it. But when I showed this to my friend, an analyst Volodia Eidlin, he asked me: "What is a gromomorphism?"

"You mean homomorphism"—I replied—"There is no such thing as gromomorphism". ("Homomorphism" is spelled and pronounced as "gomomorphism" in Russian.)

"Do you ever read anything as it is written?"—he was annoyed— "This is 'gromomorphism', black on white."

Misha Gromov with D. Burago

"Must be a misspell..."—I mumbled, but then it dawned on me. Tony's was an encoded message. He was suggesting I would immigrate from the Soviet Union to US and invited me to SUNY at Stony Brook where he worked. (We met with Tony when he visited Russia a year earlier. His visit was brief, but long enough to learn the basic conspiracy survival rules in Soviet Russia.)

Several years later I followed his suggestion. When I arrived at Stony Brook, I enjoyed Tony's hospitality as well as that of the whole mathematics department at SUNY.

The only problem I had with people was "culture shock". Everybody was very kind to me and offering their help in overcoming this mysterious "culture shock". As I could not figure out what this shock was all about and not wanting to disappoint anybody I had to invent a few shocking stories about how I missed white bears skating on the streets of Leningrad in the darkness of polar nights and a cozy family iceberg in our cellar where we kept perishable foods.

When you read a book or an article you may come across something that the author had no idea of putting in there. When you listen to a mathematician you often learn something he/she might expect you knew beforehand, something obvious from his/her perspective, something one would not dare to put on paper.

One of such "obvious" things I learned from Dima Kazhdan who remarked to me at a visit from Moscow, that Kurosh subgroup freedom theorem follows from the fact that the covering of a graph is again a graph: dimension is invariant under coverings.

Until that moment the group theory was to me a slippery formalism impossible to hold steady in my hand. But with this remark everything started slowly falling into place; very slowly—it took me about 20 years afterwords to express some other fragments of the group theory in the geometric language.

I am certain there are lots of "omitted in view of their triviality" remarks nobody ever said to me, something basic and simple I've never understood.

This equally applies to non-mathematics, you can not learn everything from the books. Only rare authors—I recall seeing this in writings by Richard Feynman (QED), Charles Cantor (The Science and Technology Behind the Human Genome Project) and Maxim Frank-Kamenetskii (Unraveling DNA)—have the clarity of mind as well as the courage to pinpoint something essential that is obvious to the initiated and with no way to guess by an outsider.

A particular frustrating instance of this happened to me with learning the French language, rather than math or science. Armed with several textbooks, tapes, etc., I dutifully followed the phonetic rules and trained myself to read aloud *ent*'s at the ends of the verbs as in *ils parl**ent***. Much later, after I've lived for ten years in Paris and have already acquired a full automatism speaking "French", I came across a textbook published in 1972 in Quebec where the author—Gilbert Taggart explained, along with lots of other things which were too late for me to learn, that this *ent* was not meant for reading.

I kept asking myself why this was kept secret in most other textbooks and eventually realized how stupid the question was: *everybody* knew this, no single person apart from myself have ever uttered "ils parl**ent**" no matter how much I tried to find one in Paris (Wouldn't it be different in Quebec?)

As I mentioned, what you learn from a mathematics paper may, especially after some time, diverge from what the author had in mind. But something opposite—kind of convergence may also occur. This once happened to me ... with a help by a burglar.

When I started studying Nash's 1956 and 1966 papers (it was at Rokhlin's seminar $\approx$1968), his proofs has stricken me as convincing as lifting oneself by the hair. Under a pressure by Rokhlin, I plodded on, and, eventually, got the gist of it: It was a seemingly circular "fixed point by iteration" argument, where the iterated maps were *forced* to contraction by adjusting the norms in the spaces involved at each step of the iteration process. The final result popped up at the end of a lengthy but a straightforward computation, which, miraculously, did lift you in the air by the hair.

I wrote an abstract version of Nash's theorem in a 1972 paper where I isolated the iteration process in the space of norms and where a part of Nash's argument was absorbed by definitions.

But when I tried to reproduce this in my book on partial differential relations, I found out that the price for the "correct formalization" was non-readability—I had to write everything anew.

It was a hard job, I was relieved when it was over and I gave the manuscript to our typist at SUNY, it was about 1979, when I still was at Stony Brook.

Next week, the secretary office was burglarized and my manuscript disappeared along with a couple of typewriters. I had to write everything again for the third time.

Trying to reconstruct the proof and being unable to do this, I found out that my "formalization by definitions" was incomplete and my argument, as stated in 1972 was invalid (for non-compact manifolds). When I simplified everything up and wrote down the proof with a meticulous care, I realized that it was almost line

In Oslo 2009 (Knut Falch/Scanpix)

for line the same as in the 1956 paper by Nash—his reasoning turned out to be a stable fixed point in the "space of ideas"! (I was neither the first nor the last to generalize/simplify/improve Nash, but his proof remains unrivaled.)

What are our ideas—"From creation to decay; Like the bubbles on a river; Sparkling, bursting, borne away" (Shelley).

Is mathematics invented or discovered?

Even if we had ever learned the answers we would be as dissatisfied as an ancient geographer if his straightforward question: —"Does the Earth rest on a whale or on the backs of four elephants?"—were befuddled by:

"Nothing exists except atoms in the void; everything else is opinion". (Leucippus? Democritus? Lucretius?)

We, mathematicians, are an equally long way from asking the right questions.

A Few Snapshots from the Work of Mikhail Gromov

D. Burago, Y. Eliashberg, M. Bestvina, F. Forstnerič, L. Guth, A. Nabutovsky, A. Phillips, J. Roe, and A. Vershik

---

Electronic supplementary material Supplementary material is available in the online version of this chapter at http://dx.doi.org/10.1007/978-3-642-39449-2_11. Videos can also be accessed at http://www.springerimages.com/videos/978-3-642-39449-2.

D. Burago
Department of Mathematics, Pennsylvania State University, 237 McAllister Building, University Park, PA 16802-6401, USA
e-mail: d.burago@math.psu.edu

Y. Eliashberg (✉)
Department of Mathematics, Stanford University, Building 380, Stanford, CA 94305, USA
e-mail: eliash@math.stanford.edu

M. Bestvina
Department of Mathematics, University of Utah, 155 South 1400 East, JWB 233, Salt Lake City, UT 84112-0090, USA
e-mail: bestvina@math.utah.edu

F. Forstnerič
Faculty of Mathematics and Physics, University of Ljubljana, Jadranska 21, 1000 Ljubljana, Slovenia
e-mail: franc.forstneric@fmf.uni-lj.si

L. Guth
Department of Mathematics, Massachusetts Institute of Technology, Headquarters Office, Building 2, Room 236, 77 Massachusetts Avenue, Cambridge, MA 02139-4307, USA
e-mail: lguth@math.mit.edu

A. Nabutovsky
Department of Mathematics, University of Toronto, Toronto, ON M5S 3G3, Canada
e-mail: alex@math.utoronto.ca

A. Phillips
Department of Mathematics, State University of New York, Stony Brook, NY 11794-3651, USA
e-mail: tony@math.sunysb.edu

H. Holden, R. Piene (eds.), *The Abel Prize 2008-2012*,
DOI 10.1007/978-3-642-39449-2_11,

This collection is the result of a collaborative work of a number of mathematicians: M. Bestvina, D. Burago, Y. Eliashberg, F. Forstnerič, L. Guth, A. Nabutovsky, A. Phillips, J. Roe and A. Vershik, coordinated and edited by D. Burago and Y. Eliashberg. Each contribution is however a single-authored paper. The papers are not unified by a common style or approach, they are indeed just snap-shots reflecting individual perception of their authors.

1 Introduction. Conceptual Thinking (by Dima Burago)

There is no particular reason why this is me who writes this introduction: there are other mathematicians who would be more appropriate for this role. It just happened that Yasha Eliashberg and myself were coordinating the effort to make this collection of papers about Misha Gromov and his work. Similarly, the collection is not nearly comprehensive in at least two respects: many more people could have made substantial contributions, and many areas of Misha's work are not discussed or even touched here. Misha wrote hundreds of pages of math, and every page needs a dozen of pages of comments and explanations. Hence making a summary of Misha's work is a silly idea. In particular, we left Misha's search for new insight in biology and algorithmics completely outside the scope of this collection. Misha himself has divided his papers into 16 topics on his personal webpage,[1] and an interested reader can see that only a small part is even mentioned here. Furthermore, there are many essays about Misha as a mathematician and his mathematics. Perhaps we should especially mention a wonderful paper by Marcel Berger "*Encounter with a Geometer I, II*" [13]. The only thing we could try to achieve here was to collect some personal perceptions of a small part of Misha's work, and perhaps how his thoughts influenced future development of math—this is probably the key point in this strange writing. There is no common style of these write-ups: We only wanted to avoid this collection of notes become boring and finding no readers. Some papers are mostly mathematical and some are rather informal, personal and impressionistic. This one is an example of the latter. Perhaps I would not risk to write in this manner if Misha had not written his autobiography, opening me the road.

I certainly met Misha many times when I was a kid and he would come to my parents' place. I have almost no memories from that time, and then he left the Soviet Union and my parents decided to stay. So I knew of Misha only as a hero from

[1] http://www.ihes.fr/~gromov/.

J. Roe
Department of Mathematics, Pennsylvania State University, 204 McAllister Building, University Park, PA 16802–6401, USA
e-mail: roe@math.psu.edu

A. Vershik
Laboratory of Representation Theory and Computational Mathematics, St. Petersburg
Department of Steklov Institute of Mathematics, 27 Fontanka, St. Petersburg 191023, Russia
e-mail: vershik@pdmi.ras.ru

legends or myths narrated to me by my dad. Already a high school boy, I would jog with my dad and he would tell me fairy tails about metrics on finitely generated groups and such. At that time, simply putting together two so seemingly far-away areas—Algebra and Geometry—had enormous influence on my naive perception of mathematics.

By the way, dates of publication of some early papers of Misha do not reflect the actual time when the results were obtained. Misha refrained from publishing papers in the Soviet Union after he had decided to leave. The manuscripts were not allowed to be taken by people who were leaving (which explains the importance of sneaking Misha's thesis to the West by Tony Phillips). So Misha had to recover everything from his memory. I think this is the case with "Almost flat manifolds". Also people leaving USSR at that time were deprived of the citizenship (against even the laws of USSR). After Perestroika, Russian citizenship had been returned to Misha, and he traveled to Russia to receive the Lobachevsky Medal (funny enough, in the nomination "for Foreign Geometers") with a Russian Passport.

As a young adult, I first met Misha in about 1989, at a conference in Muenster. This was a shock for me. I had known quite a few of excellent mathematicians, of course most of them were more mature than me and had more knowledge, but never before had I a feeling that I would never be able to achieve the same level of understanding. I am trying to figure out now what is so special about Misha. Enormous breadth of math knowledge—of course, but perhaps the key is conceptual thinking. Thinking by huge blocks—this is what differs great minds from just good scientists. Categorical thinking. I will try to illustrate this by several recollections of conversations with Misha. Disclaimer: even when I use quotation marks, everything comes from my memory, and furthermore we mostly spoke in Russian, so these cannot be precise quotations. Still, I hope they are reasonably accurate.

Some say Misha's thinking is paradoxical. I would not agree. Sometimes he says or writes something that seems totally unexpected. However, we see only the tip of the iceberg. We see only a concluding thought of a very long process of very effective thinking, and this is why it seems paradoxical. My dad told me that once he had asked Misha: "How did the idea of "*Diameter, Curvature and Betty Numbers* occur to you?" "I have been thinking about this for ten years" was Misha's response.

We recently met at a conference in Lyon. We spoke about Boltzmann and Darwin. "Theories" of both had huge flaws, they were not really scientific theories. They could not answer basic objections. Some difficulties are explained now, but many remain absolutely unanswered. Misha got very excited. We first spoke about Boltzmann, Darwin came later as another example. "This was not a scientific theory, this was much more, a way of thinking, a conceptual, a categorical approach" (I have a feeling that Misha used the two words as synonyms), "Those who objected were fools, they did not understand that these were not theorems but a method of thinking, even nowadays almost nobody understands Boltzmann. It took me years to partially understand what he meant". The same about Darwin...

I recall asking Misha about a few problems I had suggested (this was about 20 years ago). They sounded quite cute to me. He said "They lack a structure behind them, your formulations are motivated more by linguistics than mathematics". My dad recalls that Misha told him several times "All I do in mathematics is looking for rich structures". Actually, Misha said this more than two decades before his recent paper "In a Search for a Structure".

And indeed, what he has done is perhaps more about structures than individual theorems. Here are just a few examples. It was not his primary goal to get an improved version of embedding theorems by Nash; he apparently wanted to understand what was the structure behind them. Hence the h-principle, including the open, more topological one, and the whole new (geometric) approach to differential relations. In Riemannian geometry, it does not seem that Misha was so much after a particular theorem. But he changed our view, we can now think not about individual Riemannian manifolds but rather of a space of manifolds, with its rich structure, and metric structures in what used to be more differential geometry became a way more prominent. In symplectic topology, symplectic structure is locally totally flexible (Darboux's theorem; hence "topology"), however Misha has found global rigidity, which is the main starting point of a huge area now known as *hard symplectic topology*. There are many more examples, some of which are discussed in articles in this collection.

During my lecture, a quarter of a century ago, Misha interrupted me: "Are you sure that this conjecture is true?". "Yes, but I cannot prove it!" "If you are sure it is true, why are you wasting time on proving? Formulate it and someone will eventually prove it." The problem is still open. The question was basically if the geodesic flow on a surface of non-negative curvature could have positive metric entropy. (I have presented a very concrete example, but even analyzing it is still out of reach.) By the way, once I wanted to tell Misha about a result in dynamics. Misha had no idea what that was about, but his first reaction was: "Do you assume any hyperbolicity in advance? If so, I am not so much interested." He probably meant that dynamics with hyperbolicity is a structure which is relatively well understood, so he expected to hear only something incremental (and he was right), whereas dynamics without any hyperbolicity assumptions remains a rather obscure area.

Once my dad told Misha about a result of mine. Misha at first did not believe (as usual, later he said it was next to obvious). My dad suggested to explain a proof. Misha reacted: "No, one can prove anything! I want you to explain to me why this must be true"!

Actually, talking with Misha is never easy. After my first talk at IHES, also about 20 years ago, Misha said: "It was good". "What???" (During my lecture, he would object to every sentence I said!) "You resisted very well to me doing everything to ruin your talk..."

From about the same time. "Atiyah–Singer? It surprises you so much just because there is an algebraic component which you do not understand."

Almost nothing serious happening in math would nonetheless pass unnoticed by Misha. Apparently he spends a lot of time not only reading math papers but also looking for deeper understanding. So in his block thinking, he has a lot of blocks to operate with. Misha also easily gets excited about new directions, from a metric approach to P vs. NP to looking for hyperbolic structures in the inside wall of a human heart.

Not so long ago Misha sent me his paper with a comment: "You may find this rendition of your results amusing (annoying?)" I have to confess that now, after some attempts to read Misha's work I do not quite understand the ideological meaning of my own results.

When one thinks by large blocks, he certainly does not care too much about fleas. Once I asked Misha about strange enumeration of sections (such as $3\frac{3}{4}$) in his earlier papers. The explanation was remarkably simple: "When I decide to write a new section, do you think I am going to change numbers of all others?" Apparently, the density of rationals does help. Similarly, but more striking: "Misha, you formulate a conjecture in the introduction, but it is proven further in the paper and in more generality?" "Right, but when I was writing the introduction, I did not know how to prove it!" Also from a phone call. "Misha, this theorem of yours is wrong". "Nonsense, all theorems of mine are correct!" "Here is a counter example..." "Are you a fool? This is not what I had in mind! Just add a condition to rule this idiotic example out!" Also, this is not an anecdote, I indeed had a conversation with a colleague: "What are you up to nowadays?" "I am proving a theorem of Gromov" (compare with Perelman...) Remarkably, up to dull technicalities, as far as I know all Misha's theorems whose proofs were only sketched proved to be true.

Yasha Eliashberg tried to push me to write about something concrete, such as GH-convergence, collapsing, almost flat manifolds, Betty numbers and so on. But after all, with my dad and Sergei Ivanov, we wrote a 450 page textbook [29], and one of the main goals of the textbook was to serve as a bridge between researchers who are non-experts in metric geometry and Misha's "Metric structures for Riemannian and non-Riemannian spaces", [103]. A large part of the textbook is devoted to Misha's ideas. So I decided to write this eclectic essay instead.

2 Gromov's Geometry[2] (by Anatoly Vershik)

1. In the middle of the XXth century geometry and topology in Leningrad were represented by two extraordinary schools: the ones of A.D. Alexandrov and V.A. Rokhlin. Needless to say, A.D. Alexandrov—a student of B.N. Delone—obviously engaged in a geometric way of thinking. Sometimes his school was referred to as a school of visual geometry. It is certainly not quite precise. The geometric philosophy of the topologist Rokhlin is known only to those who were closely acquainted with him and his research. I remember very well a conversation I had with him in the early 60s. He stated, very emotionally (which is not typical of him), that depth and beauty of geometry cannot be compared with those of any other field of mathematics. His geometric way of thinking is clearly seen in both his topological and metric works. It may look like the triumph of algebraic topology in the 50s and 60s moved the geometric and combinatorial topology to the background, and an entirely new insight emerged. However, in reality, this triumph just shifted geometric ideas to a new level. Whereas V.A. Rokhlin promoted algebra and strongly suggested that

[2]Translated from Russian by Mariya Boyko.

his students should adopt algebraic philosophy, for him, and I think for Gromov, algebra was only one of the many languages of geometry, not the method of thinking. It is very difficult to explain this fine difference without going into details. It is easier to refer to the classic topological work of the algebraist J.P. Serre and algebraic work of the topologist J. Milnor.

One would engage here into lengthy explanations of the used terms such as geometry, geometric philosophy, geometric ideas, algebra, etc. However, I do not think that it is possible to do this in a productive way. It is better to stay at the level of vague understanding shared by most mathematicians, though with variations. I even think that clarifying the meaning of these terms is the responsibility not only of mathematicians but also of philosophers and psychologists. Even a discussion of geometric motives in modern and classical music, in poetry and, needless to say, in modern art could be meaningful.

It seems to me that the coexistence of the two schools mentioned above is remarkable and perhaps unique. Let us recall that it was A.D. Alexandrov (the Rector of the Leningrad State University in the 50s–60s) who invited V.A. Rokhlin there (on the initiative of a number of friends). Due to a variety of reasons Rokhlin was in a hard situation at that time. Taking into account Soviet reality and a unique biography of V.A. Rokhlin this was not an easy task even for a Rector. M. Gromov was one of the first topology students of V.A. Rokhlin during his first years of work in Leningrad. In my opinion Gromov was also a follower of geometric ideas of Alexandrov.

2. All junior mathematicians, no matter how talented they are, need a certain period of time to accumulate a wide supply of knowledge. Perhaps it was difficult to do that in Leningrad in the 50s and the early 60s. Despite of the presence of numerous extraordinary mathematicians, Leningrad school of mathematics was suffering from obvious narrowness (functional analysis, certain areas of algebra, classical theory of partial differential equations, Alexandrov's geometry, etc.; but it lacked modern topology, representation theory, algebraic geometry and many other areas of mathematics). The situation changed after Rokhlin's arrival. I remember the first topology course he taught (probably in 1961) which was audited by two or three professors, two or three freshmen (including Misha), and one or two Ph.D. students (including myself). The gist of the matter is not only in Rokhlin's personal intellectual investment into the broadening of academic interests of his colleagues and students. Rokhlin graduated from the Moscow State University before the war and was a student of A.N. Kolmogorov, L.S. Pontryagin, P.S. Alexandrov, A.I. Plesner, etc. at the same time. Due to Rokhlin's close relations with Moscow mathematical schools, their representatives became regular visitors of seminars in Leningrad. Rokhlin's students, in turn, became popular and were frequently invited to Moscow. Misha even received the Moscow Mathematics Society Prize, he probably was the only recipient who was not from Moscow during the entire existence of the prize. Top Moscow Mathematicians—I.M. Gelfand, V.I. Arnold, and S.P. Novikov were official opponents at Misha's doctoral defense (the latter was substituted by N.N. Uraltseva because S.P. Novikov was not able to attend). It would hardly be possible to organize this without Rokhlin.

On the other hand, after acquiring certain knowledge of the current state of mathematics (at least in one's "own" area), the main part in the formation of a researcher is played by his own effort in understanding of what has been done before, and contemplating on what is not accomplished yet. I think that Misha was very different from many of his colleagues in that way. His interests and endless curiosity, critical thinking and even lack of trust that was evident in his conversations and seminars, were indicating a constant thinking process. He continues displaying the above qualities even now. I.M. Gelfand once roughly explained the following to me: in order to avoid mistakes in his works, a mathematician must not trust himself too much, but in order to obtain significant results he must not trust others too much. The ratio of these two distrusts changes with age, and it seems that Misha finally found the perfect proportion. I, however, do not know of Misha's mistakes.

It is not appropriate to mention any personal relations here. Let me just note that I was always impressed by Misha's 100 % independence of thoughts and deeds, which illustrates the strength of his character. The Soviet regime did not favor independent and courageous people, so many research talents were not able to develop fully. However, sometimes life's circumstances and especially personal qualities of some turn out to be stronger than the routine, and then the talent obtains the full freedom of expression to the great benefit of science.

3. We discuss here only a limited circle of Gromov's ideas which play a modest role in his huge repertoire. The epigraph to Misha Gromov's book "Metric Structures for Riemannian and Non-Riemannian Spaces", [103], which we discuss below is remarkable in some respects: "Même ceux qui furent favorables à ma perception des vérités que je voulais ensuite graver dans le temple, me félicitèrent de les avoir découvertes au microscope, quand je m'étais au contraire servi d'un télescope pour apercevoir des choses, trés petites en effet, mais parce qu'elles étaient situées a une grande distance, et qui étaient chacune un monde." Marcel Proust, Le temps retrouvé (Pleiade, Paris, 1954, p. 1041).

In short, slightly roughening ornamental ligature of words by Proust, this reads as follows: "Even those who commended my perception of the truths which I wanted eventually to engrave within the temple, congratulated me on having discovered them with a microscope, when on the contrary it was a telescope that I had used to observe things which were indeed very small to the naked eye, but only because they were situated at a great distance, and which were each one of them in itself a world." (From "In search of Lost Time" translated by T. Kilmartin, revised by D.J. Enright.)

Indeed, numerous plots considered and studied by Gromov (not exclusively in this book) are not focused on the details (possibly even crucial ones) of recognized theories. They are more about new geometric realms as well as other realms which cannot be considered totally unknown, but about which we either knew almost nothing or possessed a false knowledge.

I would not undertake studying all Gromov's work from this perspective, but I think that this point of view would be useful and even necessary for those who will study his work in the future.

Among numerous geometric projects of Gromov, including a few discussed in the book in question, there is one that was especially interesting to me. However, that project perhaps was not considered especially important and interesting by mathematicians including the author himself. I am referring to the concept of mm-spaces, that is measure-metric spaces—$(X; \mu; \rho)$. Anyone who worked in analysis on manifolds, geometry and dynamical systems has encountered such triples. I would like to illustrate Proust's idea of a "telescopic" view by this concrete and important example. It is unlikely that it would occur to a traditional analyst or geometer to convert such a situation into an abstract setup and pose a question of a categorical classification of such triples where morphisms are measure-preserving isometries. It is well-known that the classification of metric spaces up to an isometry is impossible to comprehend (recently A. Kechris gave a precise meaning to this statement) and not very useful. On the contrary, the classification of standard measure spaces is trivial (V.A. Rokhlin). So, what can we say about the triples?

Let us consider a triple $\tau = (X; \mu; \rho)$, where ρ is a metric on X which turns it into a Polish space and μ a continuous Borel probability measure which is nondegenerate in the sense that every nonempty open set has a positive measure. Obviously if we randomly and independently choose n points $(x_1, x_2, \dots, x_n)$ distributed with probability μ, then the random matrix of distances $\{\rho(x_i, x_j)\}_{i,j=1}^{n}$ generates a probability measure M_n on the set of distance matrices of order n, that is, in the space of metrics on n-point sets. It is also obvious that this measure does not change under isometries that preserve the measure. Gromov posed the following question: is it true that the combination of all those measures $M_n, n = 1, 2, 3 \dots,$ fully determines the triple up to an isomorphism (that is, up to a measure-preserving isometry). The positive answer is proven (in a rather difficult way; actually by the method of momenta) by Gromov in his book (p. 120–123). This crucial fact is known as Gromov's Reconstruction Theorem. Approximately in 1997 or not much earlier M. Gromov asked me what I thought of that theorem and its proof. I conveyed to him my very simple proof based on passing to infinite sequences of independent random points. Both proofs are discussed in the book and then the author, not without a naughty trick, asks a reasonable question if the reader is puzzled by the fact that complicated analytical constructions can be replaced by a "spineless argument"? Certainly, the answer is very easy: the spine of the second proof is nontrivial though widely known. It is the individual Law of Large Numbers, and its fruitful usage requires passing to infinite sampling in infinite distance matrices. Here is the argument. For every metric triple $\tau = (X; \mu; \rho)$, let us define a map

$$F_\tau : X^N \to \mathrm{Mat}_N$$

by the formula $F_t(\{x_n\}) = \{\rho(X_i, X_j)\}(i, j)$. By D_τ we denote the F_τ-image of the Bernoulli measure M_n. We call it the matrix distribution of a metric on a space with measure μ. Obviously, this is an invariant of a triple in the sense of the equivalence in question. We show that the matrix distribution is a full invariant of the equivalence of triples and that the assertion becomes a trivial corollary of the individual law of large numbers. If the two triples $\tau = (X; \mu; \rho)$, $\tau' = (X'; \mu'; \rho')$ have the same matrix distributions $D_\tau = D_{\tau'}$ then there is a countable everywhere dense set (since the measures are non-degenerate) $X_0 \subset X$ which is isometric to some countable everywhere dense set $X_0 \subset X$. This isometry extends to an isometry X to X. The fact that this isometry preserves the measure follows from the observation that by the Law of Large Numbers, the same distance matrix determines both the measure of balls of all radii centered at points in X_0 and all measures in the algebra of sets generated by the balls. Thus the measures coincide in both spaces, since they are determined by the same matrix, and therefore the measures μ and μ' are the same. As one can see the whole argument is based on the Strong Law of Large Numbers.

This formulation demonstrates, in a somehow unexpected way, that the classification of triples is "smooth", and that the space of invariants is the standard Borel space, namely, the space of Borel probability measures on a set of metrics on integers such that the measures are ergodic and invariant under the group of infinite substitutions. At the same time a close connection with today's popular theory of infinite random matrices emerges but the distribution of matrices that arise here are not the same as the ones that are considered in that theory (GOE [Gaussian orthogonal ensemble], GUE [Gaussian unitary ensemble] etc.) This connection will undoubtedly lead to new interesting results. My work [189] proves a generalization of this theorem to a more complicated case: the matrix distribution is a complete invariant of generic measurable functions in two (and more) variables with respect to the permutation group acting by transformations that preserve measure separately in each argument.

Let me note that the concept of metric triples (Gromov's triples or mm-spaces) restores the symmetry between measure and metrics which is absent in usual analysis. Traditionally, the metric is defined first, and then one studies the variety of Borel measures on this metric space. However, independently and almost simultaneously the opposite disposition of considering various admissible metrics on a given measure space was suggested [190]. This subordination is useful in the theory of dynamical systems where the principle invariants are first defined using both the measure and metric but at the end it turns out that they are independent of the choice of the latter (for instance, in defining Kolmogorov's entropy). This opens the road for constructing new invariants. However, here it is wise to impose some conditions of compatibility of the structures of mm-spaces satisfied for the triples discussed above.

4. Another well-known and productive geometrical initiative of Gromov is the idea of a space of metric spaces with a metric on it (known as the Gromov-Hausdorff metric). It is a "nonlinear" generalization of a corresponding notion in Banach geometry, but it is much more general. It is noteworthy that introducing the concept of the Uryson universal metric space, to which several statements in the book are dedicated, allows us to simplify the definition of the Gromov-Housdorff metric. The recent application of this idea to metric triples proved to be very useful here. The Master Thesis of M. Gromov was dedicated to Banach geometry, or more precisely to applications of topological results to it. This work remains of interest even today. For instance, Gromov's results in the spirit of generalizations of the Dvoretzky theorem, Levi's principle, etc. laid a foundation for subsequent work.

A gigantic theme of Gromov "Groups as a geometric objects" has no doubt opened a new chapter in combinatorial group theory. Roughly speaking, the proposition states that the familiar word metric turns groups into metric spaces which should be approached as purely metric objects. Group properties are deeply rooted into metric structures of the spaces. This is how the initiative of introducing hyperbolic groups arises. Geometric in its nature, it allows us to speak about a reformation of our understanding of applications of group theory in dynamics, topology and analysis. Gromov's celebrated theorem stating that groups of polynomial growth are virtually nilpotent might be one of the most eminent single theorems in mathematics of the XXth century; this theorem also originates from this geometric perception of groups.

Finally, the latest paper "In a Search for a Structure" [101] conveys yet preliminary thoughts of a geometric approach to entropy. This approach is not more connected to the modern but rather quite the opposite: To quite distant in time concepts of Boltzmann and others. Simultaneously, the latest works on this topic are being analyzed. For instance, in [20] the definition of entropy is modified in a drastic and nontrivial way so that it can be applied to non-amenable groups (unlike Kolmogorov's entropy). The concept of randomness, which plays a role in Gromov's geometric repertoire, for instance, the prominent idea of random groups, is not fully supported in further work.

This circle of ideas would be quite enough for the work of one mathematician, but this is only a small part of Gromov's accomplishments.

3 The Gromomorphism $SU \to US$ (by Tony Phillips)

I first met Misha Gromov in Leningrad, in March of 1969. Sergei Novikov, my mentor at the Steklov Institute in Moscow—I was there as part of the Soviet Academy-National Academy exchange—told me: "You have to meet Gromov." The Academy arranged for my trip to Leningrad (on the express train "Krasnaya Strela") and my stay at the Bal'tiskaya Hotel. I remember standing in front of a blackboard at LOMI, the Leningrad branch of the Steklov, with Misha and with Yasha Eliashberg; Misha was a graduate student and Yasha a last-year undergrad. "What else can we tell him?" one asked the other. They told me a lot. I remember at one point Misha illustrating the crucial point in the covering-homotopy argument (that I had adapted from Smale, Thom and Poenaru and that he had enormously generalized) by drawing three points in a line on the board and an arrow dragging the middle one upward and back down on the right. Misha gave me a (carbon) copy of his thesis, the first one: *Stable mappings of foliations into manifolds*; I took it back to the Bal'tiskaya that night and in half an hour—I knew where to look—understood what he had accomplished. The next day I wrote a postcard to John Milnor, who had been my advisor, with the news. At LOMI I also met with Vladimir Rokhlin, Misha's advisor. He told me "Er ist ein fähiger Mann." Gromov is a man given to accomplishment; if he had not chosen mathematics, he would have made his mark elsewhere. (My German was much better than my Russian, and Rokhlin was fluent in German.) Rokhlin told me that he had given Misha a copy of my article on submersions, saying: "You should be able to do something with this." That night or the next night Misha invited me home, where I met Misha's friends with whom we drank. I had been speaking French with Yasha and some kind of English with Misha; now it was all Russian. After a while I realized—the only time in my life this has happened—that I did not know what language I was speaking.

That summer (1969) I spent at the University of Warwick, where a large international Symposium on Differential Equations and Dynamical Systems was under way. Along with my own work, I lectured on Misha's, and mentioned its application to André Haefliger's work on universal foliations. That year or the next André gave a series of "Lectures on the Theorem of Gromov" at the Liverpool Singularities Symposium; we were able to circumvent, in Misha's case, the usual delay at that time in getting mathematical news across the Iron Curtain.

I kept in touch with Misha after my visit. At that time, exchanging reprints was about all that was possible; communication was tricky. A postcard came from Novikov: "I met G and we spoke of you." But in 1973, as I remember, I heard from Vova Khavkin who told me that he was Misha's half-brother; he lived in Washington DC and was a Senior Associate with First Washington Associates—I still have his business card. He needed to tell me that Misha was planning to emigrate from the Soviet Union. I spoke with Jim Simons—then the Chair of our Math Department—who immediately agreed that we should try to attract Misha to Stony Brook. I crafted a letter to Misha, ostensibly pure mathematics, which could be deciphered as saying that if he could emigrate then he could count on a position at Stony Brook. In particular I referred to a "gromomorphism" (which could mean a deformation of Gromov) from *SU* (... Soviet Union) to *US*. Vova kept me up to date on the proceedings; after a few months I learned that Misha had indeed received permission to leave the Soviet Union and that his travel would be under the auspices of HIAS, the Hebrew Immigrant Aid Society, with initial destination Rome. When I finally heard from Misha himself, he was already in Rome; not too long afterwards I had the great pleasure of meeting Misha and his family at JFK International Arrivals and driving them to their new home, on Long Island.

4 The *h*-Principle (by Yasha Eliashberg)

Already as a student Misha Gromov had a number of high profile results, e.g., a solution of Banach's geometric conjecture, but the work which made him famous was his PhD dissertation which laid the ground to what is now known as the h-principle.

Many problems in Mathematics and its applications deal with partial differential equations, partial differential inequalities, of more generally with *partial differential relations*, i.e., any conditions imposed on partial derivatives of an unknown function. A solution of such a partial differential relation $\mathscr{R}$ is any function which satisfies this relation.

Any differential relation has an underlying algebraic relation which one gets by substituting all the derivatives entering the relation with new independent functions. A solution of the corresponding algebraic relation, called a *formal* solution of the original differential relation $\mathscr{R}$, is a necessary condition for the solvability of $\mathscr{R}$. Though it seems that this necessary condition should be very far from being sufficient, it was a surprising discovery in the 1950s of geometrically interesting problems where existence of a formal solution is the only obstruction for the genuine solvability. One of the first such non-trivial examples were the C^1-isometric embedding theorem of J. Nash and N. Kuiper, [124, 148] and the immersion theory of S. Smale and M. Hirsch, [113, 174]. When Gromov came to the subject, a few other interesting examples of this phenomena were found by V. Poénaru and A. Phillips (see [159, 162]).

I remember that once Misha told me that when reading someone's paper he first reads the formulation of the main result and then tries to prove it himself. In many cases he succeeds. If not, then he looks for further hints, trying to uncover the underlying reasons why the proof works. When the argument is very tricky and complicated this may mean, in his opinion, that either the proof is wrong, or that the author did not properly understand his or her own proof. As a result, in several cases when Gromov uncovered the main moving forces of the proof it allowed him to generalize and develop the results far beyond the original intention of the authors.

When Gromov studied the work of Nash-Kuiper and Smale–Hirsch–Phillips he succeeded in both cases in finding the roots. This led him to the discovery of large classes of problems satisfying the h-principle, where the letter "h" stands for *homotopy*, i.e., for which any formal solution can be deformed through formal solutions to a genuine one. In his PhD dissertation [87], paper [88] and later in his book [96], Gromov transformed original authors' ideas into two powerful general methods for solving partial differential relations: the method of *continuous sheaves* and the method of *convex integration*. The third method, called *removal of singularities*, was first introduced and explored in [90]. We discuss below these methods in some detail.

The h-principle for solutions of partial differential relations revealed the soft/hard (or flexible/rigid) dichotomy for problems formulated in terms of derivatives: a particular analytical problem is "soft" or "abides by the h-principle" if its solvability is determined by some underlying *algebraic* or *geometric* data. It is important to point out that while Gromov proved several powerful general theorems covering many geometrically interesting cases of the h-principle, the methods which he invented are even more important. They continue to be used and improved by many mathematicians to exhibit new unexpected applications in geometry and beyond.

4.1 Holonomic Approximation

The *method of holonomic approximation*, which is just a slight repackaging (see [53]) of Gromov's method of continuous sheaves, [87], is based on the iterative application of the following seemingly trivial 1-dimensional observation: *any smooth function on* $[0, \frac{1}{3}] \cup [\frac{2}{3}, 1]$ *with positive derivative can be extended to* $[0, 1]$ *with exactly two non-degenerate critical points*; moreover, this is true *parametrically*.

We will be using below a convenient Gromov's notation $\mathcal{O}p_V K$ for an unspecified neighborhood of a subset K in a manifold V. We drop V from the notation when it is clear from the context. The neighborhood may change in the process of a proof. We begin with the following problem. Given a compact set $K \subset \mathbb{R}^n$ and for each multiindex $J = (j_1, \dots, j_n)$, $j_i \geq 0$, a smooth function $f_J : \mathcal{O}pK \to \mathbb{R}$, *is it possible to simultaneously* C^0*-approximate all the functions* f_J *by derivatives of the same function, i.e., to find for any* $\varepsilon > 0$ *and an integer* $s \geq 0$ *a function* $g : \mathcal{O}pK \to \mathbb{R}$ *such that*

$$\left\| D^J g - f_J \right\|_{C^0(K)} < \varepsilon$$

for all $J = (j_1, \dots, j_n)$ *with* $|J| = \sum_1^n j_i \leq s$? Here we denoted

$$D^J g := \frac{\partial^J g}{(\partial x_1)^{j_1} \dots (\partial x_n)^{j_n}}.$$

The answer is obviously negative even for functions of one variable. Surprisingly, a slightly modified question has a positive answer:

Theorem 1 *Denote* $I^k = \{x_{k+1} = \cdots = x_n = 0, |x_j| \leq 1, j = 1, \dots, k\} \subset \mathbb{R}^n$. *For any* $\varepsilon, \delta > 0$ *and an integer* $s \geq 0$ *there exist a smooth function* $h : I^k \to [-\delta, \delta]$ *and a function* $g : \mathcal{O}pI_h \to \mathbb{R}$, *where*

$$I_h = \left\{x_{k+1} = h(x_1, \dots, x_k), x_{k+2} = \cdots = x_n = 0\right\} \subset \mathbb{R}^n$$

is the graph of the function h, *such that*

$$\left\| D^J g - f_J \right\|_{C^0(I_h)} < \varepsilon$$

for all $J = (j_1, \dots, j_n)$ *with* $|J| = \sum_1^n j_i \leq s$.

Sketch of the proof If $k = 0$ the statement is obviously true: just take as the approximating function g the Taylor polynomial at 0 of order s:

$$g(x) = \sum_1^s \frac{f_J(0)}{|J|!} x^J,$$

where we write $x^J := x_1^{j_1} \dots x_n^{j_n}$. Passing to the case $k = 1$ we view the interval I as a 1*-parametric family of points*, and construct the Taylor approximation g_u in a neighborhood of each point $u \in I = I^1$. Let us discretize this family, i.e., for a large N, which will be chosen later on, consider functions $g_{\frac{k}{N}}$, $k = 1, \dots, N$, defined on $P_k := \{\frac{k-1}{N} \leq x_1 \leq \frac{k}{N},\ 0 \leq x_j \leq \sigma,\ j = 2, \dots, n\}$, where $\sigma > 0$ is independent of N and much bigger that $\frac{1}{N}$. Note that the functions $g_{\frac{k}{N}}$ and $g_{\frac{k+1}{N}}$ disagree on the common part of the boundaries of P_k and P_{k+1}. However, if $N\sigma$ is large enough then we can find $\widetilde{g}_{\frac{2}{N}}$ on P_2 which is C^s-close to $g_{\frac{2}{N}}$, coincides with $g_{\frac{2}{N}}$ for $x_n < \frac{\sigma}{3}$ and with $g_{\frac{1}{N}}$ for $x_n > \frac{2\sigma}{3}$. Next, we find a C^s-approximation $\widetilde{g}_{\frac{3}{N}}$ of $g_{\frac{3}{N}}$ which coincides with $g_{\frac{2}{N}}$ for $x_n < \frac{\sigma}{3}$ and with $g_{\frac{3}{N}}$ for $x_n > \frac{2\sigma}{3}$, and so on. Consider the function $h(t) = 0.4\sigma(1 + \sin \pi N t)$. Then the functions $\widetilde{g}_{\frac{1}{N}} := g_{\frac{1}{N}}, \widetilde{g}_{\frac{2}{N}}, \dots, \widetilde{g}_{\frac{N-1}{N}}, \widetilde{g}_1$ define together the required smooth function g on a neighborhood of the curve $\{x_n = h(x_1),\ x_j = 0, 2 \le j \le n-1\}$. The case of a general k can be similarly deduced by induction from a 1-parametric version of the case $k-1$. □

The proof of Theorem 1 also holds in the parametric, as well as in the relative form, i.e., if the required approximation is already given on $\mathcal{O}p\,\partial I^k$ then the function h can be chosen $\equiv 0$ near ∂I_k and the function g can be kept unchanged on $\mathcal{O}p\,\partial I^k$.

To discuss applications of this theorem it is useful to introduce the language of jets, which allows one to talk about solutions of a differential relation in a geometric language. Given a (smooth) map $f\colon \mathbb{R}^n \to \mathbb{R}^q$, its s-jet at a point $x \in \mathbb{R}^n$ is the string of its partial derivatives up to order s

$$J^s(f)(x) = \{D^I f(x)\}_{|I| \le s},$$

or alternatively its Taylor polynomial of sth order.

Thus $x \mapsto (x, J^s(f)(x))$ can be viewed as a section of the trivial *s-jet bundle* $\mathbb{R}^n \times \mathbb{R}^{qN(n,s)} \to \mathbb{R}^n$, where $N(n,s) = \frac{(n+s)!}{n!s!}$ is the total number of derivatives. A fiber $x \times \mathbb{R}^{qN(n,s)}$ of this fibration is the space of all a priori possible values of jets of maps $f\colon \mathbb{R}^n \to \mathbb{R}^q$ at the point $x \in \mathbb{R}^n$. In this context the space $\mathbb{R}^n \times \mathbb{R}^{qN(n,s)}$ is called the space of s-jets of maps $\mathbb{R}^n \to \mathbb{R}^q$, or the *space of s-jets of sections* $\mathbb{R}^n \to \mathbb{R}^n \times \mathbb{R}^q$ of the trivial bundle $\mathbb{R}^n \times \mathbb{R}^q \to \mathbb{R}^n$, and denoted by $J^s(\mathbb{R}^n, \mathbb{R}^q)$. In other words, a section $J^s(f)\colon \mathbb{R}^n \to J^s(\mathbb{R}^n, \mathbb{R}^q)$ is a simultaneous graph of the map f and all its derivatives up to order s.

The property of two maps f, g to have the same s-jet at a given point is independent of the choice of coordinates in the source and target. This allows us to associate with any fiber bundle $p\colon X \to V$ its *s-jet extension* $p^{(s)} : X^{(s)} \to V$, or the space of s-jets of sections $f : V \to X$. In the case of a trivial bundle $X = V \times Q \to V$ one usually uses the notation $J^s(V, Q)$ and refers to this space as the space of s-jets of maps $V \to Q$.

Every section $f\colon V \to X$ has its s-jet extension $J^s(f) : V \to X^{(s)}$, but of course not all sections of the bundle $X^{(s)}$ are s-jet extensions of some sections of X. A section $\Phi\colon V \to X^{(s)}$ is called *holonomic* if $\Phi = J^s(f)$ for some section $f : V \to X$. Theorem 1 can be more globally reformulated as follows:

Theorem 2 *Let $X \to V$ be a smooth bundle, $A \subset V$ a closed stratified subset of positive codimension and $\Phi : \mathcal{O}p A \to X^{(s)}$ a section of the s-jet extension bundle $X^{(s)} \to V$ given over a neighborhood of A. Then there exist*

- *a C^0-small isotopy $h_t : A \to V, t \in [0,1]$, beginning at $h_0 = \mathrm{Id}$, and*
- *a section $\varphi\colon \mathcal{O}p h_1(A) \to X$*

such that $J^s(\varphi)$ C^0*-approximates* Φ *over* $h_1(A)$.

A *partial differential relation* of order s for sections $V \to X$ is any subset $\mathscr{R} \subset X^{(s)}$. Subvarieties of $X^{(s)}$ correspond to (systems of) partial differential equations, while open subsets correspond to partial differential inequalities.

A *solution* of a differential relation $\mathscr{R}$ is any section $f: V \to X$ such that $J^s(f)(V) \subset \mathscr{R}$. A *formal solution* of $\mathscr{R}$ is any section $\Phi: V \to X^{(s)}$ with $\Phi(V) \subset \mathscr{R}$. Thus (genuine) solutions of a differential relation $\mathscr{R}$ can be identified with its formal holonomic solutions, and the space $\mathrm{Sol}(\mathscr{R})$ of solutions can be viewed as a subspace of the space $\mathrm{Sol}_{\mathrm{formal}}(\mathscr{R})$ of its formal solutions. One says that a differential relation $\mathscr{R}$ *satisfies an h-principle* if the inclusion

$$j: \mathrm{Sol}_{\mathrm{formal}}(\mathscr{R}) \hookrightarrow \mathrm{Sol}(\mathscr{R})$$

is a homotopy equivalence.

The h-principle has many forms and variations. The above form of the h-principle is usually called *parametric*. However, sometimes the parametric version of the h-principle fails but the inclusion j induces a surjection on π_0. In some cases one needs to add certain additional homotopy conditions on the 0-jet part of the formal solution, (e.g., see below the discussion of the Haefliger structure obstruction for foliations). One can also talk about the relative h-principle, C^0-dense h-principle, etc.

For many fibrations $X \to V$ one can define an action of the group $\mathrm{Diff}(V)$ of diffeomorphisms of V on the space of sections of the fibration $X \to V$ (e.g., when X is the tangent bundle TV), which then naturally extends to an action on its s-jet extension $X^{(s)} \to V$. Such fibrations are called *natural*. Theorem 2 implies the following general h-principle, see [87]:

Theorem 3 *Let V be an open manifold and $X \to V$ a natural fibration. Then any open* $\mathrm{Diff}(V)$*-invariant differential relation $\mathscr{R} \subset X^{(r)}$ satisfies all the forms of the h-principle.*

Proof Indeed, for any open n-dimensional manifold V there exists an isotopy which retracts V to an arbitrary small neighborhood of its $(n-1)$-dimensional skeleton $K \subset V$. Using Theorem 2, any formal solution of $\mathscr{R}$ can be approximated by a holonomic section $J^s(\varphi)$ over a slightly deformed $(n-1)$-skeleton $\widetilde{K}$ of V. In view of the openness of $\mathscr{R}$, the section φ is a solution of $\mathscr{R}$ over $\mathcal{O}p\widetilde{K}$, and in view of the $\mathrm{Diff}\,V$-invariance of $\mathscr{R}$ the section $\widetilde{\varphi}$ remains a solution after we pull it back to V by an isotopy retracting V to a $\mathcal{O}p\widetilde{K}$. □

Among standard corollaries of Theorem 3 there are Smale–Hirsch's immersion theory, Phillips's submersion theory, mappings of open manifolds transversal to foliations, mappings of foliations regular along fibers, as well as k-mersions (i.e., maps of rank $\geq k$) and a rich variety of other applications.

Some of the above mentioned applications had either already been known before Gromov's theorem appeared, or proven independently by other authors at about the same time. Besides the unified viewpoint, essentially new were applications concerning constructions of different basic geometric *structures*, e.g., symplectic, contact, complex structures or foliations. We briefly discuss some of these applications below. Gromov's results became well known in the West thanks to Tony Phillips who visited Leningrad in 1969 and managed to sneak out Gromov's dissertation. I was later told by a mathematician who was working at that time on the h-principle type questions about almost a shock he had from reading Gromov's work. He even decided to switch the subject of his research to avoid competing with Gromov.

Let us recall that a symplectic structure on a $2n$-dimensional manifold M is a closed non-degenerate 2-form, and a contact structure on a $(2n+1)$-dimensional manifold N is a completely non-integrable hyperplane field $\xi \subset TN$. The complete non-integrability can be expressed by the Frobenius condition: if one defines ξ (locally) by a Pfaffian equation $\alpha = 0$, then the 2-form $d\alpha|_\xi$ is non-degenerate, see Sect. 6.1 for more discussion.

A formal solution for the corresponding differential relation is a non-degenerate (not necessarily closed) 2-form in the symplectic case, and a pair (α, η) of a non-vanishing 1-form α and a 2-form η such that $\eta|_{\alpha=0}$ is non-degenerate, in the contact case. In turn, existence of these formal solutions is equivalent to existence of an almost complex structure J on M in the symplectic case, and existence of a so-called *stable* almost complex structure, i.e., an almost complex structure on $N \times \mathbb{R}$ in the contact case. Theorem 3 implies the parametric h-principle for symplectic and contact structures on open manifolds. Moreover, symplectic forms can be constructed in any given cohomology class. The existence problem for closed manifolds, as well as the extension problem to top-dimensional balls is much more subtle, both in the symplectic and contact cases. We discuss it in Sect. 6.3 below.

A codimension k *foliation* on an n-dimensional manifold M can be thought of as an *integrable* $(n-k)$-dimensional distribution $\eta \subset TM$. André Haefliger discovered that there exists a homotopical obstruction for a tangent distribution η to be homotopic to an integrable one. Namely, there exists a classifying space $B\Gamma_k$ which carries a universal codimension k foliation η_{univ}. Let $\pi_{\text{univ}}\colon B\Gamma_k \to BO_k$ be the map classifying the normal bundle $TB\Gamma_k/\eta_{\text{univ}}$. If a distribution η on M is homotopic to a foliation, then the map $h\colon M \to BO_k$ classifying its normal bundle TM/η lifts to a map $\overline{h}\colon M \to B\Gamma_k$, so that $\pi_{\text{univ}} \circ \overline{h} = h$. Any h-principle type statement should take into account this obstruction. Phillips–Gromov's theorem about mappings transversal to foliations implied that the modified h-principle with the added Haefliger's condition holds for open manifolds. A surprising discovery of W.P. Thurston (see [183, 184], and also [137]) was that the h-principle also holds for closed manifolds (for codimension 1 foliations only in the absolute non-parametric case).

The situation is much worse for the problem of existence of another basic geometric structure, the complex one. The h-principle for that case is not known even for open manifolds. In fact, the best result is a theorem of Gromov–Landweber (see [88, 126]) which asserts that if M is a $2n$-dimensional open manifold which has homotopy type of an $(n+1)$-dimensional CW-complex then any almost complex structure on M is homotopic to an integrable one. The only known negative result is the failure of the extension h-principle to top-dimensional balls from their boundaries, see [47].

A modified version of Theorem 3, where the openness condition for the differential relation $\mathscr{R}$ is relaxed to the so-called *micro-flexibility*, implies a large number of further applications, such as parametric h-principles for Lagrangian, Legendrian, iso-symplectic and iso-contact (i.e., inducing the given symplectic and contact structures) immersions.

Interestingly, Theorem 1 can be also used for constructions of *embeddings*. The idea is that given an embedding one can apply Theorem 3 to sections of its small tubular neighborhood. A typical example of this kind is the problem of *directed embeddings*. Let A be a subset of a Grassmannian $G_{n,k}$ of k-planes in $\mathbb{R}^n$. For an embedding $f\colon V \to \mathbb{R}^n$ of a k-dimensional manifold V we denote by Gf the corresponding Gaussian map $M \to G_{n,k}$ which associates to each point $x \in V$ the tangent plane to $f(V)$ at the point $f(x)$. An embedding $f\colon M \to \mathbb{R}^n$ is called *A-directed* if $Gf(V) \subset A$. For instance, if $A = A_l$ consists of k-planes trivially intersecting $\mathbb{R}^l \subset \mathbb{R}^n$, $l+k \leq n$, then an embedding is A_l-directed if it projects as an immersion to $\mathbb{R}^{n-l} = \mathbb{R}^n/\mathbb{R}^l$. Similarly if $A = A_{\mathbb{R}} \subset G_{2n,n}$ is the set of totally real n-dimensional subspaces in $\mathbb{R}^{2n} = \mathbb{C}^n$, then $A_{\mathbb{R}}$-directed embeddings are exactly totally real ones, i.e., whose tangent spaces contain no complex lines.

The key role in the proof of the corresponding h-principle for A-directed embeddings plays the following lemma about approximation of a tangential homotopy.

Lemma 1 *Let $i : V \hookrightarrow \mathbb{R}^n$ be a k-dimensional, $k < n$, submanifold in $\mathbb{R}^n$, and $K \subset V$ a stratified subset of positive codimension in V. Suppose that the Gaussian map $Gi : V \to G_{n,k}$ is extended to $\mathcal{O}p_{\mathbb{R}^n} V$. Let $G_t\colon \mathcal{O}p_{\mathbb{R}^n} K \to G_{n,k}$, $t \in [0,1]$, be any homotopy beginning with $G_0 = Gi$. Then there exists an isotopy $h_t\colon \mathcal{O}p_V K \to \mathbb{R}^n$ starting with the inclusion map $h_0 = i$ and such that Gh_t C^0-approximates G_t over $\mathcal{O}p_V K$. The statement also holds in the relative form.*

Proof To deduce Lemma 1 from Theorem 1 we choose a sufficiently large integer N such that for $t \in [\frac{k-1}{N}, \frac{k}{N}]$ the planes rotate for no more than $\frac{\pi}{8}$. We inductively extend the isotopy to these intervals. Denote by τ_t the plane field $G_t(\mathcal{O}pK)$ on $\mathcal{O}pK$. If an embedding $h_{\frac{k-1}{N}}$, $k = 1, \dots, N$, is already constructed so that the tangent planes to $V_{k-1} := h_{\frac{k-1}{N}}(\mathcal{O}p_V K)$ deviate for no more than $\frac{\pi}{8}$ from $G_{\frac{k-1}{N}}$ then the distribution $\tau_{\frac{k}{N}}$ is transverse to the fibers of a sufficiently small tubular neighborhood $U(V_{k-1})$ of V_{k-1}. Hence, we can view the distribution $\tau_{\frac{k}{N}}$ along V_{k-1} as a section σ of the 1-jet extension $U(V_{k-1})^{(1)}$ of the bundle $U(V_{k-1}) \to V_{k-1}$. Applying Theorem 1 we can then construct a section $s\colon V_{k-1} \to U(V_{k-1})$ such that its 1-jet extension $J^1(s)$ approximates σ. Let $s_t\colon V_{k-1} \to U(V_{k-1})$, $t \in [\frac{k-1}{N}, \frac{k}{N}]$, be a family of sections interpolating between the 0-section and the section s. We then define $h_t := s_t \circ h_{\frac{k-1}{N}}(V)$ for $t \in [\frac{k-1}{N}, \frac{k}{N}]$, and continue the process. □

As an example of an application of Lemma 1 let us deduce the h-principle for A_l-directed embeddings: *Let M be an n-dimensional manifold and $h\colon M \to \mathbb{R}^q$ an* *embedding. Suppose there exists a homotopy* $G_t\colon TM \to G_{q,n}$, $t \in [0,1]$, *which begins at* $G_0 = Gh$ *and such that* $G_1(TM)$ *intersects trivially* $\mathbb{R}^l \subset \mathbb{R}^q$, $n+l<q$. *Then there exists a* C^0*-small isotopy* $h_t\colon M \to \mathbb{R}^q, t \in [0,1], h_0 = h$, *such that the composition* $M \xrightarrow{h_1} \mathbb{R}^n \xrightarrow{\pi} \mathbb{R}^q/\mathbb{R}^l = \mathbb{R}^{q-l}$, $n+l<q$, *is an immersion.*

Indeed, the embedding h can be extended to an embedding of a bigger open manifold $V \supset M$ of dimension $q-l$ for which there exists a homotopy of tangent planes with the same property (e.g., one can take as V the total space of the normal bundle to $\pi \circ G_1(TM)$ in $\mathbb{R}^{q-l}$). Then Lemma 1 provides the required isotopy of a neighborhood of M in V, and hence of M itself.

The h-principle for totally real embeddings can also be deduced from this lemma with some additional work. Alternatively, it follows from Gromov's method of convex integration discussed below in Sect. 4.3.

4.2 Removal of Singularities

Let $E \to M$ be a vector bundle E over a manifold M. We say that sections $\varphi_1, \dots, \varphi_q\colon M \to E$ pointwise generate E if for each $x \in M$ we have

$$\operatorname{Span}\bigl(\varphi_1(x), \dots, \varphi_q(x)\bigr) = E_x.$$

We illustrate the method of removal of singularities, see [89, 90], by its application to immersions of closed manifolds of dimension n into $\mathbb{R}^q$ for $q>n$. The h-principle in this case can be formulated as follows:

Theorem 4 (Hirsch, [113]) *Let M be a closed n-dimensional manifold. Suppose that $q>n$ and 1-forms $\alpha_1, \dots, \alpha_q$ pointwise generate $T^*(M)$. Then there exist exact forms $df_1, \dots, df_q$ with the same property. Moreover, one can choose the system of exact forms $(df_1, \dots, df_q)$ homotopic to $(\alpha_1, \dots, \alpha_q)$ through systems of generating form, as well as to C^0-approximate any given map $g := (g_1, \dots, g_q)\colon M \to \mathbb{R}^q$ by the map $f := (f_1, \dots, f_q)$.*

Sketch of the proof The proof inductively replaces the forms $\alpha_1, \dots, \alpha_q$ by exact forms. We will discuss here only the last step; the intermediate steps differ only in the notation. Suppose that we already replaced $\alpha_1, \dots, \alpha_{q-1}$ by exact forms $df_1, \dots, df_{q-1}$ such that the system of forms $(df_1, \dots, df_{q-1}, \alpha_q)$ generate T^*M. Then $\widetilde{f} := (f_1, \dots, f_{q-1})\colon M \to \mathbb{R}^{q-1}$ is a map of corank ≤ 1, and in view of the inequality $q-1 \ge n$ *generically* it has only Thom–Boardman–Morin singularities $\Sigma^{1\dots1}$. In other words, there exists a submanifold $M_1 = \Sigma^1(\widetilde{f})$ of codimension $q-n+1$ such that $\widetilde{f}|_{M \setminus M_1}$ is an immersion (i.e., $df_1, \dots, df_{q-1}$ generate T^*M over $M \setminus M_1$); the kernel $\lambda := \operatorname{Ker}|_{d\widetilde{f}} \subset TM|_{M_1}$ is a 1-dimensional bundle which is transverse to M_1 in the complement of a submanifold $M_2 := \Sigma^{11}(\widetilde{f}) \subset M_1$ of codimension $q-n+1$ in M_1; in turn $\lambda|_{M_2} \subset TM_1|_{M_2}$ is transverse to M_2 in the complement of a submanifold $M_3 \subset M_2$ of codimension $q-n+1$ in M_2 etc. Note that the bundle λ is trivial: indeed it is trivialized by the non-vanishing vector field $v \in \lambda$ defined by the equation $\alpha_q(v) = 1$. Our task is to find a function $f_q : M \to \mathbb{R}$ such that its directional derivative $df_q(v)$ along v is positive. Assume for a moment that $M_2 = \varnothing$, i.e., the vector field v is transverse to M_1. Given any smooth function $g_q : M \to \mathbb{R}$ we can find a function f_q such that $f_q = g_q$ on M_1 and outside a small neighborhood of M_1, and which satisfies the equation $df_q(v) = 1$ along M_1. If $M_2 \neq \varnothing$, but $M_3 = \varnothing$ then we first find a function f_q on $\mathcal{O}p M_2$ which is equal to g_q on M_2 and satisfies $df_q(v) = 1$ along M_2. Then $df_q(v)|_{\mathcal{O}p M_2} > 0$, and hence, we can extend f_q with this property first on $\mathcal{O}p M_1 \subset M$, and then further extend it to the whole M as equal to g_q outside a larger neighborhood of M_1. In the general case we begin with the smallest non-empty M_k and repeating the above process inductively extend the function f_q to neighborhoods of $M_k, M_{k-1}, \dots M_1$ with the property that $df_q(v) > 0$. □

A more detailed analysis of the process shows that under the above genericity assumptions one can actually solve the *equation* $df_q(v) = 1$ along M_1.

There are two advantages of this proof of Hirsch's theorem compared to the one discussed in the previous section. First, it provides much better control of the approximation of a map g by an immersion. Namely, the process gives not just C^0 but an approximation in the Sobolev norm $W^{l,p}$ with any integer l and real number p which satisfy one of the two conditions: (1) $lp < q - n - 1$, or (2) $lp = q - n - 1$, $p > 1$. This approximation, which can be shown to be the optimal one, was used in [89] to construct immersed isoperimetric films. Second, the method works in the complex analytic, and even algebraic situation. In particular, in combination with Gromov's generalization of Oka–Grauert's principle which is discussed in Forstnerič's essay, see Sect. 5, this method can be used to prove an h-principle for *holomorphic immersions of Stein manifolds* into $\mathbb{C}^q$, see [91].

4.3 Convex Integration

While the method of removal of singularities employs an induction over the dimension of the target space, gradually improving the coordinate functions of a solution, the convex integration method uses an induction over the dimension of the source manifold, one by one adjusting partial derivatives of a solution. We discuss here the method of convex integration in its simplest form, and only for differential relations of the first order, see [88, 176] for more details.

Convex integration is based on the following geometric interpretation of the integral of a vector function. For a set $A \subset \mathbb{R}^n$ we denote by $\widehat{A}$ is convex hull.

Lemma 2 *Let* $A \subset \mathbb{R}^n$ *be a path-connected subset such that* $\widehat{A} \supset \mathcal{O}p b$ *for some point* $b \in \mathbb{R}^n$. *Then for any point* $a \in A$ *there exists a path* $\Phi : [0,1] \to A$ *with* $\Phi(0) = a$ *such that* $\int_0^1 \Phi(t)dt = b$.

Proof Consider points $p_1, \dots, p_k \in A$ such that $b = \sum_1^k \lambda_j p_j$, where $\lambda_i > 0$ and $\sum_1^k \lambda_i = 1$. Consider paths $\gamma_1, \dots, \gamma_k$ connecting in A the point $a \in A$ with $p_1, \dots, p_k$. Choose $\varepsilon \in (0, \frac{1}{2k})$. Consider a path $\Phi_\varepsilon : [0,1] \to A$ which begins at a, jumps in the ε time to p_1 along γ_1, stays there the time $\lambda_1 - 2\varepsilon$, then jumps back to a in the time ε, immediately again jumps to p_2 along γ_2, stays there the time $\lambda_2 - 2\varepsilon$, jumps back to a, etc. Clearly $\int_0^1 \Phi_\varepsilon(t) \underset{\varepsilon \to 0}{\to} b$. The condition that b is an *interior* point of $\widehat{A}$ allows us when ε is sufficiently small to find the required path Φ by varying the weights λ_j. □

Similarly one can prove the following parametric version of the previous lemma.

Lemma 3 *Let $A \subset \mathbb{R}^n$ be a path-connected subset, C is a compact space and $g: C \to \mathbb{R}^n$ a continuous map such that $\widehat{A} \supset \mathcal{O}pg(C)$. Then any map $\varphi: C \to A$ can be extended to a map $\Phi: C \times [0,1] \to A$ such that $\Phi(x,0) = \varphi(x)$ and $\int_0^1 \Phi(x,t)dt = g(x)$, $x \in C$.*

The following lemma is the essence of the convex integration method.

Lemma 4 *Let $A \subset \mathbb{R}^n$ be a path-connected subset and $f: I^m \to \mathbb{R}^n$ a smooth map of the cube $I^m = \{0 \le x_j \le 1;\ j = 1, \dots, m\} \subset \mathbb{R}^m$ such that $\frac{\partial f}{\partial x_m}(I^m) \subset \operatorname{Int} \widehat{A}$. Then for any map $v: I^{m-1} = I^m \cap \{x_m = 0\} \to A$ there exists a map $g: I^m \to \mathbb{R}^n$ which is C^0-close to f, equal to f on I^{m-1} and such that $\frac{\partial g}{\partial x_m}|_{I^{m-1}} = v$, $\frac{\partial g}{\partial x_m}(x) \in A$ for all $x \in I^m$ and $\frac{\partial g}{\partial x_j}$ is C^0-close to $\frac{\partial f}{\partial x_j}$ for $j = 1 \dots, m-1$. The result also holds in the relative form, i.e., as an extension theorem.*

Proof We write $I^m = I^{m-1} \times [0,1]$, rename the last coordinate x_m into t, and denote a point of I^m as (x,t), $x \in I^{m-1}, t \in [0,1]$. We will view x as a parameter and use the notation $f'(x,t)$ for the partial derivative with respect to t. Let us divide the interval $[0,1]$ into sub-intervals of length $\frac{1}{N}$. Using Lemma 3 we find a map $\Phi: I^{m-1} \times [0, \frac{1}{N}] \to A$ such that $\Phi(x,0) = v(x)$, $x \in I^{m-1}$, and $\int_0^{\frac{1}{N}} \Phi(x,t)dt = \frac{1}{N} f'(x, \frac{1}{N})$. Again using Lemma 2 we extend the map Φ to $I^{m-1} \times [\frac{1}{N}, \frac{2}{N}]$ with the property $\int_{\frac{1}{N}}^{\frac{2}{N}} \Phi(t)dt = \frac{1}{N} f'(\frac{2}{N})$. Continuing the process we construct a continuous map $\Phi: I^{m-1} \times [0,1] \to A$ such that

$$\int_{\frac{j-1}{N}}^{\frac{j}{N}} \Phi(x,t)dt = \frac{1}{N} f'\left(x, \frac{j}{N}\right) \tag{1}$$

for all $j = 1, \dots, N$. Set

$$g(x,t) := f(x,0) + \int_0^t \Phi(x,t)dt.$$

If $t \in [\frac{j}{N}, \frac{j+1}{N}]$ then we have

$$g(x,t) = \frac{1}{N}\sum_{1}^{j} f'\left(x, \frac{i}{N}\right) + \int_{\frac{j}{N}}^{t} \Phi(x,t)dt \underset{N\to\infty}{\to} f(x,t).$$

On the other hand, $g'(x,t) = \Phi(x,t) \in A$. By differentiating integral (1) with respect to parameters x_j, $j = 1, \dots, m-1$, we get

$$\int_{\frac{j-1}{N}}^{\frac{j}{N}} \frac{\partial \Phi}{\partial x_j}(x,t)dt = \frac{1}{N}\frac{\partial f'}{\partial x_j}\left(x, \frac{j}{N}\right),$$

and thus

$$\frac{\partial g}{\partial x_j}(x,t) := \frac{\partial f}{\partial x_j}(x,0) + \int_0^t \frac{\partial \Phi}{\partial x_j}(x,t)dt$$

approximates $\frac{\partial f}{\partial x_j}(x,t)$ for $t \in [\frac{j}{N}, \frac{j+1}{N}]$. □

In order to formulate coordinate-free corollaries of Lemma 4 let us introduce some terminology. A subset $A \subset V$ of an affine space V is called *ample* if the convex envelope of each of its path connected components coincides with the whole space V. Note that the empty set is by this definition is ample as well.

Consider a smooth fibration $\pi: X \to V$. Given a point $x \in X$ we denote by Vert_x the tangent space at the point x to the fiber $\pi^{-1}(\pi(x))$ through the point x. Then the canonical projection $\pi^{(1)}: X^{(1)} \to X$ of its 1-jet extension is a vector bundle with the fiber $\mathrm{Hom}(T_yV, \mathrm{Vert}_x)$, where we set $y := \pi(x)$. The rank of this vector bundle is equal to nq, where $n = \dim V$ and $n + q = \dim X$. There are certain q-dimensional linear subspaces of its fibers $\mathrm{Hom}(T_yV, \mathrm{Vert}_x)$ which we will call *principal*. Namely, let us choose a hyperplane $L \subset T_yV$ and a homomorphism $l: L \to \mathrm{Vert}_x$. The corresponding principal subspace $P_{L,l} \subset \mathrm{Hom}(T_yV, \mathrm{Vert}_x)$ consists of homomorphisms $\lambda: T_yV \to \mathrm{Vert}_x$ such that $\lambda|_L = l$. A differential relation is called *ample in principal directions* if for every $x \in X$ and an affine subspace $Q \subset \mathrm{Hom}(T_yV, \mathrm{Vert}_x)$ parallel to a principal subspace the intersection $\mathscr{R} \cap Q$ is ample.

Theorem 5 *An open differential relation $\mathscr{R} \subset X^{(1)}$ which is ample in principal directions satisfies all forms of the h-principle.*

Proof To prove the theorem we first reduce it, by covering the manifolds X and V by coordinate neighborhoods, to the extension result for mappings of a cube in $\mathbb{R}^m$ into $\mathbb{R}^n$. Then starting with the given formal solution we use Lemma 4 to "deformalize" partial derivatives one by one. The condition of ampleness in the principal directions ensures that at each inductional step all the condition of Lemma 3 are satisfied. The crucial property that the values of partial derivatives in the parameter directions remain close guarantees that when improving a certain partial derivative we are not destroying the derivatives which were already taken care of at the previous steps of the construction. □

Theorem 5 has numerous corollaries: h-principles for a *generic* codimension two singularity, for totally real and even ε-Lagrangian (see Sect. 6.1 below) embeddings. For directed *immersions* of an n-dimensional manifolds to $\mathbb{R}^{n+1}$ it implies the following result:

Corollary 6 *Let A be any open subset of S^n such that its intersection with every great circle contains an arc of length $>\pi$. Then any stably parallelizable n-dimensional manifold M admits an immersion into $\mathbb{R}^{n+1}$ with the Gaussian image in A.*

For instance, *there exists an immersion $T^2 \to \mathbb{R}^3$ whose Gaussian image misses vertices of a regular tetrahedron.* Though the convex integration method proves existence of such an immersion, it is not easy to extract an explicit construction from the proof. An explicit effective construction of such an immersion was done by Ghomi in [66].

But the most attractive feature of the method of convex integration is that it is applicable to certain classes of differential equations. In some sense it is applicable to a *generic* "everywhere non-linear" PDE. As it was explained above, a system of partial differential equations corresponds to a submanifold (or more generally, to a stratified subset) $\mathscr{R}$ of the corresponding jet space. Under ampleness assumption the h-principle allows us to construct approximate solutions, i.e., solutions of differential relations $\mathscr{R}_\varepsilon \supset \mathscr{R}$ for a sequence of decreasing neighborhoods of $\mathscr{R}$ in the jet space. The condition which allows us to make this approximation process converging is, roughly speaking, that $\mathscr{R}$ is ample in any sufficiently small neighborhood U of any point $a \in \mathscr{R}$, i.e., for any principal direction P through the point a the convex envelope of any connected component of the intersection $\mathscr{R} \cap U \cap P$ contains $U \cap P$ (e.g., as it is the case for a curve of non-zero torsion in $\mathbb{R}^3$). The solution provided by this limiting process is C^1-smooth for differential relations in 1-jet spaces. In fact, one can even get solutions of Hölder class $C^{1,\alpha}$ for sufficiently small α depending on local differential geometric properties of the submanifold $\mathscr{R}$ (but not C^2 and even not $C^{1,\beta}$ for a sufficiently large β), see [17, 18, 35, 175]. In fact, there are cases when the above local ampleness condition is not satisfied but with an appropriate choice of an initial approximation the successive approximate solutions obtained via the convex integration method still converge to a C^1-solution. The Nash–Kuiper C^1-isometric embedding theorem is an example: *any short, i.e., distance decreasing embedding of a Riemannian manifold (M_1, g_1) into another Riemannian manifold (M_2, g_2) can be C^0-approximated by C^1-isometric embeddings, provided that* $\dim M_1 < \dim M_2$. On Fig. 1, produced by Borrelli, Jabrane, Lazarus and Thibet (project Hévéa), see [19], it is shown a C^1-isometric embedding of the flat 2-torus into $\mathbb{R}^3$, which is constructed using Gromov's convex integration scheme.

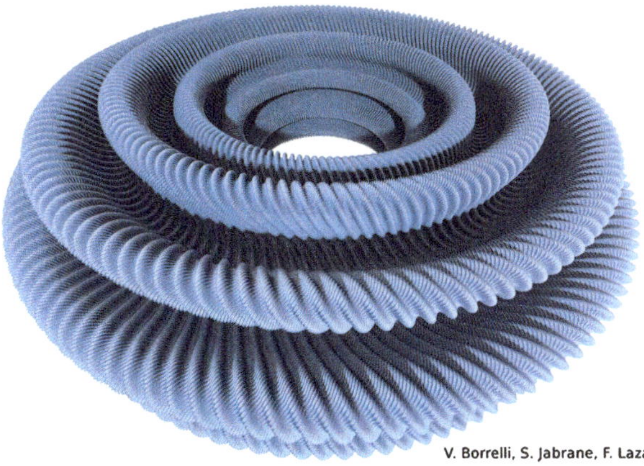

Fig. 1 C^1-isometric embedding of the flat torus in $\mathbb{R}^3$

Gromov's point of view was that though the technique of convex integration in its various versions is applicable to the large class of non-linear differential equations, the equations which are most interesting from physical applications point of view cannot exhibit an h-principle type of behavior. However, some recent development suggests that this may not be necessarily the case. In particular, Müller and Šverak, see [139] and also [38], noticed that if one drops the condition of path-connectedness in the definition of ampleness, one still can use Gromov's method for constructing non-smooth but Lipschitz solutions. Further development of this approach is an active current subject of research (see [39] for a survey) promising interesting applications, in particular in the study of weak solutions of equations of fluid dynamics and the theory of turbulence.

5 The Homotopy Principle in Complex Analysis (by Franc Forstnerič)

5.1 The Oka–Grauert Principle

The homotopy principle in complex analysis is commonly known as the *Oka Principle* after Kiyoshi Oka (1901–1978). In his series of papers during 1936–1953, Oka invented new methods of constructing global analytic objects from local ones. The Oka principle first appeared in his 1939 paper [149] where he showed that a holomorphic line bundle on a domain of holomorphy is holomorphically trivial if (and only if) it is topologically trivial.

Domains of holomorphy form a sublass of the class of *Stein manifolds* that were introduced by Karl Stein in 1951. During 1950s, Hans Grauert and Reinhold Remmert studied *Stein spaces*, complex spaces that are holomorphically convex and on which holomorphic functions separate points. In 1958, Grauert [70] extended Oka's theorem to principal fiber bundles with arbitrary complex Lie group fibers over Stein spaces, showing that the holomorphic classification of such bundles coincides with their topological classification. More precisely, if X is a Stein space and G is a complex Lie group, then the inclusion $\mathscr{O}_X^G \hookrightarrow \mathscr{C}_X^G$ of the sheaf of germs of holomorphic maps $X \to G$ into the sheaf of germs of continuous maps induces an isomorphism $H^1(X; \mathscr{O}_X^G) \cong H^1(X; \mathscr{C}_X^G)$ of the 1st Čech cohomology groups. Oka's theorem corresponds to the case when $G = \mathbb{C}^* = \mathbb{C} \setminus \{0\}$, and it says that the Picard group $\mathrm{Pic}(X) = H^1(X; \mathscr{O}^*)$ is isomorphic to the topological Picard group, and hence (via the first Chern class map) to $H^2(X; \mathbb{Z})$.

Grauert's result also pertains to fiber bundles with complex homogeneous fibers; in particular, to complex vector bundles. Interesting generalizations and applications were found by Forster and Ramspott, Henkin and Leiterer, and others (see Chap. 7 in [63]). This led to the formulation of the following heuristic principle:

Oka Principle: Analytic problems on Stein spaces which can be cohomologically formulated have only topological obstructions.

Oka's original theorem is proved by looking at the exact cohomology sequence associated to the short exact sequence $0 \to \mathbb{Z} \to \mathscr{O} \xrightarrow{\sigma} \mathscr{O}^* \to 1$, where $\sigma(f) = \exp(2\pi i f)$. This cohomological proof fails for nonabelian Lie groups, and in particular for $GL_n(\mathbb{C})$ when $n > 1$. Grauert reduced the proof to the problem of constructing holomorphic sections of an associated holomorphic fiber bundle $E \to X$ with fiber G whose transition maps are left and right multiplications by elements of G. The method is similar to the construction of global sections of coherent analytic sheaves over Stein spaces (Cartan's Theorem A). The key step consists in gluing a pair of holomorphic sections over a suitable geometric configuration (A, B) in X, called a *Cartan pair*. This is accomplished by the *Cartan lemma* on splitting a holomorphic map $f \colon A \cap B \to G$ to a complex Lie group G into a product $f = f_A f_B$ of two holomorphic maps $f_A \colon A \to G$, $f_B \colon B \to G$, each one defined on one of the larger sets A, B in our configuration.

The classical Oka–Grauert principle is limited to fiber bundles with complex homogeneous fibers. Challenging new problems in Stein geometry called for a more general Oka principle. One such case was Forster's result from 1970 [58] on the existence of proper holomorphic embeddings of a Stein manifold X^n into $\mathbb{C}^N$ for values of N well below the classical result $N = 2n + 1$ of Remmert, Bishop and Narasimhan. Forster conjectured that for $n > 1$ one can take $N = [\frac{3n}{2}] + 1$, the smallest number for which there are no topological obstructions. Forster's conjecture was only confirmed two decades later by Eliashberg and Gromov [50], using Gromov's pioneering work on Oka principle presented in the next section.

5.2 Gromov's Oka Principle

The modern development of the Oka principle started with Gromov's seminal paper of 1989 [76] in which the emphasis moved from the cohomological to the homotopy-theoretic aspect. Grauert's proof uses the exponential map on a complex Lie group for two purposes: to prove a Runge-type approximation theorem for holomorphic maps to a complex Lie group, and to linearize the gluing problem for holomorphic sections. Gromov introduced a much more flexible concept of a *dominating (holomorphic) spray* on a complex manifold Y: A triple (E, π, s) consisting of a holomorphic vector bundle $\pi : E \to Y$ and a holomorphic map $s : E \to Y$ such that for each point $y \in Y$ we have $s(0_y) = y$ and the differential $ds_{0_y} : T_{0_y}E \to T_yY$ maps the vertical subspace E_y of the tangent space $T_{0_y}E$ surjectively onto the tangent space T_yY. A complex manifold is said to be *elliptic* if it admits a dominating spray. Gromov's first main result is the following:

Theorem 7 (M. Gromov, [76]) *Maps $X \to Y$ from a Stein manifold X to an elliptic manifold Y satisfy all forms of the Oka principle. The same holds for sections $f : X \to E$ of any holomorphic fiber bundle $\pi : E \to X$ with a Stein base X and an elliptic fiber Y.*

This means that every continuous map (resp. section) is homotopic to a holomorphic one, with uniform approximation on compact holomorphically convex subsets of X and with interpolation on closed complex subvarieties of X. The analogous result holds with continuous dependence on parameters. In particular, the inclusion $\mathscr{O}(X, Y) \hookrightarrow \mathscr{C}(X, Y)$ of the space of holomorphic maps into the space of continuous sections is a weak homotopy equivalence.

Here are a few examples that were pointed out by Gromov:

(A) If a complex Lie group G acts transitively on Y by holomorphic automorphisms, we obtain a spray $s : E = Y \times \mathfrak{g} \to Y$ by taking $s(y, v) = \exp(v)y$. Here, $\mathfrak{g} = T_1G \cong \mathbb{C}^k$ ($k = \dim G$) is the Lie algebra of G.

(B) Let $V_1, \dots, V_k$ be complete holomorphic vector fields on a complex manifold Y; that is, the flow $\varphi_t^j(y)$ of V_j exists for all complex values of time t, starting at any point $y \in Y$. If the vectors $V_j(y)$ span the tangent space T_yY at each point $y \in Y$, then we get a dominating spray $s : Y \times \mathbb{C}^k \to Y$ by the formula $s(y, t_1, \dots, t_k) = \varphi_{t_1}^1 \circ \varphi_{t_2}^2 \circ \dots \circ \varphi_{t_k}^k(y)$.

(C) A dominating spray of type (B) exists on $\mathbb{C}^n \setminus A$, where A is an algebraic subvariety of $\mathbb{C}$ which does not contain any hypersurfaces.

Dominating sprays are used by Gromov in essentially the same way as sprays of type (A) in Grauert's construction; however, the details are considerably more involved. Theorem 7 also holds for the ostensibly larger class of *subelliptic* manifolds: A complex manifold Y with a finite family of holomorphic sprays (E_j, π_j, s_j) which together dominate at every point $y \in Y$, meaning that T_yY is spanned by the vector subspaces $(ds_j)_{0_y}(E_{j,y})$. For example, if $A \subset \mathbb{CP}^n$ is a projective subvariety of codimension >1, then $\mathbb{CP}^n \setminus A$ is subelliptic, but is not known to be elliptic.

Gromov considered the Oka principle in the more general context of sections of holomorphic submersions over Stein manifolds. A surjective holomorphic submersion $\pi : Z \to X$ is said to be *elliptic* if each point $x_0 \in X$ admits an open neighborhood $U \subset X$ and a family of dominating sprays s_x on the fibers Z_x, depending holomorphically on the base point $x \in U$. Similarly one defines a *subelliptic submersion*.

We introduce a *stratified (sub-) elliptic submersion* by asking that the base X, which may now be a complex space with singularities, is stratified by a descending chain of closed complex subvarieties $X = X_0 \supset X_1 \supset \cdots \supset X_m = \emptyset$, with smooth differences $S_j = X_j \setminus X_{j+1}$, such that the restriction of π to any stratum (a connected component of a difference S_j) is a (sub-) elliptic submersion. Gromov's main theorem [76, Main Theorem 4.5] is included in the following result from [59].

Theorem 8 *If X is a Stein space and $\pi : Z \to X$ is a stratified (sub-) elliptic submersion, then sections $X \to Z$ of π satisfy the Oka principle.*

Example 1 Let $\pi : E \to X$ be a holomorphic vector bundle of rank $n > 1$, and let $\Sigma \subset E$ be a complex subvariety with affine algebraic fibers $\Sigma_x = \Sigma \cap E_x \subset E_x \cong \mathbb{C}^n$ $(x \in X)$ of codimension >1. If Σ is locally uniformly tame (a condition concerning its behavior at infinity), then the restricted submersion $\pi : E \setminus \Sigma \to X$ is elliptic. If X is Stein, it follows that the Oka principle holds for sections $X \to E \setminus \Sigma$.

The Oka principle in the above example is used to construct proper holomorphic immersions and embeddings of Stein manifolds of dimension >1 into Euclidean spaces of minimal dimension; see [50] and Chap. 8 in [63]. The problem of embedding open Riemann surfaces properly holomorphically into $\mathbb{C}^2$ is still very much open since the Oka principle does not apply in this case (see Sect. 8.9 in [63] for results in this direction).

An interesting recent application of Theorem 8 was found by Ivarsson and Kutzschebauch [120] who solved the following *Gromov-Vaserstein problem*: Every null-homotopic holomorphic map $X \to SL_n(\mathbb{C})$ from a finite dimensional Stein space X to a special linear group can be factored into a finite product of upper- and lower triangular holomorphic maps into $SL_n(\mathbb{C})$.

A detailed exposition of Theorems 7 and 8 can be found in [59, 61, 62], and also in Chaps. 5 and 6 of [63].

5.3 From Elliptic Manifolds to Oka Manifolds and Oka Maps

Gromov asked in [76] whether the Oka principle for maps $X \to Y$ from Stein manifolds X to a given complex manifold Y could be characterized by a Runge approximation property for entire maps $\mathbb{C}^n \to Y$ from Euclidean spaces to Y. This conjecture was confirmed in 2006 by Forstnerič who showed that it suffices to ask for Runge approximation on a special class of compact (geometrically!) convex sets in Euclidean spaces. This condition, called CAP (the *Convex Approximation Property*), is equivalent to some dozen ostensibly different Oka properties; a complex manifold satisfying these equivalent properties is called an *Oka manifold*. (See [60] and Chap. 5 in [63] for more information.) The simple characterization of Oka manifolds by CAP paved the way to prove some functorial properties which are unknown in the class of elliptic manifolds. For example, if E and B are complex manifolds and $E \to B$ is a holomorphic fiber bundle whose fiber is an Oka manifold, then B is Oka if and only if E is Oka.

By Gromov's Theorem 7, every elliptic manifold is an Oka manifold. A partial converse, due to F. Lárusson, pertains to 'good' complex manifolds; this class includes all Stein manifolds and all quasi-projective manifolds. A good manifold Y is Oka if and only if there exists an affine holomorphic bundle $E \to Y$ whose total space E is Oka and Stein, hence elliptic.

Any natural property of objects in a given category should induce a corresponding property of morphisms. Following this philosophy, a holomorphic map $\pi : E \to B$ is said to be an *Oka map* if it is a Serre fibration and it enjoys the parametric Oka property. The latter is a parametric version of the basic Oka property of π which pertains to the possibility of deforming any continuous π-lifting $F_0 : X \to E$ of a given holomorphic map $f : X \to B$ from a Stein space X into a holomorphic lifting $F_1 : X \to E$ of f.

Finnur Lárusson explained how Oka manifolds and Oka maps naturally fit into an abstract homotopy-theoretic framework. The category of complex manifolds and holomorphic maps can be embedded into a model category such that: (a) a holomorphic map is acyclic (as a map in the ambient model category) if and only if it is a homotopy equivalence in the usual topological sense; (b) a holomorphic map is a fibration if and only if it is an Oka map. In particular, a complex manifold is fibrant if and only if it is Oka; (c) a complex manifold is cofibrant if and only if it is Stein; (d) a Stein inclusion is a cofibration. (See [127] and Sect. 7 in [60] for more information.)

A central problem is to determine the place of Oka manifolds in the classification of complex manifolds. This is well understood only in dimension one: a Riemann surface is Oka if and only if it is not Kobayashi hyperbolic. In particular, the compact Riemann surfaces that are Oka are the Riemann sphere and all elliptic curves. Already for complex surfaces the problem is difficult and to a large extent open. Whether the Oka property is preserved by modifications such as blowing up and blowing down is a closely related problem. In particular, we do not know whether an Oka manifold of dimension >1 blown up at a point, or punctured at one point, is still Oka.

6 Soft and Hard Symplectic Geometry (by Yasha Eliashberg)

This is the title of Gromov's plenary lecture at the ICM-86 in Berkeley. Symplectic geometry serves as an especially rich source of examples on both sides of the soft-hard, or as I prefer to call it, *flexible-rigid spectrum*. Flexible and rigid problems come in symplectic geometry extremely close to each other, and the development of each side towards the other one shaped and continues to shape the subject from its inception. In fact, when Gromov first entered the subject in 1967 it was unclear whether symplectic geometry has any rigid side at all.

6.1 Gromov's Alternative

To set the stage let me give some basic definitions from symplectic and contact geometries.

Symplectic geometry was born as a geometric language of classical mechanics, and similarly contact geometry is a natural set-up for geometric optics and mechanics with non-holonomic constraints.

The cotangent bundle T^*M of any smooth n-dimensional manifold M carries a canonical *Liouville* 1-form λ, usually denoted pdq, which in any local coordinates $(q_1, \dots, q_n)$ on M and dual coordinates $(p_1, \dots, p_n)$ on cotangent fibers can be written as $\lambda = \sum_1^n p_i dq_i$. The differential $\omega := d\lambda = \sum_1^n dp_i \wedge dq_i$ is called the *canonical symplectic structure* on the cotangent bundle of M. In the Hamiltonian formalism of classical mechanics the cotangent budle T^*M is viewed as the phase space of a mechanical system with the configuration space M. The p-coordinates have a mechanical meaning of momenta. The full energy of the system expressed through coordinates and momenta, i.e., viewed as a function $H\colon T^*M \to \mathbb{R}$ on the cotangent bundle (or a time-dependent family of functions $H_t\colon T^*M \to \mathbb{R}$ if the system is not conservative) is called the *Hamiltonian* of the system. The dynamics is then defined by the Hamiltonian equations $\dot{z} = X_{H_t}(z)$, $z \in T^*M$, where the Hamiltonian vector field X_{H_t} is determined by the equation $i(X_{H_t})\omega = dH_t$, and in the canonical (p, q)-coordinates has the form

$$X_{H_t} = \sum_1^n -\frac{\partial H_t}{\partial q_i}\frac{\partial}{\partial p_i} + \frac{\partial H_t}{\partial p_i}\frac{\partial}{\partial q_i}.$$

The flow of the vector field X_{H_t} preserves ω, i.e. $X_{H_t}^*\omega = \omega$. The isotopy generated by the vector field X_{H_t} is called *Hamiltonian*.

More generally, the Hamiltonian dynamics can be defined on any $2n$-dimensional manifold endowed with a *symplectic*, i.e., a closed non-degenerate differential 2-form ω. According to a theorem of Darboux any such form admits local *canonical coordinates* $p_1, \dots, p_n, q_1, \dots, q_n$ in which it can be written as $\omega = \sum_1^n dp_i \wedge dq_i$. Diffeomorphisms preserving ω are called *symplectomorphisms*. Symplectomorphisms which can be included in a time dependent Hamiltonian flow are called *Hamiltonian*. When $n = 1$ a symplectic form is just an area form, and symplectomorphisms are area preserving transformations. Though in higher dimensions symplectomorphisms are volume preserving but the subgroup of symplectomorphisms represents a small part of the group of volume preserving diffeomorphisms.

The projectivized cotangent bundle PT^*M serves as the phase space in the *geometric optics*. It can be interpreted as the *space of contact elements* of the manifold M, i.e., the space of all tangent hyperplanes to M. The form pdq does not descend to PT^*M but its kernel does, and hence the space of contact elements carries a canonical field of hyperplanes tangent to it. This field turns out to be completely non-integrable. It is called a *contact structure*. More generally, a *contact structure* on a $(2n+1)$-dimensional manifold is a completely non-integrable field of tangent hyperplanes ξ, where the complete non-integrability can be expressed by the Frobenius condition $\alpha \wedge (d\alpha)^{\wedge n} \neq 0$ for a 1-form α (locally) defining ξ by the Pfaffian equation $\alpha = 0$. Though at first glance symplectic and contact geometries are quite different, they are in fact tightly interlinked and it is useful to study them in parallel.

An important property of symplectic and contact structures is the following stability theorem due to Moser [138] in the symplectic case and to Gray [71] in the contact one:

Theorem 9 *Let ω_t, $t \in [0,1]$, be a family of symplectic (resp. contact) structures on a manifold X which coincide outside of a compact set. In the symplectic case suppose, in addition, that $\omega_t - \omega_0 = d\theta_t$, $t \in [0,1]$, where θ_t has a compact support. Then there exists an isotopy $h_t : X \to X$ with compact support which starts at the identity $h_0 = \mathrm{Id}$ and such that $h_t^*\omega_t = \omega_0$.*

Maximal integral (i.e., tangent to ξ) submanifolds of a $(2n+1)$-dimensional contact manifold (V, ξ) have dimension n and are called *Legendrian*. Their symplectic counterparts are n-dimensional submanifolds L of a $2n$-dimensional symplectic manifold (W, ω) which are isotropic for ω, i.e., $\omega|_L = 0$. They are called *Lagrangian* submanifolds. Here are two important examples of Lagrangian submanifolds. A diffeomorphism $f\colon W \to W$ of a symplectic manifold (W, ω) is symplectic if and only if its graph $\Gamma_f = \{(x, f(x));\ x \in W\} \subset (W \times W, \omega \times (-\omega))$ is Lagrangian. A 1-form θ on a manifold M viewed as a section of the cotangent bundle T^*M is Lagrangian if and only if it is closed. For instance, if $H_1(M) = 0$ then Lagrangian sections are graphs of differentials of functions, and hence the intersection points of a Lagrangian with the 0-section are critical points of the function. A general Lagrangian submanifold corresponds to a *multivalued function*, called the *front* of the Lagrangian manifold. Given a submanifold $N \subset M$ (of any codimension), the set of all hyperplanes tangent to it in TM is a Legendrian submanifold of the space of contact elements PT^*M.

It was an original idea of H. Poincaré that Hamiltonian systems should satisfies special qualitative properties. In particular, his study of periodic orbits in the so-called restricted 3-body problem led him to the following statement, now known as the "last geometric theorem of H. Poincaré": *any area preserving transformation of an annulus $S^1 \times [0,1]$ which rotates the boundary circles in opposite directions should have at least two fixed points*. Poincaré provided many convincing arguments why the statement should be true [163], but the actual proof was found by G.D. Birkhoff [16] in 1913, only after Poincaré's death. Birkhoff's proof was purely 2-dimensional and further development of Poincaré's dream of what is now called *symplectic topology* had to wait till 1960s when V. I. Arnold [4] formulated a number of conjectures formalizing this vision of Poincaré. In particular, one of Arnold's conjectures stated that the number of fixed points of a Hamiltonian diffeomorphism is bounded below by the minimal number of critical points of a function on the symplectic manifold.

At about the same time Gromov was proving his h-principle type results. Symplectic problems exhibited some remarkable flexibility. For instance, as it was already mentioned, Gromov proved that symplectic structures on open manifolds, as well as Lagrangian immersions abide an h-principle. Moreover, this is also the case for ε-Lagrangian *embeddings* (i.e., embeddings whose tangent planes deviate from Lagrangian directions by an angle $<\varepsilon$). A remarkable h-principle was proven by Gromov for the *iso-symplectic* and *iso-contact* embeddings. For instance, in the symplectic case, Gromov proved that *if* (M,ω) *and* (N,η) *are two symplectic manifolds such that* $\dim N \geq \dim M + 4$ *then any smooth embedding* $f\colon M \to N$ *which pulls back the cohomology class of the form* η *to the cohomology class of* ω, *and whose differential* df *is homotopic to a symplectic bundle isomorphism, can be* C^0*-approximated by an iso-symplectic embedding* $\widetilde{f}\colon M \to N$, *i.e.,* $\widetilde{f}^*\eta = \omega$.

Gromov formulated the following alternative: *either the group of symplectomorphisms (resp. contactomorphisms) is* C^0*-closed in the group of all diffeomorphisms, or its* C^0*-closure coincides with the group of volume preserving (resp. all) diffeomorphisms.* The proof of this alternative appeared first in his book [96] but I remember he explained to me a sketch of the proof already around 1970. One of the corollaries of Gromov's convex integration methods was that there are no additional lower bounds for the number of fixed points of a volume preserving diffeomorphism of a manifold of dimension ≥ 3. Clearly the estimates on the number of fixed points is a C^0-property, and hence, if the second part of the alternative were true this would imply that Hamiltonian diffeomorphisms of symplectic manifolds of dimension >2 have no special fixed point properties, and hence Poincaré's theorem and Arnold's conjectures reflected pure 2-dimensional phenomena. In fact, it was clear from this alternative, that all basic problems of symplectic topology are tightly interconnected. Let us list some of such problems, besides Gromov's alternative:

- *Extension of symplectic and contact structures to the ball from a neighborhood of the boundary sphere.*
- 1-parametric version of the previous question: *is it true that two structures on the ball which coincide near the boundary and which are formally homotopic relative the boundary, are isotopic?*
- *Fixed point problems for symplectomorphisms.* More generally, Lagrangian intersection problem: *Do Lagrangian manifolds under certain conditions have more intersection points than it is required by topology?*
- *Are there any non-formal obstructions to Legendrian isotopy?*

Proving an h-principle type statement in one of these problems would imply that *all* symplectic problems have soft solutions. Thus a resolution of Gromov's alternative became a question about existence of symplectic topology as a separate subject. In 1979 I proved Arnold's conjecture for 2-dimensional surfaces [46], but this still was just a two-dimensional result. The breakthrough came at the beginning of 1980s.

In 1982 Bennequin [12] proved that for certain 3-dimensional contact structures there are diffeomorphisms which could be C^0-approximated by contactomorphisms. Conley and Zehnder [34] proved Arnold's conjecture about the number of fixed points of a Hamiltonian diffeomorphism of the $2n$-torus, see the next section. Via Gromov's alternative their results implied that the group of symplectomorphisms is C^0-closed in the group of volume preserving (resp. all) diffeomorphisms, and the group of contactomorphisms of any contact 3-manifold is C^0-closed in the group of all diffeomorphisms. However, they did not observe this connection. Let me also mention that in 1981 I proved a theorem about the structure of fronts of Legendrian submanifolds which implied rigidity in both symplectic and contact cases (see [48]), but the full proof remains unpublished.

6.2 Proof of the Arnold Fixed Point Conjecture for the 2n-Torus

I sketch in this section a proof of the Arnold fixed point conjecture for the $2n$-torus, which is an adaptation of M. Gromov and myself, see [51], of the finite-dimensional reduction of Conley-Zehnder's proof given by Marc Chaperon, [32]. What is remarkable about this proof that it could be given by H. Poincaré. In fact, the first half of the proof precisely follows the first page of Poincaré's paper [163].

Theorem 10 (C. Conley and E. Zehnder [34]) *Any Hamiltonian diffeomorphism f of the $2n$-torus $(T^{2n},\omega)=(\mathbb{R}^{2n},\sum dp_i\wedge dq_i)/\mathbb{Z}^{2n}$ must have at least 2^{2n} fixed points, provided that they are non-degenerate.*

Sketch of a proof Let us lift f to a Hamiltonian diffeomorphism $F:\mathbb{R}^{2n}\to\mathbb{R}^{2n}$. We write a point of $\mathbb{R}^{2n}$ as (p,q), where $p=(p_1,\dots,p_n)$, $q=(q_1,\dots,q_n)$. We also write the symplectic form $\sum_1^n dp_i\wedge dq_i$ as $dp\wedge dq$.

We have $F(p,q)=(P(p,q),Q(p,q))$ and $dP\wedge dQ=dp\wedge dq$. If F is C^1-close to the identity then its graph

$$\Gamma_F=\big\{(p,q,P,Q)\mid P=P(p,q),\ Q=Q(p,q)\big\}\subset\mathbb{R}^{4n}$$

is graphical with respect to the splitting of $\mathbb{R}^{4n}$ into the (q,P)- and (p,Q)-coordinate subspaces, i.e.

$$\Gamma_F=\big\{p=p(q,P),\ Q=Q(q,P)\big\},$$

and hence the equation $dp\wedge dq=dP\wedge dQ$ is equivalent to the existence of a function $H(q,P)$ such that $pdq+QdP=dH$. Fixed points $p=P$, $Q=q$ of F are zeroes of the 1-form $(p-P)dq+(Q-q)dP=d(H-qP)$. In other words, fixed points are exactly the critical points of the function $G(q,P)=H(q,P)-qP$. It is not hard to check that the function G (called a generating function of F) is periodic in q, P, and hence descends to the $2n$-torus T^{2n}. Thus, in the Morse case the function G must have at least 2^{2n} critical points. Its critical points are in 1–1 correspondence with the fixed points of f, and therefore, f has as many fixed points.

As I already mentioned, the above proof almost literally repeats Poincaré's argument in his attempt to prove his "last geometric theorem". What Poincaré was missing is the following seemingly trivial idea: *an arbitrary Hamiltonian diffeomorphism is a composition of small ones.* Namely, F can be written as a composition

$F = F_N \circ \cdots \circ F_1$ of C^1-small periodic Hamiltonian symplectomorphisms. I restrict the discussion below to the case $N = 2$, the general case differs only in the notation.

The product $\Gamma := \Gamma_{F_1} \times \Gamma_{F_2} \subset \mathbb{R}^{8n}$ of the graphs of F_1 and F_2 is given by the equations

$$p_1 = p_1(q_1, P_1), \qquad Q_1 = Q_1(q_1, P_1),$$
$$p_2 = p_2(q_2, P_2), \qquad Q_2 = Q_2(q_2, P_2).$$

Furthermore, $p_i dq_i + Q_i dP_i = dH_i$ and the functions $G_i = F_i - q_i P_i$ are $\mathbb{Z}^2$-periodic, $i = 1, 2$. Set $G := G_1 + G_2$. Fixed points of F are in 1–1 correspondence with the intersection $\Gamma \cap \{p_2 = P_1, Q_1 = q_2, p_1 = P_2, Q_2 = q_1\}$, i.e., with the *zeroes of the 1-form*

$$\begin{aligned} \alpha &:= (p_1 - P_2)dq_1 + (Q_1 - q_2)dP_1 + (p_2 - P_1)dq_2 + (Q_2 - q_1)dP_2 \\ &= dG(q_1, q_2, P_1, P_2) + d\big((P_1 - P_2)(q_1 - q_2)\big). \end{aligned}$$

Changing the variables $(q_1, q_2, P_1, P_2) \mapsto (q_1, u_1 := q_2 - q_1, P_1, U_1 := P_2 - P_1)$ we get

$$\alpha = d(\widetilde{G} + u_1 U_1), \quad \text{where} \quad \widetilde{G}(q_1, u_1, P_1, U_1) := G(q_1, q_1 + u_1, P_1, P_1 + U_1).$$

One can check that the function $\widetilde{G}$ is periodic with respect to all variables, and therefore, $\widetilde{G} + u_1 U_1$ descends to a function

$$T^{2n} \times \mathbb{R}^{2n} = \mathbb{R}^{2n}/\{q_1 \sim q_1 + 1, P_1 \sim P_1 + 1\} \to \mathbb{R}.$$

Then the *stable Morse theory* implies that $\widetilde{G} + u_1 U_1$ must have at least 2^{2n} critical points, and hence f has as many fixed points. □

6.3 Advent of Holomorphic Curves

The true new era of symplectic topology started with the publication of Gromov's paper [73]. Searching for tools to establish symplectic rigidity, and, in particular, to define invariants of symplectic manifolds, Gromov turned his attention to holomorphic curves which were known to be an important tool in Algebraic Geometry. However, the environment of integrable complex structures was too rigid to be useful in Symplectic Geometry. But then, Gromov's extraordinary ability to uncover simple but deep underlying geometric ideas led him to realization that Bers–Vekua's pseudo-analytic functions could be interpreted as J-holomorphic curves in a not necessarily integrable almost complex structure J. This turned out to be precisely the right tool to deal with in Symplectic topology.

Let us recall that an almost complex structure on a $2n$-dimensional manifold W is a complex structure on its tangent bundle, i.e., an anti-involution $J: TW \to TW$, $J^2 = -\mathrm{Id}$. A map $f: (W_1, J_1) \to (W_2, J_2)$ between two almost complex manifolds of real dimension $2n_1 = \dim W_1$ and $2n_2 = \dim W_2$ is called *holomorphic* if the differential $df : TW_1 \to TW_2$ is complex linear, i.e.,

$$\overline{\partial} f := \frac{1}{2}(df \circ J_1 - J_2 \circ df) = 0. \quad (2)$$

Written in local coordinates, Eq. (2) is a system of $2n_1n_2$ equations with respect to $2n_2$ unknown functions, and hence it is overdetermined unless $n_1 = 1$, i.e., when (W_1, J_1) is a Riemann surface. Respectively, when $n_1 > 1$ then for generic non-integrable J_1 or J_2 there are no holomorphic maps $(W_1, J_1) \to (W_2, J_2)$, even locally. On the other hand, when $n_1 = 1$ Eq. (2) is an elliptic equation with the same principal symbol as the standard Cauchy-Riemann equation in the integrable case. Hence, with appropriate boundary conditions (e.g., for closed holomorphic curves, or holomorphic curves with boundaries in totally real submanifolds) this is a Fredholm problem, and assuming certain transversality we get finite-dimensional moduli spaces of solutions.

At the time when Gromov was working on his theory of J-holomorphic curves[3] there was already known an example of a remarkable application of an elliptic PDE in topology: Simon Donaldson [41] spectacularly applied in 4-dimensional topology moduli spaces of solutions of another elliptic problem, the so-called anti-self-dual Yang-Mills equation. Gromov's idea was to realize a similar scheme in symplectic topology using instead moduli spaces of holomorphic curves.

The starting problem in this scheme was to ensure compactness properties for the corresponding moduli spaces, i.e., to prove in the holomorphic curve setup an analog of Uhlenbeck's compactnes theorem [187] in the Yang-Mills theory. Gromov proved a far-going generalization of the Schwarz lemma from complex analysis which allowed him to control derivatives of a holomorphic map in terms of the diameter of its image. Combining this lemma with an ingenious use of hyperbolic geometry of Riemann surfaces Gromov proved that

Theorem 11 *Given a sequence $f_n : (S, j) \to (M, J)$ of holomorphic curves in a closed almost complex manifold such that the area* $\mathrm{Area}(f_n(S))$ *is uniformly bounded, there exists a subsequence converging to a nodal holomorphic curve with spherical bubble trees.*

Instead of giving a precise definition of a nodal holomorphic curve with spherical bubble trees, we illustrate the notion by the following example.

Example 2 Consider a sequence of holomorphic maps $f_n : \mathbb{C}P^1 \to \mathbb{C}P^1$ given in homogeneous projective coordinates by the formula

$$f_n(z_1 : z_2) = \left(\frac{z_1}{n} : z_2\right)$$

[3]Gromov used the term *pseudo*-holomorphic but I omit here this degrading prefix.

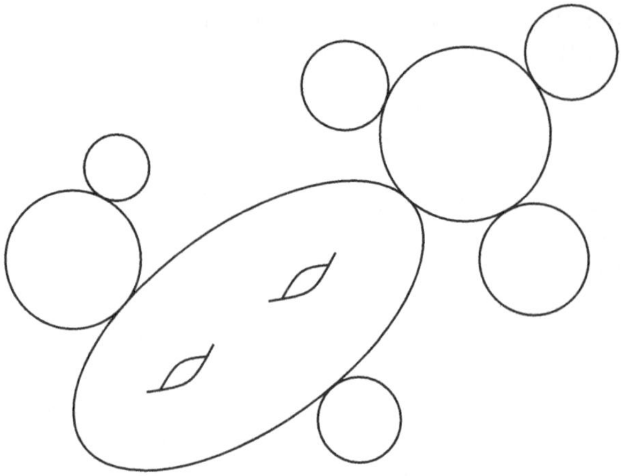

Fig. 2 Nodal holomorphic curve with spherical bubble-trees

and let $F_n\colon \mathbb{C}P^1 \to \mathbb{C}P^1 \times \mathbb{C}P^1$ be the sequence of their graphs:

$$F_n(z_1 : z_2) = \bigl(z_1 : z_2, f_n(z_1 : z_2)\bigr).$$

It is natural to think that the sequence of graphs converges to the horizontal curve $z_1 : z_2 \mapsto (z_1 : z_2, 0 : 1)$ together with the vertical spherical bubble $((1 : 0) \times CP^1) \subset \mathbb{C}P^1 \times \mathbb{C}P^1$. However, this union is not an image of $\mathbb{CP}^1$ under any holomorphic map $\mathbb{C}P^1 \to \mathbb{C}P^1 \times \mathbb{C}P^1$. What happens here is that there exists a sequence of holomorphic coordinate charts on $\mathbb{CP}^1$ centered at $(0 : 1)$ on which the sequence converges to the horizontal curve $\mathbb{C}P^1 \times (0 : 1)$, and a sequence of holomorphic coordinate charts centered at $(1 : 0)$ on which the sequence converges to the vertical curve $(0 : 1) \times \mathbb{C}P^1$.

In more general cases, in order to analyze the behavior of a sequence of holomorphic curves near a point where the gradient explodes, one may need to do some further rescaling, and as a result get a *tree* of spherical bubbles, see Fig. 2.

Gromov also analyzed the case of moduli spaces of holomorphic curves with boundaries in a totally real submanifold and proved a similar compactness theorem. Besides trees of spherical bubbles appearing at interior points there is also in this case a possibility of formation of trees of holomorphic *disc* bubbles at boundary points.

The spherical bubble formed in the above example was vertical, and this turns out to be always the case when one deals with sections of J-holomorphic bundles. Namely, suppose $\pi : W \to S$ is a smooth bundle over a Riemann surface (S, j), and J is an almost complex structure on W such that the projection $(W, J) \to (S, j)$ is holomorphic. Then, *from any sequence of holomorphic sections of the bundle π with a uniform area bound one can choose a subsequence converging to a section together with a union of vertical spherical (or disc) holomorphic bubble trees*, where the word "vertical" means that all bubble trees are contained in fibers. In particular, if we know that there are no non-constant holomorphic spheres and discs in the fibers we can conclude a conventional C^∞-convergence to a section.

The problem with applications of this compactness theorem is that in the general case it is unclear how to control the area of holomorphic curves. Here is where Gromov introduces a symplectic form into the picture. Suppose that there exists a symplectic form ω which *tames* the almost complex structure J, i.e., ω is positive on complex directions: $\omega(X, JX) > 0$ for $X \neq 0$. Then according to the Wirtinger inequality, for any holomorphic curve $f\colon S \to M$ one has $\int_S f^*\omega \geq c \operatorname{Area} f(S)$ for some positive constant c. In particular, if a Riemann surface S is closed, then the left-hand side of the inequality depends only on the homology class realized by the holomorphic curve $f\colon S \to W$, and hence when all the curves f_n belong to the same homology class the area of $f_n(S)$ is uniformly bounded, and therefore compactness in the sense of Theorem 11 holds. On the other hand, on a given symplectic manifold the space of almost complex structures which are tamed by its symplectic form is non-empty and contractible.

Similarly, one has a uniform control of the area of holomorphic curves with *Lagrangian* boundary conditions which belong to a fixed relative homology class. Gromov's compactness theorem also holds when the target manifold is non-compact, but its geometry prevents holomorphic curves to go to infinity. For instance, this is the case for holomorphic curves with boundaries in a compact Lagrangian submanifold of $\mathbb{R}^{2n}$ endowed with a symplectic form and an almost complex structures standard at infinity.

Here are a few examples of Gromov's application of holomorphic curves in symplectic geometry. A Lagrangian submanifold L of an exact symplectic manifold with a fixed primitive λ (called a *Liouville form*) is said to be *exact* if the form $\lambda|_L$ is exact. Note that if $H_1(X) = 0$ then exactness is independent of the choice of the Liouville form λ.

Theorem 12 *There are no compact exact Lagrangian submanifolds in* $\mathbb{R}^{2n}$.

Sketch of a proof We begin the sketch with Gromov's important observation that the graph of a solution of an inhomogeneous $\bar\partial$-equation $\bar\partial f = g$ for maps $(S, j) \to (M, J)$ can be interpreted as a $\widetilde{J}$-holomorphic curve in $S \times M$ for an appropriate almost complex structure $\widetilde{J}$ on $S \times M$. The almost complex structure $\widetilde{J}$ has the property that the projection $(S \times M, \widetilde{J}) \to (S, j)$ is holomorphic and the restriction of $\widetilde{J}$ to the fibers of this projection coincides with J. Let us apply this observation to the space $\mathcal{M}(C)$ of maps $f\colon (D, \partial D) \to (\mathbb{C}^n, L)$ in the trivial relative homology class, which satisfy the equation $\bar\partial f = C$ where C is a $(0, 1)$-form $\sum_1^n c_i d\bar{z}_i$ on $\mathbb{C}^n$ with constant coefficients. We note that there are no non-constant holomorphic spheres in $\mathbb{C}^n$, and exactness of the Lagrangian manifold L also implies that there are no non-constant holomorphic discs with boundary in L. Hence one can conclude that for every $(0, 1)$-form C the space $\mathcal{M}(C)$ is compact. Moreover, for a generic C the space $\mathcal{M}(C)$ is a manifold, and a generic path C_t, $t \in [0, 1]$, provides us with a cobordism between $\mathcal{M}(C_0)$ and $\mathcal{M}(C_1)$. As we already pointed out above, the space $\mathcal{M}(0)$ consists of constant discs and hence can be identified with L. Moreover, the diffeomorphism between these manifolds is given by the evaluation map $ev\colon \mathcal{M}(0) \to L$ at some fixed point $p \in \partial D$. Hence, for a generic C the evaluation map $ev\colon \mathcal{M}(C) \to L$ realizes the fundamental homology class of the manifold L, and in particular, $\mathcal{M}(C)$ is non-empty. On the other hand, when at least one of the coefficients c_j is sufficiently large then $\mathcal{M}(C) = \varnothing$. Indeed, for a constant form C the coordinate functions of any solution f of the equation $\overline{\partial} f = C$ are harmonic and hence satisfy an a priori bound for their derivatives in terms of the diameter of the Lagrangian manifold L. This contradiction proves that the Lagrangian manifold L cannot be exact. In fact, the argument proves more: there exists a non-constant J-holomorphic disc with boundary in L for any standard at infinity almost complex structure J which is tamed by the standard symplectic form on $\mathbb{C}^n$. □

Further adjusting the argument, Gromov proved that there are no exact Lagrangian submanifolds in any symplectic manifold which can be split as a product with the standard symplectic plane $\mathbb{R}^2$, and using this modification, he showed that *an exact Lagrangian submanifold can never be displaced with itself by a Hamiltonian isotopy*. An interesting application is that any closed exact Lagrangian submanifold of a cotangent bundle must intersect the 0-section. Indeed, if an exact Lagrangian submanifold $L \subset T^*M$ does not intersect the 0-section $M \subset T^*M$ then by scaling $L \to \varepsilon L$, $\varepsilon \to 0$, the submanifold L can be disjoint with itself, while remaining exact, and hence this deformation can be realized via a Hamiltonian isotopy.

The beautiful argument used by Gromov in his proof of absense of exact Lagrangian submanifolds was replicated many times in later years by several authors. In particular, it is a starting point in a remarkable theorem of Abouzaid that for certain exotic n-spheres their cotangent bundles are not symplectomorphic to T^*S^n, see [1].

As it was already mentioned above, fixed points of a symplectomorphism $f\colon (M,\omega) \to (M,\omega)$ can be interpreted as intersection points of its Lagrangian graph $\Gamma_f = \{(x, f(x))\} \subset (M,\omega) \times (M,-\omega)$ with the Lagrangian diagonal $\Delta = \{(x,x)\}$. Hence, the Lagrangian non-displaceability result implies existence of fixed points of a Hamiltonian diffeomorphism of a symplectically aspherical, i.e., containing no non-constant symplectic 2-spheres, symplectic manifold. The asphericity assumption is needed to prevent the bubbling off of holomorphic spheres.

However, this argument only shows that the set of fixed points is non-empty and does not give the lower bound predicted by Arnold's conjecture. Gromov promised a more precise estimate in his sequel paper to [73] which he referred to as [Gro2]. Unfortunately this paper had never appeared. Soon after the paper [73] was written, Andreas Floer, inspired by Gromov's work, and also motivated by Witten's work [194] and Conley–Zehnder's paper [34], came up with his Floer homology, see [56, 57], which allowed him to improve Gromov's result and to prove Arnold's conjecture for symplectically aspherical manifolds. Gromov told me that Floer suggested to him to collaborate on this project, but Gromov declined and unfortunately dropped his intended project [Gro2].

Attempts to prove Arnold's fixed point conjectures in full generality were a prominent driving force in the development of symplectic topology in the years after Gromov's work. The symplectic asphericity assumption was gradually removed (see [65]), and depending on its interpretation one can either claim that the Arnold fixed points conjecture is now fully proven, or still remains open. Indeed, Arnold formulated different versions of his conjecture. For instance, in the so-called non-degenerate case, a weaker formulation asserts that the number of fixed points of a Hamiltonian diffeomorphism of a closed symplectic manifold is bounded below by the rank of the homology of the manifold, while a more optimistic version asserts a bound in terms of the minimal number of critical points of a Morse function on the manifold. While the former version is proven, see [65], the latter one remains open for most of symplectic manifolds. The situation is even less satisfactory in the degenerate case (see, however, [51, 114, 173]).

Let us sketch a proof of another of Gromov's application of holomorphic curves:

Theorem 13 *Let (N,η) be a symplectically aspherical manifold, σ an area form on S^2, and J a tamed by $\eta\times\sigma$ almost complex structure on $(M=N\times S^2,\eta\times\sigma)$. Suppose that $\int_{S^2}\sigma$ is bounded above by the minimal positive value of η on spherical classes in N. Then each point in M belongs to a J-holomorphic sphere in M which realizes a homology class of $pt\times S^2\subset N\times S^2$.*

Sketch of a proof The space of holomorphic reparameterizations of $S^2=\mathbb{CP}^1$ is non-compact (e.g., see Example 2). To fix this problem we choose three different points $p_0,p_1,p_\infty\in S^2$ and consider the space $\mathcal{M}(J)$ of J-holomorphic maps $f\colon \mathbb{CP}_1\to S^2\times N$, such that $f(0)\in p_0\times N$, $f(1)\in p_1\times N$, $f(\infty)\in p_\infty\times N$. Then the absence of J-holomorphic curves in N guarantees compactness of $\mathcal{M}(J)$. Consider the evaluation map $ev\colon \mathcal{M}(J)\times\mathbb{CP}_1\to M$ given by the formula $ev(f,z)=f(z)$, $f\in\mathcal{M}(J)$, $z\in\mathbb{CP}^1$. When J is a split almost complex structure $\mathbb{C}P^1\times(N,\widetilde{J})$ then the map is a diffeomorphism, and therefore it has degree 1 for *any generic* J. Using compactness one more time, we conclude that it is onto for *any* J. □

Theorem 13 is the key ingredient in the proof of famous Gromov's non-squeezing theorem.

Theorem 14 *Suppose that $0<r<R$. Then there are no symplectic embeddings $D^{2n}(R)\to D^2(r)\times\mathbb{R}^{2n-2}$. Here we assume that $\mathbb{R}^{2k}$ is endowed with the standard symplectic structure and denote by D^{2k} the unit ball in $\mathbb{R}^{2k}$.*

Proof Suppose that such an embedding $f\colon D^{2n}(R)\to D^2(r)\times\mathbb{R}^{2n-2}$ does exist. Then for a sufficiently large R' we have $f(D^{2n}(R))\subset W:=D^2(r)\times D^{2n-2}(R')$. For any $\varepsilon>0$ we can find an area preserving embedding $D^2(r)$ into (S,ω) where (S,ω) is the 2-sphere with an area form ω of total area $r+\varepsilon$. We also embed $D^{2n-2}(R)$ into $(2n-2)$-torus $(T^{2n},\Theta)=(T^2,\theta)\times\cdots\times(T^2,\theta)$ with a split form, where the area $\int_{T^2}\theta$ is sufficiently large. There exists an almost complex structure J

on W tamed by the symplectic form $\omega \times \Theta$ and equal to $f_* i$ on $f(D^{2n}(R))$, where i is the standard complex structure on $\mathbb{R}^{2n} = \mathbb{C}^n$. According to Theorem 13 there is a J-holomorphic sphere C in the horizontal class passing through the image $f(0)$ of the center $0 \in D^{2n}(R)$. Denoting by ω_0 the standard symplectic structure in $\mathbb{R}^{2n}$ we have

$$\pi(r+\varepsilon)^2 = \int_C \omega \times \Theta \geq \int_{f^{-1}(C)} \omega_0 = \text{Area}\big(f^{-1}(C)\big).$$

On the other hand, the *monotonicity estimate*, see [129], implies that any properly embedded holomorphic curve passing through the center of the ball in $\mathbb{C}^n$ has its area bigger or equal than the area of a planar section trough the center of the ball. Hence, $\text{Area}(f^{-1}(C)) \geq \pi R^2$, and therefore $R \leq r$. □

This theorem for the first time established existence of specifically symplectic (e.g., different from the volume) invariants, and in particular, implied symplectic rigidity theorem (i.e., the C^0-closedness of the group of symplectomorphisms). One can also deduce from 14 that for any $n > 1$ there is a symplectic structure on $\mathcal{O}p\partial D^{2n}$ which does not extend to D^{2n}, while there are no formal obstructions for that (see Sect. 4.1). Interestingly, a similar extension problem in the contact case is still open in dimension >3.

Slightly modifying the original Gromov's definition we define swidth(U, ω) of a $2n$-dimensional manifold (U, ω), or as it is now usually called the *Gromov width* as

$$\text{swidth}(U) := \sup\big\{\pi r^2;\ B^{2n}(r) \text{ symplectically embeds into } (U, \omega)\big\}.$$

Thus, $\text{swidth}(D^2(r) \times D^{2n-2}(R)) = \pi r^2$ if $r \leq R$, and one can similarly prove that $\text{swidth}(S^2(a) \times \mathbb{R}^{2n-2}) = a$, where we denote by $S^2(a)$ the 2-sphere of area a, assume that the symplectic structure on $\mathbb{R}^{2n}$ is standard, and that the product is endowed with the split symplectic structure. It is interesting to note that this Gromov rigidity result coexists with the h-principle type observation of Polterovich (see [52]) that $\text{swidth}(T^2(a) \times \mathbb{R}^{2n-2}) = \infty$. In a similar vein is a recent result of Latschev–McDuff–Schlenk, see [128]: *the 4-torus admits an embedding of the 4-ball of full volume, and hence,* $\text{swidth}(T^2(a) \times T^2(b)) = \sqrt{2ab}$.

Using an argument somewhat similar to his proof of Theorem 13, Gromov also proved that *for any almost complex structure on $\mathbb{C}P^n$ which is tamed by the standard symplectic form one has a holomorphic sphere in the class of the generator of $H_2(\mathbb{C}P^n)$ passing through any two points*. He then applied this result (cf. the proof of the non-squeezing theorem) to show that *if there is a symplectic embedding of two disjoint balls $D^{2n}(r_1)$ and $D^{2n}(r_1)$ into $D^{2n}(1)$ then $r_1^2 + r_2^2 \leq 1$*. Indeed, if balls $D^{2n}(r_1)$ and $D^{2n}(r_1)$ embed into $D^{2n}(1)$, then they also embed into $\mathbb{C}P^n$ with the symplectic area of the generator of $H_2(\mathbb{C}P^n)$ is slightly bigger than π. Then one can choose an almost complex structure on $\mathbb{C}P^n$ tamed by the standard symplectic form and equal to the push-forward of the standard complex structure on the images of the balls. Finally, using the monotonicity theorem we conclude that the holomorphic sphere in the generator class passing through the centers of these balls has symplectic area $\geq \pi(r_1^2 + r_2^2)$, and the required inequality follows.

This result opened the whole new subject of *symplectic packing inequalities* with the most remarkable results proven in the 4-dimensional case. For instance, let us denote

$$v_k := \sup \frac{k \operatorname{vol}(D^4(r))}{\operatorname{vol} D^4(1)},$$

where the supremum is taken over all r such that $\underbrace{D^4(r) \sqcup \cdots \sqcup D^4(r)}_{k}$ symplectically embeds into $D^4(1)$. Gromov's result implies that $v_2 \le \frac{1}{2}$. In the work of Karshon [121], Traynor [185], McDuff–Polterovich [135] and Biran [14] there were computed the precise values of v_k for all k. It turns out that

k	1	2	3	4	5	6	7	8
v_k	1	1/2	3/4	1	20/25	24/25	63/64	288/289

and $v_k = 1$ for $k \ge 9$.

Inspired by Gromov's definition of invariants of symplectic domains, Helmut Hofer defined in [115] a remarkable invariant of a Hamiltonian symplectomorphism, called nowadays the *Hofer norm* (a related invariant was defined by Claude Viterbo in [192]).

To connect Hofer's and Gromov's definitions let us consider a Hamiltonian diffeotopy $h_t\colon D^{2n} \to D^{2n}$, $t \in [0,1]$, from the identity to $h_1 = h$. The isotopy is generated by a family of Hamiltonian functions $H_t\colon D^{2n} \to D^{2n}$ equal to 0 on ∂D^{2n}. Let $\Gamma_{H_t} \subset D^{2n} \times [0,1] \times \mathbb{R}$ be the graph of H_t:

$$\Gamma_{H_t} = \bigl\{u = H_t(x);\ x \in D^{2n}, t \in [0,1]\bigr\} \subset \bigl(D^{2n} \times \mathbb{R}^2, \omega + dt \wedge du\bigr),$$

where ω is the symplectic form on D^{2n}. Let us assume for a moment that H_t is positive on the interior of the ball and consider the domain

$$U_{H_t} = \bigl\{(x,t,u) \mid 0 \le u \le H_t(x),\ t \in [0,1],\ x \in D^{2n}\bigr\}.$$

Choosing a different Hamiltonian path $\widetilde{h}_t$ connecting Id with h, we observe that if the two paths are homotopic through paths with positive Hamiltonian functions then the corresponding domains U_{H_t} and $U_{\widetilde{H}_t}$ are symplectomorphic. If not the annoying positivity constraint, this would imply that symplectic invariants of the domain U_{H_t}, e.g. its Gromov width, are invariants of the Hamiltonian diffeomorphism h (or more precisely, of its lift to the universal cover of the group of Hamiltonian diffeomorphisms).

Hofer got around this difficulty and defined his conjugation invariant norm $\|h\|$ for any Hamiltonian diffeomorphism $h\colon D^{2n} \to D^{2n}$ as

$$\|h\| = \inf\bigl(\max H_t(x) - \min H_t(x)\bigr),$$

where the max and min are taken over all $(x,t) \in D^{2n} \times [0,1]$, and the infimum is taken over all Hamiltonians H_t with $H_t|_{\partial D^{2n}} = 0$ generating h. One can think of $\|h\|$ as the Gromov width of a smallest box $D^{2n} \times [0,1] \times [m,M]$ containing the graph Γ_{H_t}. The non-degeneracy of this norm, i.e., the fact that a symplectomorphism which can be generated by an arbitrary C^0-small Hamiltonian is equal to the identity, is parallel to Gromov's non-squeezing theorem. Hofer's norm can be equivalently defined by the formula $\|h\| = \inf_{H_t} \int_0^1 \|H_t\|_{C^0} dt$, where $\|H_t\|_{C^0} = \max_{x \in D^{2n}} H_t(x) - \min_{x \in D^{2n}} H_t(x)$ is the C^0-norm on the space of functions on D^{2n} equal to 0 on ∂D^{2n}, and the infimum is again taken over all Hamiltonians H_t with $H_t|_{\partial D^{2n}} = 0$ generating h. In this formulation we see that Hofer's norm is just the path-length norm on the group $\mathscr{H}$ of Hamiltonian diffeomorphisms corresponding to the Finsler metric given by the C^0-norm on the Lie algebra of the group $\mathscr{H}$. A remarkable theorem of Buhovsky and Ostrover, see [23], asserts that *any conjugation invariant Finsler (pseudo-)norm on the group of Hamiltonian diffeomorphisms that is generated by an invariant norm on the Lie algebra which is continuous with respect to the C^∞-topology, is either identically zero or equivalent to the Hofer metric.*

Hofer's norm, later generalized to all symplectic manifolds by Lalonde–McDuff [125], generates a bi-invariant metric on the group of Hamiltonian symplectomorphisms, which plays an important role in Hamiltonian Dynamics.

As Gromov demonstrated in his seminal paper, holomorphic curves are especially useful in 4-dimensional symplectic geometry due to the *positivity of intersection* property. As in the integrable case, transversely intersecting holomorphic curves in an almost complex 4-manifold intersect *positively*. Gromov sketched an argument that even in the singular case the analogy with the integrable case should hold. It turned out that the issue is quite subtle, and was settled in a series of papers of McDuff [134] and Micalleff-White [136]. As one of the applications Gromov proved the following improvement of Theorem 13 in dimension 4:

Theorem 15

1. *Let F be an orientable closed surface. Suppose that $X = S^2 \times F$ is endowed with a split symplectic structure ω, and in the case when $F = S^2$ suppose in addition that the area of the first factor is $\leq$ than the area of the second one. Then for any almost complex structure J tamed by ω the manifold X admits a foliation by J-holomorphic curves in the homology class of the first factor.*
2. *Let J be any almost complex structure on $\mathbb{C}P^2$ tamed by the standard (Fubini–Studi) symplectic form on $\mathbb{C}P^2$. Then through any two distinct points there is a unique J-holomorphic sphere in the homology class of the generator of $H_2(\mathbb{C}P^2)$, and any two such spheres intersect at 1-point. All these spheres are embedded.*
3. *Let (X,ω) be a symplectic 4-manifold with a tamed by ω almost complex structure J. Suppose that there exists an embedded J-holomorphic sphere $S \subset X$ with $S \cdot S = 1$, and there are no embedded holomorphic spheres with the self intersection equal to -1. Then (X,ω) is symplectomorphic to $\mathbb{C}P^2$ with the standard symplectic form.*

Theorem 15.3 was improved by McDuff [133]: *without assuming an absence of (-1)-curves one can conclude that X is symplectomorphic to $\mathbb{C}P^2$, possibly blown up at a few points.*

A corollary of Theorem 15.3 is that there exists a unique standard at infinity symplectic structure on $\mathbb{R}^4$, where uniqueness is understood up to symplectomorphism fixed at infinity. Another spectacular corollary is that the group of compactly supported symplectomorphisms of the standard symplectic $\mathbb{R}^4$ is contractible. This implies that the space of standard at infinity symplectic forms on $\mathbb{R}^4$ is homotopy equivalent to the group of all compactly supported diffeomorphisms. Note, however, that nothing is currently known about the topology of this group.

C.H. Taubes found a link, see [178, 179], between the Seiberg–Witten gauge theory and Gromov's theory of holomorphic curves. His result in combination with Theorem 15.3 implied that the uniqueness up to symplectomorphism result also holds for symplectic structures of a fixed total volume on $\mathbb{C}P^2$.

Though Gromov never wrote the promised second part [Gro2] of his holomorphic curves paper, which was supposed to contain a more algebraic treatment of information encoded into moduli spaces of holomorphic curves in symplectic manifolds, the subject began developing with exponentially increasing speed. We already mentioned the work of Floer who defined his Floer homology for Lagrangian intersection and fixed point problems. His theory united Gromov's holomorphic curves methods with variational techniques used by Conley–Zehnder in their proof of Arnold's conjecture for tori, and which goes back to Rabinowitz's work on mountain path method. Motivated by Physics, holomorphic curves were used to define a deformation of the product on cohomology of a symplectic manifold into a *quantum product*, and the holomorphic curve information was conveniently packaged into a generating function, which nowadays is called *Gromov–Witten potential.* Gromov–Witten theory was expanded to a relative setting as *Symplectic Field Theory*, see [49] which brought, in particular, new powerful applications in contact geometry.

There were discovered remarkable connections of Gromov–Witten theory with the theory of integrable systems. This theory is also an essential part of the mathematical theory of Mirror Symmetry, which predicts a wealth of information about symplectic manifolds and their Lagrangian submanifolds by looking at the mirror problems in complex geometry. Many of these predictions are now rigorously proven, and the picture continues to unravel.

The *Heegaard homology* theory created by P. Ozsváth and Z. Szabó [154], and more recently *embedded contact homology* of M. Hutchings [118] and C.H. Taubes [181], which are defined using J-holomorphic curves, became one of the most powerful tools in low-dimensional topology and led to solutions of several long-standing classical problems in this area.

Applications of holomorphic curves in Hamiltonian Dynamics brought us closer to the realization of Poincaré's dream of establishing qualitative properties of mechanical systems (e.g., existence and the number of periodic trajectories) without actual solving the equations of motion. In particular, the *Weinstein conjecture* asserting existence of periodic trajectories of Reeb vector fields was proven in many cases, see [116, 191], and in dimension 3 in full generality (see [180]).

Today, 27 years after Gromov's discovery, the theory of J-holomorphic curves remains one of the center-pieces of Modern Mathematics, and attracts more and more attention of Mathematicians and Physicists.

6.4 Flexible Side of Symplectic Geometry is Still Alive

We will finish this section by discussing a few advances on the flexible side of symplectic topology which happened more recently.

We already mentioned above Gromov's h-principle for iso-symplectic embeddings in codimension >2. Applying holomorphic curve technique it is not difficult to construct counter-examples to a similar h-principle in codimension 2. However, Simon Donaldson, using his theory of *almost holomorphic sections* of complex line bundles over almost complex symplectic manifolds, proved among other remarkable results the following

Theorem 16 (Donaldson, [42]) *For any closed $2n$-dimensional symplectic manifold (M, ω) with an integral cohomology class $[\omega] \in H^2(M)$ and a sufficiently large integer k there exists a codimension 2 symplectic submanifold $\Sigma \subset M$ which represents the homology class Poincaré dual to $k\omega$. Moreover, the complement $M \setminus \Sigma$ has a homotopy type of an n-dimensional cell complex* (as it is the case for complements of hyperplane sections in complex projective manifolds).

Let us denote by $P(r_1, \dots, r_n)$ the polydisc $\{|z_1| \le r_1, \dots, |z_n| \le r_n\} \subset \mathbb{C}^n$, where we assume $r_1 \le r_2 \le \dots \le r_n$. If $P(r_1, \dots, r_n)$ symplectically embeds into $P(R_1, \dots, R_n)$ then famous Gromov's non-squeezing theorem implies that $r_1 \le R_1$. We also have the volume constraint $r_1 \dots r_n \le R_1 \dots R_n$. Many people tried to prove that for a similar embedding problem for high-dimensional polydiscs there are additional constraints on radii besides the width and volume constraints. However, Larry Guth proved the following remarkable result on the flexible side, which showed that the room for additional constraints is very limited.

Theorem 17 (Guth, [107]) *There exists a constant $C(n)$ depending on the dimension n such that if $C(n)r_1 \le R_1$ and $C(n)r_1 \dots r_n \le R_1 \dots R_n$ then a polydisc $P(r_1, \dots, r_n)$ symplectically embeds into $P(R_1, \dots, R_n)$.*

While the results confirming Arnold's conjecture for intersection of two Lagrangian submanifolds remain one of centerpieces of *rigid* symplectic topology, its analog concerning lower bounds for the number of double points of a Lagrangian immersion turned out to be wrong. For instance, an "Arnold type" conjecture predicts that the minimal number $s(L)$ of transverse double points of an exact Lagrangian immersion of an orientable n-dimensional closed manifold L should satisfy the bound $s(L) \ge \frac{1}{2} \operatorname{rank} H_*(L)$, which for $L = T^n$ gives $s(L) \ge 2^{n-1}$.

However, it turns out (see [43]; for $n = 2$ the result is due to D. Sauvaget, [171]) that this conjecture is wrong:

Theorem 18 *Let L be an n-dimensional orientable manifold. If the complexified tangent bundle $T^*(L) \otimes \mathbb{C}$ is trivial*[4] *then*

$$s(L) \begin{cases} = -\frac{1}{2}\chi(L), & \text{if } n \text{ is even and } \chi(L) < 0; \\ \leq \frac{1}{2}\chi(L) + 2, & \text{if } n \text{ is even and } \chi(L) \geq 0; \\ \leq 2, & \text{if } n \text{ is odd}; \\ = 1, & \text{if } n = 3. \end{cases}$$

Here $\chi(L)$ is the Euler characteristic of L. For instance, *any 3-manifold admits a Lagrangian immersion into $\mathbb{R}^6$ with exactly 1 self-intersection point.*

It is again interesting to contrast this theorem with the following rigidity result of Ekholm and Smith [44]:

Theorem 19 *If a closed orientable $2k$-manifold L, $k > 2$, with $\chi(L) \neq -2$ admits an exact Lagrangian immersion into $\mathbb{R}^{4k}$ with one transverse double point and no other self-intersections, then L is diffeomorphic to the sphere.*

In his work Gromov approached symplectic topology both, from its flexible and rigid sides, and today the research which he has initiated is bringing both sides closer and closer to each other.

7 The Waist Inequality in Gromov's Work (by Larry Guth)

The central theme of this essay is the following inequality.

Theorem 20 (Waist inequality) *If F is a continuous map from the unit n-sphere to $\mathbb{R}^q$, then one of the fibers of F has $(n-q)$-dimensional-volume at least that of an $(n-q)$-dimensional equator. In other words,*

$$\textit{there is some } y \in \mathbb{R}^q \textit{ so that } \operatorname{Vol}_{n-q} F^{-1}(y) \geq \operatorname{Vol}_{n-q} S^{n-q}.$$

The waist inequality is a fundamental fact of Euclidean geometry. It is also a difficult theorem—it is much harder to prove than it may look at first sight. In my opinion, the waist inequality is one of the most under-appreciated theorems in geometry, and so I am excited to write about it. The waist inequality also connects with several other areas of mathematics.

Gromov began writing about the waist inequality in the early 80's, and he came back to it many times since then. When he started writing, the waist inequality could be proven as a corollary of deep work in geometric measure theory. Gromov gave several other proofs of the theorem, trying to get towards the bottom of this fundamental fact of geometry. He recognized and popularized the theorem, and gave

[4] This is a necessary and sufficient condition for existence of a Lagrangian *immersion*.

a number of applications in geometry. More recently, he wrote several papers connecting the waist inequality to other areas of mathematics, such as combinatorics and topology.

The isoperimetric inequality began as a theorem about Euclidean space. Later, people began to think about isoperimetric inequalities on other spaces, and they became a fundamental concept in geometry. Still later, people realized that many situations in different parts of mathematics are analogous to the isoperimetric inequality. Isoperimetric inequalities now play an important role in parts of group theory, graph theory, analysis, probability, computational complexity, and many other fields. In Gromov's recent work, the waist inequality is beginning to play a similar role.

7.1 Why is the Waist Inequality Hard?

The waist inequality is sharp and the optimal map is quite simple. Think of S^n as the unit sphere in $\mathbb{R}^{n+1}$. Let L be a linear map $\mathbb{R}^{n+1} \to \mathbb{R}^q$. The fibers of $L : S^n \to \mathbb{R}^q$ will be $(n-q)$-dimensional spheres, and the largest of these will be an $(n-q)$-dimensional equator.

The waist inequality for maps $F : S^n \to \mathbb{R}$ follows easily from the isoperimetric inequality on the sphere. One special case of the isoperimetric inequality says that if $U \subset S^n$ has half the volume of S^n, then the boundary of U has $(n-1)$-volume at least as big as an equator. Now we choose a value y so that the set $\{x : F(x) < y\}$ has exactly half the volume of S^n. The boundary of this set is exactly the fiber $F^{-1}(y)$, and the isoperimetric inequality tells us that it has $(n-1)$-volume at least as big as S^{n-1}.

But this approach is hard to generalize to maps $F : S^n \to \mathbb{R}^q$ for $q \geq 2$. When $q \geq 2$, a fiber of the map F does not divide S^n into regions, so there is no analogue of the method we used to choose y. A key difficulty is that it is not clear which value $y \in \mathbb{R}^q$ we should look at.

There is another important difference between the case $q = 1$ and the case $q \geq 2$. If $q = 1$, then most of the fibers of F are pretty big, but when $q \geq 2$, they may be almost all tiny. We can make this precise as follows. For any map $F : S^n \to \mathbb{R}^q$, we let $U_F(w)$ denote the union of all the fibers with $(n-q)$-volume at least w. When $q = 1$, the same argument we used to prove the waist inequality delivers the following stronger information. If $L : S^n \to \mathbb{R}$ is a linear map and $F : S^n \to \mathbb{R}$ is any continuous map, then $\operatorname{Vol} U_F(w) \geq \operatorname{Vol} U_L(w)$ for every w! The waist inequality (for maps $S^n \to \mathbb{R}$) is a corollary of this stronger inequality. And the stronger inequality fails dramatically for larger q. For any $q \geq 2$ and any $n > q$, there are maps $F_\varepsilon : S^n \to \mathbb{R}^q$ so that $U_{F_\varepsilon}(\varepsilon)$ has volume less than ε. (In other words, all but a tiny volume of the sphere S^n is covered by tiny fibers of F_ε.)

If $q \geq 2$, then there is no obvious candidate $y \in \mathbb{R}^q$ which should have a large fiber, and if we wander randomly around the domain S^n, we may see only tiny fibers. We will have to overcome these difficulties even to prove the waist inequality with a non-sharp constant.

Proving the waist inequality with a sharp constant is much harder. To get some perspective, we note that the corresponding problem for the unit cube is wide open! Conjecturally, any map $F : [0, 1]^n \to \mathbb{R}^q$ should have a fiber with $(n - q)$-volume at least 1, but no one knows how to prove it. The conjecture is true for linear maps by a theorem of Vaaler from 1979. One great advantage of a linear map is that we know which fiber should be big: the fiber through the center of the cube. Vaaler proved [188] that any k-plane through the center of a unit cube intersects the cube in k-volume at least 1. The sharp waist inequality for the unit cube would be a major non-linear generalization of Vaaler's theorem.

7.2 *A Quick History of the Waist Inequality, Part 1*

The first proof of the waist inequality is essentially due to Almgren. In [3], Almgren developed a new set of tools in minimal surface theory, allowing him to find minimal surfaces by using minimax arguments. Almgren's minimax arguments can be used to prove the waist inequality under mild hypotheses about the regularity and genericity of the map F. We call this first proof the minimax proof of the waist inequality. The proof is hard. I do not know the details of the proof. Almgren's paper [3] is over 100 pages long, and it requires a good background in geometric measure theory. (The full proof may require an additional regularity result like Allard's regularity theorem...) Suppose that a student wanted to learn the proof after taking good first-year graduate courses in differential geometry, PDE, analysis, and topology. They would probably have to read over 200 pages of math.

In the early 80's, Gromov gave a short proof of the waist inequality with a non-sharp constant ([94], page 134). The proof is only a few pages long. We give a detailed sketch of it at the end of this essay. The sharp constant is certainly interesting, but the waist inequality is important even with a non-sharp constant, and having a proof of reasonable length makes a big difference.

In [81], Gromov gave a new proof of the waist inequality (with the sharp constant). The proof does not use geometric measure theory, and instead it uses a lot of algebraic topology. One key ingredient is a generalization of the Borsuk–Ulam theorem, proven using characteristic classes. Another key ingredient is (a cousin of) the Brunn-Minkowski inequality. This proof is also pretty hard. In my opinion, it is a little shorter than the minimax proof, but that may just be subjective. Our imaginary student will probably still have to read over 100 pages of math to learn the proof.

There are now at least three different proofs of the waist inequality: the minimax proof, the short proof based on the isoperimetric inequality (with non-sharp constant), and the Borsuk–Ulam proof. The proofs with the sharp constant are surprisingly long and difficult. I tried my best to explain why the waist inequality is hard to prove, but I do not feel I can account for the length and difficulty of these proofs. There may be more proofs yet to be discovered, and it is not at all clear that we have found the simplest proof.

In the last few years, Gromov has been coming back to the idea of waists in a sequence of papers on a wide range of topics. These papers apply the idea of the waist inequality to questions in geometry, topology, and combinatorics. A main goal of this essay is to explain some of this work.

7.3 Combinatorial Analogues of the Waist Inequality

Suppose that we have N points in $\mathbb{R}^q$. There are $\binom{N}{q+1}$ different q-simplices with vertices among these points. It is interesting to try to understand how much these simplices have to overlap with each other. In [11], Barany proved that there is always a point which lies in a definite fraction of all these simplices. This result is sometimes called the point selection theorem.

Theorem 21 (Point selection, [11]) *For any N points in $\mathbb{R}^q$, there is some other point in $\mathbb{R}^q$ which lies in at least $c(q)\binom{N}{q+1}$ of the q-simplices that they determine.*

The point selection theorem can be rephrased in a way that makes it look like the waist inequality. Given N points in $\mathbb{R}^q$, we can define a linear map L from the $(N-1)$-simplex Δ^{N-1} to $\mathbb{R}^q$ sending the vertices of the simplex to the given points. Now for a point y in $\mathbb{R}^q$, the fiber $L^{-1}(y)$ is a subset of Δ^{N-1}, and we define the size $|L^{-1}(y)|$ to be the number of q-faces of Δ^{N-1} which $L^{-1}(y)$ intersects. The point selection theorem can now be rephrased as follows:

For any linear map $L : \Delta^{N-1} \to \mathbb{R}^q$, there is a point $y \in \mathbb{R}^q$ so that $|L^{-1}(y)| \geq c(q)\binom{N}{q+1}$.

Using this analogy, Gromov adapted ideas from the waist inequality to give a new proof of the point selection theorem. The short proof of the waist inequality adapts smoothly to this combinatorial setting. Moreover, the argument automatically gives the following more general result:

Theorem 22 ([83]) *For any continuous map $F : \Delta^{N-1} \to \mathbb{R}^q$, there is a point $y \in \mathbb{R}^q$ so that $|F^{-1}(y)| \geq c(q)\binom{N}{q+1}$.*

This story is an example of why it is useful to have many proofs of fundamental theorems. The minimax and Borsuk–Ulam proofs of the waist inequality do not adapt to the combinatorial setting (as far as we know...) The minimax and Borsuk–Ulam proofs use a lot of specific information about the geometry of S^n, which makes them more difficult to adapt to other situations. The short proof, although it gives a weaker result, is more adaptable.

One important open problem in this area is to understand the asymptotic behavior of the constant $c(q)$. Gromov and Barany both give a constant $c(q)$ which is approximately $1/q!$. On the other hand, in the worst known examples, the constant $c(q)$ is approximately e^{-q}. In particular, it would be interesting to know whether $c(q)$ decays exponentially or super-exponentially.

The analogy with the waist inequality gives a new perspective on this problem. Gromov's proof of point selection is based on the short proof of the waist inequality. The short proof of the waist inequality gives a non-sharp constant which is too small by a factor $\sim 2^q/q!$. This gap roughly matches the gap between the known constants and the worst examples in the point selection problem. Perhaps a better understanding of the waist inequality will some day help us to understand the asymptotic behavior of the point selection problem.

7.4 Topological Analogues of the Waist Inequality

We begin with an image from everyday life. Imagine we have sewn up a tear in a pair of pants. At the end we take out the sewing needle, and there is some thread sticking out. We have to tie the thread up in a knot to make sure that it does not go back through the fabric and undo the stitches. What kind of knot should we tie? The knot has to be thick to prevent it from pulling back through the hole in the fabric. For example, a long string of trefoil knots will not work, because they may pull through the fabric one at a time. We want our knot to be thicker than the hole. And we want the knot to stay thick even if it gets jostled. How can we make such a knot?

A friend who sews described to me one way of doing this in practice. Take the loose end of string, wrap it several times around your finger, then roll it gradually off your finger, pressing it into a ball and pulling it tight at the same time. When she does it, the result is a tightly knotted knob of string too thick to go back through the hole in the fabric. Even if I squeeze it, it stays round and resists squishing into a narrow tube. Mathematically, it is not clear to me why this works. What is the geometry/topology that keeps this knot from passing through the hole in the fabric?

Let us formulate a mathematical question in a similar spirit. We define the waist of a knot $K \subset \mathbb{R}^3$ to be the smallest W so that we can isotope K to a position where it meets each horizontal plane in $\leq W$ points. For example, consider a torus knot $T_{p,q}$ with $p < q$. If we take a standard representative and orient it in a sensible way, it meets each horizontal plane in $\leq 2p$ points. Can we isotope it to some strange position to reduce this number $2p$? Recently in [156], Pardon gave an elegant proof that the waist of $T_{p,q}$ is exactly $2p$. This estimate about the waist was the first step in Pardon's solution to a problem about the distortion of knots that Gromov posed in the early 80's.

Let us digress a little to explain this problem. If $K \subset \mathbb{R}^n$, we recall that the intrinsic distance between two points is the length of the shortest curve between the two points in K. The extrinsic distance is the distance between the two points in $\mathbb{R}^n$. The distortion of K is the largest value of the ratio between intrinsic distance and extrinsic distance. If the distortion is large, it means that there are two points of K which are close together, but the shortest path between them in K is long. In the early 80's, Gromov asked whether there are isotopy classes of knots that require arbitrarily large distortion, and the question was open for almost thirty years. One key difficulty is that knots with distortion <100 can be extremely complex: there are infinitely many isotopy classes, and they can have arbitrarily large values of many (all?) standard knot invariants. Pardon estimated the distortion of torus knots: he proved that if $2 \leq p < q$, then any $T_{p,q}$ torus knot has distortion $\gtrsim p$. In particular, there are torus knots that require arbitrarily large distortion. Pardon's key idea was to connect estimates for distortion with estimates for waists, and he combined this with his estimate for the waist of $T_{p,q}$.

Returning to the theme of waists, we next consider a closed 3-manifold M. We say that the waist of M is the smallest W so that we can find a map $M \to \mathbb{R}^3$ where each fiber is a surface of total genus at most W. (The total genus of a disconnected surface is the sum of the genus of each component.) Gromov explored the waists of 3-manifolds in [100]. The most interesting 3-manifolds considered are arithmetic hyperbolic 3-manifolds. These 3-manifolds are defined in an algebraic way, and the definition is simple and natural from the point of view of algebra. From the point of view of geometry and topology, they are interesting and complex. Gromov showed that an arithmetic hyperbolic 3-manifold triangulated with N simplices has waist $\sim N$. Any 3-manifold triangulated with N simplices has waist $\lesssim N$, and so the arithmetic hyperbolic 3-manifolds are in some sense as complicated as possible.

From this starting point, Gromov showed that arithmetic hyperbolic 3-manifolds are topologically complicated in many other ways. The most interesting result has to do with the "Morse theory" of maps from M to $\mathbb{R}^2$. The standard Morse theory connects the topology of a manifold M with the critical points of a smooth generic map $M \to \mathbb{R}$. For example, if M is topologically complicated, then the Morse theory proves that any smooth generic function must have many critical points. It is natural and interesting to try to replace the target $\mathbb{R}$ with something more general, but it turns out to be very difficult to formulate an interesting analogue of Morse theory with a higher-dimensional target. If we take a generic smooth map $M \to \mathbb{R}^2$, then the set of critical points in M will not be discrete—it will typically be 1-dimensional. In nice cases, it will be a 1-dimensional manifold—a union of circles. Next we may ask how the topology of M is connected with the set of critical points of a generic smooth map $M \to \mathbb{R}^2$. If M is a topologically complicated 3-manifold, does it imply that the set of critical points must have many circles? Surprisingly, the answer is no. In [45], Eliashberg proved that any closed orientable 3-manifold admits a generic smooth map to $\mathbb{R}^2$ where the set of critical points consists of 4 small (unknotted) circles! (Eliashberg's construction is a special case of his "h-principle for folded maps". It is a part of the theory of h-principles, which Gromov played a big role in developing. This aspect of Gromov's work is discussed in Eliashberg's essay in this volume.)

At this point, it may look as though there are no analogues of the Morse inequalities for maps from M^3 to $\mathbb{R}^2$. But Gromov observed that interesting inequalities appear when we switch our attention from the set of critical points (in M) to the set of critical values (in $\mathbb{R}^2$). For nice maps $M^3 \to \mathbb{R}^2$, the set of critical values will be an immersed curve. And if M is complicated topologically, then Gromov proved that this curve must have a large number of self-intersections. In particular, Gromov proved that if M is an arithmetic hyperbolic 3-manifold triangulated with N simplices, then the number of self-intersections of the curve of critical values must be $\gtrsim N^2$. Any 3-manifold triangulated with N simplices admits a map to $\mathbb{R}^2$ where the curve of critical values has $\lesssim N^2$ self-intersections, and so the arithmetic hyperbolic 3-manifolds are again as complicated as possible. One key step in the proof is to see how this Morse inequality problem is connected with the waist of a 3-manifold, and another key step is to estimate the waists of arithmetic hyperbolic 3-manifolds.

The point that I would like to make here is that the waist of a knot or a 3-manifold is a useful definition. The problem on the distortion of knots was pretty old. And the problem of finding some type of Morse inequalities for maps to $\mathbb{R}^2$ is even older, although it did not have a precise formulation. The two theorems we talked about required several new ideas, but the idea of waists played a key role in both.

So far, we have only talked about waists for maps from a 3-manifold to $\mathbb{R}$. This corresponds to the "easy" case of the waist inequality: the case of maps to $\mathbb{R}$. For maps to $\mathbb{R}^q$ with $q \geq 2$, much less is known, and there are many interesting open problems described in [100] and [83]. For example, if we consider a map from a 7-dimensional arithmetic hyperbolic manifold to $\mathbb{R}^2$, does one of the fibers need to be topologically complicated? This looks difficult. In fact, the situation is unclear even for high-dimensional tori. If we consider a continuous map from the n-torus T^n to $\mathbb{R}^q$, what can we say about the complexity of the fibers? It is straightforward to construct maps where the most complicated fiber consists of two $(n-q)$-tori. In this case, the sum of the Betti numbers of each fiber is at most 2^{n-q+1}. Does every continuous map $T^n \to \mathbb{R}^q$ have a fiber with the sum of the Betti numbers at least 2^{n-q+1}? This is unknown. For $q = 1$, Gromov proved a nearly sharp lower bound on the topological complexity of fibers in [83]. For $q \geq 2$, the best lower bound [100] is still far from the upper bound of 2^{n-q+1}.

The waist inequality gives an interesting perspective on topological complexity, inspired by geometry. Let us compare a large arithmetic hyperbolic 3-manifold with the connected sum of many 3-dimensional tori. In some ways, they are both topologically complicated: both have large homology groups, both have fundamental groups that require many generators, and both require many simplices to triangulate. But the arithmetic hyperbolic 3-manifold is far more difficult to understand or to imagine. In some fundamental way, the arithmetic hyperbolic 3-manifold is much more complex. The waist gives one perspective for describing this complexity.

7.5 A Quick History of the Waist Inequality, Part 2

The waist inequality has taken a long time to get recognition as something important. I want to try here to address the history of people writing about the waist inequality. Actually I know of very few examples of people writing about the waist inequality, and I will mention all the ones that I know. It is hard to be sure if I missed something—if I did, I would definitely like to hear about it.

Who first posed the question of the waist inequality? I have no idea. Plausibly it could be a hundred years old. But as far as I personally know, it might not have been posed until the 70's or even the 80s?

Almgren's paper on varifolds [3] from the early 60's contains the tools for the minimax proof of the waist inequality. But Almgren did not state the waist inequality in that paper. He had a different goal: he proved that every closed Riemannian manifold contains a minimal surface of every dimension. I suspect that the waist inequality was known in the geometric measure theory community in the 60's and 70's as a folk theorem, but I do not know of any place that it was published. The most well-known source on minimax techniques in minimal surface theory is Pitts's book [160], which also does not mention the waist inequality. Even today, I do not know of any published source that states the waist inequality and then explains in detail how it follows from Almgren's minimax methods.

The first place that I know where something like the waist inequality appears in writing is Gromov's paper [94] (on pages 106 and 133–135). Gromov describes Almgren's work, discusses some possible connections between the waist inequality and systolic geometry, and gives his short proof of the waist inequality with a non-sharp constant.

I would be very interested to know what people thought about the waist inequality in the 60's, 70's, 80's, 90's. Since I know so little about what other people were thinking, I thought I might mention my first experience with the waist inequality. I first learned about the waist inequality reading Gromov's book *Metric Structures* when I was a graduate student. (The waist inequality is described in a paragraph in Section $2.12\frac{1}{2}$.) When I first read about it, it did not make much impression on me. It did not seem surprising, and I thought incorrectly that it was the sort of thing that had been known for a long time. But then I needed to use it in my thesis in many places, and over time, I gradually developed a great respect for it.

In 2003, Gromov wrote an entire paper about the waist inequality [81]. I believe that this was the first time anyone wrote a paper about the waist inequality. From then until the present, it has played an important role in much of his work. During that time, Gromov popularized the waist inequality as something important and worth knowing.

7.6 Quantitative Topology

All of the proofs of the waist inequality use some topology. They all use degree theory, and one uses Borsuk–Ulam and characteristic classes. The waist inequality is closely connected to topology, more so than its cousin the isoperimetric inequality. One reason is that the waist inequality implies the topological invariance of dimension. To see this, suppose that we had a homeomorphism (or just a continuous injective map) $\Phi : \mathbb{R}^{q+1} \to \mathbb{R}^q$. Let $L : S^n \to \mathbb{R}^{q+1}$ be a linear map, so that the fibers of L are $(n-q-1)$-dimensional spheres. Now consider the composition $F : \Phi \circ L : S^n \to \mathbb{R}^q$. The composition F is continuous, and so the waist inequality should apply. Since Φ is injective, each fiber of F would be an $(n-q-1)$-dimensional sphere. These fibers would be one dimension lower than they "should" be! In particular, each fiber would have $(n-q)$-volume equal to zero. So we see that the waist inequality (even with a non-sharp constant) implies topological invariance of dimension.

One useful way of thinking about the waist inequality is as a quantitative version of the topological invariance of dimension. Now there is already an important quantitative version of the topological invariance of dimension, going back to the very beginning of the subject.

Lebesgue covering lemma *If U_i are open sets covering the n-cube $[0, 1]^n$, and if each point of the cube lies in $\leq n$ of the sets U_i, then one of the U_i has diameter ≥ 1.*

The Lebesgue covering lemma has the following corollary, which looks analogous to the waist inequality.

Corollary *If $n > q$, and $F : [0, 1]^n \to \mathbb{R}^q$ is a continuous map, then one of the fibers of F has diameter ≥ 1.*

This corollary should be compared to the waist inequality for the unit cube. As we mentioned above, the sharp constant in the waist inequality for the unit cube is unknown, but the waist inequality for the sphere implies the following non-sharp result:

Theorem 23 (Waist inequality for the unit cube) *If $n > q$, and $F : [0, 1]^n \to \mathbb{R}^q$ is a continuous map, then one of the fibers of F has $(n - q)$-volume $\geq c(q, n) > 0$. (Conjecturally, we should be able to take $c(q, n) = 1$ for all q, n.)*

Comparing the last two results, we see that the waist inequality is like the Lebesgue covering lemma with diameter replaced by $(n - q)$-volume. I wonder if people working in topological dimension theory asked about the waist inequality, but I do not know of any evidence that they did. It seems to me that the waist inequality plays a natural role in topological (or geometrical) dimension theory. I like to view this story as an attempt to move the understanding of dimension from linear algebra into topology and geometry. Here are some highlights of the story in chronological order—each grouped around one fundamental fact from linear algebra.

Linear Algebra Theorem 1 *If $n > q$, there is no surjective linear map from $\mathbb{R}^q$ to $\mathbb{R}^n$.*

The topological analogue of this theorem is false! In the 1870's, Peano constructed a surjective continuous map from $\mathbb{R}^q$ to $\mathbb{R}^n$. This important example showed that people's intuition can be wrong and emphasized the need to be careful in topology. After Peano's example, it might have seemed like a bad idea to keep trying to generalize ideas from linear algebra into topology and geometry. Remarkably, some very good generalizations followed.

Linear Algebra Theorem 2 *If $n > q$, there is no injective linear map from $\mathbb{R}^n$ to $\mathbb{R}^q$.*

This time, the topological analogue of the theorem is true! In 1909, Brouwer proved that there is no injective continuous map from $\mathbb{R}^n$ to $\mathbb{R}^q$. Brouwer's proof was based on his discovery of the degree of a map. Lebesgue proposed his covering lemma almost immediately, and the lemma was proven by Brouwer a few years later. The covering lemma implies a quantitative geometric version of the theorem: If $n > q$, and $F : [0,1]^n \to \mathbb{R}^q$ is a continuous map, then one of the fibers of F has diameter ≥ 1.

Linear Algebra Theorem 3 *If $n > q$, then any linear map from $\mathbb{R}^n$ to $\mathbb{R}^q$ has kernel of dimension at least $n - q$.*

This theorem is a stronger version of Linear Algebra Theorem 2. In particular, if n is much bigger than q, then the fibers of a linear map $\mathbb{R}^n \to \mathbb{R}^q$ must be very large. The Lebesgue covering lemma and its corollary do not provide a good geometric analogue for this stronger theorem. They say that a non-linear map $F : [0,1]^n \to \mathbb{R}^q$ must have a fiber with a large diameter, but this fiber may look more like a hair than like an $(n - q)$-dimensional plane. The waist inequality says that any map from $[0,1]^n$ to $\mathbb{R}^q$ has a fiber which is large in an $(n - q)$-dimensional sense. The waist inequality is a good geometric analogue of this fundamental theorem of linear algebra. (But one may ask for a fiber which is large in a different way besides volume, and it is not clear to me what is the qualitatively strongest thing we can say about one of the fibers…)

The geometric results that we have mentioned here are examples of quantitative topology. They are quantitive geometric versions of qualitative theorems from topology. They are geometric estimates with close ties to topology, estimates whose proofs are powered by topology. This is a young area pioneered by Gromov. (See Nabutovsky's essay in this chapter for other material related to quantitative topology.) So far, there are only a handful of general methods in the area. Some of the fundamental tools go back to Gromov's short proof of the waist inequality in the early 80's. In the next section, we try to give a flavor of this area via a detailed sketch of this proof.

7.7 Gromov's Short Proof of the Waist Inequality

In the early 80's, Gromov gave a short proof of the waist inequality with a non-sharp constant. We describe the proof here, slightly adapted to the present context. I think it is a good idea to try to include a real proof in a survey like this. This theorem is not nearly as deep the non-squeezing theorem or the groups of polynomial growth theorem, but it is one of my favorites of Gromov's short proofs. I think it should be accessible to many people.

This proof has had a significant influence, and I think it will continue to be influential in the future. The proof combines topological arguments and quantitative geometric estimates. This can sometimes feel (to me) like trying to mix water and oil, and this proof gives a great example of how to get them to work together.

Another reason that the proof is influential is that it is very robust. I have used it and adapted it in several of my papers on quantitative topology ([106, 108]). Gromov adapted it to the combinatorial setting of the point selection theorem in [83]. We will describe that adaptation in the next section.

For any $X \subset S^n$, let $\mathrm{Cov}_r(X)$ denote the minimal number of balls of radius r which are needed to cover X. We give here an estimate for the covering numbers of fibers, analogous to the waist inequality.

Theorem 24 (Essentially contained in [94], page 134) *For each $n > q$, there exists a constant $\beta_{n,q} > 0$ so that the following holds. Suppose $F : S^n \to \mathbb{R}^q$ is a continuous map. For each $r > 0$, there is some $y \in \mathbb{R}^q$ so that the fiber $F^{-1}(y) \subset S^n$ has* $\mathrm{Cov}_r(F^{-1}(y)) \geq \beta_{n,q} r^{-(n-q)}$.

The theorem says that each continuous map $S^n \to \mathbb{R}^q$ has a fiber with covering size comparable to an $(n-q)$-dimensional equator. At the very end we explain how to estimate the volumes of fibers by similar methods.

The first main ingredient of the proof is the following fundamental theorem of topology.

Ingredient 1 (Brouwer) *The identity map $S^n \to S^n$ is not homotopic to a constant map.*

The second main ingredient of the proof is an isoperimetric-type inequality which is essentially due to Federer and Fleming.

Ingredient 2 *Suppose that $X \subset S^n$ is a proper subset of S^n. Then X is contained in a contractible set $Y \subset S^n$ with* $\mathrm{Cov}_r(Y) \leq C_n r^{-1} \mathrm{Cov}_r(X)$.

We sketch the proof of this isoperimetric inequality. The set $Y \supset X$ will be a cone. Recall that the cone $C_p X$ is defined to be the union of all the minimal geodesics which start at p and end at a point of X. If X contains the antipode to p, then $C_p X$ is the entire sphere S^n; otherwise, $C_p X$ is a contractible proper subset of S^n. We have to choose p so that we can control $\mathrm{Cov}_r(C_p X)$. Federer and Fleming had the remarkable idea to choose p randomly and estimate the average value of $\mathrm{Cov}_r(C_p X)$.

Let us see how to estimate the average value of $\mathrm{Cov}_r(C_p X)$, averaged over all $p \in S^n$. Let B denote an r-ball centered at a point x. If B lies in the hemisphere centered at p, then $C_p B$ may easily be covered by $\lesssim r^{-1}$ balls of radius r. However, if B lies near to the antipode of p, then $C_p B$ may be much larger. Let us write $\bar{p}$ for the antipode of p. A short calculation shows that $\mathrm{Cov}_r(C_p B) \lesssim r^{-1} \mathrm{dist}(x, \bar{p})^{-(n-1)}$. The exponent $-(n-1)$ is very important. The key point is that the function $\mathrm{dist}(x, \bar{p})^{-(n-1)}$ is integrable in p! Therefore, if we move p randomly on S^n, then the average value of $\mathrm{Cov}_r(C_p B)$ is $\lesssim r^{-1}$. The same holds true for each r-ball used to cover X, and we see that the average of $\mathrm{Cov}_r\, C_p X$ is $\lesssim r^{-1} \mathrm{Cov}_r\, X$.

I think this averaging trick of Federer and Fleming is a wonderful idea. For any particular choice of p, it is hard to calculate what is going on. Naive choices for p, such as the point at the greatest distance from X, do not work. But if we average over p, then everything becomes transparent.

Notice that one of main ingredients is a fundamental result of topology and the other is an isoperimetric-type inequality, a fundamental quantitative estimate from geometry. Now we are going to sketch how these two ingredients work together in Gromov's proof of the waist inequality. The proof is by contradiction. Suppose that we have a map $F : S^n \to \mathbb{R}^q$ and that, for some $r > 0$, every fiber of F can be covered by $\beta r^{-(n-q)}$ balls of radius r. We get to choose the constant $\beta = \beta_{n,q} > 0$ as small as we like. Using the structure from F, we will construct a homotopy from the identity map to a constant map. This homotopy contradicts ingredient 1, proving our theorem.

We are going to construct a homotopy from the identity to a constant map: we will construct a map $H : S^n \times [0, 1] \to S^n$, which is the identity at time 0 and constant at time 1. We just defined the map H on $S^n \times \{0\} \cup S^n \times \{1\}$, and now we have to extend H to the rest of $S^n \times [0, 1]$. We will use the map F to help us construct this homotopy. The map F allows us to organize the sphere S^n into small pieces which overlap nicely. We choose a fine triangulation of $\mathbb{R}^q$, and we consider $F^{-1}(\Delta)$ for different simplices Δ in the triangulation. Now we construct the homotopy in small steps:

Step 0. Define H on $F^{-1}(v) \times [0, 1]$ for each vertex v of the fine triangulation.
Step 1. Define H on $F^{-1}(\Delta^1) \times [0, 1]$ for each 1-simplex of our triangulation.
...
Step q. Define H on $F^{-1}(\Delta^q) \times [0, 1]$ for each q-simplex of our triangulation.

When we get to step j, we have already defined H on the boundary of $F^{-1}(\Delta^j) \times [0, 1]$ for each j-simplex Δ^j in our triangulation. Let us take a minute to think about the boundary of $F^{-1}(\Delta^j) \times [0, 1]$. It has two parts:

- The top and bottom: $F^{-1}(\Delta) \times \{0\}$ and $F^{-1}(\Delta) \times \{1\}$.
- The sides: $F^{-1}(f^{j-1}) \times [0, 1]$ where f^{j-1} is a hyperface of Δ^j.

We defined H on the top and bottom at the very outset, and we defined H on the sides at step $j - 1$. So in step j, we have to extend H from the boundary of $F^{-1}(\Delta^j) \times [0, 1]$ to all of $F^{-1}(\Delta^j) \times [0, 1]$.

How do we know there is any extension at all? If the map H from the boundary of $F^{-1}(\Delta) \times [0, 1]$ does not cover all of S^n, then the map from the boundary is contractible and we can use this to build an extension. So the key point is that the image of the boundary does not cover all of S^n. In order to control this, we try to make the image of H as small as possible at each step, and we prove estimates about the size of the image of H. We will construct H so that it obeys the following estimates:

1. The image of the boundary of $F^{-1}(\Delta^j) \times [0, 1]$ has Cov_r-size $\lesssim \beta r^{-n+q-j}$.
2. The image of $F^{-1}(\Delta^j) \times [0, 1]$ has Cov_r-size $\lesssim \beta r^{-n+q-j-1}$.

The top and bottom of $\partial(F^{-1}(\Delta^j) \times [0,1])$ have small images because of our hypotheses. Because H is the identity at time 0, the image of the top is exactly $F^{-1}(\Delta^j) \subset S^n$. Since our triangulation is fine, this lies in a small neighborhood of $F^{-1}(y)$ for y the center of Δ^j, and so the image of the top has covering size $\lesssim \beta r^{-n+q}$. Because H is constant at time 1, the image of the bottom is just a point! This establishes estimate 1 for $j = 0$. Also, the sides at step j are controlled by estimate 2 at step $j-1$. Now we can do a proof by induction, and we just have to check the following claim:

Suppose we have defined H on the boundary of $F^{-1}(\Delta^j) \times [0,1]$, and the image of the boundary has covering size $\lesssim \beta r^{-n+q-j}$. Then we can extend H to $F^{-1}(\Delta^j) \times [0,1]$ so that its image has covering size $\lesssim \beta r^{-n+q-j-1}$.

This claim follows directly from the isoperimetric inequality. We define X to be the image of the boundary. By the isoperimetric inequality, X is contained in a contractible Y with $\mathrm{Cov}_r(Y) \lesssim r^{-1}\mathrm{Cov}_r(X) \lesssim \beta r^{-n+q-j-1}$. Since Y is contractible, we can extend H to a map from $F^{-1}(\Delta^j) \times [0,1]$ into Y.

When we use the isoperimetric inequality, we have to know that X is a proper subset of S^n. By choosing β small enough, we can arrange that $\mathrm{Cov}_r(X) \lesssim \beta r^{-n+q-j}$ is always strictly smaller than $\mathrm{Cov}_r(S^n) \sim r^{-n}$. This criterion determines the constant β.

(Technical remarks: The theorem implies that one of the fibers is fairly large in terms of covering by balls. If the fibers are decently regular, then we can let $r \to 0$, and we see that one of the fibers has $(n-q)$-volume at least $c_{n,q} > 0$. More generally, with a small amount of extra record-keeping, we can deal with coverings by balls of varying radii and estimate the Hausdorff content. In this way, it follows that any continuous map $S^n \to \mathbb{R}^q$ has a fiber with $(n-q)$-dimensional Hausdorff measure at least $c(n,q) > 0$. For a general continuous map, one still does not know the sharp waist inequality for the Hausdorff measure of the fibers, but Gromov proved in [81] that the sharp waist inequality holds for the Minkowski volume of a fiber.)

7.8 Gromov's Proof of Point Selection

The great thing about the short proof of the waist inequality is how flexible it is. As an example, we explain here how to adapt it to the combinatorial setting of the point selection theorem. Gromov's generalization of the point selection theorem goes as follows:

Theorem 25 ([83]) *For any continuous map $F : \Delta^N \to \mathbb{R}^q$, there is a point $y \in \mathbb{R}^q$ so that the fiber $F^{-1}(y)$ intersects at least a fraction $c(q)$ of all the q-faces of Δ^N. (The constant $c(q)$ does not depend on N—it is approximately $1/q!$.)*

We will focus attention on the boundary of Δ^N, which is homeomorphic to the sphere S^{N-1}. Basically, we are just going to adapt the proof of the waist inequality from a round sphere S^{N-1} to $\partial\Delta^N$. The first main ingredient of the proof is exactly the same as above.

Ingredient 1 (Brouwer) *The identity map $\partial\Delta^N \to \partial\Delta^N$ is not homotopic to a constant map.*

We follow the same outline of proof. Suppose that we have a map $F : \partial\Delta^N \to \mathbb{R}^q$ and every fiber of F intersects only a small fraction of the q-faces of Δ^N. We will construct a homotopy $H : \partial\Delta^N \times [0, 1] \to \partial\Delta^N$, which is the identity at time 0 and constant at time 1. We choose a fine triangulation of $\mathbb{R}^q$, and then we construct the homotopy H in small steps:

Step 0. Define H on $F^{-1}(v) \times [0, 1]$ for each vertex v of the fine triangulation.
Step 1. Define H on $F^{-1}(e) \times [0, 1]$ for each edge e of our triangulation.
...
Step q. Define H on $F^{-1}(f^q) \times [0, 1]$ for each q-face f^q of our triangulation.

As before, the boundary of $F^{-1}(f^j) \times [0, 1]$ has two parts:

- The top and bottom: $F^{-1}(f^j) \times \{0\}$ and $F^{-1}(f^j) \times \{1\}$.
- The sides: $F^{-1}(f^{j-1}) \times [0, 1]$ where f^{j-1} is a hyperface of f^j. There are $j + 1$ sides.

We defined H on the top and bottom at the very outset, and we defined H on the sides at step $j - 1$. So in step j, we have to extend H from the boundary of $F^{-1}(f^j) \times [0, 1]$ to all of $F^{-1}(f^j) \times [0, 1]$. As long as the map H from the boundary of $F^{-1}(f) \times [0, 1]$ does not cover all of $\partial\Delta^N$, then the map from the boundary is contractible and we can extend it to $F^{-1}(f) \times [0, 1]$. The key point is that the image of the boundary does not cover all of $\partial\Delta^N$. In order to control this, we need to estimate the size of the image at every step.

At this moment, we need to adapt our idea of size to the situation. We do not know anything about the volumes of the fibers of F. Instead, we know that each fiber of F intersects only a small fraction of the q-simplices of F. We make this the basis of our notion of size. After fine-tuning a little bit, Gromov settled on the following definition.

For a subset $X \subset \partial\Delta^N$, we let $\|X\|_j$ denote the probability that X intersects the face spanned by $j + 1$ randomly chosen vertices $v_0, \dots, v_j$ of Δ^N. It may happen that the vertices $v_0, \dots, v_j$ are not distinct, in which case this face has dimension $<j$. But in our situation, we will have N much larger than j, in which case $\|X\|_j$ is essentially the probability that X intersects a random j-face of the simplex. The hypothesis of our proof by contradiction is that every fiber obeys $\|F^{-1}(y)\|_q \le \beta$ for an appropriate β which is close to $c(q)$.

With this language, we can describe our estimates about the map H. We will construct H so that, for each j-face f^j in our triangulation of $\mathbb{R}^q$, the following holds:

1. The image of the boundary of $F^{-1}(f^j) \times [0, 1]$ has j-norm $\lesssim\beta$.
2. The image of $F^{-1}(f^j) \times [0, 1]$ has $(j - 1)$-norm $\lesssim\beta$.

The top and bottom of $\partial(F^{-1}(f^j) \times [0,1])$ have small images because of our hypotheses. Because H is the identity at time 0, the image of the top is exactly $F^{-1}(f^j) \subset \partial \Delta^N$. Since our triangulation is fine, this lies in a small neighborhood of $F^{-1}(y)$ for y the center of f^j, and so the image of the top has q-norm $\leq \beta$. It also has j-norm $\leq \beta$ for all $j < q$. Because H is constant at time 1, the image of the bottom is just a point! This establishes estimate 1 for $j = 0$. Also, the sides at step j are controlled by estimate 2 at step $j-1$. Now we can do a proof by induction, and we just have to check the following claim:

Suppose we have defined H on the boundary of $F^{-1}(f^j) \times [0,1]$, and the image of the boundary has j-norm <1. Then we can extend H to $F^{-1}(f^j) \times [0,1]$ so that its image has $(j-1)$-norm obeying the following bound:

$$\left\| H\left(F^{-1}(f^j) \times [0,1]\right) \right\|_{j-1} \leq \left\| H\left(\partial\left(F^{-1}(f^j) \times [0,1]\right)\right) \right\|_j .$$

This claim follows from an isoperimetric inequality adapted to this situation.

Ingredient 2 *Suppose that $X \subset \partial \Delta^N$ with $\|X\|_j < 1$. Then X is contained in a contractible set $Y \subset \partial \Delta^N$ with $\|Y\|_{j-1} \leq \|X\|_j$.*

The proof of this ingredient is closely modeled on the argument of Federer and Fleming using random cones. The first task is to give an appropriate definition of a cone. Let p denote the center of one of the $(N-1)$-faces of $\partial \Delta^N$, and let $\bar{p}$ denote the opposite vertex. Given any point $x \in \partial \Delta^N$ which is not $\bar{p}$, then we can define the ray from x to p as follows. If x lies in the same $(N-1)$-face as p, then draw the ordinary Euclidean ray from x to p. If not, imagine the simplex sitting on a table, with the face containing p face down. Draw the line from $\bar{p}$ through x, and follow it until it hits the base of the simplex at some point x'. The ray from x to p is given by the line segment from x to x' and then the line segment from x' to p. If $\bar{p}$ is not in X, then the cone $C_p(X)$ is the union of all the rays from $x \in X$ to p. As long as $\bar{p} \notin X$, the cone $C_p(X)$ is contractible. (If $\bar{p} \in X$, then we can define $C_p X$ to be all of $\partial \Delta^N$.)

Now we consider $Y = C_p(X)$ where p is the center of a random $(N-1)$-face. We have to estimate the $(j-1)$-norm of Y. Because of the geometry of our set up, the face spanned by $v_0, \ldots, v_{j-1}$ intersects Y if and only if the face spanned by $v_0, \ldots, v_{j-1}, p$ intersects X. Therefore, we get the following simple equation:

$$Average_p \left\| C_p(X) \right\|_{j-1} = \|X\|_j .$$

We choose p so that $Y = C_p(X)$ enjoys $\|Y\|_{j-1} \leq \|X\|_j < 1$. Evidently, Y is not all of $\partial \Delta^N$, and so we must have $\bar{p} \notin X$ and Y is contractible. This finishes the proof of Ingredient 2.

Finally, we just need to choose β small enough so that we have $\|X\|_j < 1$ at every step. Doing a careful book-keeping, it turns out that $\beta \sim c(q) \sim \frac{1}{q!}$ works.

8 Quantitative Topology and Quantitative Geometric Calculus of Variations (by Alex Nabutovsky)

This is a survey of some of the original work of Misha Gromov on quantitative topology and curvature-free geometric inequalities in Riemannian geometry as well as some recent advances in these areas motivated by the problems that he posed and involving approaches that he invented. In particular, we discuss upper bounds for measures of various extremal objects on a Riemannian manifold in terms of the volume or the diameter of the underlying manifold, and various problems about fillings and slicings of Riemannian manifolds.

8.1 Quantitative Topology

A number of fundamental works of Misha Gromov are devoted to Quantitative Topology. Its object of study is quantitative properties of maps between Riemannian manifolds, (or, more generally, length spaces) that are known to exist for non-constructive topological reasons sometimes involving homotopy-theoretic arguments. In particular, one is interested in complexity of "optimal" homotopies, extensions, liftings, etc. (cf. [92, 99, 103]). There are many diverse ways to measure the complexity of a map $f : X \longrightarrow Y$. One of the most natural measures of complexity of a Lipschitz map is its Lipschitz constant or dilatation $\operatorname{dil} f = \sup_{x_1,x_2\in X} \frac{\operatorname{dist}_Y(f(x_1),f(x_2))}{\operatorname{dist}_X(x_1,x_2)}$. For example, if X is a circle of length 1, and we reparametrize f proportionally to the arclength, then the dilatation of f coincides with the length of the closed curve $f(X)$ in Y. More generally, dilatation measures how much lengths of curves in X can increase under f. Gromov notices that one can also define the complexity as k-dilation dil_k that measures the maximal increase of the k-area of k-dimensional subsets of X under f. (Here k is any positive integer number that does not exceed $\dim X$.) If X and Y are Riemannian manifolds, and f is C^1, then dil_k is the C^0-norm of the kth exterior power of the derivative df. Of course, 1-dilation coincides with the dilatation. One can also consider L^p-norms of the kth exterior power of $\Lambda^k df$ for $p < \infty$ (cf. [97]). If f is an embedding, then Gromov suggests that its distortion $\max\{\operatorname{dil} f, \operatorname{dil} f^{-1}|_{f(X)}\}$ provides yet another natural measure of complexity.

Once an appropriate complexity functional had been chosen, one can ask, for example, how many homotopy classes of the considered maps are representable by maps with complexity $\leq x$, or what is the complexity of "optimal" homotopies contracting a given contractible map of complexity $\leq x$.

8.1.1 Fundamental Group and Homotopies Contracting Null-Homotopic Curves

Assume that X is a circle of radius one, and we have replaced f by a homotopic map parametrized proportionally to the arclength. The two basic questions are: (1) How long are the generators of $\pi_1(M^n)$? and (2) Given a contractible closed curve γ of length $\leq l$ in a closed Riemannian manifold Y what is the minimal value of L such that γ can be contracted via closed curves of length $\leq L$? Can one also estimate the Lipschitz constant of an "optimal" contracting homotopy? The answer for the first question is provided by the following theorem of Gromov ([103], Proposition 5.28): For each closed Riemannian manifold M^n of diameter d, each $p \in M^n$ there exists a finite presentation of $\pi_1(M^n)$, where all generators g_i can be represented by loops of length $\leq 2d$ based at p and all relations are of the form $g_i g_j g_k^{-1} = e$. (Gromov also proved that one can choose loops representing a system of generators of $\pi_1(M^n)$ so that the lengths of these loops are strictly less than $2d$, unless M^n is diffeomorphic to a real projective space and has a very special metric.) This frequently quoted theorem is not difficult to prove: Take a sufficiently fine triangulation of M^n, connect the base point with the vertices v_i of the triangulation by minimizing geodesics g_{v_i}, and for every 1-simplex $v_i v_j$ of the triangulation consider the loop formed by g_{v_i}, $v_i v_j$, and g_{v_j} travelled in the opposite direction towards the base point. These loops will be the generators of $\pi_1(M^n)$, and each 2-simplex of the triangulation will yield a relation. The generators are represented by loops of length $\leq 2d + \varepsilon$, where ε can be made arbitrarily small by choosing a sufficiently fine triangulation. A compactness argument finishes the proof.

Recently, A. Petrunin noticed that this theorem has the following unexpected implication: Consider a Riemannian manifold M of diameter 1 with a finite fundamental group of order k. It is obvious that the universal covering $\tilde{M}$ of M with the pullback Riemannian metric has diameter $\leq 2k$. It is not difficult to improve the upper bound to k ([141]), but Petrunin noticed that if $\pi_1(M^n)$ is a cyclic group of order $k \longrightarrow \infty$, then the diameter of $\tilde{M}$ is bounded by $O(\log k)$. He asked if, in fact, the diameter of $\tilde{M}$ is always bounded by a sublinear function of k, where $O(\log^{const} k)$ seems a reasonable guess [158].

The second question (about lengths of closed curves in an "optimal" homotopy) is partially answered by a result by Gromov ([103], Proposition 2.26) that deals with the simply-connected case. In particular, Gromov proved that for each closed simply-connected manifold M^n there exists a constant $c(M^n)$ such that each closed curve $f : S^1 \longrightarrow M^n$ can be contracted to a point via curves of length $\leq c(M^n)\,\mathrm{length}(f)$. Moreover, this upper bound can be improved to $(1 + o(1))\,\mathrm{length}(f)$, as $\mathrm{length}(f) \longrightarrow \infty$. Furthermore, for each $n \geq 2$ this assertion can be generalized to the situation, when S^1 is replaced by any metric polyhedron of dimension $k \leq n - 1$, providing that M^n is assumed to be k-connected. In this case Gromov proved that each mapping $P \longrightarrow M^n$ of dilatation x is contractible to a constant map via maps of dilatation $\leq (1 + o(1))x$, as $x \longrightarrow \infty$.

In the case, when the fundamental group of M^n is infinite, the behaviour of the maximal length of closed curves arizing in "optimal" homotopies contracting closed curves of length x is determined by the Dehn function of $\pi_1(M^n)$. (The Dehn function measures the number of times that one needs to apply relations to demonstrate that a word of length $\leq N$ representing the trivial element, indeed, represents the trivial element. It turns out that its growth does not depend on a particular choice of a finite presentation of a group.)

Returning to the simply connected case note that the result of Gromov that we have just quoted is of asymptotic nature, as it does not provide any estimate for the constant $c(M^n)$ and, thus, cannot be used as is for specific closed curves γ. It had been noticed in [140] that even if M^n is diffeomorphic to the n-sphere, there is no computable upper bound on $c(M^n)$ in terms of various Riemannian functionals of M^n such as volume, diameter, various norms of the curvature tensor, injectivity radius, etc. The reason is that despite the simply-connectedness of M^n, the geometry of M^n can "strongly resemble" the geometry of a smooth homology sphere such that its fundamental group has an unsolvable word problem, and, therefore, non-computably fast growing Dehn function. This result explains why most existing (and expected) results in quantitative homotopy topology are either of asymptotic nature or involve specific Riemannian metrics on considered manifolds. Two notable exceptions are the cases of low dimensions ($n = 2$, and possibly, $n = 3$), and the case, when one a priori knows that all closed curves in M^n can be "easily" contracted. This last case is relevant for study of periodic geodesics on M^n and, more generally, the geometry of loop spaces on M^n. It turns out that in this case one usually can at least theoretically obtain the desired upper bounds even for maps of manifolds of high dimensions (see [146] for some first results in this direction). Now we would like to make a digression and to discuss a seemingly trivial issue, namely "optimal" homotopies contracting the boundaries of Riemannian 2-discs.

8.1.1.A. How Difficult is it to Contract the Boundary of a Riemannian 2-Disc?

In the first version of [77] Gromov posed the following question: Given a Riemannian 2-disc of diameter d with a boundary of length L, is it possible to contract the boundary to a point through closed curves of length $\leq const(L + d)$, where $const$ is an absolute constant, e.g., 10^{10}? By the time his paper was ready for publication S. Frankel and M. Katz found counterexamples showing that, in general, no such bound is possible [64]. Then they asked if there exists an upper bound for length of curves in an "optimal" contracting homotopy in terms of d, L and the area A of the discs. Twenty years later a positive answer for proven in a paper by Y. Liokumovich, the author and R. Rotman [132]: One can bound the lengths of the curves by $L + 200d \max\{1, \ln \frac{\sqrt{A}}{d}\}$.

8.1.2 Quantitative Homotopy Theory for Higher Dimensions

Let us review the foundational results of the theory in the case when $X = S^n$. (Gromov observes that his results and conjectures probably generalize for the general case of simply-connected X, see [97].) Theorem 7.10 [103] asserts that if Y is a simply-connected Riemannian manifold, then the number of different homotopy classes of maps $X \longrightarrow Y$ of dilatation $\leq z$ grows at most as $O(z^{c(Y)})$, where $c(Y)$ depend only of the rational homotopy type of Y; the proof involves Chen–Sullivan minimal models theory. The basic idea is that the norm of a rational homotopy class represented by a given map f of a sphere can be represented as an integral of a certain differential form associated with f, and the norm of this form can be majorized in terms of the dilatation of f. Gromov asks if the number of homotopy classes of maps $S^n \longrightarrow Y$ of dilatation $\leq z$ is always asymptotic to z^r for some r providing that $\pi_1(Y)$ is finite. He also asks what is the behaviour of the minimal possible dilatation of a homotopy contracting a contractible map f of dilatation $\leq z$ between two simply-connected Riemannian manifolds X, Y. In particular, Gromov asks if the minimal possible dilatation of a contracting homotopy always bounded by $const(X, Y)z^\alpha$ for an appropriate α depending on a rational homotopy type of X and Y. A well-known (and open) particular case of this problem involves contractible maps $f : S^N \longrightarrow S^n$ between two round spheres of radius one. Gromov notes that one can prove a straightforward quantitative version of Serre's theorem asserting the finiteness of homotopy groups of spheres $\pi_N(S^n)$ for $N > 2n-1$ yielding an upper bound of the form $\exp(\exp(\dots(z)\dots))$, where the height of the tower of exponentials is determined by N and asks if the optimal upper bound is, in fact, polynomial (or even linear) in z. This question is highly relevant for upper bounds in various filling problems via cobordism theory (see Chap. 5 of [78]). More specifically, let M^n be an orientable manifold triangulated with N n-simplices so that not more than $c(n)$ n-simplices are adjacent to any vertex. Assume that M^n bounds an orientable manifold W^{n+1}. One would like to find W^{n+1} triangulated with the smallest possible number of simplices so that not more than $C(n)$ simplices meet at any vertex. Consider the smallest number of $(n+1)$ simplices as a function of N. What is its behaviour as $N \longrightarrow \infty$? Does it admit a polynomial bound? Cobordism theory can be used to rewrite the property of being orientably null cobordant as vanishing of appropriate elements in certain homotopy groups, and then to obtain an upper bound that behaves as a tower of exponentials of N ([78], Chap. 5).

Gromov has also similar results about embeddings. In [92] he used Haefliger's reduction of embedding theory $V \longrightarrow W$ to homotopy theory (for a simply connected W) in metastable range ($\dim W > \frac{3}{2}\dim V + 2$) to show that the number of smooth isotopy classes of such embeddings of distortion $\leq x$ is bounded by a polynomial of x. He also noticed that the existence of infinitely many (smooth) Haefliger knots $S^{4k-1} \subset S^{6k}$ implies that this dimension restriction is necessary. Yet in a recent paper [54] S. Ferry and S. Weinberger observe that for each x the number of *topological* isotopy classes of embeddings $V \longrightarrow W$ with distortion $\leq x$ is finite as long as $\dim W - \dim V \neq 2$. They pose an interesting problem of estimating this number of topological isotopy classes as a function of x. In particular, they ask whether or not this number can be bounded by a polynomial function (as in the results by Gromov described at the beginning of this section).

We also would like to mention a recent work by M. Krcal, J. Matousek, F. Sergeraert on fast computations of homotopy groups and, more generally, Postinkov towers of simply connected simplicial complexes ([123], see also [30]). They hope to prove that homotopy groups of simply-connected finite simplicial complexes are computable in a polynomial time. The flavour of their work is different from Gromov's point of view. To illustrate the difference consider the problem of computing $\pi_2(X)$ for a simply-connected finite simplicial complex X. Hurewicz theorem implies that $\pi_2(X)$ is isomorphic to $H_2(X)$, and the problem of calculation of $H_2(X)$

is essentially a problem of linear algebra that can be solved by means of a very fast algorithm. Yet, if $\pi_2(X) \neq 0$, then complexity of the simplest non-contractible 2-sphere in X does not admit any *computable* upper bound in terms of the number of simplices of X ([146], Theorem 0.5). Informally speaking, the reason why this is so is that, in general, we do not have any "evidence" of simply-connectedness of X.

8.1.3 k-Dilations

Recall, that k-dilation of a map f between two Riemannian manifolds measures the maximal amount of increase of k-areas under f, and that dilatation can be also regarded as 1-dilation. What happens if $k > 1$? We refer the reader to [105] for the review of the old results as well as new results about relationships between homotopy properties of maps and their k-dilation. The foundational result relating homotopy properties of maps and their k-dilation asserts that for even n the absolute value of the Hopf invariant of a map $f : S^{2n-1} \longrightarrow S^n$ between round spheres of radius one does not exceed $c(n)\,\mathrm{dil}_n^2(f)$, where $\mathrm{dil}_n(f)$ denotes the n-dilation of f. A stronger form of this result of Gromov can be found in [97] on p. 222. Furthermore, Gromov proved that any non-contractible map $S^m \longrightarrow S^2$ has a 2-dilation $>c(m) > 0$ (see p. 230 of [97] for a stronger result). Tsui and Wang [186] proved that one can take here $c(m) = 1$. Most recently, L. Guth [105] proved that non-contractible maps $S^m \longrightarrow S^{m-1}$ can have arbitrarily small k-dilations if and only if $k > \frac{m+1}{2}$. In particular, there exist non-contractible maps $S^4 \longrightarrow S^3$ with arbitrarily small 3-dilation. In the same paper Guth proved that in some homotopy groups of spheres and for some values of k there exist non-trivial elements representable by maps of arbitrarily small k-dilation , as well as some other elements that are not. (This happens, for example, for $\pi_8(S^5) = Z_{24}$, and $k = 4$.)

Numerous results about relationships between homotopy properties of maps f and Lebesque norms of Df (or $\Lambda^k Df$) can be found in Chaps. 2 and 3 of [97].

8.1.4 Quantitative Morse Theory

In the previous section we discussed representations of *individual* maps $f : V \longrightarrow W$ by homotopic maps of small dilatation. An alternative point of view suggested by M. Gromov in [92] is to consider the dilatation (or any other interesting measure of complexity) as a functional on a space of Lipschitz maps $V \longrightarrow W$, and to study different asymptotic aspects of topology of the inclusion maps of sublevel sets $\mathrm{dil}^{-1}([0, x])$ in the whole space. Earlier, we were discussing the number $N(x)$ of connected components of the space of maps that have a non-zero intersection with $\mathrm{dil}^{-1}([0, x])$, but one can also ask what is the maximal number $m = m(x)$ such that every map of an arbitrary polyhedron K of dimension $\leq m$ into the space of Lipschitz maps $V \longrightarrow W$ is homotopic to a map into $\mathrm{dil}^{-1}([0, x])$. This last question is already interesting enough, when we consider the spaces of based or free loops on a simply-connected Riemannian manifold W, but [92] discusses this problem for a variety of different complexity functionals on different spaces of maps. For spaces of either free or based loops on simply-connected closed Riemannian manifolds Gromov proved that $m(x)$ has a linear growth with x; see Chap. 7.A in [103]. Then he applied this result to prove strong lower bounds for the number of periodic geodesics on W of length $\leq x$. Gromov did not estimate the constant C in the upper bound Cx for $m(x)$, but his proof implies that it can be estimated in terms of the maximal *width* $w(W)$ of the optimal homotopy contracting a closed curve of length $\leq 2d$, where d denotes the diameter of W. (The width of a homotopy is the maximal length of a trajectory of a point during the homotopy. For every closed curve of length $\leq 2d$ one takes the infimum of widths over the set of homotopies contracting the curve. Then one defines $w(W)$ as the supremum of the resulting quantity over the set of all closed curves of length $\leq 2d$.) In [141] we prove that one can explicitly majorize the constant C for the space of based loops in terms of $S(W)$ defined as the minimal x such that each closed curve of length $\leq 2d$ is contractible via closed curves of length $\leq x$. This is stronger than Gromov's result as one can easily majorize $S(W)$ in terms of $w(W)$, but not vice versa. The meaning of our result is that if the length functional on a space of based loops on W has a very deep critical point of index >0 of finite depth (i.e., corresponding to a null-homologous cycle), then it has a very deep local minimum. In the same paper we prove that if the length functional on M^n has a very deep local minimum of finite depth, then it must have many deep local minima. We do not know if these or similar results hold for spaces of free loops or 1-cycles.

8.2 *Quantitative Geometric Calculus of Variations*

One of the central results of Gromov's paper [94] asserts that the length l of the shortest *non-contractible* periodic geodesic on an essential closed n-dimensional Riemannian manifold M^n does not exceed $c(n)\,\mathrm{vol}(M^n)^{\frac{1}{n}}$ (Gromov's systolic inequality). Here $c(n)$ is an explicit constant that depends only on the dimension of the manifold. A manifold M^n is called essential if the natural map f from M^n to the classifying space of its fundamental group $K(\pi_1(M^n), 1)$ has the following property: The image of the fundamental homology class of M^n under the homomorphism induced by f is a non-zero element of the nth homology group of $K(\pi_1(M^n), 1)$. (Here one uses homology groups with coefficients in Z, if M^n is orientable, and in Z_2, if M^n is non-orientable.) Note that the essentiality is a homotopy property. All essential manifolds are nonsimply-connected; the examples include real projective spaces, closed surfaces of genus >0, and, more generally, closed aspherical manifolds. A result by I. Babenko [7] implies that no generalizations of Gromov's upper bound for the length of a shortest *non-contractible* periodic geodesic is possible if the underlying manifold is not essential. So, in this respect the result of Gromov is optimal.

In the same paper, Gromov asks if a similar upper bound for the length l of a shortest non-trivial periodic geodesic in terms of the volume holds for all closed (possibly simply-connected) Riemannian manifolds. (The periodic geodesic does not need to be non-contractible anymore. Recall that the celebrated theorem proven by A. Fet and L. Lyusternik asserts the existence of a non-trivial periodic geodesic on every Riemannian manifold [55].) Gromov conjectured that such an estimate should exist, if $n=2$ (where the only remaining case was the case of M^n is diffeomorphic to S^2). Indeed, soon afterwards C. Croke proved that the length of a shortest non-trivial periodic geodesic on a Riemannian 2-sphere M does not exceed $31\sqrt{\mathrm{Area}(M)}$ [36]. This upper bound was later improved by Rotman to $4\sqrt{2}\sqrt{\mathrm{Area}(M)}$ [164], which is still more than three times higher than the conjectured by M. Gromov and E. Calabi optimal bound $12^{\frac{1}{4}}\sqrt{\mathrm{Area}(M)}$ (that corresponds to the limit case of 2-spheres glued out of two flat equilateral triangles). This question of Gromov remains open for all non-essential manifolds of dimension ≥ 3. One can extend it also to complete non-compact manifolds of finite volume, where one does not even know if a non-trivial periodic geodesic always exist. (However, if $n=2$, the existence of a non-trivial periodic geodesic (and even an infinite set of distinct periodic geodesics) was proven by V. Bangert [10] and G. Thorbergsson [182]; Croke proved that the smallest length of a non-trivial periodic geodesic can be majorized by a multiple of the square root of the area as in the compact case [36].)

One can also ask if the smallest length l of a non-trivial periodic geodesic can be majorized by $c(n)d$, where d denotes the diameter of the Riemannian manifold, and $c(n)$ depends only on its dimension. It is well-known and easy to prove that if the manifold is not simply-connected, then $l \leq 2d$, so the question is interesting only for simply-connected manifolds. The question remains completely open for all simply-connected manifolds of dimension ≥ 3, but, again, Croke proved the desired upper bound for $n=2$ with $c(n)=9$ ([Cr]). Later this upper bound was independently improved to $4d$ by the author and Rotman [142] and S. Sabourau [168]. On the other hand a natural conjecture that the length of a shortest periodic geodesic on a Riemannian 2-sphere does not exceed $2d$ (so that the round spheres are optimal) unexpectedly turned out to be false. This fact had been proven by F. Balacheff, Croke and M. Katz in [8]. However, no one has any idea of what are the Riemannian metrics of diameter one on S^2, where the length of a shortest non-trivial periodic geodesic is the maximal (or nearly maximal) possible.

One had better luck with establishing curvature-free upper bounds for the smallest length of a non-trivial closed geodesic net. Closed geodesic nets in Riemannian manifolds are homological "cousins" of periodic geodesics. They are the critical points of the length functional on the space of immersed finite (multi)graphs. A closed *geodesic net* in a Riemannian manifold M^n is an immersed finite (multi)graph such that each edge is immersed as a geodesic and with the following additional property: For each vertex the sum of unit tangent vectors to all edges of the graph adjacent to the considered vertex and directed from it is equal to zero. All periodic geodesics are closed geodesic nets, but closed geodesic nets are still "rare". In particular, a geodesic loop is a closed geodesic net only if it is a periodic geodesic. It is also extremely difficult to prove the existence of closed geodesic nets that are not unions of periodic geodesics with [111] being the only known to me work in this direction. Yet in [143] it was proven that the length of a shortest closed geodesic net on a closed Riemannian manifold M^n can be majorized by $c(n)d$, where d denote the diameter of M^n and also by $c(n)\,\mathrm{vol}^{\frac{1}{n}}(M^n)$. Here $c(n)$ is an explicit constant. Later, Rotman [165, 167] found much better values for $c(n)$ and demonstrated that there always exists a closed geodesic net satisfying these estimates that looks like a wedge of geodesic loops based at the same point. Misha Gromov once asked me another interesting question about geodesic nets which is still open: Is it always true that the union of all closed geodesic nets on a closed surface endowed with a Riemannian metric is an everywhere dense subset of this surface?

Of course, one can ask whether or not there exist curvature-free upper bounds for the smallest area (or volume) of a closed minimal surface in a closed Riemannian manifold. Here the theory of systolic freedom developed by I. Babenko, M. Katz and A. Suciu (see [103], Appendix D and references there) that is based on the first examples of systolic freedom found by Gromov [98] implies that the topology of the manifold does not seem to help. (This contrasts with the situation with the length of a shortest periodic geodesic, where the essentiality of the manifold guarantees the existence of the upper bound in terms of the volume.) In [144] we proved upper bounds for volumes of embedded minimal hypersurfaces in manifolds of dimension between 3 and 7 as well as stationary k-varifolds of arbitrary dimensions and codimensions (a quantitative version the existence results proven by F. Almgren and J. Pitts; see [161]). Our upper bounds are valid only for manifolds with vanishing homology groups in all dimensions up to $k-1$ and involve the homological filling functions for these dimensions in addition to either the diameter or the volume of the underlying manifold. An obvious question is whether the triviality of homology groups is really necessary here and whether or not one can get rid of the filling functions in these upper estimates. In a conversation with the author Misha Gromov observed that one probably cannot hope for the estimates in terms of only the diameter without the filling functions, and one can look for the counterexamples that are obtained as approximations of appropriate Carnot–Carathéodory metrics by Riemannian metrics (see [97]). The work of D. Burago and S. Ivanov [24] seems to provide another source of the potential counterexamples. The question about the existence of the upper bounds in terms of the volume of the ambient manifold that do not involve the filling function is widely open.

Finally, recall that Serre's theorem asserts that any two points on a closed Riemannian manifold can be connected by infinitely many geodesics. If these two points coincide, then the resulting geodesics are geodesic loops. Sabourau proved that for each Riemannian manifold M^n there exists a geodesic loop based at some point of M^n of length $\leq c(n)\,\mathrm{vol}(M^n)^{\frac{1}{n}}$ [169]. The example of a long thin ellipsoid of revolution shows that one cannot have such a bound for the smallest length of a non-trivial geodesic loop based at each point of the manifold. Yet Rotman [166] proved the existence of a non-trivial geodesic loop of length $\leq 2nd$ based at any prescribed point of each closed n-dimensional Riemannian manifold of diameter d. The same estimate holds for the length of the second shortest geodesic between any pair of points on the manifold [145]. Further, for every closed Riemannian manifold M^n , each pair of points $p, q \in M^n$ and each k there exist k distinct geodesics between p and q of length $\leq 4nk^2d$ [141]. Note that the proof of Serre's theorem given by A. Schwartz [172] implies that the lengths of geodesics connecting a fixed pair of points grow at most linearly with k. It is interesting question whether or not one has a linear upper bound of the form $c(n)kd$ for the length of the first k geodesics. Recently, the author and Rotman proved that this is the case for $n = 2$, where the only non-trivial case is when M is diffeomorphic to the 2-sphere. (The proof in the particular case, when the two points coincide, had been published as [147].) Another interesting open question is whether or not one can eliminate the dependance on the dimension n in these upper bounds.

8.3 Gromov's Filling Technique

The upper bounds discussed in Sect. 8.2 are effective versions of topological existence theorems proven using the Morse theory. In all such proofs one constructs a non-trivial homology class in a loop space or a space of cycles on considered manifold. This cycle can be represented by a map of a sphere or a torus to the ambient manifold. A "sweep-out" of this cycle by loops can yield the desired cycle in the loop space. Thus, one is led to the following question: Given a cycle in M^n representing a non-trivial homology class h in the loop space of M^n is it possible to sweep it out by loops of length bounded in terms of the diameter or the volume of the underlying manifold and the information about $[h] \in H_*(\Omega M^n)$? One can also pose this question for spaces of free loops as well as for spaces of cycles in M^n of various dimensions. The answers for these questions are almost invariably negative or (for some spaces of cycles) unknown. (The only known exceptions happen, when $n = 2$. They are discussed in Sect. 4.) Therefore, this approach does not immediately lead to the results mentioned in Sect. 2. However, one frequently uses different versions of the following technique invented by Gromov and used to prove the systolic inequality (cf. [94], pp. 9–10): One embeds M^n in $L^\infty(M^n)$ using the Kuratowski embedding. Then one uses a famous theorem proven by Gromov asserting that one can represent the embedded M^n as a boundary of a $(n+1)$-chain W^{n+1} in $L^\infty(M^n)$ so that the distance from every point of W^{n+1} to M^n does not exceed $a(n)\,\mathrm{vol}^{\frac{1}{n}}(M^n)$ for some explicit constant $a(n)$. (The minimum of $\max_{x \in W^{n+1}} \mathrm{dist}(x, M^n)$ over all fillings W^{n+1} of M^n in $L^\infty(M^n)$ is called the filling radius on M^n and is denoted *FillRad*(M^n). The just mentioned Gromov's theorem is usually written as *FillRad*$(M^n) \leq a(n)\,\mathrm{vol}^{\frac{1}{n}}(M^n)$.) Afterwards one attempts to construct an extension of the classifying map $f : M^n \longrightarrow K$, where K denotes $K(\pi_1(M^n), 1)$, to W^{n+1}. Note that the essentiality of M^n implies that such an extension cannot exist. Assuming that W^{n+1} is endowed with a very finite triangulation, one maps the vertices of this triangulation to the closest points in M^n, and 1-simplices to minimizing geodesics between the images of the endpoints (of length $\leq 2c(n)\,\mathrm{vol}^{\frac{1}{n}}(M^n)$). The next step consists in contracting the already constructed maps of the boundaries of 2-simplices into M^n. These boundaries are represented by closed curves of length $\leq 6a(n)\,\mathrm{vol}^{\frac{1}{n}}(M^n)$. Assume that all closed curves of length $\leq a(n)\,\mathrm{vol}^{\frac{1}{n}}(M^n)$ are contractible. Then this step of the extension becomes trivial. We are not going to have any problems with a further extension of this map to simplices of W^{n+1} of higher dimension as all homotopy groups $\pi_l(K)$, $l > 1$ vanish. So, the desired extension exists, and the resulting contradiction refutes our assumption that all closed curves of length $\leq 6a(n)\operatorname{vol}(M^n)^{\frac{1}{n}}$ are contractible. Now one can take a short non-contractible curve and shrink it to a non-contractible periodic geodesic.

This technique of Gromov can be modified or generalized in many ways. One can summarize the main idea of this technique as gaining a control over geometry of a non-contractible map by attempting an impossible extension. One can try to extend different maps that are known to be non-extendable or try to fill nonnull-homologous cycles by chains. Various versions of Gromov's technique were used to prove many results mentioned in Sect. 8.2. However, note that all modifications of this technique work only if one is capable to contract fairly general maps of spheres in an effective way, or at least has/assumes the existence of contracting homotopies with some good properties.

We would like to finish this section by noting that the proofs of many results mentioned in Sect. 8.2 work like that: One organizes the extension process (in some version of the original Gromov non-existing extension argument) so that the desired "nice" extension to discs/simplices can be blocked only if there exist a desired extremal object with "small" measure. Looking from a different perspective, one separates here between the case when the Riemannian manifold M^n has a "nice" shape, (where the desired extensions are possible, and a desired minimal object exists as a corollary of the original existence proof), and the case, when the Riemannian manifold M^n is "rugged", and this "ruggedness" somehow implies the existence of the desired minimal objects.

8.4 Slicing Riemannian Manifolds

In [94], Appendix A, Gromov defines a number of slicing invariants of Riemannian manifolds and proves several theorems about them (see also [75]). For example, given a Riemannian manifold M and a natural number k one can define the Urysohn width (or diameter), $\operatorname{wid}_k(M)$, (also denoted $\operatorname{diam}_k(M)$), as the infimum of x such that there exists a continuous map f into a k-dimensional polyhedron K such that for each point $p \in K$ the diameter of the inverse image $f^{-1}(p)$ of p under f does not exceed x. Recently L. Guth [104] proved a difficult theorem asserting that for each closed n-dimensional manifold M^n its $(n-1)$-width $\operatorname{wid}_{n-1}(M^n)$ does not exceed $c(n)\operatorname{vol}(M^n)^{\frac{1}{n}}$ for an appropriate constant $c(n)$. This result solves a well-known problem of Gromov (posed in [94]) and is obtained using a refining of Guth's technique from [110]. Note that the Urysohn widths $\operatorname{wid}_i(M^n)$ decrease with i. Earlier G. Perelman proved that one has a stronger inequality $\prod_{i=0}^{n-1}\operatorname{wid}_i(M^n) \leq C(n)\operatorname{vol}(M^n)$ as well as the opposite inequality $\prod_{i=0}^{n-1}\operatorname{wid}_i(M^n) \geq c(n)\operatorname{vol}(M^n)$ for appropriate positive constants $c(n)$ and $C(n)$ in the case when M^n has non-negative sectional curvature. This result of Perelman had been previously conjectured by Gromov in [75].

Gromov observed that Urysohn k-widths measure the (Gromov-Hausdorff) distance between M^n and the nearest k-dimensional metric polyhedron. One of his most well-known open problems involves complete Riemannian manifolds of scalar curvature ≥ 1 and asks whether or not any such manifold has $\mathrm{Wid}_{n-2}(M^n) \leq c(n)$ for some $c(n)$, that is, whether or not each manifold of scalar curvature ≥ 1 is $c(n)$-close to a $(n-2)$-dimensional metric polyhedron ([94], p. 130, [75]).

A different kind of widths, $W_k(M^n)$, is defined by slicing M^n into k-cycles (or, relative k-cycles, if M^n is a Riemannian manifold with boundary), measuring the supremum of k-volumes of the cycles, and passing to the infimum with respect to all such slicings of M^n. In [106] Guth noticed that one cannot estimate $W_{n-1}(M^n)$ in terms of n and $\mathrm{vol}(M^n)$ alone. If $n \geq 3$, this is impossible even if M^n is diffeomorphic to S^n. Moreover, in this case one even cannot majorize the minimal volume of a $(n-1)$-cycle cutting M^n into two domains of volume between $\frac{1}{4}\mathrm{vol}(M^n)$ and $\frac{3}{4}\mathrm{vol}(M^n)$ by $c(n)\,\mathrm{vol}(M^n)^{\frac{n-1}{n}}$. The proof uses the results of Burago and Ivanov from [24]. Yet in the same paper Guth proved that if U is a domain in R^n, then $W_k(U) \leq c(n)\,\mathrm{vol}^{\frac{k}{n}}(M^n)$.

One of the most interesting problems in this direction was posed by Guth in [109]. He asked whether or not $W_1(M^3)$ can be bounded by $\mathit{const}\,\mathrm{vol}^{\frac{1}{3}}(M^3)$, when M^3 is diffeomorphic to the 3-torus, and where *const* can be any constant. In particular, is there always a map from a Riemannian 3-torus of volume one to the plane such that the inverse images of all points of the plane are (not necessarily connected) curves of length less than 10^{10}?

A problem of finding *lower* bounds for $W_k(M^n)$ is also interesting. In particular, Gromov waist inequality implies the optimal lower bound $\mathrm{vol}(S^k)$ for $W_k(S^n)$ where S^n and S^k are round spheres of radius 1 (see [81] and its generalizations in [108]).

When $n=2$, one has many positive results about cutting and slicing of surfaces and fewer unsolved problems. First, Balacheff and Sabourau proved that each closed surface M of genus g can be sliced into (disjoint) 1-cycles (not necessarily connected closed curves) of length $\leq 10^8\sqrt{g+1}\sqrt{\mathrm{Area}(M)}$ [9]. The dependance on g is optimal here. Easy examples show that there are no similar upper bounds if one wants all 1-cycles to be connected, even if M is diffeomorphic to S^2. Liokumovich used Riemannian metrics constructed by Frankel and Katz to prove that, there is no constant c such that each Riemannian 2-sphere M can be sliced or even swept-out by loops of length $\leq c\,\mathrm{diam}(M)$ [131]. Very recently he extended this result to sweep-outs by 1-cycles. (The difference between slicing and sweeping-out is that the intersections are allowed during a sweep-out. One can formally define a sweep-out of a 2-sphere M by free loops as the family of images of parallels of a round S^2 under a map $S^2 \longrightarrow M$ of non-zero degree.) On the other hand it had been recently proven in [132] that each 2-sphere can be sliced into simple closed curves of length $\leq 200d\max\{1, \ln\frac{\sqrt{A}}{d}\}$, where A and d denote its area and diameter; this upper bound is optimal up to a multiplicative constant.

The most well-known open problem about slicing of surfaces (by L. Guth and P. Buser) asks whether one can slice a closed Riemannian surface of genus g and area 1 into not necessarily connected closed curves so that the length of each *connected component* of each curve has length $\leq$*const*, where *const* is an absolute constant that does *not* depend on g. (Of course, the quoted result of [9] implies the existence of a slicing, where the total length of all connected components of each curve does not exceed $10^8\sqrt{g+1}$.)

8.5 Filling Riemannian Manifolds

In Sect. 8.1.2 we have briefly discussed the posed by Gromov problem of finding the "sufficiently small" orientable manifolds bounded by a given orientably null-cobordant manifold. Another class of filling problems of interest for Gromov involves representing Riemannian manifolds M^n as boundaries of Riemannian manifolds or even metric polyhedra W^{n+1} so that the induced distance function on the boundary of W^{n+1} coincides with the original distance function on M^n. We have already mentioned a deep theorem by Gromov asserting that one can find W^{n+1} such that the distance from any point of W^{n+1} to M^n does not exceed $c(n)\,\mathrm{vol}^{\frac{1}{n}}(M^n)$. Further, M. Katz [122] proved that one can choose W so that for each point $x \in W$ $\mathrm{dist}(x, M)$ does not exceed $\frac{1}{3}d$, where d denotes the diameter of M (and the constant $\frac{1}{3}$ is optimal here!) Note that we can isometrically embed M^n into $L^\infty(M^n)$ using the Kuratowski embedding, and observe that arbitrary fillings of M^n inside $L^\infty(M^n)$ have the desired property that the restriction of the distance function of the filling to its boundary coincides with the distance function of M^n. A well-known open problem also posed by Gromov [94] asks whether or not the minimal area of a surface filling a circle of radius 1 in the considered sense is equal to 2π. In other words, is the filling of a circle by the round hemisphere the most "economical", if we would like to ensure that no path between two points of the boundary circle through a filling surface is shorter than a path along the boundary? In this connection we would like to mention a recent result by Burago and Ivanov [27] asserting that for each normed vector space H and any bounded domain in an affine 2-plane inside H, this domain has the smallest 2-dimensional Hausdorff measure among all surfaces in H with the same boundary. This result solves a problem that had been open for fifty years, and it is still not known if a similar result holds for affine k-subspaces with $k > 2$. (See also an earlier paper [25] by the same authors. In particular, Theorems 2 and 4 there illustrate why this and related questions are much more complicated than what one might initially think.)

A related problem is when the distance function on the boundary M^n of a Riemannian manifold W^{n+1} determines the Riemannian metric on W^{n+1}. A Riemannian manifold W^{n+1} with boundary is called *boundary rigid* if each Riemannian manifold U^{n+1} with the same boundary and the same distance function on the boundary is isometric to W^{n+1} via a boundary preserving isometry.

It is not difficult to construct examples of manifolds with boundary that are not boundary rigid, as the distance function is, obviously, not sensitive to any changes in the geometry in the domain of W^{n+1} located at the distance $> \mathrm{diam}(\partial W^{n+1})$ from the boundary. Yet is natural to conjecture that so called *simple* manifolds are boundary rigid (Michel's conjecture). A manifold with boundary is called simple if its boundary is strictly convex, all pairs of points on the boundary are connected by unique geodesics, and these geodesics do not have conjugate points. Michel's conjecture had been proven in dimension two by L. Pestov and G. Uhlmann [157], who used analytical methods, but is widely open for higher dimensions. Recently, Burago and Ivanov proved the boundary rigidity of Riemannian manifolds with boundary that are C^3-close to compact domains with smooth boundaries in either (flat) Euclidean spaces or the hyperbolic space H^n [26, 28]. Their methods are geometric and, in particular, use ideas from [94], but the proofs are quite involved.

9 Geometric Group Theory (by Mladen Bestvina)

Traditionally, combinatorial group theory studies finitely presented groups from the point of view of the combinatorics of the presentation. In his work, Gromov introduced a metric point of view in group theory.

We quote [80] for Gromov's thoughts on this.

> *The idea of negative curvature was injected into group theory by Dehn and grew into small cancellation theory. In the course of development, the geometric roots were forgotten and the role of curvature was reduced to a metaphor. Algebraists do not trust geometry. It eventually turned out that the geometric language of Dehn and Alexandrov accomplishes many needs of combinatorial group theory more efficiently than the combinatorial language. Summing up, geometry furnishes a proper language, while combinatorial group theory (especially random groups) provides a pool of objects for a meaningful usage of this language.*

Gromov's papers are best enjoyed by opening up a random page and thinking about the mathematics there. In this overview we will attempt to give the reader a flavor of Gromov's approach to group theory.

The basic concept in what follows is that of the word metric on a finitely generated group and the associated Cayley graph. Let G be a group and A a finite generating set. It will be convenient to assume that A is symmetric, i.e., that $A^{-1} = A$. The *word length* $|g|$ of $g \in G$ is the smallest n such that $g = a_1 a_2 \cdots a_n$ for some $a_i \in A$. The *word metric* on G is defined as $d(g, g') = |g^{-1}g'|$; it is invariant under left translations. The associated *Cayley graph* X has vertex set G and two vertices are joined by an edge if the distance between them is 1. Assigning length 1 to each edge turns X into a geodesic metric space (the distance between two points equals the length of a shortest path joining them; such paths are *geodesics*).

9.1 Groups of Polynomial Growth

A major theme in Gromov's work is rescaling and passing to (Gromov–Hausdorff) limit. These ideas play a major role in the stunning theorem that can be viewed as the birth of modern geometric group theory.

Theorem 26 [93] *Let G be a finitely generated group of polynomial growth. Then G contains a nilpotent subgroup of finite index.*

The assumption on polynomial growth means that there is a polynomial $P(x)$ such that for every $R \geq 1$ the number of vertices in the ball of radius R in the Cayley graph X is bounded by $P(R)$. Changing the finite generating set may require replacing P by another polynomial. It was known earlier (a theorem of Wolf) that finitely generated nilpotent groups have polynomial growth, and that finite index subgroups have equivalent growth functions.

We will outline a proof below. The following are the main ingredients:

- (Milnor–Wolf) If Γ is finitely generated and solvable and has subexponential growth, then Γ is virtually nilpotent.
- If Γ is finitely generated and has subexponential growth, and

$$1 \to K \to \Gamma \to \mathbb{Z} \to 1$$

  is exact, then K is finitely generated. If in addition Γ has polynomial growth of degree $\leq d$, then K has polynomial growth of degree $\leq d-1$.
- Let Γ be finitely generated and have polynomial growth. Then there is a metric space Y and a homomorphism $\ell : \Gamma \to \mathrm{Isom}(Y)$ to the isometry group of Y so that:

  (1) Y is homogeneous, i.e., $\mathrm{Isom}(Y)$ acts transitively on Y.
  (2) Y is connected, locally connected, complete, locally compact, finite dimensional. The metric on Y is proper (closed balls are compact).
  (3) If $\ell(\Gamma)$ is finite and Γ is not virtually Abelian, then $\Gamma' = \mathrm{Ker}(\ell)$ has, for each neighborhood U of the identity in $\mathrm{Isom}(Y)$, a representation $\rho : \Gamma' \to \mathrm{Isom}(Y)$ so that $\rho(\Gamma') \cap U \neq \{id\}$.

- (Gleason–Montgomery–Zippin) The above $\mathrm{Isom}(Y)$ is a Lie group with finitely many components.
- (Tits alternative) A finitely generated subgroup of $GL_n(\mathbb{R})$ either contains a non-abelian free group (and hence has exponential growth) or is virtually solvable (and hence is either virtually nilpotent or has exponential growth).
- (Jordan's theorem) For every n there is q so that every finite subgroup of $GL_n(\mathbb{R})$ contains an Abelian subgroup of index $\leq q$.

Proof (Outline of proof of Gromov's theorem) By the fifth bullet, there is nothing more to be done if Γ is a linear group. In fact, by Milnor–Wolf, induction on degree, and the second bullet, it suffices to construct an epimorphism $\Gamma' \to \mathbb{Z}$ for a finite index subgroup $\Gamma' \subset \Gamma$. If $\ell(\Gamma)$ is infinite, it is virtually nilpotent by the fifth bullet, and it virtually maps onto $\mathbb{Z}$ by definition of nilpotent groups. Suppose $\ell(\Gamma)$ is finite. Let L be the identity component of $\mathrm{Isom}(Y)$. We now have:

Claim *There is a subgroup $\Delta \subset \Gamma$ of finite index that admits homomorphisms to L with arbitrarily large images.*

To prove the claim, the key is that for all N there is a neighborhood $U \subset L$ of the identity so that no nontrivial element in U has order $\leq N$. Then apply (3) to get homomorphisms $\Gamma' \to \mathrm{Isom}(Y)$ with arbitrarily large images. Some subgroup Δ of index $\leq$ the number of components of $\mathrm{Isom}(Y)$ maps to L infinitely often.

To finish the argument, if $L \subset GL_n(\mathbb{R})$ is a linear group, Jordan's theorem implies that Δ has infinite abelianization and we are done. If L is not linear, replace it with $L/Z(L)$. □

9.2 Gromov–Hausdorff Limits

There are three ways of passing from the discrete group $\mathbb{Z}^n$ to Euclidean space $\mathbb{R}^n$. An algebraist might tensor with $\mathbb{R}$. A topologist would "connect the dots", i.e., build a $\mathbb{Z}^n$-invariant contractible complex by attaching cells of dimension >0. Finally, a geometer, following Gromov, would rescale and pass to the limit. It is tempting to think that this procedure can be discovered by thinking about the Bieberbach theorem that any torsion free group that contains $\mathbb{Z}^n$ as a subgroup of finite index acts freely and cocompactly by isometries on $\mathbb{R}^n$.

Suppose we start with $\Gamma = \mathbb{Z}^2$ and the usual metric on Γ. Now "stand twice as far" from Γ, i.e., consider the same group but with metric divided by 2. Continue "moving away" i.e., rescaling the metric. The result is a sequence of denser and denser lattices in $\mathbb{R}^2$, and the limiting object is $\mathbb{R}^2$.

This is Gromov's signature construction, going from discrete to continuous, and we proceed to describe it in more detail.

Let X be a metric space. When $A \subset X$ we denote

$$N_r(A) = \{x \in X \mid d(x, A) < r\}$$

and when $A, B \subset X$ the *Hausdorff distance* is

$$d_H(A, B) = \inf\{r > 0 \mid N_r(A) \supset B, N_r(B) \supset A\}.$$

If X is a compact metric space, the space of nonempty closed subsets with Hausdorff metric is a compact metric space.

Now let A, B be two compact metric spaces. The *Gromov–Hausdorff distance* $d_{GH}(A, B)$ is defined to be the infimum of numbers $R > 0$ such that there are isometric embeddings of A, B into some metric space X with $d_H(A, B) < R$. It turns out that $d_{GH}(A, B) = 0$ implies that A and B are isometric. We say a sequence A_i converges to A if $d_{GH}(A_i, A) \to 0$.

As an example, consider the Hopf fibration $h : S^3 \to S^2$. Choose a Riemannian metric on S^2 and on S^3 so that at every point the derivative h_* is an isometry when restricted to the subspace orthogonal to the fiber. Then for every $t > 0$ define the metric g_t on S^3 by scaling the fiber direction by t. As $t \to 0$ the sequence (S^3, g_t) converges to S^2. In a similar way, one can construct a sequence of metrics on the Möbius band converging to a closed interval.

It is not true that every sequence of compact metric spaces has a subsequence that converges. For example, every convergent sequence has uniformly bounded diameters. There is a more subtle obstruction to convergence. Recall that a metric space is *totally bounded* if for every $\varepsilon > 0$ there is a finite cover by ε-balls. We say that a sequence of metric spaces is *uniformly totally bounded* if for every $\varepsilon > 0$ there is $N = N(\varepsilon)$ such that every space in the sequence admits a cover by N ε-balls. For example, any convergent sequence is uniformly totally bounded.

Theorem 27 (Gromov's Compactness Criterion) *A uniformly totally bounded sequence of compact metric spaces has a convergent subsequence.*

There is a version of Gromov–Hausdorff convergence with basepoints. The distance $d_{GH}((A,a),(B,b))$ is the infimum of numbers $R > 0$ such that there are isometric embeddings $A, B \hookrightarrow X$ so that $d_H(A, B) < R$ and $d(a,b) < R$.

We have to deal with noncompact metric spaces, but with a preferred basepoint. If Y_i, Y are metric spaces in which closed balls are compact and $y_i \in Y_i$, $y \in Y$ are basepoints, define $(Y_i, y_i) \to (Y, y)$ to mean that for every $R > 0$ the sequence of balls $B(y_i, R)$ converges to $B(y, R)$ (as compact metric spaces with basepoints).

For example, letting Y_i be $\frac{1}{i}\mathbb{Z}^n$, i.e., $\mathbb{Z}^n$ with the standard word metric scaled by $1/i$, we obtain a sequence that converges to $\mathbb{R}^n$ with ℓ^1-metric (with origin as basepoints).

More interestingly, consider the (discrete) Heisenberg group H, the group of upper triangular 3×3 matrices with integer entries and 1's on the diagonal. This is a nilpotent group, a lattice in a nilpotent Lie group diffeomorphic to $\mathbb{R}^3$. Rescaling a word metric as above produces a convergent sequence, and the limit is a metric space Y homeomorphic to $\mathbb{R}^3$ but with the *Carnot–Caratheodory metric*, whose Hausdorff dimension is 4. To understand this phenomenon better, it is useful to note that the center of H, i.e., the group

$$Z = \begin{pmatrix} 1 & 0 & * \\ 0 & 1 & 0 \\ 0 & 0 & 1 \end{pmatrix}$$

is *quadratically distorted*, that is, the length (in H) of the element that has N in the upper right corner is $\sim \sqrt{|N|}$. For example, the commutator of $\begin{pmatrix} 1 & k & 0 \\ 0 & 1 & 0 \\ 0 & 0 & 1 \end{pmatrix}$ and $\begin{pmatrix} 1 & 0 & 0 \\ 0 & 1 & k \\ 0 & 0 & 1 \end{pmatrix}$ is $\begin{pmatrix} 1 & 0 & k^2 \\ 0 & 1 & 0 \\ 0 & 0 & 1 \end{pmatrix}$.

The quotient H/Z is isomorphic to $\mathbb{Z}^2$. It is then not surprising that the limiting space Y fibers over $\mathbb{R}^2$ with fiber $\mathbb{R}$ equipped with the metric $d(x, y) = \sqrt{|x - y|}$ whose Hausdorff dimension is 2, causing Y to have Hausdorff dimension 4.

On the other hand, if X is the infinite 3-valent tree or hyperbolic plane it is not hard to see that the sequence of rescalings $\frac{1}{i}X$ does not have a convergent subsequence. This follows from volume considerations. In both examples, volume of a ball of radius R is exponential in R, so the number of balls of radius R required to cover a ball of radius $2R$ is $\sim\frac{e^{2R}}{e^R} = e^R$, and so the sequence of 2-balls in $\frac{1}{i}X$ is not uniformly totally bounded.

When G is a group of polynomial growth, Gromov shows that there is a sequence of carefully chosen scales r_i so that the sequence $\frac{1}{r_i}G$ converges to a metric space Y. We will not go into the details of the third bullet.

9.3 Groups as Metric Spaces and Quasi-isometries

A *quasi-isometric embedding* of one metric space to another is a function $f : X \to Y$ such that for some $K \geq 1$, $L \geq 0$ the following inequality holds for all $x, x' \in X$:

$$\frac{1}{K}d(x,x') - L \leq d(f(x), f(y)) \leq Kd(x,x') + L.$$

We say f is a *quasi-isometry* if, in addition, there is $R > 0$ so that every R-ball in Y intersects the image of f. Thus when x, x' are at a small distance the above carries essentially no information. On large scales f behaves like a K-Lipschitz map. Note that there is no requirement that f is continuous. For example, $\mathbb{Z}$ is quasi-isometric to $\mathbb{R}$, and two different word metrics on the same group are quasi-isometric. "Being quasi-isometric" is an equivalence relation. In his ICM address in Warsaw in 1983 [95] (see also [72]) Gromov proposes to study groups up to quasi-isometry, e.g., to look for invariants of quasi-isometries or to classify finitely generated groups quasi-isometric to some model space. Here are some examples, given by Gromov in [95]:

- Any group quasi-isometric to a virtually nilpotent group is virtually nilpotent (this follows from Gromov's polynomial growth theorem since "having polynomial growth" is a quasi-isometry invariant).
- Any group quasi-isometric to $\mathbb{Z}^n$ is virtually $\mathbb{Z}^n$.
- Any torsion free group quasi-isometric to a free group is free (follows from the celebrated Stallings' splitting theorem about groups with infinitely many ends).
- Any torsion free group quasi-isometric to $\mathbb{H}^n$, $n \geq 2$, acts cocompactly and properly by isometries on $\mathbb{H}^n$ (follows from the work of Mostow, Sullivan, Tukia).

In [77] Gromov offers a rich sample of the subject and fertile ideas that have served as inspiration to many researchers.

9.4 CAT(−1) and CAT(0) Spaces

A theme in Gromov's work is a study of singular spaces. Traditionally, one seeks to resolve singularities by smooth objects. Gromov's philosophy is to live with them and take advantage of them. The notion of negative curvature for singular spaces (i.e., not manifolds) is crucial in geometric group theory.

Gromov used comparison inequalities from differential geometry to *define* curvature for complete geodesic metric spaces (those where any two points are joined by a geodesic, i.e., path of length equal to the distance between the endpoints). Given such a space X, say that it is $CAT(0)$ (or $CAT(-1)$) if every geodesic triangle abc satisfies the *comparison inequality*: let $a'b'c'$ be a triangle in Euclidean (or hyperbolic) plane with same side lengths as abc. Let $p \in [a,b]$, $q \in [a,c]$ and let $p' \in [a',b']$, $q' \in [a',c']$ be the corresponding points in the comparison triangle. Then $d(p,q) \leq d(p',q')$. Thus triangles are thinner in X than in the comparison plane. If X is a complete simply connected Riemannian manifold, then X is $CAT(0)$ ($CAT(-1)$) precisely when the sectional curvature is ≤ 0 (≤ -1). Thus, this notion captures local negative curvature, while δ-hyperbolicity is an *asymptotic* concept. For more information about *CAT* geometry see [22]. The acronym *CAT* stands for Cartan–Aleksandrov–Toponogov (in French, one adds Hadamard and *CAT* becomes *CHAT*). A version of the Cartan–Hadamard theorem is that $CAT(0)$ spaces are contractible.

9.5 Hyperbolization of Polyhedra

Important examples are provided by cubical complexes. Such a complex X is $CAT(0)$ if and only if it is simply connected and all links (which are naturally simplicial complexes) are *flag*, i.e., if the 1-skeleton of a simplex is in the link, the whole simplex is in the link (or, the way Warren Dicks likes to put it, every non-simplex contains a non-edge). Gromov offered several simple constructions (going by the name "hyperbolization") that turn a connected finite polyhedron to another polyhedron P_h, whose universal cover is $CAT(0)$. Here is one such construction. Assume P is a finite cubical complex. If $\dim P \leq 1$ then let $P_h = P$. If $\dim P = 2$ replace each 2-cell $\square^2$ (a square) by the Möbius band $\square^2_h = (\partial\square^2 \times [-1,1])/\mathbb{Z}_2$ where $\mathbb{Z}_2$ acts diagonally and by $x \mapsto -x$ in each coordinate. The boundary of $\square^2_h$ can be identified with $\partial\square^2$ and we use the same attaching map. So for example, this will turn the boundary of the 3-cube to a (nonorientable) surface of Euler characteristic $\chi = -4$. This surface has a natural fixed point free involution induced by the antipodal map on the cube, and one continues inductively on dimension. Define

$$\square^k_h = \left(\left(\partial\square^k\right)_h \times [-1,1]\right)/\mathbb{Z}_2$$

so for example in dimension 3 one obtains a compact 3-manifold M whose boundary is the above surface with $\chi = -4$ and it is totally geodesic in M. If P has dimension k, first hyperbolize the $(k-1)$-skeleton P^{k-1}, and then for each k-cube in P glue in $\square^k_h$ to the hyperbolized boundary of the cube.

Here is a topological application. It was the first general construction of closed aspherical manifolds (another construction can be made, à la Mike Davis, using Coxeter groups).

Theorem 28 *Every closed PL manifold M is cobordant to an aspherical manifold.*

Proof Let $P = cM$ be the cone on M and cubulate P. The hyperbolization P_h is a manifold away from the cone point c, whose link is M, and the boundary of P is totally geodesic, so it is aspherical. Removing a small neighborhood of c gives the desired cobordism. □

Gromov also has hyperbolization procedures that preserve orientability of manifolds. This construction should be compared to the earlier Kan–Thurston asphericalization, which does not preserve manifolds (but does preserve homology).

9.6 Hyperbolic Groups

Before Gromov's paper [74] one studied fundamental groups of closed Riemannian manifolds of negative curvature. It was becoming increasingly clear, particularly in the work of Cannon [31] that there is a more abstract concept, depending only on the Cayley graph of a group. Gromov made the definition in [74], which evolved from his ICM Warsaw address [95] in 1983 (see also [72, 6.4]). For more information, see [67] and [22].

The Cayley graph X of a group G (and the group G) is *δ-hyperbolic* for some $\delta > 0$ if for any three vertices $a, b, c \in X$ and any three geodesics $[a, b]$, $[b, c]$, $[a, c]$ between them, every vertex of $[a, c]$ is within δ of some vertex in $[a, b] \cup [b, c]$. If a group is δ-hyperbolic with respect to one finite set of generators, then for any other choice of generators the Cayley graph is δ'-hyperbolic for a suitable $\delta' > 0$, so it makes sense to talk about a finitely generated hyperbolic group, without specifying a generating set (or δ). Fundamental groups of negatively curved closed manifolds are hyperbolic, and so are free groups, while $\mathbb{Z}^2$ is not. Likewise, if G acts as a cocompact covering group on a $CAT(-1)$ space X then G is hyperbolic.

Gromov sketches a whole theory of hyperbolic groups. A driving force is the "Morse lemma" that every quasi-geodesic in X is a uniform distance away from a geodesic. This property fails in Euclidean plane.

For example, any hyperbolic group acts properly and cocompactly on a simplicial complex. Gromov attributes the following construction to Rips. Fix $d \geq 4\delta + 1$, and let $P_d(X)$ be the simplicial complex with vertex set G and with $g_1, \ldots, g_n$ spanning a simplex if $d(g_i, g_j) \leq d$ for all i, j (the distance is in X). Then $P_d(X)$ is contractible and G acts by left multiplication cocompactly and properly discontinuously. In particular, G is finitely presented (and if it is torsion-free it has a compact classifying space).

Every hyperbolic group G has a *space at infinity*, or a *boundary* ∂G. A point of ∂G is an equivalence class of geodesic rays in X, where two rays are equivalent if they are within bounded distance of each other. There is a natural topology on the boundary, making it into a compact metrizable space. In the case of negatively curved closed manifolds one always gets a sphere as the space at infinity, but in the world of general hyperbolic groups there is a rather rich supply of spaces that can occur. This may be a particularly attractive feature of the subject that persuaded so many topologists to study it.

Gromov's motivation for thinking about hyperbolic groups may have been his desire to understand, from a geometric point of view, why Burnside groups $B(n) = \langle x, y \mid w^n = 1\rangle$ where $w = w(x, y)$ runs over all words in x, y, are infinite for large n (work of Novikov-Adyan and Ivanov). In [74] he gives a construction of infinite torsion groups, as follows. Start with a hyperbolic group, e.g., a nonabelian free group. Pass to the quotient by killing a suitable high power of an infinite order element, and argue that it is also hyperbolic. Repeat the process infinitely many times; the group in the limit (given correct choices of words to kill) is a finitely generated infinite torsion group. Gromov laments that geometry is not yet developed sufficiently to construct such groups with order of elements *bounded*. This was done a bit later by Olshanskii [152] and Ivanov–Olshanskii [119], but it is likely that only [40] put the matter to rest in Gromov's mind.

A variation of this construction is to kill "sufficiently long, random" words instead of high powers. With this, one can show for example that every hyperbolic group (which is not virtually cyclic) has an infinite quotient which is hyperbolic and has Kazhdan's property (T). See [153] for details.

Here is the intuition. Say X is a $CAT(-1)$ space and G a deck group of a covering map $X \to X/G$ with quotient compact. Let $g \in G$ be an element of infinite order, and represent g as closed geodesic ℓ in X/G. Assume in addition that ℓ passes through a singular vertex $v \in X/G$ so that in the link $Lk(v)$ the distance between the two directions determined by ℓ are at distance $>\pi$. For $n > 1$ form a cone by taking a geodesic triangle ABC in $\mathbb{H}^2$ such that the angle at A is $2\pi/n$ and the sides AB and AC have lengths 1. Then identify AB to AC. This is a cone whose base is not convex, but the angle at $B = C$ can be made arbitrarily close to π by making n large. Scale the metric up so that the base has the same length as ℓ and glue it to ℓ so that $B = C$ gets glued to v. The resulting space is no longer $CAT(-1)$, but it is $CAT(-\varepsilon)$ for a suitable $\varepsilon > 0$. It is also not a polyhedron, but an *orbihedron*, that is, the cone point corresponds to a point fixed by $\mathbb{Z}_n$. Its orbihedron fundamental group is $G/\langle\langle g^n\rangle\rangle$. For a modern version of this kind of a construction, see [37].

Hyperbolic groups, cube complexes and the construction above (in the form of *Dehn fillings*) play a crucial role in the recent solution of the virtual Haken conjecture about 3-manifolds by Wise and Agol, see [2].

9.7 Isoperimetric Functions

Suppose G is a finitely presented group with generators $a_1, \dots, a_n$ and relators $R_1, \dots, R_m$. It is convenient to assume that inverses of generators (relators) are generators (relators). A "loop" of length L is any word $W = a_{i_1} a_{i_2} \cdots a_{i_L}$ that represents the identity in G. Thus one has an identity in the free group

$$W = \prod_{s=1}^{M} w_s R_{j_s} w_s^{-1}$$

for some M and some choice of relators and words w_s. The smallest possible M is called the *area* of W. By $\varphi(L)$ denote the largest area of any loop of length $\leq L$. Then φ is called an *isoperimetric function* or a *Dehn function* for G. If φ' is an isoperimetric function of G with respect to a different presentation, then

$$\varphi'(L) \leq a\varphi(bL+c)+d$$

for suitable constants $a, b, c, d > 0$. This leads to a natural equivalence relation among functions $\mathbb{N} \to [0, \infty)$, and one talks e.g., about linear, or quadratic, or polynomial, or exponential isoperimetric functions. For example, $\mathbb{Z} \times \mathbb{Z}$ has a quadratic isoperimetric function.

It is essentially due to Dehn that hyperbolic groups have a linear isoperimetric function. Gromov proved the converse, that a finitely presented group G is hyperbolic if and only if it has a linear isoperimetric function. What's more, if G has an isoperimetric function of the form $aL^2 + b$ for a sufficiently small $a > 0$, then G is hyperbolic. There are countably many exponents α so that L^α is (equivalent to) an isoperimetric function; the set of these α is 1 together with a dense subset of $[2, \infty)$ (see [15, 21, 170]). The interval $(1, 2)$ is the *Gromov gap*. Gromov also has a sharp version of the statement that can be used to algorithmically detect δ-hyperbolicity. The following is Papasoglu's formulation [155].

Theorem 29 *Let $G = \langle a_1, \ldots, a_n \mid R_1, \ldots, R_m \rangle$ be a triangular presentation of G (i.e., all relators have length 3). Assume that for some integer $K > 0$ every word w with area in the interval $[K^2/2, 240K^2]$ has area bounded by $\frac{1}{20000}L^2$, where L is the length of w. Then K^2L is un upper bound for the isoperimetric function.*

Constants such as 20000 are the so called *Gromov constants*. They appear in statements where one is not particularly concerned with sharp inequalities. They are useful when different constants in an argument appear with different orders of magnitude, making it easier to follow proofs.

Isoperimetric functions in groups measure the complexity of the word problem: find an algorithm that decides whether a given a word in the generators represents the trivial element of the group. If G is the fundamental group of a closed Riemannian manifold M, then the universal cover $\tilde{M}$ also has a natural isoperimetric function ψ, namely $\psi(L)$ is the smallest number such that every loop of length $\leq L$ bounds a disk of area $\leq \psi(L)$. What makes this concept particularly appealing is that ψ is equivalent to the combinatorial isoperimetric function.

9.8 L_2-Cohomology

There are many equivalent definitions of amenability. For the purposes of this section, we say a group G that acts cocompactly and properly discontinuously on a connected (simplicial) complex is *amenable* if there is a sequence F_j of finite subcomplexes such that

$$\frac{|\partial F_j|}{|F_j|} \to 0$$

where $|A|$ denotes the number of cells in a finite complex A. For example, $\mathbb{Z}^2$ is amenable since it acts on $\mathbb{R}^2$ in the usual way, and for F_j one can take a $j \times j$ square, when $|F_j|$ grows like j^2 while $|\partial F_j|$ grows like j. Likewise, it is not hard to see that nonabelian free groups are not amenable. For example, in a 3-valent tree any finite subtree T must have more vertices of valence 1 than of valence 3 leading to the inequality $\frac{|\partial T|}{|T|} > \frac{1}{2}$.

Theorem 30 [33] *Let Y be a finite complex with universal cover X contractible and with $G = \pi_1(Y)$ amenable. Then the Euler characteristic $\chi(Y) = 0$.*

A theorem of this nature was proved by Gottlieb [69]: If instead of amenability one assumes that G has nontrivial center the same conclusion holds. Cheeger-Gromov go on to prove a generalization of Gottlieb's theorem: if G contains an infinite normal amenable subgroup then $\chi(Y) = 0$.

Let

$$0 \to C_2^0(X) \to C_2^1(X) \to \cdots$$

denote the cochain complex of square summable cochains on X, i.e., $c \in C_2^p(X)$ is a real valued function on p-cells e^p of X such that $\sum_{e^p} c(e^p)^2 < \infty$. Each $C_2^p(X)$ is a Hilbert space and G acts by orthogonal transformations. The coboundary operator $d : C_2^p(X) \to C_2^{p+1}(X)$ has an adjoint $d^* : C_2^{p+1}(X) \to C_2^p(X)$ (i.e., $\langle dg, h\rangle = \langle g, d^*h\rangle$). A cocycle u is *harmonic* if $d^*(u) = 0$. The subspace of harmonic p-cocycles is denoted $\mathscr{H}^p$. It is canonically isometric to L^2-cohomology

$$\mathrm{Ker}\big(d : C_2^p(X) \to C_2^{p+1}(X)\big)/\overline{\mathrm{Im}\big(d : C_2^{p-1}(X) \to C_2^p(X)\big)}.$$

Any G-invariant closed subspace V of $l^2(G)^n$ has a *von Neumann dimension* $\dim_G(V)$, a real number in $[0, n]$, defined as the trace of the orthogonal projection $\Pi_V : l^2(G)^n \to l^2(G)^n$ onto V. Each $C_2^p(X)$ is a Hilbert space of the form $l^2(G)^n$ and so $\mathscr{H}^p$ has a dimension. By the standard Hopf argument the alternating sum $\sum_p (-1)^p \dim_G \mathscr{H}^p = \chi(X)$.

The trace of a G-equivariant bounded operator $F : l^2(G)^n \to l^2(G)^n$ is simply $\sum_{i=1}^n \langle 1_i, F(1_i)\rangle$ where 1_i denotes $1 \in l^2(G)$ in the ith coordinate (the function that assigns $1 \in \mathbb{R}$ to $1 \in G$ and 0 to all other $g \in G$).

Proof We will argue that $\dim_G \mathscr{H}^p = 0$ for all p. Fix a (strict) fundamental domain $D \subset X$ and Følner tiles F_j. Assume for simplicity that F_j is a union of say N_j

translates of D. Then, denoting by χ_{e^p} the characteristic function of e^p,

$$\dim_G \mathscr{H}^p = \mathrm{tr}(\Pi_{\mathscr{H}^p}) = \sum_{e^p \in D} \langle \Pi_{\mathscr{H}^p}(\chi_{e^p}), \chi_{e^p} \rangle = \frac{1}{N_j} \sum_{e^p \in D} \langle \Pi_{\mathscr{H}^p}(\chi_{e^p}), \chi_{e^p} \rangle$$
$$\leq \frac{1}{N_j} \dim \mathrm{Im}(g_j)$$

where $g_j : C^p(F_j) \to C^p(F_j)$ is the composition of $\Pi_{\mathscr{H}^p}$ with the restriction to F_j, and the last inequality comes from the fact that $|g_j| \leq 1$.

Now the key claim is that cochains in this image are completely determined by their values on p-cells that intersect the boundary of F_j. This implies that $\dim_G \mathscr{H}^p \leq \frac{n_{pj}}{N_j}$ where n_{pj} is the number of p-cells that intersect the boundary of F_j, and this number goes to 0 by the choice of Følner tiles.

To prove the claim, assume that h is a harmonic p-cocycle that vanishes on p-cells that intersect the boundary of F_j. We will also assume that $h = 0$ outside F_j and argue that $h = 0$. Since X is contractible, $h = dg$ for some (not necessarily square summable) $(p-1)$-cochain g. Let g' be a p-cochain of finite support that agrees with g on all cells that intersect F_j. Then

$$\langle h, h \rangle = \langle dg, h \rangle = \langle dg', h \rangle = \langle g', d^*h \rangle = \langle g', 0 \rangle = 0$$

so $h = 0$. □

9.9 *Random Groups*

In [74] Gromov states that most groups are hyperbolic. He develops the subject further in [77]. New applications are found in [102] and [82]. For more information on this subject see [68] and [150].

Perhaps the main application of the theory of random groups (other than making phrases such as "random groups are hyperbolic" precise) is a construction of groups with surprising properties. The best example of this is Gromov's construction of a finitely presented group G that does not admit a uniform embedding in Hilbert space. This means that there is no $f : G \to \ell^2$ and a function $\Phi : \{0, 1, 2, 3, \ldots\} \to [0, \infty)$ such that $\Phi(d(x, y)) \leq |f(x) - f(y)| \leq d(x, y)$ for all $x, y \in G$ and so that $\Phi(n) \to \infty$ as $n \to \infty$. By a theorem of Yu groups that admit uniform embedding in ℓ^2 satisfy the coarse Baum-Connes conjecture, and in fact Gromov's groups provide counterexamples to certain generalizations of this conjecture. See [112].

9.9.1 Model with a Fixed Number of Relators

The most straightforward concept of a "random group" is the following. Fix two numbers $k \geq 2$ and $r \geq 1$ and for l large consider presentations

$$G = \langle a_1, a_2, \ldots, a_k \mid R_1, R_2, \ldots, R_r \rangle$$

where R_i are *reduced cyclic words* in $a_i^{\pm 1}$ of length l. Denote the number of such presentations by $N(k,r,l)$. Let $N_{hyp}(k,r,l)$ be the number of such presentations that define hyperbolic groups.

Theorem 31 [74, 153]

$$\lim_{l\to\infty} \frac{N_{hyp}(k,r,l)}{N(k,r,l)} = 1.$$

In fact there is a version of this theorem where one allows lengths of relators to go to infinity at different rates.

9.9.2 Density Model

A more sophisticated model of randomness appears in [77], the so called "density model". When X is a finite set and A a subset, we say the *density* of A in X is

$$dens(A) = \frac{\log |A|}{\log |X|}$$

and the *codensity* is

$$codens(A) = 1 - dens(A).$$

Then it turns out that for randomly chosen sets $A, B \subset X$ we have

$$codens(A \cap B) = codens(A) + codens(B)$$

(where $codens(A) > 1$ is interpreted as $A = \emptyset$). For example, two random subsets of density $<\frac{1}{2}$ are disjoint, meaning: if $d < \frac{1}{2}$ and X has size l, the probability $P(l,d)$ that two subsets of X of size $\sim l^d$ are disjoint goes to 1 as $l \to \infty$. The "birthday paradox" can be interpreted in this language as follows: If $A \subset X$ has density $<\frac{1}{2}$ a random function $A \to X$ is injective; if $d > \frac{1}{2}$ it is not injective.

Now fix $k \geq 2$ as before and let $S(l)$ be the set of reduced words in $a_i^{\pm 1}$ of length l. Fix d, $0 < d < 1$, and consider a random subset $\mathscr{R} \subset S(l)$ of density d (of course, $|S(l)|^d$ is likely not an integer; just take a nearest integer and let $\mathscr{R}$ have that cardinality).

We say that a property P of groups *occurs with overwhelming probability at density* d if its probability goes to 1 as $l \to \infty$. There are also variations on the above model. For example, instead of considering relators of length exactly l one could look at some interval $(l-c, l+c)$ for a fixed c.

Theorem 32 *Let $G = \langle a_1, a_2, \ldots, a_k \mid \mathscr{R} \rangle$ with R randomly chosen as above with density d. Then with overwhelming probability:*

- *G is trivial or $\mathbb{Z}_2$, if $d > \frac{1}{2}$,*

- *G is hyperbolic and not virtually cyclic, if* $d < \frac{1}{2}$,
- *G satisfies Kazhdan's property (T), if* $d > \frac{1}{3}$,
- *G does not satisfy property (T), if* $d < \frac{1}{5}$,
- *G is a-T-menable and acts freely and cocompactly on a CAT(0) cubical complex, if* $d < \frac{1}{6}$,
- *G is small cancellation* $C'(1/6)$, *if* $d < \frac{1}{12}$ *G is small cancellation* $C'(1/6)$ *(see below).*

The first two statements are due to Gromov, the third to Żuk [195], and the rest to Ollivier and Wise (see [150] for a detailed discussion).

We only indicate the proof of the first statement. Consider the map on the set of relators that forgets the last letter. Since $d > \frac{1}{2}$ this map is not injective (with overwhelming probability). It follows that in G the last two letters are identified. In fact, with overwhelming probability, all of $a_i^{\pm 1}$ are identified, giving the result.

9.9.3 Graph Model

First recall the classical small cancellation theory. Given a group presentation $G = \langle a_1, \ldots, a_k \mid \mathscr{R} \rangle$ one says a *piece* P is a word in $a_i^{\pm 1}$ such that P appears more than once (with either orientation) in cyclic words in $\mathscr{R}$. For $\lambda \in (0,1)$ the presentation is $C'(\lambda)$ if whenever p is a piece contained in some R_i then $|P| < \lambda |R_i|$. Groups with a $C'(1/6)$ presentation are all hyperbolic.

If one has a presentation of the form $\langle a, b, c \mid R_1 = R_2 = R_3 \rangle$ where R_i are say of length 100, one could rewrite it as $\langle a, b, c \mid R_1 R_2^{-1}, R_1 R_3^{-1} \rangle$ but this will usually not be of type $C'(1/6)$. There is a natural extension of small cancellation theory, introduced by Gromov in [82]; here is a special case of that extension. Let Γ be a finite graph (for the above example take the theta graph) and label edges by words in $a_i^{\pm 1}$ (with R_1, R_2, R_3 in the above example). We are imagining subdividing each edge of Γ and labeling each subdivision edge with $a_i^{\pm 1}$ so that the given word reads along the original edge (and the inverse of the word reads in the opposite direction). This labeling satisfies $C'(\lambda)$ if whenever a labeled segment is embedded in two different ways in Γ, its length is $< \lambda r_0$ where r_0 is the injectivity radius of Γ. We are tacitly assuming that the labeling is *reduced* i.e., every embedded path reads a reduced word. It now turns out, not surprisingly, that

- for any $\lambda > 0$ random labeling by sufficiently long words (say of roughly the same length) will be $C'(\lambda)$,
- if the labeling is $C'(1/6)$ then the associated group (obtained from the free group by moding out words read along closed paths) is hyperbolic,
- moreover, with overwhelming probability the graph Γ will be isometrically embedded in the Cayley graph of the quotient group.

Now one wants to continue the process and pass to a further quotient. It is necessary to first generalize small cancellation theory to allow labeling with elements of a hyperbolic group, and we will not go into the details. The interested reader should consult [68] and [151]. The result of an infinite sequence of steps is a finitely generated group G whose Cayley graph contains quasi-isometrically embedded copies of a pre-chosen infinite family of finite graphs. Choosing this family to be a sequence of expanders, one obtains a group that does not admit a uniform embedding in Hilbert space. With care one can arrange that this group is a subgroup of a finitely presented group, which then does not uniformly embed in Hilbert space.

10 Gromov's Work on Manifolds of Positive Scalar Curvature (by John Roe)

10.1 Introduction

On a Riemannian manifold of dimension n, the curvature tensor is a rather complicated geometric object, having $n^2(n^2-1)/12$ components at each point. The Riemann tensor determines and is determined by the *sectional curvatures* of the manifold along the various tangent 2-planes. Two contractions of the curvature tensor, the *Ricci curvature* and the *scalar curvature*, are specially important. The scalar curvature at a point is simply (up to a scaling factor) the average of sectional curvature over the Grassmannian of tangent 2-planes at that point; the scalar curvature operator can also be characterized abstractly as the unique (up to a scaling factor) Diff-equivariant quasilinear second order differential operator from metrics to smooth functions.

Its characterization as "average sectional curvature" makes it easy to build examples of manifolds that carry metrics of positive scalar curvature. For instance, whatever X is (provided it has sectional curvatures bounded below), the product metric on $X \times S^2$ will have positive scalar curvature provided that one takes the S^2 factor small enough. This simple example shows that the *global* consequences of positive scalar curvature are going to be delicate. In the world of positive *Ricci* curvature, for example, local estimates of geodesic divergence can be integrated to yield global bounds on volume and diameter (Bonnet-Myers theorem). But $X \times S^2$ can have scalar curvature as large as you please and still have exponential volume growth on scales larger than the diameter of the S^2-factor. In fact, it is not readily apparent that scalar curvature provides *any* geometric or topological constraint on the manifold M, and this impression is reinforced by results showing that metrics with *negative* scalar curvature are exceedingly common on any manifold (Kazdan–Warner, Lohkamp).

Misha Gromov's work with Blaine Lawson [78, 79, 84–86] addresses the questions:

- Which manifolds admit metrics of positive, or of uniformly positive, scalar curvature?
- For a manifold admitting such a metric, can we classify such metrics (for instance, up to an appropriate notion of isotopy or cobordism)?

- Can one describe geometrically the "large scale" consequences of positive scalar curvature?

To take the last question first, one of Gromov's guiding ideas is that an n-manifold with scalar curvature greater than ε^{-2} (where ε is a positive constant) should be "$(n-2)$-dimensional on scales greater than ε". After the fact, one can see how this idea originates from consideration of the $X \times S^2$ examples above. *Formulating* the idea, and using it to generate conjectures which can be turned into provable theorems, is of course a quite different matter.

10.2 Simply-Connected Compact Manifolds

Let M be a compact even-dimensional spin manifold. The Atiyah–Singer index theorem [5, 6] identifies the index of the Dirac operator on M with a certain Pontrjagin number, the $\widehat{A}$-genus of M. (In fact, Atiyah has written that "accounting for" the integrality of $\widehat{A}(M)$ for spin manifolds was a key motivation in the proof of the index theorem.)

In 1963, Lichnerowicz [130] computed the square of the Dirac operator in terms of the associated connection Laplacian. His formula,

$$D^2 = \nabla^* \nabla + \frac{1}{4}\kappa,$$

where κ is the scalar curvature, shows that the $\widehat{A}$-genus vanishes for a compact spin manifold admitting a metric with $\kappa > 0$. (In making this argument, Lichnerowicz was following a path traversed earlier by Bochner, who considered the Laplacian on 1-forms and used Hodge theory to proved that positive *Ricci* curvature implies the vanishing of $H^1(M; \mathbb{Q})$, a fact that can also be deduced from the Bonnet-Myers theorem.) Since it is not hard to give examples of compact spinable manifolds with non-vanishing $\widehat{A}$-genus, Lichnerowicz' theorem shows that there are *topological obstructions* to the existence of positive scalar curvature metrics.

It is then natural to ask *how many* such obstructions there are. Amazingly, it can be shown that in the simply-connected case, the Lichnerowicz obstruction is the only one! More precisely,

Theorem 33 *Let M be a simply-connected compact manifold of dimension ≥ 5. Then*

(a) *If M is spinable, then it has a positive scalar curvature metric if and only if $\widehat{A}(M) = 0$.*
(b) *If M is not spinable then it* always *has a positive scalar curvature metric.*

Historically, part (b) came first and is due to Gromov and Lawson; part (a), based on similar techniques but requiring more elaborate input from algebraic topology, was proved by Stefan Stolz.

The dimension restriction is a hint that *surgery* is going to get involved somewhere. What is surgery?

It is not too hard to believe the following statement.

Lemma 5 *Let M and N be n-manifolds of positive scalar curvature, $n \geq 3$. Then their connected sum $M\#N$ also carries a positive scalar curvature metric.*

The reason this is plausible is that $M\#N$ is smoothly glued together from three pieces—M minus a disc, N minus a disc, and a tube $S^{n-1} \times [0, 1]$ joining them—each of which individually carry positive scalar curvature metrics (since $n - 1 \geq 2$). Making the disc really small, so that the positive scalar curvature on the tube is large, one expects that there is enough wiggle room to carry out the necessary smoothings while preserving the positive scalar curvature condition.

Now to form a connected sum one excises a pair of discs—that is, a space $D^n \times S^0$—from the disjoint union $M \sqcup N$, and replaces it by a cylinder $S^{n-1} \times D^1$. More generally one can consider a *surgery* where one excises a region diffeomorphic to $D^q \times S^{n-q}$ and replaces it by $S^{q-1} \times D^{n-q+1}$, which has the same boundary $S^{q-1} \times S^{n-q}$. This process is called a *surgery*. Generalizing the connected sum example above, Gromov and Lawson showed (by a delicate construction) that if the *codimension* q is at least 3, then surgery on a positive scalar curvature manifold yields a new manifold which also admits a positive scalar curvature metric.

Repeated surgeries generate an equivalence relation, which is a form of bordism. To prove (b), Gromov and Lawson used this idea to show that the existence of a positive scalar curvature metric on M (non-spinable, simply connected) depends only on the class of M in the oriented bordism ring Ω_*^{SO}. Explicit generators for this ring are known from the work of Wall [193], and they all admit positive scalar curvature metrics. This finishes the proof. (For case (a), the relevant bordism theory is Ω_*^{Spin}, and explicit generators for this ring are *not* known. This is the reason for the greater complexity of Stolz' proof [177] in this case.)

10.3 Beyond Simple Connectivity

Consider the *non*-simply-connected manifold $M = \mathbb{T}^n$, the n-torus. Since the tangent bundle admits a flat connection, all characteristic classes—and the $\widehat{A}$-genus in particular—vanish. For $n = 2$, the Gauss-Bonnet theorem tells us that there is no metric of positive scalar (= Gaussian) curvature. What about higher dimensions?

Remember that a guiding principle for Gromov is that an n-manifold having scalar curvature greater than ε^{-2} should be "$(n - 2)$-dimensional on scales greater than $O(\varepsilon)$". Now suppose that our torus M is equipped with a metric of positive scalar curvature. Consider its universal cover $X = \widetilde{M}$. Then X is also equipped with a metric of (uniformly) positive scalar curvature and therefore is expected to be $(n - 2)$-dimensional on large scale. On the other hand, the large scale geometry of X is the same as that of the fundamental group $\mathbb{Z}^n = \pi_1(M)$, or equivalently of $\mathbb{R}^n$, and this is n-dimensional. Contradiction! This line of thought leads one to expect that the torus should *not* admit a positive scalar curvature metric, a theorem which was proved first by Schoen and Yau (for $n \le 7$, using minimal hypersurface methods) and then shortly afterwards by Gromov and Lawson (in all dimensions, using the Dirac operator). In both cases, the results apply to a much wider family of examples, but it is helpful for concreteness to focus on the torus.

How are Dirac operator methods to be applied to the torus, where the index of Dirac clearly vanishes? Observe that an elliptic operator D on a compact manifold M defines not simply an integer invariant (its index) but in fact an *index homomorphism*

$$i_D\colon K(M) \to \mathbb{Z}, \qquad [E] \mapsto \operatorname{Index} D_E$$

from the K-theory of M to the integers—in the language of Kasparov, this is the pairing of K-theory with the K-homology class $[D]$ of the operator D. An explicit representative for D_E, the "operator with coefficients", can be constructed by making use of a Hermitian connection on the complex vector bundle E.

Positive scalar curvature implies the ordinary index of the Dirac operator is zero, but it does not in general imply that the more refined invariant i_D is zero. This is because the Lichnerowicz–Weitzenbock formula for D_E,

$$(D_E)^2 = \nabla^*\nabla + \frac{1}{4}\kappa + R_E,$$

contains an additional term R_E which is a contraction of the curvature tensor of E. What we can therefore see straight away is that $i_D([E])$ vanishes when E is flat. More generally, if E is "almost" flat—if it has a connection whose curvature is small enough compared to the lower bound on κ—then $i_D([E])$ will still be zero.

The wonderful idea of Gromov and Lawson is then to exploit the geometry of *covering spaces* of M to produce K-theory representatives that are almost flat in the appropriate sense. In earlier work they made use of towers of *finite* covering spaces, but they soon developed new analytical tools that would permit them to work with the universal covering directly. That is what I will describe below.

Definition 1 [86, Definition 5.1] A complete oriented Riemannian n-manifold X is *ε-hyperspherical*, for some $\varepsilon > 0$, if there is a smooth map $f\colon X \to S^n$ which is constant outside a compact set, of nonzero degree, and *ε-contracting* (that is, the pull-back $T^*_{f(x)}S^n \to T^*_x X$ shrinks the length of covectors by a factor at least ε). A compact manifold M is *enlargeable* if for every $\varepsilon > 0$ it admits a covering that is spin and ε-hyperspherical.

For example, by considering the universal cover one sees that $\mathbb{T}^n$ is enlargeable (in any metric—the property does not depend on the choice of Riemannian metric and is obviously true in the flat metric). Similarly, any compact manifold that admits a metric of non-positive sectional curvature is enlargeable (again, consider the universal cover, and use the inverse of the Riemannian exponential map to build the desired contractions f).

Theorem 34 [86] *A compact enlargeable manifold cannot carry a metric of positive scalar curvature.*

To prove this, Gromov and Lawson introduced the notion of the *relative index*. Let X be a complete spin manifold carrying a metric of uniformly positive scalar curvature. Let E be a Hermitian vector bundle that is flat outside a compact set. Then the Lichnerowicz-Weitzenbock formula shows that D_E^2 is (uniformly) "positive outside a compact set". It turns out that this is enough to ensure that D_E is a Fredholm operator. Moreover, just as we argued above, if the curvature of E is sufficiently small, then this operator is invertible and thus has index zero.

Suppose now that E_0 and E_1 are *two* such hermitian vector bundles and that they are identified by an isomorphism outside a compact set. In this circumstance, one can define the *relative index*

$$\operatorname{rel\,Index}(D; E_0, E_1) := \operatorname{Index} D_{E_0} - \operatorname{Index} D_{E_1}.$$

Gromov–Lawson show that there is a topological expression, analogous to the usual Atiyah–Singer theorem, for this relative index: it is

$$\operatorname{rel\,Index}(D; E_0, E_1) = \big\langle \widehat{A}(M) \smile \operatorname{ch}\big([E_0) - [E_1]\big), [M]\big\rangle,$$

where $([E_0] - [E_1])$ is the class in K-theory $K(X)$ defined (via the "clutching construction") by the bundles E_0, E_1 and the isomorphism between them at ∞. Again, one can understand this process as a pairing between K-theory and K-homology, but now one must remember that the relevant homology theory is the "locally finite" version of K-homology and its dual is the "compactly supported" version of K-cohomology.

To prove Theorem 34, now, it is enough to show that for each $\varepsilon > 0$ one can find such a pair of bundles E_0, E_1 on some covering space X of M, whose curvature is less than ε, and such that the n-dimensional component of the Chern character

$$\operatorname{ch}^n\big([E_0) - [E_1]\big) \in H_c^n(X; \mathbb{Q})$$

is a nonzero multiple of the orientation class. And the definitions of "enlargeable" and "hyperspherical" are set up exactly to permit such bundles to be pulled back from a *fixed* "Bott generator" on the sphere S^n.

10.4 Macroscopic Dimension and K-Area

The argument above (which of course has many generalizations and extensions that I have not discussed) relates the existence of a positive scalar curvature metric on a compact M to the macroscopic (large scale) geometry of $\widetilde{M}$ or equivalently of the fundamental group $\pi_1(M)$. Other articles in this volume explain how the theme of *macroscopic geometry of groups* has recurred throughout Gromov's work. In fact, enlargeability for M can be understood as a formulation of the idea that $\pi_1(M)$ is large-scale n-dimensional.

There is still a gap between this result, though, and the intuition that if M has positive scalar curvature then $\widetilde{M}$ should have dimension $n-2$ (at most) on the large scale. This intuition suggests that we can allow *one* "non-large" direction and still obstruct positive scalar curvature metrics. It is a beautiful fact that the Gromov–Lawson method described above is exactly adapted to this intuition. Control of the curvature terms R_E in the Bochner–Weitzenbock formula comes from the contractive property of f^* on the cotangent space. But since the curvature is actually a 2-form, we actually need only the strictly weaker property of contractivity on $\bigwedge^2 T^*$ to make the arguments work. And this contractivity can be obtained even if f has one non-contracting direction:

Definition 2 Let us define $\bigwedge^2$*-hyperspherical* and $\bigwedge^2$*-enlargeable* in the same way as in Definition 1, but with the contracting condition on covectors replaced by the corresponding condition on 2-covectors (elements of $\bigwedge^2 T^* S^n$).

In contrast to Definition 1, however, it is convenient to consider the possibility that a *non*-compact manifold M should be $\bigwedge^2$-enlargeable; in this case the conditions are required to hold for *any* Riemannian metric on M. A basic example is $N \times \mathbb{R}$ for any enlargeable (compact) manifold N. Using functorial properties one can create many more examples, e.g., any finite volume hyperbolic manifold is $\bigwedge^2$-enlargeable.

Analogous to Theorem 34 one has.

Theorem 35 *No $\bigwedge^2$-enlargeable manifold can carry a complete metric of positive scalar curvature.*

The proof uses the same idea as that of the earlier theorem. In more recent work [78], Gromov has reformulated things in terms of a numerical invariant, the K-area:

Definition 3 Let M be an oriented Riemannian manifold. The *K-area* of M is

$$\sup_E \left\{ \|R_E\|^{-1} : E \text{ is homologically significant} \right\}.$$

Here E is a Hermitian vector bundle over M, $\|R_E\|$ is the operator norm of its curvature, and "homologically significant" means that the subring of $H^*(M;\mathbb{Q})$ generated by the Chern classes of E contains a nonzero multiple of the fundamental class.

The K-area has various natural functorial properties and Theorem 35 follows from these together with the index-theoretic statement that

$$\text{K-area}(M) \le C_n A^{-1}$$

if $A > 0$ is a lower bound for the scalar curvature of M.

Remark 1 Gromov (and others) have given categorical descriptions of the "large scale structure" of a space or group and what one might mean by "dimension" in the context of such structures [77]. It is worth emphasizing that the above methods do not prove the still-conjectural result that if M^n (compact, spin) admits positive scalar curvature, then the asymptotic dimension (in Gromov's sense) of $\pi_1(M)$ is at most $n-2$. Indeed, the K-area definition involves the tangent bundle in what seems to be an unavoidable way. Although Gromov is equipped with a very satisfactory categorical formulation of what "large scale geometry" is, he is still free to reach beyond that formulation as the problem at hand demands. As he says in this context in [78], "one should be ready to modify the definitions if the geometry calls for it". The results show how fruitful this approach can be.

References

[1] Abouzaid, M.: On the wrapped Fukaya category and based loops. J. Symplectic Geom. **10**(1), 27–79 (2012)

[2] Agol, I., Groves, D., Manning, J.: The virtual Haken conjecture. arXiv:1204.2810

[3] Almgren, F.J.: The Theory of Varifolds—A Variational Calculus in the Large for the k-Dimensional Area Integrated. Mimeographed Notes (1965)

[4] Arnold, V.I.: Sur une propriété topologique des applications globalement canoniques de la méchanique classique. C. R. Acad. Paris **261**, 3719–3722 (1965)

[5] Atiyah, M.F., Singer, I.M.: The index of elliptic operators I. Ann. Math. **87**, 484–530 (1968)

[6] Atiyah, M.F., Singer, I.M.: The index of elliptic operators III. Ann. Math. **87**, 546–604 (1968)

[7] Babenko, I.K.: Asymptotic invariants of smooth manifolds. Izv. Akad. Nauk SSSR, Ser. Mat. **56**(4), 707–751 (1992)

[8] Balacheff, F., Croke, C., Katz, M.G.: A Zoll counterexample to a geodesic length conjecture. Geom. Funct. Anal. **19**(1), 1–10 (2009)

[9] Balacheff, F., Sabourau, S.: Diastolic and isoperimetric inequalities on surfaces. Ann. Sci. Éc. Norm. Supér. (4) **43**(4), 579–605 (2010)

[10] Bangert, V.: Closed geodesics on complete surfaces. Math. Ann. **251**(1), 83–96 (1980)

[11] Bárány, I.: A generalization of Carathéodory's theorem. Discrete Math. **40**(2–3), 141–152 (1982)

[12] Bennequin, D.: Entrelacements et équations de Pfaff. In: Third Schnepfenried Geometry Conference, Vol. 1 (Schnepfenried, 1982). Astérisque, vol. 107, pp. 87–161. Soc. Math., Paris (1983)

[13] Berger, M.: Encounter with a geometer, I, II. Not. Am. Math. Soc. **47**(2, 3), 183–194, 326–340 (2000)

[14] Biran, P.: A stability property of symplectic packing. Invent. Math. **136**(1), 123–155 (1999)

[15] Birget, J.-C., Ol'shanskii, A.Yu., Rips, E., Sapir, M.V.: Isoperimetric functions of groups and computational complexity of the word problem. Ann. of Math. (2) **156**(2), 467–518 (2002)

[16] Birkhoff, G.D.: Proof of Poincaré's geometric theorem. Trans. Am. Math. Soc. **14**(1), 14–22 (1913)

[17] Borisov, Ju.F.: $C^{1,\alpha}$-isometric immersions of Riemannian spaces. Dokl. Akad. Nauk SSSR **163**, 11–13 (1965)

[18] Borisov, Yu.F.: Irregular surfaces of the class $C^{1,\beta}$ with an analytic metric. Sib. Mat. Zh. **45**(1), 25–61 (2004)

[19] Borrelli, V., Jabrane, S., Lazarus, F., Thibert, F.: Flat tori in three-dimensional space and convex integration. Proc. Natl. Acad. Sci. USA **109**, 7218–7223 (2012)

[20] Bowen, L.P.: A measure-conjugacy invariant for free group actions. Ann. of Math. (2) **171**(2), 1387–1400 (2010)
[21] Brady, N., Bridson, M.R.: There is only one gap in the isoperimetric spectrum. Geom. Funct. Anal. **10**(5), 1053–1070 (2000)
[22] Bridson, M.R., Haefliger, A.: Metric Spaces of Non-positive Curvature. Grundlehren der Mathematischen Wissenschaften [Fundamental Principles of Mathematical Sciences], vol. 319. Springer, Berlin (1999)
[23] Buhovsky, L., Ostrover, Y.: On the uniqueness of Hofer's geometry. Geom. Funct. Anal. **21**(6), 1296–1330 (2011)
[24] Burago, D., Ivanov, S.: On asymptotic isoperimetric constant of tori. Geom. Funct. Anal. **8**(5), 783–787 (1998)
[25] Burago, D., Ivanov, S.: On asymptotic volume of Finsler tori, minimal surfaces in normed spaces, and symplectic filling volume. Ann. of Math. (2) **156**(3), 891–914 (2002)
[26] Burago, D., Ivanov, S.: Area minimizers and boundary rigidity of almost hyperbolic metrics. arXiv:1011.1570
[27] Burago, D., Ivanov, S.: Minimality of planes in normed spaces. arXiv:1204.1543
[28] Burago, D., Ivanov, S.: Boundary rigidity and filling volume minimality of metrics close to a flat one. Ann. of Math. (2) **171**(2), 1183–1211 (2010)
[29] Burago, D., Burago, Y., Ivanov, S.: A Course in Metric Geometry. Graduate Studies in Mathematics. Amer. Math. Soc., Providence (2001)
[30] Cadek, M., Krcal, M., Matousek, J., Sergeraert, F., Vokrinek, L., Wagner, U.: Computing all maps into a sphere. arXiv:1105.6257
[31] Cannon, J.W.: The combinatorial structure of cocompact discrete hyperbolic groups. Geom. Dedic. **16**(2), 123–148 (1984)
[32] Chaperon, M.: Une idée du type "géodésique brisée" pour les systémes hamiltoniens. C. R. Acad. Sci. **298**(6), 293–296 (1984)
[33] Cheeger, J., Gromov, M.: L_2-cohomology and group cohomology. Topology **25**(2), 189–215 (1986)
[34] Conley, C.C., Zehnder, E.: The Birkhoff-Lewis fixed point theorem and a conjecture of V.I. Arnol'd. Invent. Math. **73**(1), 33–49 (1983)
[35] Conti, S., De Lellis, C., Székelyhidi, L. Jr.: h-principle and rigidity for c^1-isometric embeddings. In: The Abel Symposium 2010, pp. 8–116. Amer. Math. Soc., Providence (2012)
[36] Croke, C.B.: Area and the length of the shortest closed geodesic. J. Differ. Geom. **27**(1), 1–21 (1988)
[37] Dahmani, F., Guirardel, V., Osin, D.: Hyperbolically embedded subgroups and rotating families in groups acting on hyperbolic spaces. arXiv:1111.7048
[38] De Lellis, C., Székelyhidi, L. Jr.: The Euler equations as a differential inclusion. Ann. of Math. (2) **170**(3), 1417–1436 (2009)
[39] De Lellis, C., Székelyhidi, L. Jr.: The h-principle and the equations of fluid dynamics. Bull. Am. Math. Soc. **49**(3), 347–375 (2012)
[40] Delzant, T., Gromov, M.: Courbure mésoscopique et théorie de la toute petite simplification. J. Topol. **1**(4), 804–836 (2008)
[41] Donaldson, S.K.: An application of gauge theory to four-dimensional topology. J. Differ. Geom. **18**(2), 279–315 (1983)
[42] Donaldson, S.K.: Symplectic submanifolds and almost-complex geometry. J. Differ. Geom. **44**(4), 666–705 (1996)
[43] Ekholm, T., Eliashberg, Y., Murphy, E., Smith, I.: Constructing Lagrangian immersions with few double points. Geom. and Funct. Anal. doi:10.1007/s00039-013-0243-6
[44] Ekholm, T., Smith, I.: Exact Lagrangian immersions with a single double point. arXiv:1111.5932
[45] Eliashberg, Ja.M.: Surgery of singularities of smooth mappings. Izv. Akad. Nauk SSSR, Ser. Mat. **36**, 1321–1347 (1972)
[46] Eliashberg, Ja.M.: Estimates on of the Number of Fixed Points of Area-Preserving Transformations of Surfaces (1979). VINITI, Syktyvkar University

[47] Eliashberg, Ja.M.: Complex structures on a manifold with boundary. Funkc. Anal. Prilozh. **14**(1), 89–90 (1980)
[48] Eliashberg, Ja.M.: The wave fronts structure theorem and its applications to symplectic topology. Funkc. Anal. Prilozh. **21**(3), 65–72 (1987)
[49] Eliashberg, Y., Givental, A., Hofer, H.: Introduction to symplectic field theory. In: Special Volume–GAFA2000, pp. 560–673. Birkhäuser, Basel (2000)
[50] Eliashberg, Y., Gromov, M.: Embeddings of Stein manifolds of dimension n into the affine space of dimension $3n/2+1$. Ann. of Math. (2) **136**(1), 123–135 (1992)
[51] Eliashberg, Y., Gromov, M.: Lagrangian intersection theory: finite-dimensional approach. In: Geometry of Differential Equations. American Mathematical Society Translations—Series 2, vol. 186, pp. 27–128. Amer. Math. Soc., Providence (1998)
[52] Eliashberg, Y., Polterovich, L.: Unknottedness of Lagrangian surfaces in symplectic 4-manifolds. Int. Math. Res. Not. **11**, 295–301 (1993)
[53] Eliashberg, Y.M., Mishachev, N.M.: Holonomic approximation and Gromov's h-principle. In: Essays on Geometry and Related Topics, Vols. 1, 2. Monogr. Enseign. Math., vol. 38, pp. 271–285. Enseignement Math, Geneva (2001)
[54] Ferry, S., Weinberger, S.: Quantitative algebraic topology I: Lipschitz homotopy I, www.pnas.org/cgi/doi/10.1073/pnas.1208041110
[55] Fet, A.I.: Variational problems on closed manifolds. Transl. Am. Math. Soc. **1953**(90), 61 (1953)
[56] Floer, A.: Morse theory for Lagrangian intersections. J. Differ. Geom. **28**(3), 513–547 (1988)
[57] Floer, A.: The unregularized gradient flow of the symplectic action. Commun. Pure Appl. Math. **41**(6), 775–813 (1988)
[58] Forster, O.: Plongements des variétés de Stein. Comment. Math. Helv. **45**, 170–184 (1970)
[59] Forstnerič, F.: The Oka principle for sections of stratified fiber bundles. Pure Appl. Math. Q. **6**(3), 843–874 (2010). Special Issue: In honor of Joseph J. Kohn. Part 1
[60] Forstnerič, F., Lárusson, F.: Survey of Oka theory. N.Y. J. Math. **17A**, 11–38 (2011)
[61] Forstnerič, F., Prezelj, J.: Oka's principle for holomorphic fiber bundles with sprays. Math. Ann. **317**(1), 117–154 (2000)
[62] Forstnerič, F., Prezelj, J.: Oka's principle for holomorphic submersions with sprays. Math. Ann. **322**(4), 633–666 (2002)
[63] Forstnerič, F.: Stein Manifolds and Holomorphic Mappings, 3. Folge. Ergebnisse der Mathematik und ihrer Grenzgebiete, vol. 56. Springer, Berlin (2011)
[64] Frankel, S., Katz, M.: The Morse landscape of a Riemannian disk. Ann. Inst. Fourier (Grenoble) **43**(2), 503–507 (1993)
[65] Fukaya, K., Ono, K.: Arnold conjecture and Gromov–Witten invariant for general symplectic manifolds. In: The Arnoldfest, Toronto, ON, 1997. Fields Inst. Commun., vol. 24, pp. 173–190. Amer. Math. Soc., Providence (1999)
[66] Ghomi, M.: Directed immersions of closed manifolds. Geom. Topol. **15**(2), 699–705 (2011)
[67] Ghys, É., de la Harpe, P. (eds.): Sur les groupes hyperboliques d'après Mikhael Gromov. Progress in Mathematics, vol. 83. Birkhäuser Boston (1990). Papers from the Swiss Seminar on Hyperbolic Groups held in Bern, 1988
[68] Ghys, É.: Groupes aléatoires (d'après Misha Gromov, ...). Astérisque **294**(viii), 173–204 (2004)
[69] Gottlieb, D.H.: A certain subgroup of the fundamental group. Am. J. Math. **87**, 840–856 (1965)
[70] Grauert, H.: Analytische Faserungen über holomorph-vollständigen Räumen. Math. Ann. **135**, 263–273 (1958)
[71] Gray, J.W.: Some global properties of contact structures. Ann. of Math. (2) **69**, 421–450 (1959)
[72] Gromov, M.: Hyperbolic manifolds, groups and actions. In: Riemann Surfaces and Related Topics: Proceedings of the 1978 Stony Brook Conference, State Univ. New York,

Stony Brook, N.Y., 1978. Ann. of Math. Stud., vol. 97, pp. 183–213. Princeton Univ. Press, Princeton (1981)

[73] Gromov, M.: Pseudoholomorphic curves in symplectic manifolds. Invent. Math. **82**(2), 307–347 (1985)

[74] Gromov, M.: Hyperbolic groups. In: Essays in Group Theory. Math. Sci. Res. Inst. Publ., vol. 8, pp. 75–263. Springer, New York (1987)

[75] Gromov, M.: Width and related invariants of Riemannian manifolds. Astérisque (163–164):6, 93–109, 282 (1989), 1988. On the geometry of differentiable manifolds (Rome, 1986).

[76] Gromov, M.: Oka's principle for holomorphic sections of elliptic bundles. J. Am. Math. Soc. **2**, 851–897 (1989)

[77] Gromov, M.: Asymptotic invariants for infinite groups. In: Niblo, G.A., Roller, M.A. (eds.) Geometric Group Theory. LMS Lecture Notes, vol. 182, pp. 1–295. Cambridge University Press, Cambridge (1993)

[78] Gromov, M.: Positive curvature, macroscopic dimension, spectral gaps and higher signatures. In: Functional Analysis on the Eve of the 21st Century, Vol. II, New Brunswick, NJ, 1993. Progr. Math., vol. 132, pp. 1–213. Birkhäuser, Boston (1996)

[79] Gromov, M.: Spaces and questions. Geom. Funct. Anal. **Part I**, 118–161 (2000). Special Volume, GAFA 2000 (Tel Aviv, 1999)

[80] Gromov, M.: CAT(κ)-spaces: construction and concentration. Zap. Nauč. Semin. POMI **280**(7), 100–140 (2001). 299–300, Geom. Topol.

[81] Gromov, M.: Isoperimetry of waists and concentration of maps. Geom. Funct. Anal. **13**(1), 178–215 (2003)

[82] Gromov, M.: Random walk in random groups. Geom. Funct. Anal. **13**(1), 73–146 (2003)

[83] Gromov, M.: Singularities, expanders and topology of maps. Part 2: From combinatorics to topology via algebraic isoperimetry. Geom. Funct. Anal. **20**(2), 416–526 (2010)

[84] Gromov, M., Lawson, H.B.: The classification of simply-connected manifolds of positive scalar curvature. Ann. Math. **111**, 423–434 (1980)

[85] Gromov, M., Lawson, H.B.: Spin and scalar curvature in the presence of a fundamental group. Ann. Math. **111**, 209–230 (1980)

[86] Gromov, M., Lawson, H.B.: Positive scalar curvature and the Dirac operator. Publ. Math. Inst Hautes Études Sci. **58**, 83–196 (1983)

[87] Gromov, M.L.: Stable mappings of foliations into manifolds. Izv. Akad. Nauk SSSR, Ser. Mat. **33**, 707–734 (1969)

[88] Gromov, M.L.: Convex integration of differential relations. I. Izv. Akad. Nauk SSSR, Ser. Mat. **37**, 329–343 (1973)

[89] Gromov, M.L., Eliashberg, Ja.M.: Construction of nonsingular isoperimetric films. Tr. Mat. Inst. Steklova **116**, 18–33 (1971) and 235. Boundary value problems of mathematical physics, 7

[90] Gromov, M.L., Eliashberg, Ja.M.: Elimination of singularities of smooth mappings. Izv. Akad. Nauk SSSR, Ser. Mat. **35**, 600–626 (1971)

[91] Gromov, M.L., Eliashberg, Ja.M.: Nonsingular mappings of Stein manifolds. Funct. Anal. Priložen. **5**(2), 82–83 (1971)

[92] Gromov, M.: Homotopical effects of dilatation. J. Differ. Geom. **13**(3), 303–310 (1978)

[93] Gromov, M.: Groups of polynomial growth and expanding maps. Inst. Hautes Études Sci. Publ. Math. **53**, 53–73 (1981)

[94] Gromov, M.: Filling Riemannian manifolds. J. Differ. Geom. **18**(1), 1–147 (1983)

[95] Gromov, M.: Infinite groups as geometric objects. In: Proceedings of the International Congress of Mathematicians, Vols. 1, 2, Warsaw, 1983, pp. 385–392. PWN, Warsaw (1984)

[96] Gromov, M.: Partial Differential Relations. Ergebnisse der Mathematik und ihrer Grenzgebiete (3) [Results in Mathematics and Related Areas (3)], vol. 9. Springer, Berlin (1986)

[97] Gromov, M.: Carnot-Carathéodory spaces seen from within. In: Sub-Riemannian Geometry. Progr. Math., vol. 144, pp. 79–323. Birkhäuser, Basel (1996)

[98] Gromov, M.: Systoles and intersystolic inequalities. In: Actes de la Table Ronde de Géométrie Différentielle, Luminy, 1992. Sémin. Congr., vol. 1, pp. 291–362. Soc. Math. France, Paris (1996)
[99] Gromov, M.: Quantitative homotopy theory. In: Prospects in Mathematics, Princeton, NJ, 1996, pp. 45–49. Amer. Math. Soc., Providence (1999)
[100] Gromov, M.: Singularities, expanders and topology of maps. I. Homology versus volume in the spaces of cycles. Geom. Funct. Anal. **19**(3), 743–841 (2009)
[101] Gromov, M.:. In a search for a structure. Preprint http://www.ihes.fr/~gromov/topics/recent.html
[102] Gromov, M.: Spaces and questions. Geom. Funct. Anal. **Part I**, 118–161 (2000). Special Volume, GAFA 2000 (Tel Aviv, 1999)
[103] Gromov, M.: Metric Structures for Riemannian and Non-Riemannian Spaces. Modern Birkhäuser Classics. Birkhäuser, Boston (2007). English edition. Based on the 1981 French original, with appendices by M. Katz, P. Pansu and S. Semmes, Translated from the French by Sean Michael Bates
[104] Guth, L.: In preparation
[105] Guth, L.: Contraction of areas vs. topology of maps. Preprint
[106] Guth, L.: The width-volume inequality. Geom. Funct. Anal. **17**(4), 1139–1179 (2007)
[107] Guth, L.: Symplectic embeddings of polydisks. Invent. Math. **172**(3), 477–489 (2008)
[108] Guth, L.: Minimax problems related to cup powers and Steenrod squares. Geom. Funct. Anal. **18**(6), 1917–1987 (2009)
[109] Guth, L.: Metaphors in systolic geometry. In: Proceedings of the International Congress of Mathematicians. Volume II, pp. 745–768. Hindustan Book Agency, New Delhi (2010)
[110] Guth, L.: Volumes of balls in large Riemannian manifolds. Ann. of Math. (2) **173**(1), 51–76 (2011)
[111] Hass, J., Morgan, F.: Geodesic nets on the 2-sphere. Proc. Am. Math. Soc. **12**, 3843–3850 (1996)
[112] Higson, N., Lafforgue, V., Skandalis, G.: Counterexamples to the Baum-Connes conjecture. Geom. Funct. Anal. **12**(2), 330–354 (2002)
[113] Hirsch, M.W.: Immersions of manifolds. Trans. Am. Math. Soc. **93**, 242–276 (1959)
[114] Hofer, H.: Lusternik–Schnirelman-theory for Lagrangian intersections. Ann. Inst. Henri Poincaré, Anal. Non Linéaire **5**(5), 465–499 (1988)
[115] Hofer, H.: On the topological properties of symplectic maps. Proc. R. Soc. Edinb., Sect. A, Math. **115**(1–2), 25–38 (1990)
[116] Hofer, H.: Pseudoholomorphic curves in symplectizations with applications to the Weinstein conjecture in dimension three. Invent. Math. **114**(3), 515–563 (1993)
[117] Hutchings, M.: Recent progress on symplectic embedding problems in four dimensions. arXiv:1101.1069
[118] Hutchings, M.: Embedded contact homology and its applications. In: Proceedings of the International Congress of Mathematicians. Volume II, pp. 1022–1041. Hindustan Book Agency, New Delhi (2010)
[119] Ivanov, S.V., Ol'shanskiĭ, A.Yu.: Hyperbolic groups and their quotients of bounded exponents. Trans. Am. Math. Soc. **348**(6), 2091–2138 (1996)
[120] Ivarsson, B., Kutzschebauch, F.: Holomorphic factorization of mappings into $SL_n(\mathbb{C})$. Ann. of Math. (2) **175**(1), 45–69 (2012)
[121] Karshon, Y.: Invent. Math. **115**(1), 431–434 (1994). Appendix to D. McDuff and L. Polterovich
[122] Katz, M.: The filling radius of two-point homogeneous spaces. J. Differ. Geom. **18**(3), 505–511 (1983)
[123] Krcal, M., Matousek, J., Sergeraert, F.: Polynomial-time homology for simplicial Eilenberg-Maclane spaces. arXiv:1201.6222
[124] Kuiper, N.H.: On C^1-isometric imbeddings. I, II. Nederl. Akad. Wetensch. Proc. Ser. A. 58 = Indag. Math. **17**, 545–556 (1955), 683–689

[125] Lalonde, F., McDuff, D.: The geometry of symplectic energy. Ann. of Math. (2) **141**(2), 349–371 (1995)
[126] Landweber, P.S.: Complex structures on open manifolds. Topology **13**, 69–75 (1974)
[127] Lárusson, F.: Model structures and the Oka principle. J. Pure Appl. Algebra **192**(1–3), 203–223 (2004)
[128] Latschev, J., McDuff, D., Schlenk, F.: The Gromov width of 4-dimensional tori. arXiv:1111.6566
[129] Lawson, H.B. Jr.: Lectures on Minimal Submanifolds. Vol. I, 2nd edn. Mathematics Lecture Series, vol. 9. Publish or Perish, Wilmington (1980)
[130] Lichnerowicz, A.: Spineurs harmoniques. C. R. Math. Acad. Sci. **257**, 7–9 (1963)
[131] Liokumovich, Y.: Spheres of small diameter with long sweep-outs. Proc. Amer. Math. Soc. **141**(1), 309–312 (2013)
[132] Liokumovich, Y., Nabutovsky, A., Rotman, R.: Contracting the boundary of a Riemannian 2-disc. arXiv:1205.5474
[133] McDuff, D.: The structure of rational and ruled symplectic 4-manifolds. J. Am. Math. Soc. **3**(3), 679–712 (1990)
[134] McDuff, D.: Singularities and positivity of intersections of J-holomorphic curves. In: Holomorphic Curves in Symplectic Geometry. Progr. Math., vol. 117, pp. 191–215. Birkhäuser, Basel (1994). With an appendix by Gang Liu
[135] McDuff, D., Polterovich, L.: Symplectic packings and algebraic geometry. Invent. Math. **115**(3), 405–434 (1994). With an appendix by Yael Karshon
[136] Micallef, M.J., White, B.: The structure of branch points in minimal surfaces and in pseudoholomorphic curves. Ann. of Math. (2) **141**(1), 35–85 (1995)
[137] Mishachev, N.M., Eliashberg, Ja.M.: Surgery of singularities of foliations. Funct. Anal. Priložen. **11**(3), 43–53 (1977). 96
[138] Moser, J.: On the volume elements on a manifold. Trans. Am. Math. Soc. **120**, 286–294 (1965)
[139] Müller, S., Šverák, V.: Convex integration for Lipschitz mappings and counterexamples to regularity. Ann. of Math. (2) **157**(3), 715–742 (2003)
[140] Nabutovsky, A.: Disconnectedness of sublevel sets of some Riemannian functionals. Geom. Funct. Anal. **6**, 703–725 (1996)
[141] Nabutovsky, A., Rotman, R.: Lengths of geodesics and quantitative Morse theory on loop spaces. Submitted. http://www.math.toronto.edu/alex/morse09.pdf
[142] Nabutovsky, A., Rotman, R.: The length of the shortest closed geodesic on a 2-dimensional sphere. Int. Math. Res. Not. **23**, 1211–1222 (2002)
[143] Nabutovsky, A., Rotman, R.: Volume, diameter and the minimal mass of a stationary 1-cycle. Geom. Funct. Anal. **14**(4), 748–790 (2004)
[144] Nabutovsky, A., Rotman, R.: Curvature-free upper bounds for the smallest area of a minimal surface. Geom. Funct. Anal. **16**(2), 453–475 (2006)
[145] Nabutovsky, A., Rotman, R.: Length of geodesics on a two-dimensional sphere. Am. J. Math. **131**(2), 549–569 (2009)
[146] Nabutovsky, A., Rotman, R.: Upper bounds on the length of a shortest closed geodesic and quantitative Hurewicz theorem. J. Eur. Math. Soc. **5**(3), 203–244 (2003)
[147] Nabutovsky, A., Rotman, R.: Linear bounds for lengths of geodesic loops on Riemannian 2-spheres. J. Differ. Geom. **89**(2), 217–232 (2011)
[148] Nash, J.: C^1 isometric imbeddings. Ann. of Math. (2) **60**, 383–396 (1954)
[149] Oka, K.: Sur les fonctions des plusieurs variables. III: Deuxième problème de cousin. J. Sci. Hiroshima Univ. **9**(1–3), 7–19 (1939)
[150] Ollivier, Y.: A January 2005 Invitation to Random Groups. Ensaios Matemáticos [Mathematical Surveys], vol. 10. Soc. Brasileira de Matemática, Rio de Janeiro (2005)
[151] Ollivier, Y.: On a small cancellation theorem of Gromov. Bull. Belg. Math. Soc. Simon Stevin **13**(1), 75–89 (2006)
[152] Ol'shanskiĭ, A.Yu.: Periodic quotient groups of hyperbolic groups. Mat. Sb. **182**(4), 543–567 (1991)

[153] Ol'shanskiĭ, A.Yu.: Almost every group is hyperbolic. Int. J. Algebra Comput. **2**(1), 1–17 (1992)
[154] Ozsváth, P., Szabó, Z.: Holomorphic disks and three-manifold invariants: properties and applications. Ann. of Math. (2) **159**(3), 1159–1245 (2004)
[155] Papasoglu, P.: An algorithm detecting hyperbolicity. In: Geometric and Computational Perspectives on Infinite Groups, Minneapolis, MN and New Brunswick, NJ, 1994. DIMACS Ser. Discrete Math. Theoret. Comput. Sci., vol. 25, pp. 193–200. Amer. Math. Soc., Providence (1996)
[156] Pardon, J.: On the distortion of knots on embedded surfaces. Ann. of Math. (2) **174**(1), 637–646 (2011)
[157] Pestov, L., Uhlmann, G.: Two dimensional compact simple Riemannian manifolds are boundary distance rigid. Ann. of Math. (2) **161**(2), 1093–1110 (2005)
[158] Petrunin, A.: Diameter of universal cover. http://mathoverflow.net/questions/8534/diameter-of-universal-cover
[159] Phillips, A.: Submersions of open manifolds. Topology **6**, 171–206 (1967)
[160] Pitts, J.T.: Existence and regularity of minimal surfaces on Riemannian manifolds. Bull. Am. Math. Soc. **82**(3), 503–504 (1976)
[161] Pitts, J.T.: Regular minimal hypersurfaces exist on manifolds in dimensions up to six. In: Seminar on Minimal Submanifolds. Ann. of Math. Stud., vol. 103, pp. 199–205. Princeton Univ. Press, Princeton (1983)
[162] Poenaru, V.: On regular homotopy in codimension 1. Ann. of Math. (2) **83**, 257–265 (1966)
[163] Poincaré, H.: Sur une théorème de géométrie. Rend. Circ. Mat. Palermo **33**, 375–507 (1912)
[164] Rotman, R.: The length of a shortest closed geodesic and the area of a 2-dimensional sphere. Proc. Am. Math. Soc. **134**(10), 3041–3047 (2006) (electronic)
[165] Rotman, R.: The length of a shortest geodesic net on a closed Riemannian manifold. Topology **46**(4), 343–356 (2007)
[166] Rotman, R.: The length of a shortest geodesic loop at a point. J. Differ. Geom. **78**(3), 497–519 (2008)
[167] Rotman, R.: Flowers on Riemannian manifolds. Math. Z. **269**(1–2), 543–554 (2011)
[168] Sabourau, S.: Filling radius and short closed geodesics of the 2-sphere. Bull. Soc. Math. Fr. **132**(1), 105–136 (2004)
[169] Sabourau, S.: Global and local volume bounds and the shortest geodesic loops. Commun. Anal. Geom. **12**(5), 1039–1053 (2004)
[170] Sapir, M.V., Birget, J.-C., Rips, E.: Isoperimetric and isodiametric functions of groups. Ann. of Math. (2) **156**(2), 345–466 (2002)
[171] Sauvaget, D.: Curiosités lagrangiennes en dimension 4. Ann. Inst. Fourier (Grenoble) **54**(6), 1997–2020 (2005). 2004
[172] Schwartz, A.S.: Geodesic arcs on Riemann manifolds. Usp. Mat. Nauk **13**(6 (84)):181–184 (1958)
[173] Schwarz, M.: A quantum cup-length estimate for symplectic fixed points. Invent. Math. **133**(2), 353–397 (1998)
[174] Smale, S.: The classification of immersions of spheres in Euclidean spaces. Ann. of Math. (2) **69**, 327–344 (1959)
[175] Spring, D.: On the regularity of solutions in convex integration theory. Invent. Math. **104**(1), 165–178 (1991)
[176] Spring, D.: Convex Integration Theory. Monographs in Mathematics, vol. 92. Birkhäuser, Basel (1998). Solutions to the h-principle in geometry and topology
[177] Stolz, S.: Simply connected manifolds of positive scalar curvature. Ann. Math. **136**(3), 511–540 (1992). ArticleType: research-article/Full publication date: Nov., 1992 / Copyright © 1992 Annals of Mathematics
[178] Taubes, C.H.: SW $\Rightarrow$ Gr: from the Seiberg–Witten equations to pseudo-holomorphic curves. J. Am. Math. Soc. **9**(3), 845–918 (1996)
[179] Taubes, C.H.: Gr $\Longrightarrow$ SW: from pseudo-holomorphic curves to Seiberg–Witten solutions. J. Differ. Geom. **51**(2), 203–334 (1999)

[180] Taubes, C.H.: The Seiberg–Witten equations and the Weinstein conjecture. II. More closed integral curves of the Reeb vector field. Geom. Topol. **13**(3), 1337–1417 (2009)
[181] Taubes, C.H.: Embedded contact homology and Seiberg–Witten Floer cohomology I, II, III, IV, V. Geom. Topol. **14**(5), 2497–3000 (2010)
[182] Thorbergsson, G.: Closed geodesics on non-compact Riemannian manifolds. Math. Z. **159**(3), 249–258 (1978)
[183] Thurston, W.P.: Existence of codimension-one foliations. Ann. of Math. (2) **104**(2), 249–268 (1976)
[184] Thurston, W.: The theory of foliations of codimension greater than one. Comment. Math. Helv. **49**, 214–231 (1974)
[185] Traynor, L.: Symplectic packing constructions. J. Differ. Geom. **41**(3), 735–751 (1995)
[186] Tsui, M.-P., Wang, M.-T.: Mean curvature flows and isotopy of maps between spheres. Commun. Pure Appl. Math. **57**(8), 1110–1126 (2004)
[187] Uhlenbeck, K.K.: Connections with L^p bounds on curvature. Commun. Math. Phys. **83**(1), 31–42 (1982)
[188] Vaaler, J.D.: A geometric inequality with applications to linear forms. Pac. J. Math. **83**(2), 543–553 (1979)
[189] Vershik, A.M.: Classification of measurable functions of several arguments, and invariantly distributed random matrices. Funkc. Anal. Prilozh. **36**(2), 12–27 (2002). 95
[190] Vershik, A.M.: Dynamics of metrics in measure spaces and their asymptotic invariants. Markov Process. Relat. Fields **16**(1), 169–184 (2010)
[191] Viterbo, C.: A proof of Weinstein's conjecture in $\mathbf{R}^{2n}$. Ann. Inst. Henri Poincaré, Anal. Non Linéaire **4**(4), 337–356 (1987)
[192] Viterbo, C.: Symplectic topology as the geometry of generating functions. Math. Ann. **292**, 685–710 (1992)
[193] Wall, C.T.C.: Determination of the cobordism ring. Ann. Math. **72**(2), 292–311 (1960). ArticleType: research-article/Full publication date: Sep., 1960 / Copyright © 1960 Annals of Mathematics
[194] Witten, E.: Supersymmetry and Morse theory. J. Differ. Geom. **17**(4), 661–692 (1983). 1982
[195] Żuk, A.: Property (T) and Kazhdan constants for discrete groups. Geom. Funct. Anal. **13**(3), 643–670 (2003)

List of Publications for Mikhail Leonidovich Gromov

1967

[1] A geometrical conjecture of Banach. *Math. USSR, Izv.*, 1:1055–1064. Also available in *Izv. Akad. Nauk SSSR, Ser. Mat.*, 31:1105–1114 (in Russian).

1968

[2] On the number of simplexes of subdivisions of finite complexes. *Math. Notes*, 3:326–332. Also available in *Mat. Zametki*, 3:511–522 (in Russian).

[3] The mappings of the foliations in the manifolds equipped with extra structure. *Vestnik Leningrad. Univ.*, 23(19):167.

[4] Transversal mappings of foliations. *Sov. Math., Dokl.*, 9:1126–1129.

1969

[5] On simplexes inscribed in a hypersurface. *Math. Notes*, 5:52–56.

[6] Stable mappings of foliations into manifolds. *Math. USSR, Izv.*, 3:671–694. Also available in *Izv. Akad. Nauk SSSR, Ser. Mat.*, 33:707–734 (in Russian).

1970

[7] Isometric imbeddings and immersions. *Sov. Math., Dokl.*, 11:794–797.

[8] (with V.A. Rokhlin). Embeddings and immersions in Riemannian geometry. *Russ. Math. Surv.*, 25(5):1–57. Also available in *Usp. Mat. Nauk* 25, No. 5(155):3–62 (in Russian).

1971

[9] (with Y. Eliashberg). Elimination of singularities of smooth mappings. *Math. USSR, Izv.*, 5:615–639. Also available in *Izv. Akad. Nauk SSSR, Ser. Mat.*, 35:600–626 (in Russian).

[10] (with Y. Eliashberg). Nonsingular mappings of Stein manifolds. *Funct. Anal. Appl.*, 5:156–157. Also available in *Funkcional. Anal. i Priložen.*, 5(2):82–83 (in Russian).

[11] A topological technique for the construction of solutions of differential equations and inequalities. In *Actes Congr. internat. Math. (Nice 1970)*, 2:221–225. Gauthier–Villars, Paris.

H. Holden, R. Piene (eds.), *The Abel Prize 2008-2012*,
DOI 10.1007/978-3-642-39449-2_12,

[12] (with Y. Eliashberg). Construction of nonsingular isoperimetric films. *Proc. Steklov Inst. Math.*, 116:13–28.

1972

[13] Smoothing and inversion of differential operators. *Mat. Sb. (N.S.)*, 88(130): 382–441.

1973

[14] (with A.V. Bubennikov). *Darstellende Geometrie. (Nacertatel'naja geometrija.) Lehrbuch. 2. überarb. und erg. Aufl.* Moskau: "Vyssaja Skola". 416 S. R. 1.18.
[15] Degenerate smooth mappings. *Mat. Zametki*, 14:509–516.
[16] (with Y. Eliashberg). Construction of a smooth mapping with a prescribed Jacobian. I. *Functional Anal. Appl.*, 7:27–33. Also available in *Funkcional. Anal. i Priložen.*, 7(1):33–40 (in Russian).
[17] Convex integration of differential relations. I. *Math. USSR, Izv.*, 7(2):329–343. Also available in *Izv. Akad. Nauk SSSR Ser. Mat.*, 37:329–343 (in Russian).

1978

[18] Manifolds of negative curvature. *J. Differ. Geom.*, 13(2):223–230.
[19] Almost flat manifolds. *J. Differ. Geom.*, 13(2):231–241.
[20] Homotopical effects of dilatation. *J. Differ. Geom.*, 13(3):303–310.

1980

[21] Synthetic geometry in Riemannian manifolds. In *Proc. int. Congr. Math. (Helsinki 1978)*, pages 415–419, Helsinki. Acad. Sci. Fennica.
[22] (with H.B. Lawson, Jr.). Spin and scalar curvature in the presence of a fundamental group. I. *Ann. of Math. (2)*, 111(2):209–230.
[23] (with H.B. Lawson, Jr.). The classification of simply connected manifolds of positive scalar curvature. *Ann. of Math. (2)*, 111(3):423–434.
[24] (with M. Brin). On the ergodicity of frame flows. *Invent. Math.*, 60(1):1–7.

1981

[25] Hyperbolic manifolds according to Thurston and Jorgensen. Semin. Bourbaki, 32e annee, Vol. 1979/80, Exp. 546, Lect. Notes Math. 842, 40–53.
[26] Curvature, diameter and Betti numbers. *Comment. Math. Helv.*, 56:179–195.
[27] Hyperbolic manifolds, groups and actions. In *Riemann surfaces and related topics: Proc. 1978 Stony Brook Conf.*, volume 97 of *Ann. Math. Stud.*, pages 183–213. Princeton Univ. Press, Princeton, N.J.
[28] Groups of polynomial growth and expanding maps. Appendix by J. Tits. *Inst. Hautes Études Sci. Publ. Math.*, 53:53–73.
[29] *Structures métriques pour les variétés riemanniennes. Redige par J. Lafontaine et P. Pansu.* Textes Mathematiques, 1. Paris: Cedic/Fernand Nathan. VII, 152 p.

1982

[30] (with J. Cheeger and M. Taylor). Finite propagation speed, kernel estimates for functions of the Laplace operator, and the geometry of complete Riemannian manifolds. *J. Differ. Geom.*, 17(1):15–53.

[31] Volume and bounded cohomology. *Inst. Hautes Études Sci. Publ. Math.*, 56:5–99.

1983

[32] Filling Riemannian manifolds. *J. Differ. Geom.*, 18(1):1–147.

[33] (with H.B. Lawson, Jr.). Positive scalar curvature and the Dirac operator on complete Riemannian manifolds. *Inst. Hautes Études Sci. Publ. Math.*, 58:83–196.

[34] Asymptotic geometry of homogeneous spaces. In *Differential geometry on homogeneous spaces*, Conf. Torino/Italy 1983, Rend. Semin. Mat., Torino, Fasc. Spec., pages 59–60.

1984

[35] Infinite groups as geometric objects. In *Proc. Int. Congr. Math. (Warszawa 1983)*, Vol. 1, pages 385–392.

[36] (with V.D. Milman). Brunn theorem and a concentration of volume phenomena for symmetric convex bodies. Lindenstrauss, Joram (ed.) et al., Israel seminar on geometrical aspects of functional analysis (GAFA) 1983–84. Tel Aviv: Tel Aviv University, Paper V, 12 p.

1985

[37] (with W. Ballmann and V. Schroeder). *Manifolds of nonpositive curvature*, volume 61 of *Progress in Mathematics*. Birkhäuser, Boston, MA.

[38] Pseudo holomorphic curves in symplectic manifolds. *Invent. Math.*, 82(2): 307–347.

[39] (with J. Cheeger). On the characteristic numbers of complete manifolds of bounded curvature and finite volume. In *Differential geometry and complex analysis*, pages 115–154. Springer, Berlin.

[40] Isometric immersions of Riemannian manifolds. The mathematical heritage of Élie Cartan, Semin. (Lyon 1984), Astérisque, No.Hors Sér. 1985, 129–133.

[41] (with J. Cheeger). Bounds on the von Neumann dimension of L^2-cohomoloy and the Gauss–Bonnet theorem for open manifolds. *J. Differ. Geom.*, 21(1):1–34.

1986

[42] (with J. Cheeger). L_2-cohomology and group cohomology. *Topology*, 25(2): 189–215.

[43] Large Riemannian manifolds. In *Curvature and topology of Riemannian manifolds (Katata 1985)*, volume 1201 of *Lecture Notes Math.*, pages 108–121. Springer, Berlin.

[44] (with J. Cheeger). Collapsing Riemannian manifolds while keeping their curvature bounded. I. *J. Differ. Geom.*, 23(3):309–346.
[45] *Partial differential relations*, volume 9 of *Ergebnisse der Mathematik und ihrer Grenzgebiete*. Springer-Verlag, Berlin.
[46] *Soft and hard symplectic geometry. A plenary address presented at the international congress of mathematicians, Berkeley, CA, USA, August 1986. Videotape (NTSC; 60 min. VHS)*. ICM Series. Providence, RI: American Mathematical Society (AMS).

1987

[47] Entropy, homology and semialgebraic geometry. *Astérisque*, 145/146:225–240. Sémin. Bourbaki, 38ème année, Vol. 1985/86, Exposé 663.
[48] (with W. Thurston). Pinching constants for hyperbolic manifolds. *Invent. Math.*, 89(1):1–12.
[49] (with V.D. Milman). Generalization of the spherical isoperimetric inequality to uniformly convex Banach spaces. *Compos. Math.*, 62(3):263–282.
[50] Monotonicity of the volume of intersection of balls. In *Geometrical aspects of functional analysis (1985–1986)*, volume 1267 of *Lecture Notes in Math.*, pages 1–4, Springer, Berlin.
[51] Hyperbolic groups. In *Essays in group theory*, volume 8 of *Math. Sci. Res. Inst. Publ.*, pages 75–263. Springer, New York. Also available in *Institut Komp'yuternykh Issledovanij.*, 2002 (in Russian).
[52] Cauchy–Riemann equation in Lagrange intersection theory. In *Periodic solutions of Hamiltonian systems and related topics (Il Cioco, 1986)*, volume 209 of *NATO Adv. Sci. Inst. Ser. C Math. Phys. Sci.*, pages 175–176. Reidel, Dordrecht.
[53] Soft and hard symplectic geometry. In *Proc. Int. Congr. Math. (Berkeley 1986)*, Vol. 1, pages 81–98. Amer. Math. Soc., Providence, RI.

1988

[54] (with I. Piatetski-Shapiro). Non-arithmetic groups in Lobachevsky spaces. *Inst. Hautes Études Sci. Publ. Math.*, 66:93–103.
[55] Rigid transformations groups. In *Géométrie différentielle (Paris 1986)*, volume 33 of *Trav. en Cours*, pages 65–139. Hermann, Paris.
[56] Dimension, nonlinear spectra and width. In *Geometric aspects of functional analysis (1986/1987)*, volume 1317 of *Lect. Notes in Math.*, pages 132–184. Springer, Berlin.
[57] Width and related invariants of Riemannian manifolds. *Astérisque* 163/164: 93–109.
[58] (with H.B. Lawson, Jr. and W. Thurston). Hyperbolic 4-manifolds and conformally flat 3-manifolds. *Inst. Hautes Études Sci. Publ. Math.*, 68:27–45.

1989

[59] Sur le groupe fondamental d'une variété kählérienne. *C. R. Acad. Sci. Paris Sér. I Math.*, 308(3):67–70.

[60] (with J. Bourgain). Estimates of Bernstein widths of Sobolev spaces. In *Geometric aspects of functional analysis (1987–1988)*, volume 1376 of *Lect. Notes in Math.*, pages 176–185. Springer, Berlin.
[61] Oka's principle for holomorphic sections of elliptic bundles. *J. Amer. Math. Soc.*, 2(4):851–897.
[62] Soft differential equations. In *IXth International Congress on Mathematical Physics (Swansea, 1988)*, pages 374–376. Hilger, Bristol.

1990

[63] (with A. Connes and H. Moscovici). Conjecture de Novikov et fibrés presque plats. *C. R. Acad. Sci. Paris Sér. I Math.*, 310(5):273–277.
[64] (with J. Cheeger). Collapsing Riemannian manifolds while keeping their curvature bounded. II. *J. Differ. Geom.*, 32(1):269–298.
[65] Convex sets and Kähler manifolds. In *Advances in differential geometry and topology*, pages 1–38. World Sci. Publ., Teaneck, NJ.

1991

[66] Kähler hyperbolicity and L_2-Hodge theory. *J. Differ. Geom.*, 33(1):263–292.
[67] (with J. Cheeger). Chopping Riemannian manifolds. In *Differential geometry*, volume 52 of *Pitman Monogr. Surverys Pure Appl. Math.*, pages 85–94. Longman Sci. Tech., Harlow.
[68] (with Y. Eliashberg). Convex symplectic manifolds. In *Several complex variables and complex geometry, Part 2 (Santa Cruz, CA, 1989)*, volume 52 of *Proc. Symp. Pure Math.* pages 135–162. Amer. Math. Soc., Providence, RI.
[69] (with G. D'Ambra). Lectures on transformation groups: geometry and dynamics. In *Surveys in differential geometry (Cambridge, MA 1990)*, pages 19–111. Lehigh Univ., Bethlem, PA.
[70] Foliated Plateau problem. II: Harmonic maps of foliations. *Geom. Funct. Anal.*, 1(3):253–320.
[71] (with M.A. Shubin). Von Neumann spectra near zero. *Geom. Funct. Anal.*, 1(4):375–404. Erratum, *Geom. Funct. Anal.*, 5(4):729, 1995.
[72] (with P. Pansu). Rigidity of lattices: An introduction. In *Geometric topology: recent developments (Montecatini Terme 1990)*, volume 1504 of *Lect. Notes in Math.*, pages 39–137. Springer, Berlin.
[73] Sign and geometric meaning of curvature. *Rend. Semin. Mat. Fis. Milano*, 61:9–123. Also available in *Izhevsk*, 1999 (in Russian).

1992

[74] (with J. Cheeger and K. Fukaya). Nilpotent structures and invariant metrics on collapsed manifolds. *J. Amer. Math. Soc.*, 5(2):327–372.
[75] (with M.A. Shubin). The Riemann–Roch theorem for general elliptic operators. *C. R. Acad. Sci. Paris Sér. I Math.*, 314(5):363–367.
[76] (with Y. Eliashberg). Embeddings of Stein manifolds of dimension n into the affine space of dimension $3n/2+1$. *Ann. of Math. (2)*, 136(1):123–135.
[77] Spectral geometry of semi-algebraic sets. *Ann. Inst. Fourier*, 42(1–2):249–274.

[78] Stability and pinching. In *Geometry Seminars. Sessions on Topology and Geometry of Manifolds (Bologna 1990)*, pages 55–97. Univ. Stud. Bologna, Bologna.

[79] (with M.A. Shubin). Near-cohomology of Hilbert complexes and topology of non-simply connected manifolds. *Astérisque*, 210:283–294. Méthodes semi-classiques, Vol. 2 (Nantes, 1991).

[80] (with Y. Burago and G. Perelman). A. D. Alexandrov spaces with curvature bounded below. *Russ. Math. Surv.*, 47(2):1–58. Also available in *Uspekhi Mat. Nauk*, 47(2(284)):3–51 (in Russian).

[81] (with R. Schoen). Harmonic maps into singular spaces and p-adic superrigidity for lattices in groups of rank one. *Inst. Hautes Études Sci. Publ. Math.*, 76:165–246.

1993

[82] (with A. Connes and H. Moscovici). Group cohomology with Lipschitz control and higher signatures. *Geom. Funct. Anal.*, 3(1):1–78.

[83] (with M.A. Shubin). The Riemann–Roch theorem for elliptic operators. In *I.M. Gelfand Seminar*, volume 16 of *Adv. Soviet Math.*, pages 211–241. Amer. Math. Soc., Providence, RI.

[84] (with M.A. Shubin). The Riemann–Roch theorem for elliptic operators and solvability of elliptic equations with additional conditions on compact subsets. In *Journ. Équ. Dériv. Partielles (St.-Jean-de-Monts, 1993)*, 13 pages. École Polytech., Palaiseau.

[85] Asymptotic invariants of infinite groups. In *Geometric group theory. Vol. 2 (Sussex, 1991)*, volume 182 of *London Math. Soc. Lecture Note Ser.*, pages 1–295. Cambridge Univ. Press, Cambridge.

[86] Metric invariants of Kähler manifolds. In *Differential geometry and topology (Alghero, 1992)*, pages 90–116. World Sci. Publ., River Edge, NJ.

1994

[87] (with M.A. Shubin). The Riemann–Roch theorem for elliptic operators and solvability of elliptic equations with additional conditions on compact subsets. *Invent. Math.*, 117(1):165–180.

1995

[88] Geometric reflections on the Novikov conjecture. In *Novikov conjectures, index theorems and rigidity, Vol. 1 (Oberwolfach, 1993)*, volume 226 of *London Math. Soc. Lecture Note Ser.*, pages 164–173. Cambridge Univ. Press, Cambridge.

1996

[89] Carnot–Carathéodory spaces seen from within. In *Sub-Riemannian geometry*, volume 144 of *Progr. Math.*, pages 79–323. Birkhäuser, Basel.

[90] Systoles and intersystolic inequalities. Besse, Arthur L. (ed.), Actes de la table ronde de géométrie différentielle en l'honneur de Marcel Berger, Luminy,

France, 12–18 juillet, 1992. Paris: Société Mathématique de France. Sémin. Congr. 1, 291–362.

[91] Positive curvature, macroscopic dimension, spectral gaps and higher signatures. In *Functional analysis on the eve of the 21st century, Volume II (Brunswick, 1993)*, volume 132 of *Prog. Math.*, 1–213. Birkhäuser Boston, Boston, MA.

1997

[92] (with Y. Eliashberg). Lagrangian intersections and the stable Morse theory. *Boll. Un. Mat. Ital. B (7)*, 11(2):289–326.

1998

[93] (with G. Henkin and M. Shubin). L^2 holomorphic functions on pseudoconvex coverings. In *Operator theory for complex and hypercomplex analysis (Mexico City, 1994)*, volume 212 of *Contemp. Math.*, pages 81–94. Amer. Math. Soc., Providence, RI.

[94] (with G. Henkin and M. Shubin). Holomorphic L^2 functions on coverings of pseudoconvex manifolds. *Geom. Funct. Anal.*, 8(3):552–585.

[95] (with Y. Eliashberg). Lagrangian intersection theory: finite-dimensional approach. In *Geometry of differential equations*, volume 186 of *Amer. Math. Soc. Transl. Ser. 2*, pages 27–118. Amer. Math. Soc., Providence, RI.

[96] Possible trends in mathematics in the coming decades. *Notices Amer. Math. Soc.*, 45(7):846–847. Also in *Sviti Mat.*, 7(1):3–5, 2001 (in Ukrainian).

1999

[97] Quantitative homotopy theory. In *Prospects in mathematics (Princeton, NJ, 1996)*, pages 45–49. Amer. Math. Soc., Providence, RI.

[98] Endomorphisms of symbolic algebraic varieties. *J. Eur. Math. Soc. (JEMS)*, 1(2):109–197.

[99] *Metric structures for Riemannian and non-Riemannian spaces*, volume 152 of *Progress in Mathematics*. Birkhäuser Boston Inc., Boston, MA. Also available as *Modern Birkhäuser Classics*, 2007.

[100] Topological invariants of dynamical systems and spaces of holomorphic maps. I. *Math. Phys. Anal. Geom.*, 2(4):323–415.

2000

[101] A. Carbone, M. Gromov and P. Prusinkiewicz, editors. *Pattern formation in biology, vision and dynamics. Including papers form the conference on pattern formation, Bures-sur-Yvette, France, December 2–6, 1997.* World Sci. Publ., Singapore.

[102] N. Alon, J. Bourgain, A. Connes, M. Gromov and V. Milman, editors. *GAFA 2000. Visions in mathematics—Towards 2000. Proceedings of a meeting, Tel Aviv, Israel, August 25–September 3, 1999. Part I. Special volume of the journal Geometric and Functional Analysis.* Birhäuser, Basel. Also available as *Modern Birkhäuser Classics*, Birkhäuser, 2010.

[103] N. Alon, J. Bourgain, A. Connes, M. Gromov and V. Milman, editors. *GAFA 2000. Visions in mathematics—Towards 2000. Proceedings of a meeting, Tel Aviv, Israel, August 25–September 3, 1999. Part II. Special volume of the journal Geometric and Functional Analysis.* Birkhäuser, Basel. Also available as *Modern Birkhäuser Classics*, Birkhäuser, 2010.

[104] Three remarks on geodesic dynamics and fundamental group. *Enseign. Math. (2)*, 46(3–4):391–402.

[105] Spaces and questions. *Geom. Funct. Anal.* (Special Volume, part I):118–161. GAFA 2000 (Tel Aviv, 1999).

2001

[106] (with A. Carbone). Mathematical slices of molecular biology. *Gaz. Math.*, 88:80.

[107] Possible trends in mathematics in the coming decades. In *Mathematics unlimited—2001 and beyond*, pages 525–527. Springer, Berlin.

[108] Mesoscopic curvature and hyperbolicity. In *Global differential geometry: the mathematical legacy of Alfred Gray (Bilbao, 2000)*, volume 288 of *Contemp. Math.*, pages 58–69. Amer. Math. Soc., Providence, RI.

[109] (with J.-P. Bourguignon, E. Calabi, J. Eells and O. García-Prada). Where does geometry go? A research and education perspective. In *Global differential geometry: the mathematical legacy of Alfred Gray (Bilbao, 2000)*, volume 288 of *Contemp. Math.*, pages 442–457. Amer. Math. Soc., Providence, RI.

[110] Small cancellation, unfolded hyperbolicity, and transversal measures. In *Essays on geometry and related topics*, volume 38 of *Monogr. Enseign. Math.*, pages 371–399. Enseignment Math., Geneva.

2003

[111] (with A. Carbone). Functional labels and syntactic entropy on DNA strings and proteins. *Theor. Comput. Sci.*, 303(1):35–51.

[112] Isoperimetry of waists and concentration of maps. *Geom. Funct. Anal.*, 13(1):178–215.

[113] On the entropy of holomorphic maps. *Enseign. Math. (2)*, 49(3–4):217–235.

[114] Random walk in random groups. *Geom. Funct. Anal.*, 13(1):73–146.

2004

[115] S. Donaldson and Y. Eliashberg and M. Gromov, editors. *Different faces of geometry.* International Mathematical Series (New York) 3. New York, NY: Kluwer Academic/Plenum Publishers.

[116] (with M. Bertelson). Dynamical Morse entropy. In *Modern dynamical systems and applications*, pages 27–44. Cambridge Univ. Press, Cambridge.

[117] CAT(κ)-spaces: construction and concentration. *J. Math. Sci., New York*, 119(2):178–200. Also available in *Zap. Nauchn. Sem. S.-Peterburg. Otdel. Mat. Inst. Steklov (POMI)*, 280:100–140, 2001 (in Russian).

2005

[118] (with T. Delzant). Cuts in Kähler groups. In *Infinite groups: geometric, combinatorial and dynamical aspects*, volume 248 of *Progr. Math.*, pages 31–55. Birkhäuser, Basel.

2008

[119] Entropy and isoperimetry for linear and non-linear group actions. *Groups Geom. Dyn.*, 2(4):499–593.
[120] (with N.V. Evtushenko and A.V. Kolomeets). On the synthesis of adaptive tests for nondeterministic finite state machines. *Program. Comput. Softw.*, 34(6):322–329.
[121] (with T. Delzant). Mesoscopic curvature and very small cancellation theory. (Courbure mésoscopique et théorie de la toute petite simplification.). *J. Topol.*, 1(4):804–836.
[122] Mendelian dynamics and Sturtevant's paradigm. In *Geometric and probabilistic structures in dynamics*, volume 469 of *Contemp. Math.*, pages 227–242. Amer. Math. Soc., Providence, RI.

2009

[123] Singularities, expanders and topology of maps. I: Homology versus volume in the spaces of cycles. *Geom. Funct. Anal.*, 19(3):743–841.

2010

[124] Singularities, expanders and topology of maps. II: From combinatorics to topology via algebraic isoperimetry. *Geom. Funct. Anal.*, 20(2):416–526.

2011

[125] Crystals, proteins, stability and isoperimetry. *Bull. Am. Math. Soc., New Ser.*, 48(2):229–257.

2012

[126] (with L. Guth). Generalizations of the Kolmogorov–Barzdin embedding estimates. *Duke Math. J.*, 161(13):2549–2603.
[127] (with J. Fox, V. Lafforgue, A. Naor, and J. Pach). Overlap properties of geometric expanders. *J. Reine Angew. Math.*, 671:49–83.
[128] Super stable Kählerian horseshoe? In *Essays in mathematics and its applications*, pages 151–229. Springer, Heidelberg.
[129] Hilbert volume in metric spaces. Part 1. *Cent. Eur. J. Math.*, 10(2):371–400.

2013

[130] V. Capasso, M. Gromov, A. Harel-Bellan, N. Morozova and L. Pritchard, editors. *Pattern formation in morphogenesis. Problems and mathematical issues.*, volume 15 of *Springer Proceedings in Mathematics*. Springer, Berlin.

Curriculum Vitae for Mikhail Leonidovich Gromov

Born:	December 23, 1943 in Boksitogorsk, USSR
Degrees/education:	Master, Leningrad University, 1965
	PhD, Leningrad University, 1969
	Post-Doctoral Thesis, Leningrad University, 1973
Positions:	Assistant Professor, Leningrad University, 1967–1974
	Professor, State University of New York at Stony Brook, 1974–1981
	Professor, Université de Paris VI, 1981–1982
	Permant Professor, Institut des Hautes Études Scientifiques at Bures-sur-Yvette, 1982–
	Professor, University of Maryland, 1991–1996
	Jay Gould Professor of Mathematics, Courant Institute of Mathematical Sciences, New York University, 1996–
Memberships:	National Academy of Sciences (USA), 1989
	American Academy of Arts and Sciences, 1989
	Académie des sciences, France (Associé étranger: 1989, titulaire: 1997)
	Norwegian Academy of Science and Letters, 2009
	Hungarian Academy of Sciences, 2011
	Royal Society, 2011
	Russian Academy of Sciences (Foreign member), 2011
Awards and prizes:	Prize of the Mathematical Society of Moscow, 1971
	Oswald Veblen Prize in Geometry, 1981
	Élie Cartan Prize, 1984
	Prix des Assurances de Paris, 1989
	Wolf Prize, 1993
	Lobachevsky Medal, 1997
	Leroy P. Steele Prize for Seminal Contribution to Research, 1997
	Balzan Prize, 1999
	Kyoto Prize in Basic Sciences, 2002

H. Holden, R. Piene (eds.), *The Abel Prize 2008-2012*,
DOI 10.1007/978-3-642-39449-2_13, © Springer-Verlag Berlin Heidelberg 2014

	Frederic Esser Nemmers Prize in Mathematics, 2004
	János Bolyai Prize, 2005
	Abel Prize, 2009
Honorary degrees:	University of Geneva, 1992
	Tel Aviv University, 2009
	University of Neuchâtel, 2009

2010

John Torrence Tate

"for his vast and lasting impact on the theory of numbers"

ABEL
PRISEN

Autobiography

John Tate

What follows is a sketch of some incidents in my education and early career, followed by a very brief summary of the last 50 years of my mathematical life to retirement.

Before beginning, I would like to thank Jim Milne for his willingness to take on the job of writing about my work, and for the remarkably thorough account he has given of it, with indications of its place in different aspects of the development of arithmetic geometry during the last half of the 20th century.

1925–1937

I was born on March 13, 1925, in Minneapolis, an only child. My father was an experimental physicist at the University of Minnesota who was Dean of the College of Arts and Sciences for a few years before he was called to lead the US antisubmarine research effort during WWII. The Physics building at the University of Minnesota is named the Tate Laboratory of Physics after him. His father was a doctor in rural Iowa, descended from Scotch Presbyterian ministers who had been in the US for several generations, moving west with the frontier. I know little about his mother, except that she was of Irish descent and died when my father was about 12.

My mother had a thorough knowledge of the classics. Before I was born she taught English in high school. Her father had come to the US from Germany as a teenager, settled in Lincoln, Nebraska, and eventually became head of the Germanic Languages department at the University of Nebraska. Her mother had come to Lincoln from Sweden with her family as a child.

I learned about negative numbers at an early age from the mercury thermometer mounted outside our kitchen window. That instrument also illustrated for me the

Electronic supplementary material Supplementary material is available in the online version of this chapter at http://dx.doi.org/10.1007/978-3-642-39449-2_14. Videos can also be accessed at http://www.springerimages.com/videos/978-3-642-39449-2.

J. Tate (✉)
Department of Mathematics, Harvard University, 1 Oxford Street, Cambridge MA 02138, USA
e-mail: tate@math.utexas.edu

H. Holden, R. Piene (eds.), *The Abel Prize 2008-2012*,
DOI 10.1007/978-3-642-39449-2_14, © Springer-Verlag Berlin Heidelberg 2014

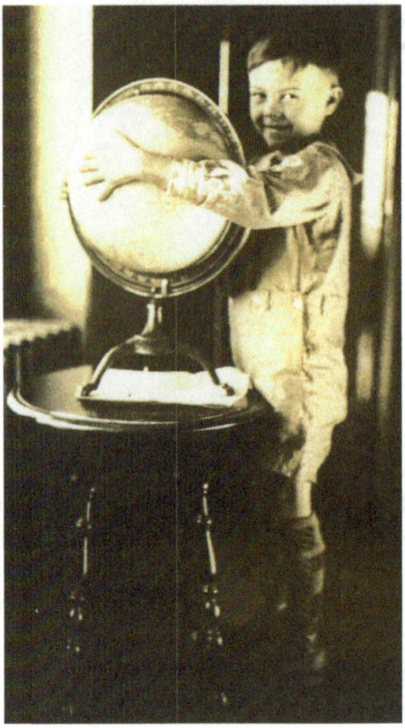

John Tate, age 4

concepts of linear function and coordinate change, for it was marked in Fahrenheit on one side and in Celcius on the other. The point marked −40 on both sides was near the bottom of the scale. The mercury was always above that, but it did fall below −30 F on a few winter mornings, which was a welcome event, for it meant no school.

I loved puzzles. I liked the Pastime jigsaw puzzles cut from plywood which my mother rented at a department store. Even more, I liked to read my father's puzzle books by Henry Dudeney. Most of the puzzles were too difficult for me to solve, but I enjoyed just contemplating the questions, which were usually of a mathematical or logical nature.

From my father I had a good idea of what science was about. He did not push me, but would sometimes explain some basic fact, such that the distance a body falls in x seconds is about $16.1x^2$ feet, or how something worked, like locks in a canal. I liked math and science, but was not very good at arithmetic, and especially hated the long division drills in fourth grade.

1937–1946

For my secondary education I attended Saint Paul Academy, a private day school, where I liked my first math teacher, Max Sporer. Somehow we understood each other. Once when I asked a question he said "your problem, Tate, is that you are trying to think". Another time we were doing a class exercise in which he would write a number on the board and ask one of us to declare it prime or to factor it. I enjoyed this until he gave me 91. By then I was 12 or 13 and my interest in mathematics was becoming clearer. I still remember Mr. Sporer's proof of the quadratic formula, which I thought was quite elegant because he multiplied by $4a$ before completing the square.

After I learned that $n!$ means the product of the numbers from 1 to n, I decided I would privately denote the sum of the numbers from 1 to n by $n?$. I soon realized that this was rather silly since my $n?$ is equal to $(n^2+n)/2$. However, this led eventually to my first attempt at mathematical research. After learning about the formulas for the sum of the first N squares and the sum of the first N cubes, and playing with sequences and their difference sequences for some time, I managed to convince myself that there were polynomials $P_n(x) = \frac{x^{n+1}}{n+1} + \frac{x^n}{2} + \cdots + C_n x$ such that $\sum_{i=1}^{N} i^n = P_n(N)$, and that these polynomials without constant term could be computed inductively from the relation $P_n'(x) = nP_{n-1}(x) + C_n$, if only one knew the constants C_n. I assumed that there would be a simple formula for C_n in terms of n and set out to try to guess it by computing the first few C_n's. It seemed clear that $C_n = 0$ for odd $n > 1$, but finding $C_2 = \frac{1}{6}$, $C_4 = -\frac{1}{30}$, $C_6 = \frac{1}{42}$, I was baffled, and gave up. Had I gone farther, I would have been even more baffled, because the sequence continues $-1/30, 5/66, -691/2730, 7/6, \ldots$. The C_n's are the Bernoulli numbers.

E.T. Bell's book *Men of Mathematics*, made a big impression on me. Reading it gave me my first knowledge of the history of mathematics in the West and of some great theorems. From the chapter on Fermat I learned that if p is a prime, then p divides $n^p - n$ for every integer n, and that if $p = 4m + 1$ then p is the sum of two squares. From the chapter on Gauss I learned that one could construct a regular 17-gon with ruler and compass, and also learned the law of quadratic reciprocity, which Bell makes vivid by giving a few explicit examples. This law fascinated me. I tried to figure out why it was so, of course with no success.

I graduated from S.P.A. in May of 1942, and went with no break to Harvard, since it was then on a full time schedule, including a summer term, because of the war. I had five consecutive terms at Harvard, the first three as a civilian, the last two as an apprentice seaman in the Navy's V-12 officer training program for which I volunteered when I became eligible for the draft. I then spent three terms at M.I.T. being trained as a meteorologist, and a term in midshipman school at Cornell University, graduating as an Ensign in the US Navy with a special knowledge of meteorology. But by that time the war in Europe was over, and it had progressed so far in the Pacific that more meteorologists were no longer needed. I was assigned to minesweeping research until I was discharged a year later, having been in the Navy for three years without leaving the east coast of the USA.

1946–1953

During that time I managed to graduate from Harvard. My degree was in mathematics for convenience, but I decided to go to graduate school in physics. This was a strange decision. Although I had always liked both subjects, I was really much more

John T. Tate Jr. with John T. Tate Sr. in New York in 1944

interested in math and had shown more talent for it. But from reading Bell's book I had the idea that to do valuable research in mathematics one had to be a genius like the people he wrote about, whereas from my father's example I saw that with intelligence and hard work one could make a difference in physics. I knew I was no Gauss or Galois, but thought I was reasonably intelligent and could be diligent.

So I began graduate school in physics in Princeton in fall, 1946. At that time physics and math shared a common room in the old Fine Hall. One day a fellow student pointed to a man across the room and said, "That's Artin!" "Who's Artin?" I asked. I was surprised to be told "He's the great algebraist", for I had never heard of him. Only later did I notice that my favorite math book, *Moderne Algebra*, by Van der Waerden, was based on lectures by E. Artin and E. Noether.

As a Navy veteran, I took advantage of a feature of the G.I. Bill of Rights. If a professor certified that a book I wanted would be useful to me in my studies, the US government paid for it. Early in the spring term I realized that I had acquired about twenty mathematics books in this way, but only two in physics. Also, in reading Von Neuman's *Mathematical Foundations of Quantum Mechanics*, I found the first part to be a marvellously clear axiomatic characterization of Hilbert space, but the rest was not clear at all. Deciding finally that I should switch from physics to mathematics, I asked permission from Lefschetz who was then head of the math department. He told me that too many people had been wanting to make the change I requested. Did we think the math prelims, which were oral, were easier than those in physics, which were written? He said I could put my application in with all the other applications to the math department and take my chances.

I did so, and also started to sit in on a couple of math courses. One was a course of Artin's in which he was developing measure theory. At the end of one lecture he stated a lemma and challenged us try to prove it. Highly motivated to show I was a worthy applicant for admission to math, I thought about it many hours with no success, but finally saw the trick in the middle of the night before the next class. In the morning when Artin asked who had found a proof I was the only one to raise

John Tate in his office, Cambridge, MA, 1956

my hand. When he told me to go to the board and explain it to the class I was so nervous that I could barely speak or write. To my great relief, as soon as he saw that I did have a proof, he took over and explained it clearly. Soon after, I was happy to hear that I had been admitted to the math department.

It was a phenomenal bit of luck that I ended up in the department with Emil Artin, the man who had proved the ultimate generalization of my favorite theorem, the law of quadratic reciprocity, and who was also a great teacher and mentor. I became his student, learned a great deal of algebra and number theory from him, and owe to him the suggestion of a wonderful thesis topic, proving by abstract harmonic analysis on the adèle ring the functional equation for Hecke's L-functions which Hecke had proved by classical Fourier analysis. In a sense this was simply a big exercise, but I think I gave a good solution. I soon realized that the topic was in the air at the time; Iwasawa and Weil had the same idea.

After earning my Ph.D. I stayed on at Princeton for three years as an Instructor. During the second year, 1951–1952, I helped Artin in a seminar doing class field theory by cohomological methods. We were very fortunate that Serge Lang, who had just finished his Ph.D. with Artin, took notes and wrote them up into what eventually became the main part of the book *Class Field Theory*, by E. Artin and me, recently republished by AMS Chelsea.

1953–1959

After a year at Columbia University, in which I was very happy to become acquainted with Bernie Dwork, I accepted a tenure track offer at Harvard. Fortunately, Harvard's enlightened policy of offering sabbatical years to tenure candidates enabled me to spend the year 1957–1958 in Paris. That was a great year for me. I attended Serre's lectures, witnessed the birth of the theory of schemes in Grothendieck's seminar, and became friends with those two amazing individuals who were to have such an important influence in my mathematical life. I found new directions for my research which I will briefly describe.

I had become interested in the arithmetic of elliptic curves and, thanks to Lang's influence, also Abelian varieties. The key results of class field theory are equivalent to knowing the Galois cohomology of the of the multiplicative group over local and global fields and their interrelations, in dimensions 1 and 2. I hoped my background in that area would help me study the local and global Galois cohomology of Abelian varieties. Just before arriving in Paris, I saw that, over a p-adic field k, there is a natural pairing between the group of torsors of an elliptic curve over k and the group of k-rational points on the curve, with values in the Brauer group of k. By local class field theory that Brauer group is canonically isomorphic to $\mathbb{Q}/\mathbb{Z}$, and I was able to prove that the pairing gives a perfect Pontrjagin duality between the compact profinite group of points and the discrete group of torsors. Lang, who was also in Paris that fall, insisted that I should extend the result to Abelian varieties and taught me what I needed to know in order to do so.

Having such success with the local theory I optimistically started thinking about the global situation. My optimism was unfounded. It soon became clear that the cohomological picture would be very nice if the group which is now called the Tate-Shafarevich group and denoted by Ш, is finite, and would be a mess if it is not. It also became clear that the global cohomology just gave a very general way of looking at the classical theory of descent and that the finiteness question was related to the effectiveness of the descent procedure. My naive hope that the cohomological machinery might enable me to find a proof of the finiteness of Ш gradually faded. In fact, it was not until almost 30 years later, with the addition of completely new methods by Thaine, Rubin, Kolyvagin, Gross, Zagier, that the finiteness was proved in special situations. The general case is still a complete mystery. Michael Artin has remarked that the question whether Ш is finite is a special case of a more general question: Is the Brauer group of every scheme proper over $\mathrm{Spec}(\mathbb{Z})$ finite?

From Grothendieck I learned that the most important thing in Galois cohomology is the cohomology of the absolute Galois group G_k of a field k, rather than that of the finite Galois groups of which G_k is the projective limit, and that G_k often has a finite cohomological dimension. For example for global and local fields k the cohomological dimension of G_k is 2 (with a grain of salt involving the real field and the prime 2). Thus the salient fact about the higher dimensional groups of class field theory which I had worked so hard to determine earlier is that they die under inflation and are trivial for G_k.

During that year in Paris I also tried to learn more about the structure of the group of points $E(k)$ of an elliptic curve E over a p-adic field k. Results of Dieudonné and Lazard on formal groups gave information about the subgroup $E_1(k)$ of points reducing to 0 over the residue field. In case of good ordinary reduction, the result was especially striking. I regret never having published it, but I did think to send the statement as a challenge to Dwork, who I saw as the world's greatest p-adic analyst. Almost by return mail he sent a proof completely different from mine, for the Legendre curve, using the hypergeometric function $F(\frac{1}{2}, \frac{1}{2}, 1, \lambda)$. He saw aspects of the situation which I had not dreamed of.

That sabbatical year in Paris which was so crucial to my career ended with a complete surprise, an invitation to collaborate with Bourbaki. This pleased me very

Barry Mazur (*right*) and John Tate (*left*)

much, for I admired his work. By the next train I went to the little village of Pelvoux in the French Alps to join his collaborators at their June congress. Working with them was a unique privilege.

A year later, still thinking about the $E(k)$ for p-adic k, I was thrilled to realize that some series expansions occurring in the classical theory of theta functions, which involve only integer coefficients, make sense over any field k complete with respect to an absolute value function $x \mapsto |x|$, and that using them one can construct, for each $q \in k$ such that $0 < |q| < 1$, an elliptic curve E_q defined over k, together with a k-analytic homomorphism $k^* = \mathbb{G}_m(k) \to E_q(k)$ which is surjective, with kernel $q^{\mathbb{Z}}$. The j invariant of E_q is given by the classical Fourier expansion of the j-function:

$$j = q^{-1} + 744 + 196884\,q + 21493760\,q^2 + \cdots$$

For $k = \mathbb{C}$ or $\mathbb{R}$, every elliptic curve over k can be obtained in this way, but for k nonarchimedean, e.g. p-adic, as the displayed formula shows, the j-invariant of E_q satisfies $|j| = |q|^{-1} > 1$, and $q = j^{-1} + 744j^{-2} + 750420j^{-3} + \cdots$ is uniquely determined by j. Moreover, every elliptic curve over k with such a non-integral j-invariant is a quadratic twist of the curve E_q which has the same j.

1959–1989

In fall 1959 I was given tenure at Harvard. I continued teaching at Harvard for the next thirty years, except for three more sabbatical years in Paris visiting the I.H.E.S. and/or Orsay and a half year visit to Berkeley. My research owes much to direct contact with colleagues in Cambridge and Paris. I think especially of Michael Artin, A. Grothendieck, David Mumford, J.-P. Serre and, later, Benedict Gross and Barry Mazur. Above all, an extensive postal correspondence with Serre was invaluable in helping me get my ideas straight.

Teaching at all levels, from calculus classes to mentoring Ph.D. students, has always been a pleasure for me. A good way to learn a topic is to teach it. On the

John Tate, Park City, 2000

other hand, giving graduate courses and talking with Ph.D. students is often a source of new ideas and cannot be separated from research efforts, contrary to the thinking of those who promote effort reporting. I have had more than 40 Ph.D. students, several of whom have become good friends.

1989–2009

In the late 1980's the offer of a Chair at the University of Texas at Austin prompted me to think about trying something new. This was appealing, for various reasons, and I felt fortunate to have such an opportunity. The U.T. Math. Department and the University itself were huge compared with what I had known, but size was the least of it. Nearly everything was new. I taught there for almost 20 years before retiring. During that time, although there was a great disparity among positions, the department was run democratically, and the atmosphere was very positive. I enjoyed being part of it and working with my colleagues in number theory, Vaaler, Villegas, and Voloch, from whom I learned a lot.

It was a time of change in Austin as the city grew rapidly, doubling in size in twenty years. The general spirit of the university seemed to be evolving along with the skyline of its home city. One sign of this evolution, to my delight, touched me directly, in connection with the Abel award. The 300 foot tower on the main building of the university, which is centrally located on high ground, is visible from almost anywhere in the city. By tradition it is illuminated in UT's distinctive burnt-orange color to signal an athletic victory. Under University President Bill Powers, academic achievement is honored in the same way.

2012

Looking back on my research, I take satisfaction in having proved or helped prove several important theorems, no one of which is so great that it stands out above the others. Some of my best ideas have been conjectures on which I could make little

(*From left to right*) Steve Leslie (Provost), Mary Ann Rankin (Dean), John Tate, President Bill Powers, University of Texas at Austin, 2010

progress, though others have. I liked trying to imagine what should be true, not only by making conjectures of my own, but in trying to generalize and clarify those of others, such as those of Birch and Swinnerton-Dyer, and of Stark.

Although I was born on a Friday the thirteenth, I feel I've had more than my share of good fortune. Two specific examples are my meeting, by pure chance, Emil Artin, the mentor from whom I learned so much, and my having had the opportunity to spend an especially valuable year 1958–1959 in Paris, thanks to Harvard's policy of offering junior sabbaticals. I did not write easily, and am thankful that my colleagues included unpublished results of mine in their papers and books, crediting me fully. More generally, I have had the support of family in spite of my obsession with mathematics, and have enjoyed good health to date. Finally, I feel fortunate to have been inclined toward mathematics, a field in which cooperation is so much more common than competition, and in which I could earn a living by doing what I most liked to do.

The Work of John Tate

J.S. Milne

Tate helped shape the great reformulation of arithmetic and geometry which has taken place since the 1950s.
Andrew Wiles.[1]

This is an exposition of Tate's work, written on the occasion of the award to him of the Abel prize. True to the epigraph, I have attempted to explain it in the context of the "great reformulation".

Notations

We speak of the primes of a global field where others speak of the places.

$M_S = S \otimes_R M$ for M an R-module and S and R-algebra.

$|S|$ is the cardinality of S.

$X_n = \mathrm{Ker}(x \mapsto nx\colon X \to X)$ and $X(\ell) = \bigcup_{m \geq 0} X_{\ell^m}$ (ℓ-primary component, ℓ a prime).

$\mathrm{Gal}(K/k)$ or $G(K/k)$ denotes the Galois group of K/k.

$\mu(R)$ is the group of roots of 1 in R.

$R^\times$ denotes the group of invertible elements of a ring R (apologies to Bourbaki).

$V^\vee$ denotes the dual of a vector space or the contragredient of a representation.

K^{al}, K^{sep}, K^{ab}, K^{un} ... denote an algebraic, separable, abelian, unramified... closure of a field K.

$\mathcal{O}_K$ denotes the ring of integers in a local or global field K.

[1] Introduction to Tate's talk at the conference on the Millenium Prizes (2000).

Electronic supplementary material Supplementary material is available in the online version of this chapter at http://dx.doi.org/10.1007/978-3-642-39449-2_15. Videos can also be accessed at http://www.springerimages.com/videos/978-3-642-39449-2.

J.S. Milne (✉)
Department of Mathematics, University of Michigan, 2074 East Hall, 530 Church Street, Ann Arbor, MI 48109-1043, USA
e-mail: jmilne@umich.edu

H. Holden, R. Piene (eds.), *The Abel Prize 2008-2012*,

DOI 10.1007/978-3-642-39449-2_15, © Springer-Verlag Berlin Heidelberg 2014

1 Hecke L-Series and the Cohomology of Number Fields

1.1 Background

Kronecker, Weber, Hilbert, and Ray Class Groups For every abelian extension L of $\mathbb{Q}$, there is an integer m such that L is contained in the cyclotomic field $\mathbb{Q}[\zeta_m]$; it follows that the abelian extensions of $\mathbb{Q}$ are classified by the subgroups of the groups $(\mathbb{Z}/m\mathbb{Z})^\times \simeq G(\mathbb{Q}[\zeta_m]/\mathbb{Q})$ (Kronecker–Weber). On the other hand, the unramified abelian extensions of a number field K are classified by the subgroups of the ideal class group C of K (Hilbert). In order to be able to state a common generalization of these two results, Weber introduced the ray class groups. A modulus $\mathfrak{m}$ for a number field K is the formal product of an ideal $\mathfrak{m}_0$ in $\mathcal{O}_K$ with a certain number of real primes of K. The corresponding ray class group $C_\mathfrak{m}$ is the quotient of the group of ideals relatively prime to $\mathfrak{m}_0$ by the principal ideals generated by elements congruent to 1 modulo $\mathfrak{m}_0$ and positive at the real primes dividing $\mathfrak{m}$. For $\mathfrak{m} = (m)\infty$ and $K = \mathbb{Q}$, $C_\mathfrak{m} \simeq (\mathbb{Z}/m\mathbb{Z})^\times$. For $\mathfrak{m} = 1$, $C_\mathfrak{m} = C$.

Takagi and the Classification of Abelian Extensions Let K be a number field. Takagi showed that the abelian extensions of K are classified by the ray class groups: for each modulus $\mathfrak{m}$, there is a well-defined "ray class field" $L_\mathfrak{m}$ with $G(L_\mathfrak{m}/K) \approx C_\mathfrak{m}$, and every abelian extension of K is contained in a ray class field for some modulus $\mathfrak{m}$. Takagi also proved precise decomposition rules for the primes in an extension L/K in terms of the associated ray class group. These would follow from knowing that the map sending a prime ideal to its Frobenius element gives an isomorphism $C_\mathfrak{m} \to G(L_\mathfrak{m}/K)$, but Takagi didn't prove that.

Dirichlet, Hecke, and L-Series For a character χ of $(\mathbb{Z}/m\mathbb{Z})^\times$, Dirichlet introduced the L-series

$$L(s,\chi) = \prod_{(p,m)=1} \frac{1}{1-\chi(p)p^{-s}} = \sum_{(n,m)=1} \chi(n)n^{-s}$$

in order to prove that each arithmetic progression, a, $a+m$, $a+2m$, ... with a relatively prime to m has infinitely many primes. When χ is the trivial character, $L(s,\chi)$ differs from the zeta function $\zeta(s)$ by a finite number of factors, and so has a pole at $s=1$. Otherwise $L(s,\chi)$ can be continued to a holomorphic function on the entire complex plane and satisfies a functional equation relating $L(s,\chi)$ and $L(1-s,\bar{\chi})$.

Hecke proved that the L-series of characters of the ray class groups $C_\mathfrak{m}$ had similar properties to Dirichlet L-series, and noted that his methods apply to the L-series of even more general characters, now called Hecke characters (Hecke 1918, 1920). The L-series of Hecke characters are of fundamental importance. For example, Deuring (1953) showed that the L-series of an elliptic curve with complex multiplication is a product of two Hecke L-series.

Artin and the Reciprocity Law Let K/k be an abelian extension of number fields, corresponding to a subgroup H of a ray class group $C_{\mathfrak{m}}$. Then

$$\zeta_K(s)/\zeta_k(s) = \prod_{\chi} L(s,\chi) \quad \text{(up to a finite number of factors)} \tag{1}$$

where χ runs through the nontrivial characters of $C_{\mathfrak{m}}/H$. From this and the results of Dirichlet and Hecke, it follows that $\zeta_K(s)/\zeta_k(s)$ is holomorphic on the entire complex plane. In the hope of extending this statement to nonabelian extensions K/k, Artin (1923) introduced what are now called Artin L-series.

Let K/k be a Galois extension of number fields with Galois group G, and let $\rho\colon G \to \mathrm{GL}(V)$ be a representation of G on a finite dimensional complex vector space V. The Artin L-series of ρ is

$$L(s,\rho) = \prod_{\mathfrak{p}} \frac{1}{\det(1-\rho(\sigma_{\mathfrak{p}})\mathbb{N}\mathfrak{P}^{-s} \mid V^{I_{\mathfrak{P}}})}$$

where $\mathfrak{p}$ runs through the prime ideals of K, $\mathfrak{P}$ is a prime ideal of K lying over $\mathfrak{p}$, $\sigma_{\mathfrak{p}}$ is the Frobenius element of $\mathfrak{P}$, $\mathbb{N}\mathfrak{P} = (\mathcal{O}_K : \mathfrak{P})$, and $I_{\mathfrak{P}}$ is the inertia group.

Artin observed that his L-series for one-dimensional representations would coincide with the L-series of characters on ray class groups if the following "theorem" were true:

> for the field L corresponding to a subgroup H of a ray class group $C_{\mathfrak{m}}$, the map $\mathfrak{p} \mapsto (\mathfrak{p}, L/K)$ sending a prime ideal $\mathfrak{p}$ not dividing $\mathfrak{m}$ to its Frobenius element induces an isomorphism $C_{\mathfrak{m}}/H \to G(L/K)$.

Initially, Artin was able to prove this statement only for certain extensions. After Chebotarev had proved his density theorem by a reduction to the cyclotomic case, Artin (1927) proved the statement in general. He called it the reciprocity law because, when K contains a primitive mth root of 1, it directly implies the classical mth power reciprocity law.

Artin noted that $L(s,\rho)$ can be analytically continued to a meromorphic function on the whole complex plane if its character χ can be expressed in the form

$$\chi = \sum_i n_i \operatorname{Ind} \chi_i, \quad n_i \in \mathbb{Z}, \tag{2}$$

with the χ_i one-dimensional characters on subgroups of G, because then

$$L(s,\rho) = \prod_i L(s,\chi_i)^{n_i}$$

with the $L(s,\chi_i)$ abelian L-series. Brauer (1947) proved that the character of a representation can always be expressed in the form (1), and Brauer and Tate found what is probably the simplest known proof of this fact (see p. 323).

To complete his program, Artin conjectured that, for every nontrivial irreducible representation ρ, $L(s,\rho)$ is holomorphic on the entire complex plane. This is called the Artin conjecture. It is known to be true if the character of ρ can be expressed in the form (2) with $n_i \geq 0$, and in a few other cases.

Chevalley and Idèles Chevalley gave a purely local proof of local class field theory, and a purely algebraic proof of global class field theory, but probably his most lasting contribution was to reformulate class field theory in terms of idèles.

An idèle of a number field K is an element $(a_v)_v$ of $\prod_v K_v^\times$ such that $a_v \in \mathcal{O}_v^\times$ for all but finitely many primes v. The idèles form a group J_K, which becomes a locally compact topological group when endowed with the topology for which the subgroup

$$\prod_{v|\infty} K_v^\times \times \prod_{v \text{ finite}} \mathcal{O}_v^\times$$

is open and has the product topology.[2]

Let K be number field. In Chevalley's reinterpretation, global class field theory provides a homomorphism $\phi\colon J_K/K^\times \to G(K^{\mathrm{ab}}/K)$ that induces an isomorphism

$$J_K/\big(K^\times \cdot \mathrm{Nm}\, J_L\big) \longrightarrow G(L/K)$$

for each finite abelian extension L/K. For each prime v of K, local class field theory provides a homomorphism $\phi_v\colon K_v^\times \to G(K_v^{\mathrm{ab}}/K_v)$ that induces an isomorphism

$$K_v^\times/\mathrm{Nm}\, L^\times \to G(L/K_v)$$

for each finite abelian extension L/K_v. The maps ϕ_v and ϕ are related by the diagram:

$$\begin{array}{ccc} K_v^\times & \xrightarrow{\phi_v} & G(K_v^{\mathrm{ab}}/K_v) \\ \downarrow{\scriptstyle i_v} & & \downarrow \\ J_K & \xrightarrow{\phi} & G(K^{\mathrm{ab}}/K). \end{array}$$

Beyond allowing class field theory to be stated for infinite extensions, Chevalley's idèlic approach greatly clarified the relation between the local and global reciprocity maps.

1.2 Tate's Thesis and the Local Constants

The modern definition is that a Hecke character is a quasicharacter of $J/K^\times$, i.e., a continuous homomorphism $\chi\colon J \to \mathbb{C}^\times$ such that $\chi(x) = 1$ for all $x \in K^\times$. We

[2]The original topology defined by Chevalley is not Hausdorff. It was Weil who pointed out the need for a topology in which the Hecke characters become the characters on J (Weil 1936). By the time of Tate's thesis, the correct definition seems to have been common knowledge.

explain how to interpret χ as a map on a group of ideals, which is the classical definition.

For a finite set S of primes, including the infinite primes, let J^S denote the subgroup of J_K consisting of the idèles $(a_v)_v$ with $a_v = 1$ for all $v \in S$, and let I^S denote the group of fractional ideals generated by those prime ideals not in S. There is a canonical surjection $J^S \to I^S$. For each Hecke character χ, there exists a finite set S such that χ factors through $J^S \xrightarrow{\text{can.}} I^S$, and a homomorphism $\varphi\colon I^S \to \mathbb{C}^\times$ arises from a Hecke character if and only if there exists an integral ideal $\mathfrak{m}$ with support in S, complex numbers $(s_\sigma)_{\sigma \in \mathrm{Hom}(K,\mathbb{C})}$, and integers $(m_\sigma)_{\sigma \in \mathrm{Hom}(K,\mathbb{C})}$ such that

$$\varphi\big((\alpha)\big) = \prod_{\sigma \in \mathrm{Hom}(K,\mathbb{C})} \sigma(\alpha)^{m_\sigma} \left|\sigma(\alpha)\right|^{s_\sigma}$$

for all $\alpha \in K^\times$ with $(\alpha) \in I^S$ and $\alpha \equiv 1 \pmod{\mathfrak{m}}$.

Hecke's Proof The classical proof uses that $\mathbb{R}^n$ is self-dual as an additive topological group, and that the discrete subgroup $\mathbb{Z}^n$ of $\mathbb{R}^n$ is its own orthogonal complement under the duality. The Poisson summation formula follows easily[3] from this: for any Schwartz function f on $\mathbb{R}^n$ and its Fourier transform $\hat{f}$,

$$\sum_{m \in \mathbb{Z}^n} f(m) = \sum_{m \in \mathbb{Z}^n} \hat{f}(m).$$

Write the L-series as a sum over integral ideals, and decompose it into a finite family of sums, each of which is over the integral ideals in a fixed element of an ideal class group. The individual series are Mellin transforms of theta series, and the functional equation follows from the transformation properties of the theta series, which, in turn, follow from the Poisson summation formula.

Tate's Proof An adèle of K is an element $(a_v)_v$ of $\prod K_v$ such that $a_v \in \mathcal{O}_v$ for all but finitely many primes v. The adèles form a ring $\mathbb{A}$, which becomes a locally compact topological ring when endowed with its natural topology.

Tate proved that the ring of adèles $\mathbb{A}$ of K is self-dual as an additive topological group, and that the discrete subgroup K of $\mathbb{A}$ is its own orthogonal complement under the duality. As in the classical case, this implies an (adèlic) Poisson summation formula: for any Schwartz function f on $\mathbb{A}$ and its Fourier transform $\hat{f}$

$$\sum_{\gamma \in K} f(\gamma) = \sum_{\gamma \in K} \hat{f}(\gamma).$$

[3] Let f be a Schwartz function on $\mathbb{R}$, and let $\hat{f}$ be its Fourier transform on $\hat{\mathbb{R}} = \mathbb{R}$. Let ϕ be the function $x + \mathbb{Z} \mapsto \sum_{n \in \mathbb{Z}} f(x+n)$ on $\mathbb{R}/\mathbb{Z}$, and let $\hat{\phi}$ be its Fourier transform on $\widehat{\mathbb{R}/\mathbb{Z}} = \mathbb{Z}$. A direct computation shows that $\hat{f}(n) = \hat{\phi}(n)$ for all $n \in \mathbb{Z}$. The Fourier inversion formula says that $\phi(x) = \sum_{n \in \mathbb{Z}} \hat{\phi}(n)\chi(x)$; in particular, $\phi(0) = \sum_{n \in \mathbb{Z}} \hat{\phi}(n) = \sum_{n \in \mathbb{Z}} \hat{f}(n)$. But, by definition, $\phi(0) = \sum_{n \in \mathbb{Z}} f(n)$.

Let χ be a Hecke character of K, and let χ_v be the quasicharacter $\chi \circ i_v$ on $K_v^\times$. Tate defines local L-functions $L(\chi_v)$ for each prime v of K (including the infinite primes) as integrals over K_v, and proves functional equations for them. He writes the global L-function as an integral over J, which then naturally decomposes into a product of local L-functions. The functional equation for the global L-function follows from the functional equations of the local L-functions and the Poisson summation formula.

Although, the two proofs are superficially similar, in the details they are quite different. Once Tate has developed the harmonic analysis of the local fields and of the adèle ring, including the Poisson summation formula, "an analytic continuation can be given at one stroke for all of the generalized ζ-functions, and an elegant functional equation can be established for them ... without Hecke's complicated theta-formulas"(Tate 1967b, pp. 305–306).

One consequence of Tate's treating all primes equally, is that the Γ-factors arise naturally as the local zeta functions of the infinite primes. By contrast, in the classical treatment, their appearance is more mysterious.

As Kudla (2003) writes:

> Tate provides an elegant and unified treatment of the analytic continuation and functional equation of Hecke L-functions. The power of the methods of abelian harmonic analysis in the setting of Chevalley's adèles/idèles provided a remarkable advance over the classical techniques used by Hecke. ... In hindsight, Tate's work may be viewed as giving the theory of automorphic representations and L-functions of the simplest connected reductive group $G = \mathrm{GL}(1)$, and so it remains a fundamental reference and starting point for anyone interested in the modern theory of automorphic representations.

Tate's thesis completed the re-expression of the classical theory in terms of idèles. In this way, it marked the end of one era, and the start of a new.

Notes Tate completed his thesis in May 1950. It was widely quoted long before its publication in 1967. Iwasawa obtained similar results about the same time as Tate, but published nothing except for the brief notes Iwasawa (1952, 1992).

Local Constants Let χ be a Hecke character, and let $\Lambda(s, \chi)$ be its completed L-series. The theorem of Hecke and Tate says that $\Lambda(s, \chi)$ admits a meromorphic continuation to the whole complex plane, and satisfies a functional equation

$$\Lambda(1 - s, \chi) = W(\chi) \cdot \Lambda(s, \bar{\chi})$$

with $W(\chi)$ a complex number of absolute value 1. The number $W(\chi)$ is called the root number or the epsilon factor. It is a very interesting number. For example, for a Dirichlet character χ with conductor f, it equals $\tau(\chi)/\sqrt{\pm f}$ where $\tau(\chi)$ is the Gauss sum $\sum_{a=1}^{f} \chi(a)\mathbf{e}(a/f)$. An importance consequence of Tate's description of the global functional equation as a product of local functional equations is that he obtains an expression

$$W(\chi)=\prod_{v} W(\chi_v) \tag{3}$$

of $W(\chi)$ as a product of (explicit) local root numbers $W(\chi_v)$.

Langlands pointed out[4] that his conjectural correspondence between degree n representations of the Galois groups of number fields and automorphic representations of GL_n requires that there be a similar decomposition for the root numbers of Artin L-series, or, more generally, for the Artin-Hecke L-series that generalize both Artin and Hecke L-series (see p. 268). For a Hecke character, the required decomposition is just that of Tate. Every expression (2), p. 261, of an Artin character χ as a sum of monomial characters gives a decomposition of its root number $W(\chi)$ as a product of local root numbers—the problem is to show that the decomposition is *independent* of the expression of χ as a sum.[5]

For an Artin character χ, Dwork (1956) proved that there exists a decomposition (3) of $W(\chi)$ well-defined up to signs; more precisely, he proved that there exists a well-defined decomposition for $\chi(-1)W(\chi)^2$. Langlands completed Dwork's work and thereby found a local proof that there exists a well-defined decomposition for $W(\chi)$. However, he abandoned the writing up of his proof when Deligne (1973) found a simpler global proof.

Tate (1977b) gives an elegant exposition of these questions, including a proof of (3) for Artin root numbers by a variant of Deligne's method, and a proof of a theorem of Fröhlich and Queyrut that $W(\chi)=1$ when χ is the character of a representation that preserves a quadratic form.

1.3 The Cohomology of Number Fields

Tate Cohomology With the action of a group G on an abelian group M, there are associated homology groups $H_r(G,M)$, $r\geq 0$, and cohomology groups $H^r(G,M)$, $r\geq 0$. When G is finite, the map $m\mapsto\sum_{\sigma\in G}\sigma m$ defines a homomorphism

$$H_0(G,M)\stackrel{\text{def}}{=}M_G\xrightarrow{\mathrm{Nm}_G}M^G\stackrel{\text{def}}{=}H^0(G,M),$$

---

[4]See his "Notes on Artin L-functions" and the associated comments at http://publications.ias.edu/rpl/section/22.

[5]In fact, this is not quite true, but is true for "virtual representations" with "virtual degree 0". The decomposition of the root number of the character χ of a Galois representation is obtained by writing it as $\chi=(\chi-\dim\chi\cdot 1)+\dim\chi\cdot 1$.

and Tate defined cohomology groups $\hat{H}^r(G,M)$ for all integers r by setting

$$\hat{H}^r(G,M) \stackrel{\text{def}}{=} \begin{cases} H_{-r-1}(G,M) & r < -1 \\ \operatorname{Ker}(\operatorname{Nm}_G) & r = -1 \\ \operatorname{Coker}(\operatorname{Nm}_G) & r = 0 \\ H^r(G,M) & r > 0. \end{cases}$$

The diagram

$$\begin{array}{ccccccccc}
 & & \hat{H}^{-1}(G,M') & \to & \hat{H}^{-1}(G,M) & \to & \hat{H}^{-1}(G,M'') & & \\
 & & \downarrow & & \downarrow & & \downarrow & & \\
\cdots \to H_1(G,M'') & \to & H_0(G,M') & \to & H_0(G,M) & \to & H_0(G,M'') & \to & 0 \\
 & & \downarrow \scriptstyle{\operatorname{Nm}_G} & & \downarrow \scriptstyle{\operatorname{Nm}_G} & & \downarrow \scriptstyle{\operatorname{Nm}_G} & & \\
0 & \to & H^0(G,M') & \to & H^0(G,M) & \to & H^0(G,M'') & \to & H^1(G,M') \to \cdots \\
 & & \downarrow & & \downarrow & & \downarrow & & \\
 & & \hat{H}^0(G,M') & \to & \hat{H}^0(G,M) & \to & \hat{H}^0(G,M'') & &
\end{array}$$

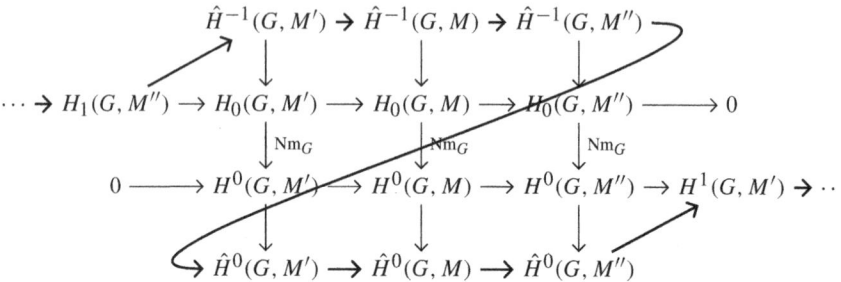

shows that a short exact sequence of G-modules gives an exact sequence of cohomology groups infinite in both directions. These groups are now called the Tate cohomology groups. Most of the usual constructions for cohomology groups (except the inflation maps) extend to the Tate groups.

Notes Tate's construction was included in Serre (1952/1953), Cartan and Eilenberg (1956), and elsewhere. Farrell (1977/78) extended Tate's construction to infinite groups having finite virtual cohomological dimension (Tate-Farrell cohomology), and others have defined an analogous extension of Hochschild cohomology (Tate-Hochschild cohomology).

The Cohomology Groups of Algebraic Number Fields Let G be a finite group, let C be a G-module, and let u be an element of $H^2(G,C)$. Assume that $H^1(H,C)=0$ for all subgroups H of G and that $H^2(H,C)$ is cyclic of order $(H:1)$ with generator the restriction of u. Then Tate (1952c) showed that cup product with u defines an isomorphism

$$x \mapsto x \cup u \colon \hat{H}^r(G,\mathbb{Z}) \to \hat{H}^{r+2}(G,C) \tag{4}$$

for all $r \in \mathbb{Z}$. He proves this by constructing an exact sequence

$$0 \to C \to C(\varphi) \to \mathbb{Z}[G] \to \mathbb{Z} \to 0,$$

depending on the choice of a 2-cocycle φ representing u, and showing that

$$\hat{H}^r\big(G, C(\varphi)\big) = 0 = \hat{H}^r\big(G, \mathbb{Z}[G]\big)$$

for all $r \in \mathbb{Z}$. Now the double boundary map is an isomorphism $\hat{H}^r(G,\mathbb{Z}) \to \hat{H}^{r+2}(G,C)$.

On taking G to be the Galois group of a finite extension L/K of number fields, C to be the idèle class group of L, and u the fundamental class of L/K,[6] one obtains for $r = -2$ the inverse of the Artin reciprocity map

$$G/[G,G] \xrightarrow{\simeq} C^G.$$

Let L/K be a finite Galois extension of global fields (e.g., number fields) with Galois group G. There is an exact sequence of G-modules

$$1 \to L^\times \to J_L \to C_L \to 1 \tag{5}$$

where J_L is the group of idèles of L and C_L is the idèle class group. Tate determined the cohomology groups of the terms in this sequence by relating them to those in the much simpler sequence

$$0 \to X \to Y \to Z \to 0. \tag{6}$$

Here Y is the free abelian group on the set of primes of L (including the infinite primes) with G acting through its action on the primes, and Z is just $\mathbb{Z}$ with G acting trivially; the map $Y \to Z$ is $\sum n_P P \mapsto \sum n_P$, and X is its kernel. Tate proved that there is a canonical isomorphism of doubly infinite exact sequences

$$\begin{array}{ccccccccc}
\cdots & \longrightarrow & \hat{H}^r(G,X) & \longrightarrow & \hat{H}^r(G,Y) & \longrightarrow & \hat{H}^r(G,Z) & \longrightarrow & \cdots \\
 & & \downarrow\simeq & & \downarrow\simeq & & \downarrow\simeq & & \\
\cdots & \longrightarrow & \hat{H}^{r+2}(G,L^\times) & \longrightarrow & \hat{H}^{r+2}(G,J_L) & \longrightarrow & \hat{H}^{r+2}(G,C_L) & \longrightarrow & \cdots.
\end{array} \tag{7}$$

Tate announced this result in his Short Lecture at the 1954 International Congress, but did not immediately publish the proof.

The Tate-Nakayama Theorem Nakayama (1957) generalized Tate's isomorphism (4) by weakening the hypotheses—it suffices to require them for Sylow subgroups—and strengthening the conclusion—cup product with u defines an isomorphism

$$x \mapsto x \cup u \colon \hat{H}^r(G,M) \to \hat{H}^{r+2}(G, C \otimes M)$$

provided C or M is torsion-free. Building on this, Tate (1966c) proved that the isomorphism (7) holds with each of the sequences (5) and (6) replaced by its tensor product with M. In other words, he replaced the torus $\mathbb{G}_m$ implicit in (7) with an arbitrary torus defined over K. He also proved the result for any "suitably large" set of primes S—the module $L^\times$ is replaced with the group of S-units in L and J_L is

[6]Which had been discovered by Nakayama and Weil, cf. Artin and Tate, 2009, p. 189.

replaced by the group of idèles whose components are units outside S. This result is usually referred to as the Tate-Nakayama theorem, and is widely used, for example, throughout the Langlands program including in the proof of the fundamental lemma.

Abstract Class Field Theory: Class Formations Tate's theorem (see (4) above) shows that, in order to have a class field theory over a field k, all one needs is, for each system of fields

$$k^{\text{sep}} \supset L \supset K \supset k, \quad [L\colon k] < \infty, \quad L/K \text{ Galois},$$

a $G(L/K)$-module C_L and a "fundamental class" $u_{L/K} \in H^2(G(L/K), C_L)$ satisfying Tate's hypotheses; the pairs $(C_L, u_{L/K})$ should also satisfy certain natural conditions when K and L vary. Then Tate's theorem then provides "reciprocity" isomorphisms

$$C_L^G \xrightarrow{\simeq} G/[G,G], \quad G = G(L/K),$$

Artin and Tate (1961, Chap. 14) formalized this by introducing the abstract notion of a class formation.

For example, for any nonarchimedean local field k, there is a class formation with $C_L = L^\times$ for any finite extension L of k, and for any global field, there is a class formation with $C_L = J_L/L^\times$. In both cases, $u_{L/K}$ is the fundamental class.

Let k be an algebraic function field in one variable with algebraically closed constant field. Kawada and Tate (1955a) show that there is a class formation for unramified extensions of K with C_L the dual of the group of divisor classes of L. In this way they obtain a "pseudo class field theory" for k, which they examine in some detail when $k = \mathbb{C}$.

The Weil Group Weil was the first to find a common generalization of Artin L-series and Hecke L-series. For this he defined what is now known as the Weil group. The Weil group of a finite Galois extension of number fields L/K is an extension

$$1 \to C_L \to W_{L/K} \to \mathrm{Gal}(L/K) \to 1$$

corresponding to the fundamental class in $H^2(G_{L/K}, C_L)$. Each representation of $W_{L/K}$ has an L-series attached to it, and the L-series arising in this way are called Artin-Hecke L-series. Weil (1951) constructed these groups, thereby discovering the fundamental class, and proved the fundamental properties of Weil groups. Artin and Tate (1961, Chap. XV) developed the theory of the Weil groups in the abstract setting of class formations, basing their definition on the existence of a fundamental class. In the latest (2009) edition of the work, Tate expanded their presentation and included a sketch of Weil's original construction (pp. 185–189).

Summary The first published exposition of class field theory in which full use of the cohomology theory is made is Chevalley (1954). There Chevalley writes:

> One of the most baffling features of classical class field theory was that it appeared to say practically nothing about normal extensions that are not abelian. It was discovered by A. Weil and, from a different point of view, T. Nakayama that class field theory was actually much richer than hitherto suspected; in fact, it can now be formulated in the form of statements about normal extensions without any mention whatsoever of abelian extensions. Of course, it is true that it is only in the abelian case that these statements lead to laws of decomposition for prime ideals of the subfield and to the law of reciprocity. Nevertheless, it is clear that, by now, we know something about the arithmetic of non abelian extensions. In fact, since the work of J. Tate, it may be said that we know almost everything that may be formulated in terms of cohomology in the idèle class group, and generally a great deal about everything that can be formulated in cohomological terms.

Notes Tate was not the first to make use of group cohomology in class field theory. In a sense it had always been there, since crossed homomorphisms and factor systems had long been used. Weil and Nakayama independently discovered the fundamental class, Weil by constructing the Weil group, and Nakayama as a consequence of his work (partly with Hochschild) to determine the cohomology groups of number fields in degrees 1 and 2. Tate's contribution was to give a remarkably simple description of all the basic cohomology groups of number fields, and to construct a general isomorphism that, in the particular case of an abelian extension and in degree -2, became the Artin reciprocity isomorphism.

1.4 The Cohomology of Profinite Groups

Krull (1928) showed that, when the Galois group of an infinite Galois extension of fields Ω/F is endowed with a natural topology, there is a Galois correspondence between the intermediate fields of Ω/F and the *closed* subgroups of the Galois group. The topological groups that arise as Galois groups are exactly the compact groups G whose open normal subgroups U form a fundamental system $\mathcal{N}$ of neighbourhoods of 1. Tate described such topological groups as being "of Galois-type", but we now say they are "profinite".

For a profinite group G, Tate (1958d) considered the G-modules M such that $M=\bigcup_{U\in\mathcal{N}}M^U$. These are the G-modules M for which the action is continuous relative to the discrete topology on M. For such a module, Tate defined cohomology groups $H^r(G,M)$, $r\geq 0$, using continuous cochains, and he showed that

$$H^r(G,M)=\varinjlim_{U\in\mathcal{N}}H^r\big(G/U,M^U\big)$$

where $H^r(G/U,M^U)$ denotes the usual cohomology of the (discrete) finite group G/U acting on the abelian group M^U. In particular, $H^r(G,M)$ is torsion for $r>0$.

The cohomological dimension and strict cohomological dimension of a profinite group G relative to a prime number p are defined by the conditions:

$$\mathrm{cd}_p(G) \leq n \iff H^r(G, M)(p) = 0 \text{ whenever } r > n \text{ and } M \text{ is torsion;}$$

$$\mathrm{scd}_p(G) \leq n \iff H^r(G, M)(p) = 0 \text{ whenever } r > n.$$

Here (p) denotes the p-primary component. The (strict) cohomological dimension of a field is the (strict) cohomological dimension of its absolute Galois group. Among Tate's theorems are the following statements:

(a) A pro p-group G is free if and only if $\mathrm{cd}_p(G) = 1$. (A pro p-group is a profinite group G such that G/U is a p-group for all $U \in \mathcal{N}$; it is free if it is of the form $\varprojlim F/N$ where F is the free group on symbols $(a_i)_{i \in I}$, say, and N runs through the normal subgroups of G containing all but finitely many of the a_i and such G/N is a finite p-group.)
(b) If k is a local field other than $\mathbb{R}$ or $\mathbb{C}$, then $\mathrm{scd}_p(k) = 2$ for all $p \neq \mathrm{char}(k)$.
(c) Let $K \supset k$ be an extension of fields of transcendence degree n. Then

$$\mathrm{cd}_p(K) \leq \mathrm{cd}_p(k) + n,$$

with equality if K is finitely generated over k, $\mathrm{cd}_p(k) < \infty$, and $p \neq \mathrm{char}(k)$. In particular, if k is algebraically closed, then the p-cohomological dimension of a finitely generated K is equal to its transcendence degree over k ($p \neq \mathrm{char}(k)$).

According to Tate (1958d), statement (c) "historically arose at [the theory's] beginning. Its conjecture and the sketch of its proof are due to Grothendieck". Indeed, from Grothendieck's point of view, the cohomology of the absolute Galois group of a field k should be interpreted as the étale cohomology of Spec k, and the last statement of (c) is suggested by the weak Lefschetz theorem in étale cohomology.

Notes Tate explained the above theory in his 1958 seminar at Harvard.[7] Douady reported on Tate's work in a Bourbaki seminar in 1959, and Lang included Tate's unpublished article (1958d) as Chap. VII of his 1967 book (Lang 1967). Serre included the theory in his course at the Collège de France, 1962–1963; see Serre (1964a). Tate himself published only the brief lectures Tate (2001).

1.5 Duality Theorems

In the early 1960s, Tate proved duality theorems for modules over the absolute Galois groups of local and global fields that have become an indispensable tool in Iwasawa theory, the theory of abelian varieties, and in other parts of arithmetic geometry. The main global theorem was obtained independently by Poitou, and is now referred to as the Poitou-Tate duality theorem.

[7] See Shatz, Math Reviews 0212073.

Throughout, K is a field, $\bar{K}$ is a separable closure of K, and G is the absolute Galois group $\mathrm{Gal}(\bar{K}/K)$. All G-modules are discrete (i.e., the action is continuous for the discrete topology on the module). The dual M' of such a module is $\mathrm{Hom}(M, \bar{K}^{\times})$.

Local Results Let K be a nonarchimedean local field, i.e., a finite extension of $\mathbb{Q}_p$ or $\mathbb{F}_p((t))$. Local class field theory provides us with a canonical isomorphism $H^2(G, \bar{K}^{\times}) \simeq \mathbb{Q}/\mathbb{Z}$. Tate proved that, for every finite G-module M whose order m is not divisible by characteristic of K, the cup-product pairing

$$H^r(G, M) \times H^{2-r}(G, M') \to H^2(G, \bar{K}^{\times}) \simeq \mathbb{Q}/\mathbb{Z} \tag{8}$$

is a perfect duality of finite groups for all $r \in \mathbb{N}$. In particular, $H^r(G, M) = 0$ for $r > 2$. Moreover, the following holds for the Euler-Poincaré characteristic of M:

$$\frac{|H^0(G, M)||H^2(G, M)|}{|H^1(G, M)|} = \frac{1}{(\mathcal{O}_K : m\mathcal{O}_K)}.$$

A G-module M is said to be unramified if the inertia group I in G acts trivially on M. When M is unramified and its order is prime to the residue characteristic, Tate proved that the submodules $H^1(G/I, M)$ and $H^1(G/I, M')$ of $H^1(G, M)$ and $H^1(G, M')$ are exact annihilators in the pairing (8).

Let $K = \mathbb{R}$. In this case, there is a canonical isomorphism $H^2(G, \bar{K}^{\times}) \simeq \frac{1}{2}\mathbb{Z}/\mathbb{Z}$. For any finite G-module M, the cup-product pairing of Tate cohomology groups

$$\hat{H}^r(G, M) \times \hat{H}^{2-r}(G, M') \to H^2(G, \bar{K}^{\times}) \simeq \frac{1}{2}\mathbb{Z}/\mathbb{Z}$$

is a perfect pairing for all $r \in \mathbb{Z}$. Moreover, $\hat{H}^r(G, M)$ is a finite group, killed by 2, whose order is independent of r.

Global Results Let K be a global field, and let M be a finite G-module whose order is not divisible by the characteristic of K. Let

$$H^r(K_v, M) = \begin{cases} H^r(G_{K_v}, M) & \text{if } v \text{ is nonarchimedean} \\ \hat{H}^r(G_{K_v}, M) & \text{otherwise.} \end{cases}$$

The local duality results show that $\prod_v H^0(K_v, M)$ is dual to $\bigoplus_v H^2(K_v, M')$ and that $\prod'_v H^1(K_v, M)$ is dual to $\prod'_v H^1(K_v, M')$—the $'$ means that we are taking the restricted product with respect to the subgroups $H^1(G_{K_v}/I_v, M^{I_v})$.

In the table below, the homomorphisms at right are the duals of the homomorphisms at left with M replaced by M', i.e., $\beta^r(M) = \alpha^{2-r}(M')^*$ with $-^* =$

Hom(−, ℚ/ℤ):

$$\begin{array}{ll} H^0(K,M) \xrightarrow{\alpha^0} \prod_v H^0(K_v,M) & \bigoplus_v H^2(K_v,M) \xrightarrow{\beta^2} H^0(K,M')^* \\ H^1(K,M) \xrightarrow{\alpha^1} \prod'_v H^1(K_v,M) & \prod'_v H^1(K_v,M) \xrightarrow{\beta^1} H^1(K,M')^* \\ H^2(K,M) \xrightarrow{\alpha^2} \bigoplus_v H^2(K_v,M) & \prod_v H^0(K_v,M) \xrightarrow{\beta^0} H^2(K,M')^*. \end{array}$$

The Poitou-Tate duality theorem states that there is an exact sequence

$$\begin{array}{ccccccc} 0 \longrightarrow & H^0(K,M) & \xrightarrow{\alpha^0} & \prod_v H^0(K_v,M) & \xrightarrow{\beta^0} & H^2(K,M')^* & \\ & & & & & \downarrow & \\ & H^1(K,M')^* & \xleftarrow{\beta^1} & \prod'_v H^1(K_v,M) & \xleftarrow{\alpha^1} & H^1(K,M) & \\ & \downarrow & & & & & \\ & H^2(K,M) & \xrightarrow{\alpha^2} & \bigoplus_v H^2(K_v,M) & \xrightarrow{\beta^2} & H^0(K,M')^* & \longrightarrow 0 \end{array} \tag{9}$$

(with explicit descriptions for the unnamed arrows). Moreover, for $r \geq 3$, the map

$$H^r(K,M) \to \prod_{v \text{ real}} H^r(K_v,M)$$

is an isomorphism. In fact, the statement is more general in that one replaces the set of all primes with a nonempty set S containing the archimedean primes in the number field case (there is then a restriction on the order of M).

For the Euler-Poincaré characteristic, Tate proved that

$$\frac{|H^0(G,M)||H^2(G,M)|}{|H^1(G,M)|} = \frac{1}{|M|^{r_1+2r_2}} \prod_{v|\infty} |M^{G_v}| \tag{10}$$

where r_1 and r_2 are the numbers of real and complex primes.

Notes Tate announced the above results (with brief indications of proof) in his talk at the 1962 International Congress except for the last statement on the Euler-Poincaré characteristic, which was announced in Tate (1966e). Tate's proofs of the local statements were included in Serre (1964b). Later Tate (1966f) proved a duality theorem for an abstract class formation, which included both the local and global duality results, and in which the exact sequence (9) arises as a sequence of Exts. This proof, as well as proofs of the formulas for the Euler-Poincaré characteristics, are included in Milne (1986).

1.6 Expositions

The notes of the famous Artin-Tate seminar on class field theory have been a standard reference on the topic since they first became available in 1961. They have recently been republished in slightly revised form by the American Mathematical Society. Tate made important contributions, both in his article on global class field theory and in the exercises, to another classic exposition of algebraic number theory, namely, the proceeding of the 1965 Brighton conference (Cassels and Fröhlich 1967). His talk on Hilbert's ninth problem, which asked for "a proof of the most general reciprocity law in any number field", illuminates the problem and the work done on it (Tate 1976a). Tate's contribution to the proceedings of the Corvallis conference, gave a modern account of the Weil group and an explanation of the hypothetical nonabelian reciprocity law in terms of the more general Weil-Deligne group (Tate 1979).

2 Abelian Varieties and Curves

In the course of proving the Riemann hypothesis for curves and abelian varieties in the 1940s, Weil rewrote the foundations of algebraic geometry, including the theory of abelian varieties. This made it possible to do algebraic geometry in a rigorous fashion over arbitrary base fields. In the late 1950s, Grothendieck rewrote the foundations again, developing the more natural and flexible language of schemes.

2.1 The Riemann Hypothesis for Curves

After Hasse proved the Riemann hypothesis for elliptic curves over finite fields in 1930, he and Deuring realized that, in order to extend the proof to curves of higher genus, one should replace the endomorphisms of the elliptic curve by correspondences. However, they regarded correspondences as objects in a double field, and this approach didn't lead to a proof until Roquette (1953) (Roquette was a student of Hasse). In the meantime Weil had realized that everything needed for the proof could be found already in the work of the Italian geometers on correspondences, at least in characteristic zero. In order to give a rigorous proof, he laid the foundations for algebraic geometry over arbitrary fields,[8] and completed the proof of the Riemann hypothesis for all curves over finite fields in 1945 (Weil 1945).[9]

[8]The main lacunae at the time were a rigorous intersection theory taking account of the phenomenon of pure inseparability and the construction of the Jacobian variety in nonzero characteristic.

[9]Much has been written on these events. I've found the following particularly useful: Schappacher (2006).

The key point of Weil's proof is that the inequality of Castelnuovo-Severi continues to hold in characteristic p, i.e., for a divisor D on the product of two complete nonsingular curves C and C' over an algebraically closed field,

$$[D \cdot D] \leq 2dd' \tag{11}$$

where $d = [D \cdot (P \times C')]$ and $d' = [D \cdot (C \times P')]$ are the degrees of D over C and C' respectively. Mattuck and Tate (1958a) showed that it is possible to derive (11) directly and easily from the Riemann-Roch theorem for surfaces, for which they were able to appeal to Zariski (1952) or to a sheaf-theoretic proof of Serre which is sketched in Zariski (1956).

The Mattuck-Tate proof is the most attractive geometric proof of Weil's theorem. Grothendieck (1958) simplified it further by showing that the Castelnuovo-Severi inequality can most naturally be derived from the Hodge index theorem for surfaces, which itself can be derived directly from the Riemann-Roch theorem.

Hodge proved his index theorem for smooth projective varieties over $\mathbb{C}$. That it should hold for such varieties in nonzero characteristic is known as Grothendieck's "Hodge standard conjecture", whose proof Grothendieck calls one of the "most urgent tasks in algebraic geometry" (Grothendieck 1969). In the more than forty years since Grothendieck formulated the conjecture, almost no progress has been made towards its proof—even in characteristic zero, there exists no algebraic proof in dimensions greater than 2.

The Tate Module of an Abelian Variety Let A be an abelian variety over a field k. For a prime l, let $A(l) = \bigcup A(k^{\mathrm{sep}})_{l^n}$ where $A(k^{\mathrm{sep}})_{l^n} = \mathrm{Ker}(A(k^{\mathrm{sep}}) \xrightarrow{l^n} A(k^{\mathrm{sep}}))$. Then $A \rightsquigarrow A(l)$ is a functor from abelian varieties over k to l-divisible groups equipped with an action of $\mathrm{Gal}(k^{\mathrm{sep}}/k)$. When $l \neq \mathrm{char}(k)$, $A(l) \simeq (\mathbb{Q}_\ell/\mathbb{Z}_\ell)^{2\dim A}$, and Weil used $A(l)$ to study the endomorphisms of A. Tate observed that it is more convenient to work with

$$T_l A = \varprojlim A\left(k^{\mathrm{sep}}\right)_{l^n},$$

which is a free $\mathbb{Z}_l$-module of rank $2\dim A$ when $l \neq \mathrm{char}(k)$—this is now called the Tate module of A.

2.2 *Heights on Abelian Varieties*

The Néron-Tate (Canonical) Height Let K be a number field, and normalize the absolute values $|\cdot|_v$ of K so that the product formula holds:

$$\prod_v |a|_v = 1 \quad \text{for all } v \in K^\times.$$

The logarithmic height of a point $P = (a_0 : \ldots : a_n)$ of $\mathbb{P}^n(K)$ is defined to be

$$h(P) = \log\Bigl(\prod_v \max\bigl\{|a_0|_v, \ldots, |a_n|_v\bigr\}\Bigr).$$

The product formula shows that this is independent of the representation of P.

Let X be a projective variety. A morphism $f: X \to \mathbb{P}^n$ from X into projective space defines a height function $h_f(P) = h(f(P))$ on X. In a Short Communication at the 1958 International Congress, Néron conjectured that, in certain cases, the height is given by a quadratic form.[10] Tate proved this for abelian varieties by a simple direct argument.

Let A be an abelian variety over a number field K. A nonconstant map $f: A \to \mathbb{P}^n$ of A into projective space is said to be symmetric if the inverse image D of a hyperplane is linearly equivalent to $(-1)^*D$. For a symmetric embedding f, Tate proved that there exists a unique *quadratic* map $\hat{h}: A(K) \to \mathbb{R}$ such that $\hat{h}(P) - h_f(P)$ is bounded on $A(K)$. To say that $\hat{h}$ is quadratic means that $\hat{h}(2P) = 4\hat{h}(P)$ and that the function

$$P, Q \mapsto \frac{1}{2}\bigl(\hat{h}(P+Q) - \hat{h}(P) - \hat{h}(Q)\bigr) \tag{12}$$

is bi-additive on $A(K) \times A(K)$.

Note first that there exists at most one function $\hat{h}: A(K) \to \mathbb{R}$ such that (a) $\hat{h}(P) - h_f(P)$ is bounded on $A(K)$, and (b) $\hat{h}(2P) = 4\hat{h}(P)$ for all $P \in A(K)$. Indeed, if $\hat{h}$ satisfies (a) with bound B, then

$$\bigl|\hat{h}(2^n P) - h_f(2^n P)\bigr| \leq B$$

for all $P \in A(K)$ and all $n \geq 0$. If in addition it satisfies (b), then

$$\left|\hat{h}(P) - \frac{h_f(2^n P)}{4^n}\right| \leq \frac{B}{4^n}$$

for all n, and so

$$\hat{h}(P) = \lim_{n\to\infty} \frac{h_f(2^n P)}{4^n}. \tag{13}$$

Tate used Eq. (13) to define $\hat{h}$, and applied results of Weil on abelian varieties to verify that it is quadratic.

Let A' be the dual abelian variety to A. For a map $f: A \to \mathbb{P}^n$ corresponding to a divisor D, let $\varphi_f: A(K) \to A'(K)$ be the map sending P to the point on A

---

[10] Only the title, *Valeur asymptotique du nombre des points rationnels de hauteur bornée sur une courbe elliptique*, of Néron's communication is included in the Proceedings. The sentence paraphrases one from: Lang, Serge. Les formes bilinéaires de Néron et Tate. Séminaire Bourbaki, 1963/64, Fasc. 3, Exposé 274.

represented by the divisor $(D+P)-D$. Tate showed that there is a unique bi-additive pairing

$$\langle\,,\rangle\colon\quad A'(K)\times A(K)\to\mathbb{R} \tag{14}$$

such that, for every symmetric f, the function $\langle\varphi_f(P),P\rangle+2h_f(P)$ is bounded on $A(K)$.

Néron (1965) found his own construction of $\hat{h}$, which is much longer than Tate's, but which has the advantage of expressing $\hat{h}$ as a sum of local heights. The height function $\hat{h}$, is now called the Néron-Tate, or canonical, height. It plays a fundamental role in arithmetic geometry.

Notes Tate explained his construction in his course on abelian varieties at Harvard in the fall of 1962, but did not publish it. However, it was soon published by others (Lang 1964; Manin 1964).[11]

Variation of the Canonical Height of a Point Depending on a Parameter Let T be an algebraic curve over $\mathbb{Q}^{\mathrm{al}}$, and let $E\to T$ be an algebraic family of elliptic curves parametrised by T. Let $P\colon T\to E$ be a section of E/T, and let $\hat{h}_t$ be the Néron-Tate height on the fibre E_t of E/T over a closed point t of T. Tate (1983a) proves that the map $t\mapsto\hat{h}_t(P_t)$ is a height function on the curve T for a certain divisor class $q(P)$ on T; moreover, the degree of $q(P)$ is the Néron-Tate height of P regarded as a point on the generic fibre of E/T.

As Tate noted "The main obstacle to extending the theorem in this paper to abelian varieties seems to be the lack of a canonical compactification of the Néron model in higher dimensions." After Faltings compactified the moduli stack of abelian varieties (Faltings 1985), one of his students, William Green, extended Tate's theorem to abelian varieties (Green 1989).

Height Pairings Via Biextensions Let A be an abelian variety over a number field K, and let A' be its dual. The classical Néron-Tate height pairing is a pairing

$$A(K)\times A'(K)\to\mathbb{R}$$

whose kernels are precisely the torsion subgroups of $A(K)$ and $A'(K)$. In order, for example, to state a p-adic version of the conjecture of Birch and Swinnerton-Dyer, it is necessary to define a $\mathbb{Q}_p$-valued height pairing,

$$A(K)\times A'(K)\to\mathbb{Q}_p.$$

When A has good ordinary or multiplicative reduction at the p-adic primes, Mazur and Tate (1983b) use the expression of the duality between A and A' in terms of biextensions, and exploit the local splittings of these biextensions, to define such pairings. They compare their definition with other suggested definitions. It is not known whether the pairings are nondegenerate modulo torsion.

[11] Also see footnote 10.

2.3 The Cohomology of Abelian Varieties

The Local Duality for Abelian Varieties Let A be an abelian variety over a field k. A principal homogeneous space over A is a variety V over k together with a regular map $A \times V \to V$ such that, for every field K containing k for which $V(K)$ is nonempty, the pairing $A(K) \times V(K) \to V(K)$ makes $V(K)$ into a principal homogeneous space for $A(K)$ in the usual sense. The isomorphism classes of principal homogeneous spaces form a group, which Tate (1958b) named the Weil-Châtelet group, and denoted $\mathrm{WC}(A/k)$.

For a finite extension k of $\mathbb{Q}_p$, local class field theory provides a canonical isomorphism $H^2(k, \mathbb{G}_m) \simeq \mathbb{Q}/\mathbb{Z}$. Tate (ibid.) defines an "augmented" cup-product pairing

$$H^r(k, A) \times H^{1-r}(k, A') \to H^2(k, \mathbb{G}_m) \simeq \mathbb{Q}/\mathbb{Z}, \tag{15}$$

and proves that it is a perfect duality for $r = 1$. In other words, the discrete group $\mathrm{WC}(A/k)$ is canonically dual to the compact group $A'(k)$. Later, he showed that (15) is a perfect duality for all r. In the case $k = \mathbb{R}$, he proved that $H^1(\mathbb{R}, A)$ is canonically dual to $A'(\mathbb{R})/A'(\mathbb{R})^\circ = \pi_0(A'(\mathbb{R}))$.

Notes The above results are proved in Tate (1958b, 1959b, or 1962d). The analogous statements for local fields of characteristic p are proved in Milne (1970).

Principal Homogeneous Spaces over Abelian Varieties Lang and Tate (1958c) explain the relation between the set $\mathrm{WC}(A/k)$ of isomorphism classes of principal homogeneous spaces over a group variety A and the Galois cohomology group $H^1(k, A)$. Briefly, there is a canonical injective map $\mathrm{WC}(A/k) \to H^1(k, A)$ which Weil's descent theorems show to be surjective. This generalizes results of Châtelet.

Let K be a field complete with respect to a discrete valuation with residue field k, and let A be an abelian variety over K with good reduction to an abelian variety $\bar{A}$ over k. Then, for any integer m prime to the characteristic of k, Lang and Tate (ibid.) prove that there is a canonical exact sequence

$$0 \to H^1(k, \bar{A})_m \to H^1(k, A)_m \to \mathrm{Hom}\big(\mu_m(k), \bar{A}\big(k^{\mathrm{sep}}\big)_m\big) \to 0.$$

In the final section of the article, they study abelian varieties over global fields. In particular, they prove the weak Mordell-Weil theorem.

As Cassels wrote,[12] the article Tate (1962b) provides "A laconic but useful review of the existing state of knowledge [on principal homogeneous spaces for abelian varieties] for different types of groundfield."

The Conjecture of Birch and Swinnerton-Dyer For an elliptic curve A over $\mathbb{Q}$, Mordell showed that the group $A(\mathbb{Q})$ is finitely generated. It is easy to compute the

[12] See his Math. Reviews 0138625.

torsion subgroup of $A(\mathbb{Q})$, but there is at present no proven algorithm for computing its rank $r(A)$. Computations led Birch and Swinnerton-Dyer to conjecture that $r(A)$ is equal to the order of the zero at 1 of the L-series of A, and further work led to a more precise conjecture. Tate (1966e) formulated the analogues of their conjectures for an abelian variety A over a global field K.

Let v be a nonarchimedean prime of K, and let $\kappa(v)$ be the corresponding residue field. If A has good reduction at v, then it gives rise to an abelian variety $A(v)$ over $\kappa(v)$. The characteristic polynomial of the Frobenius endomorphism of $A(v)$ is a polynomial $P_v(T)$ of degree $2d$ with coefficients in $\mathbb{Z}$ such that, when we factor it as $P_v(T)=\prod_i(1-a_iT)$, then $\prod_i(1-a_i^m)$ is the number of points on $A(v)$ with coordinates in the finite field of degree m over $\kappa(v)$. For any finite set S of primes of K including the archimedean primes and those where A has bad reduction, we define the L-series $L_S(s,A)$ by the formula

$$L_S(A,s)=\prod_{v\notin S}P_v\left(A,Nv^{-s}\right)^{-1}$$

where $Nv=[\kappa(v)]$. The product converges for $\Re(s)>3/2$, and it is conjectured that $L_S(A,s)$ can be analytically continued to a meromorphic function on the whole complex plane. This is known in the function field case, and over $\mathbb{Q}$ for elliptic curves. The analogue of the first conjecture of Birch and Swinnerton-Dyer for A is that

$$L_S(A,s)\text{ has a zero of order } r(A) \text{ at } s=1. \tag{16}$$

Let ω be a nonzero global differential d-form on A. As $\Gamma(A,\Omega_A^d)$ has dimension 1, ω is uniquely determined up to multiplication by an element of $K^\times$. For each nonarchimedean prime v of K, let μ_v be the Haar measure on K_v for which $\mathcal{O}_v$ has measure 1, and for each archimedean prime, take μ_v to be the usual Lebesgue measure on K_v. Define

$$\mu_v(A,\omega)=\int_{A(K_v)}|\omega|_v\mu_v^d$$

Let μ be the measure $\prod\mu_v$ on the adèle ring $\mathbb{A}_K$ of K, and set $|\mu|=\int_{\mathbb{A}_K/K}\mu$. For any finite set S of primes of K including all archimedean primes and those nonarchimedean primes for which A has bad reduction or such that ω does not reduce to a nonzero differential d-form on $A(v)$, we define

$$L_S^*(s,A)=L_S(s,A)\frac{|\mu|^d}{\prod_{v\in S}\mu_v(A,\omega)}.$$

The product formula shows that this is independent of the choice of ω. The asymptotic behaviour of $L_S^*(s,A)$ as $s\to 1$, which is all we are interested in, doesn't depend on S. The analogue of the second conjecture of Birch and Swinnerton-Dyer is that

$$\lim_{s\to 1}\frac{L_S^*(s,A)}{(s-1)^{r(A)}}=\frac{[\text{Ш}(A)]\cdot|D|}{[A'(K)_{\text{tors}}]\cdot[A(K)_{\text{tors}}]} \tag{17}$$

where $Ш(A)$ is the Tate-Shafarevich group of A,

$$Ш(A) \stackrel{\text{def}}{=} \operatorname{Ker}\Big(H^1(K, A) \to \prod_v H^1(K_v, A)\Big),$$

which is conjectured to be finite, and D is the discriminant of the height pairing (14), which is known to be nonzero.

Global Duality In his talk at the 1962 International Congress, Tate stated the local duality theorems reviewed above (p. 277), and he announced some global theorems which we now discuss.

In their computations, Birch and Swinnerton-Dyer found that the order of the Tate-Shafarevich group predicted by (17) is always a square. Cassels and Tate conjectured independently that the explanation for this is that there exists an alternating pairing

$$Ш(A) \times Ш(A) \to \mathbb{Q}/\mathbb{Z} \tag{18}$$

that annihilates only the divisible subgroup of $Ш(A)$. Cassels (1962) proved this for an elliptic curve over a number field. For an abelian variety A and its dual abelian variety A', Tate proved that there exists a canonical pairing

$$Ш(A) \times Ш(A') \to \mathbb{Q}/\mathbb{Z} \tag{19}$$

that annihilates only the divisible subgroups; moreover, for a divisor D on A and the homomorphism $\varphi_D\colon A \to A'$, $a \mapsto [D_a - D]$, it defines, the pair $(\alpha, \varphi_D(\alpha))$ maps to zero under (19) for all $\alpha \in Ш(A)$. The pairing (19), or one of its several variants, is now called the Cassels-Tate pairing.

For an elliptic curve A over a number field k such that $Ш(A)$ is finite, Cassels determined the Pontryagin dual of the exact sequence

$$0 \to Ш(A) \to H^1(k, A) \to \bigoplus_v H^1(k_v, A) \to Б(A) \to 0 \tag{20}$$

(regarded as a sequence of discrete groups). Assume that $Ш(A)$ is finite. Using Tate's local duality theorem (see p. 277) for an elliptic curve, Cassels (1964) showed that the dual of (20) takes the form

$$0 \leftarrow Ш \leftarrow \Theta \leftarrow \prod_v A(k_v)' \leftarrow \widetilde{A(k)} \leftarrow 0 \tag{21}$$

for a certain explicit Θ and with $\widetilde{A(k)}$ equal to the closure of $A(k)$ in $\prod_v A(k_v)'$. Tate proved the same statement for abelian varieties over number fields, except that, in (21), it is necessary to replace A with its dual A'. So modified, the sequence (21) is now called the Cassels-Tate dual exact sequence.

Let A and B be isogenous elliptic curves over a number field. Then $L_S(s, A) = L_S(s, B)$ and $r(A) = r(B)$, and so the first conjecture of Birch and Swinnerton-Dyer is true for A if and only if it is true for B. Cassels (1965) proved the same statement for the second conjecture. This amounts to showing that a certain product of terms doesn't change in passing from A to B (even though the individual terms may change). Using his duality theorems and the formula (10), p. 272, for the Euler-Poincaré characteristic, Tate (1966e, 2.1) proved the same result for abelian varieties over number fields.

Tate's global duality theorems were widely used, even before there were published proofs. Since 1994, the duality theorems have been used in cryptography.

Notes Tate's results are more general and complete than stated above; in particular, he works with a nonempty set S of primes of k (not necessarily the complete set). Proofs of the theorems of Tate in this subsection can be found in Milne (1986).

2.4 Serre-Tate Liftings of Abelian Varieties

In a talk at the 1964 Woods Hole conference, Tate discussed some results of his and Serre on the lifting of abelian varieties from characteristic p.

For an abelian scheme A over a ring R, let A_n denote the kernel of $A \xrightarrow{p^n} A$ regarded as a finite group scheme over R, and let $A(p)$ denote the direct system

$$A_1 \hookrightarrow A_2 \hookrightarrow \cdots \hookrightarrow A_n \hookrightarrow \cdots$$

of finite group schemes. Let R be an artinian local ring with residue field k of characteristic $p \neq 0$. An abelian scheme A over R defines an abelian variety $\bar{A}$ over k and a system of finite group schemes $A(p)$ over R. Serre and Tate prove that the functor

$$A \rightsquigarrow \left(\bar{A}, A(p)\right)$$

is an equivalence of categories (Serre-Tate theorem). In particular, to lift an abelian variety A from k to R amounts to lifting the system of finite group schemes $A(p)$.

This has many important consequences.

- Let A and B be abelian schemes over a complete local noetherian ring R with residue field a field k of characteristic $p \neq 0$. A homomorphism $f\colon \bar{A} \to \bar{B}$ of abelian varieties over k lifts to a homomorphism $A \to B$ of abelian schemes over R if and only if $f(p)\colon \bar{A}(p) \to \bar{B}(p)$ lifts to R. For the artinian quotients of R, this is part of the above statement, and the statement for R follows by passing to the limit over the artinian quotients of R and applying a theorem of Grothendieck.
- Let A be an abelian variety over a perfect field k of characteristic $p \neq 0$. If A is ordinary, then

$$A_n \approx \left(\mathbb{Z}/p^n\mathbb{Z}\right)^{\dim A} \times (\mu_{p^n})^{\dim A},$$

  and so each A_n has a canonical lifting to a finite group scheme over the ring $W(k)$ of Witt vectors of k. Thus A has a canonical lifting to an abelian scheme

over $W(k)$ (at least formally, but the existence of polarizations implies that the formal abelian scheme is an abelian scheme). Deligne (1969) has used this to give a "linear algebra" description of the category of ordinary abelian varieties over a finite field similar to the classical description of abelian varieties over $\mathbb{C}$.
- Over a ring R in which p is nilpotent, the infinitesimal deformation theory of A is equivalent to the infinitesimal deformation theory of $A(p)$. For example, when A is ordinary, this implies that the local deformation space of an ordinary abelian variety A over k has a natural structure of a formal torus over $W(k)$ of relative dimension $\dim(A)^2$.

Notes Lifting results were known to Hasse and Deuring for elliptic curves. The canonical lifting of an ordinary abelian variety was found by Serre, prompting Tate to prove the general result. Lubin, Serre, and Tate (1964b) contains a sketch of the proofs. The liftings of A obtained from liftings of the system $A(p)$ are sometimes called Serre-Tate liftings, especially in the ordinary case. Messing (1972) includes a proof of the Serre-Tate theorem.

2.5 Mumford-Tate Groups and the Mumford-Tate Conjecture

In 1965, Mumford gave a talk at the AMS Summer Institute (Mumford 1966) whose results he described as being "partly joint work with J. Tate". In it, he attached a reductive group to an abelian variety, and stated a conjecture. The first is now called the Mumford-Tate group, and the second is the Mumford-Tate conjecture.

Let A be a complex abelian variety of dimension g. Then $V \stackrel{\text{def}}{=} H_1(A, \mathbb{Q})$ is a $\mathbb{Q}$-vector space of dimension $2g$ whose tensor product with $\mathbb{R}$ acquires a complex structure through the canonical isomorphism

$$H_1(A, \mathbb{Q})_{\mathbb{R}} \simeq \mathrm{Tgt}_0(A).$$

Let $u\colon U^1 \to \mathrm{GL}(V_{\mathbb{R}})$ be the homomorphism describing this complex structure, where $U^1 = \{z \in \mathbb{C} \mid |z| = 1\}$. The Mumford-Tate group of A is defined to be the smallest algebraic subgroup H of GL_V such that $H(\mathbb{R})$ contains $u(U^1)$.[13] Then H is a reductive algebraic group over $\mathbb{Q}$, which acts on $H^*(A^r, \mathbb{Q})$, $r \in \mathbb{N}$, through the isomorphisms

$$H^*(A^r, \mathbb{Q}) \simeq \bigwedge^{*} H^1(A^r, \mathbb{Q}),$$

$$H^1(A^r, \mathbb{Q}) \simeq r H^1(A, \mathbb{Q}),$$

$$H^1(A, \mathbb{Q}) \simeq \mathrm{Hom}(V, \mathbb{Q}).$$

[13] Better, it should be thought of as the pair (H, u).

It can be characterized as the algebraic subgroup of GL_V that fixes exactly the Hodge tensors in the spaces $H^*(A^r, \mathbb{Q})$, i.e., the elements of the $\mathbb{Q}$-spaces

$$H^{2p}(A^r, \mathbb{Q}) \cap \bigoplus H^{p,p}(A^r).$$

Let A be an abelian variety over a number field k, and, for a prime number l, let

$$V_l A = \mathbb{Q}_l \otimes_{\mathbb{Z}_l} T_l A$$

where $T_l A$ is the Tate module of A (p. 274). Then

$$V_l A \simeq \mathbb{Q}_l \otimes_{\mathbb{Q}} H_1(A_{\mathbb{C}}, \mathbb{Q}).$$

The Galois group $G(k^{\mathrm{al}}/k)$ acts on $A(k^{\mathrm{al}})$, and hence there is a representation

$$\rho_l \colon G(k^{\mathrm{al}}/k) \to \mathrm{GL}(V_l A)$$

The Zariski closure $H_l(A)$ of $\rho_l(G(k^{\mathrm{al}}/k))$ is an algebraic group in $\mathrm{GL}_{V_l A}$. Although $H_l(A)$ may change when k is replaced by a finite extension, its identity component $H_l(A)^\circ$ does not and can be thought of as determining the image of ρ_l up to finite groups. The Mumford-Tate conjecture states that

$$H_l(A)^\circ = (\text{Mumford–Tate group of } A_{\mathbb{C}})_{\mathbb{Q}_l} \text{ inside } \mathrm{GL}_{V_l A} \simeq (\mathrm{GL}_{H_1(A_{\mathbb{C}}, \mathbb{Q})})_{\mathbb{Q}_l}.$$

In particular, it posits that the $\mathbb{Q}_l$-algebraic groups $H_l(A)^\circ$ are independent of l in the sense that they all arise by base change from a single algebraic group over $\mathbb{Q}$. In the presence of the Mumford-Tate conjecture, the Hodge and Tate conjectures for A are equivalent. Much is known about the Mumford-Tate conjecture.

Let H be the Mumford-Tate group of an abelian variety A, and let $u \colon U^1 \to H(\mathbb{R})$ be the above homomorphism. The centralizer K of u in $H(\mathbb{R})$ is a maximal compact subgroup of $H(\mathbb{R})$, and the quotient manifold $X = H(\mathbb{R})/K$ has a unique complex structure for which $u(z)$ acts on the tangent space at the origin as multiplication by z. With this structure X is isomorphic to a bounded symmetric domain, and it supports a family of abelian varieties whose Mumford-Tate groups "refine" that of A. The quotients of X by congruence subgroups of $H(\mathbb{Q})$ are connected Shimura varieties.

The notion of a Mumford-Tate group has a natural generalization to an arbitrary polarizable rational Hodge structure. In this case the quotient space X is a homogeneous complex manifold, but it is not necessarily a bounded symmetric domain. The complex manifolds arising in this way were called Mumford-Tate domains by Green et al. (2010). As these authors say: "Mumford-Tate groups have emerged as the principal symmetry groups in Hodge theory."

2.6 Abelian Varieties over Finite Fields (Weil, Tate, Honda Theory)

Consider the category whose objects are the abelian varieties over a field k and whose morphisms are given by

$$\mathrm{Hom}^0(A, B) \stackrel{\text{def}}{=} \mathrm{Hom}(A, B) \otimes \mathbb{Q}.$$

Weil's results (Weil 1946) imply that this is a semisimple abelian category whose endomorphism algebras are finite dimensional $\mathbb{Q}$-algebras. Thus, to describe the category up to equivalence, it suffices to list the isomorphism classes of simple objects and, for each class, describe the endomorphism algebra of an object in the class. This the theory of Weil, Tate, and Honda does when k is finite. Briefly: Weil showed that there is a well-defined map from isogeny classes of simple abelian varieties to conjugacy classes of Weil numbers, Tate proved that the map is injective and determined the endomorphism algebra of each simple class, and Honda used the theory of Shimura and Taniyama to prove that the map is surjective.

In more detail, let k be a field with $q = p^a$ elements. Each abelian variety A over k admits a Frobenius endomorphism π_A, which acts on the k^{al}-points of A as $(a_0 : a_1 : \ldots) \mapsto (a_0^q : a_1^q : \ldots)$. Weil proved that the image of π_A in $\mathbb{C}$ under any homomorphism $\mathbb{Q}[\pi_A] \to \mathbb{C}$ is a Weil q-integer, i.e., it is an algebraic integer with absolute value $q^{\frac{1}{2}}$ (this is the Riemann hypothesis). Thus, attached to every simple abelian variety A over k, there is a conjugacy class of Weil q-integers. Isogenous simple abelian varieties give the same conjugacy class.

Tate (1966b) proved that a simple abelian A is determined up to isogeny by the conjugacy class of π_A, and moreover, that $\mathbb{Q}[\pi_A]$ is the centre of $\mathrm{End}^0(A)$. Since $\mathrm{End}^0(A)$ is a division algebra with centre the field $\mathbb{Q}[\pi_A]$, class field theory shows that its isomorphism class is determined by its invariants at the primes v of $\mathbb{Q}[\pi_A]$. These Tate determined as follows:

$$\mathrm{inv}_v\big(\mathrm{End}^0(A)\big) = \begin{cases} \frac{1}{2} & \text{if } v \text{ is real,} \\ \frac{\mathrm{ord}_v(\pi_A)}{\mathrm{ord}_v(q)} [\mathbb{Q}[\pi_A]_v : \mathbb{Q}_p] & \text{if } v | p, \\ 0 & \text{otherwise.} \end{cases}$$

Moreover,

$$2 \dim A = \big[\mathrm{End}^0(A) : \mathbb{Q}[\pi_A]\big]^{\frac{1}{2}} \cdot \big[\mathbb{Q}[\pi_A] : \mathbb{Q}\big].$$

The abelian varieties of CM-type over $\mathbb{C}$ are classified up to isogeny by their CM-types, and every such abelian variety has a model over $\mathbb{Q}^{\mathrm{al}}$. When we choose a p-adic prime of $\mathbb{Q}^{\mathrm{al}}$, an abelian variety A of CM-type over $\mathbb{Q}^{\mathrm{al}}$ specializes to an abelian variety $\bar{A}$ over a finite field of characteristic p. The Shimura-Taniyama formula determines $\pi_{\bar{A}}$ up to a root of 1 in terms of the CM-type of A. Using this, Honda (1968) proved that every Weil q-number arises from an abelian variety, possibly after a finite extension of the base field. An application of Weil restriction of scalars completes the proof.

2.7 Good Reduction of Abelian Varieties

The language of Weil's foundations of algebraic geometry is ill-suited to the study of algebraic varieties in mixed characteristic. For example, it makes it cumbersome to prove even that an algebraic variety over a number field has good reduction at almost all primes of the field (Shimura 1955, Theorem 26). Serre and Tate (1968a) use schemes and Néron's theory of minimal models (Néron 1964) to simplify and sharpen known results for abelian varieties, and to extend some statements from elliptic curves to abelian varieties.

Let R be a discrete valuation ring with field of fractions K and perfect residue field k. For an abelian variety A over K, Néron proved that the functor sending a smooth R-scheme X to $\mathrm{Hom}(X_K, A)$ is represented by a smooth group scheme $\tilde{A}$ of finite type over R. Using this, Serre and Tate prove the following criterion:

> If A has good reduction, then the $\mathrm{Gal}(K^{\mathrm{sep}}/K)$-module $A(K^{\mathrm{sep}})_m$ is unramified for all integers m prime to $\mathrm{char}(k)$; conversely, if $A(K^{\mathrm{sep}})_m$ is unramified for infinitely many m prime to $\mathrm{char}(k)$, then A has good reduction.

The necessity was known earlier, and the sufficiency was known to Ogg and Shafarevich. in the case of elliptic curves. Because it is a direct consequence of the existence of Néron's models, Serre and Tate call it the "Néron-Ogg-Shafarevich criterion". It is of fundamental importance.

Serre and Tate say that an abelian variety has *potential good reduction* if it acquires good reduction after a finite extension of the base field, and they prove a number of results about such varieties. For example, when R is strictly henselian, there is a smallest extension L of K in K^{al} over which such an abelian variety A has good reduction, namely, the extension of K generated by the coordinates of the points of order m for any $m \geq 3$ prime to $\mathrm{char}(k)$. Moreover, just as for elliptic curves, the notion of the conductor of an abelian variety is well defined.

Let (A, i) be an abelian variety over a number field K with complex multiplication by E. By this we mean that E is a CM field of degree $2\dim A$ over $\mathbb{Q}$, and that i is a homomorphism of $\mathbb{Q}$-algebras $E \to \mathrm{End}^0(A)$. Serre and Tate apply their earlier results to show that such an abelian variety A acquires good reduction everywhere over a cyclic extension L of K; moreover, L can be chosen to have degree m or $2m$ where m is the least common multiple of the images of the inertia groups acting on the torsion points of A.

Let (A, i) and E be as in the last paragraph, and let C_K be the idèle class group of K. Shimura and Taniyama (1961, 18.3) show there exists a (unique) homomorphism $\rho\colon C_K \to (\mathbb{R} \otimes_{\mathbb{Q}} E)^{\times}$ with the following property: for each $\sigma : E \to \mathbb{C}$, let χ_σ be the Hecke character

$$\chi_\sigma : C_K \xrightarrow{\rho} (\mathbb{R} \otimes_{\mathbb{Q}} E)^{\times} \xrightarrow{1\otimes\sigma} \mathbb{C}^{\times};$$

then the L-series $L(s, A)$ coincides with the product $\prod_\sigma L(s, \chi_\sigma)$ of the L-series of the χ_σ, except possibly for the factors corresponding to a finite number of primes of K.[14] Serre and Tate make this more precise by showing that the conductor of A is the product of the conductors of the χ_σ (which each equals the conductor of ρ). In particular, the support of the conductor of each χ_σ equals the set of primes where A has bad reduction, from which it follows that $L(s, A)$ and $\prod_\sigma L(s, \chi_\sigma)$ coincide exactly.

2.8 CM Abelian Varieties and Hilbert's Twelfth Problem

A CM-type on a CM field E is a subset Φ of $\mathrm{Hom}(E, \mathbb{C})$ such that $\Phi \sqcup \bar{\Phi} = \mathrm{Hom}(E, \mathbb{C})$. For $\sigma \in \mathrm{Aut}(\mathbb{C})$, let $\sigma\Phi = \{\sigma \circ \varphi \mid \varphi \in \Phi\}$. Then $\sigma\Phi$ is also a CM-type on E. The reflex field of (E, Φ) is the subfield F of $\mathbb{C}$ such that an automorphism σ of $\mathbb{C}$ fixes F if and only if $\sigma\Phi = \Phi$. It is easy to see that F is a CM-subfield of $\mathbb{Q}^{\mathrm{al}} \subset \mathbb{C}$.

Let (A, i) be an abelian variety over $\mathbb{C}$ with complex multiplication by E. Then E acts on the tangent space of A at 0 through a CM-type Φ, and (A, i) is said to be of CM-type (E, Φ). For $\sigma \in \mathrm{Aut}(\mathbb{C}/\mathbb{Q})$, $\sigma(A, i)$ is of CM-type $\sigma\Phi$, and it follows that $\sigma(A, i)$ is isogenous to (A, i) if and only if σ fixes the reflex field F. Fix a polarization λ of A whose Rosati involution acts as complex multiplication on E. For an integer $m \geq 1$, let $\mathcal{S}(m)$ be the set of isomorphism classes of quadruples (A', λ', i', η) such that (A', λ', i') is isogenous to (A, λ, i) and η is a level m-structure on (A', i'). According to the preceding observation, $\mathrm{Aut}(\mathbb{C}/F)$ acts on the set $\mathcal{S}(m)$. Shimura and Taniyama prove that this action factors through $\mathrm{Aut}(F^{\mathrm{ab}}/F)$, and they describe it explicitly. In this way, they generalized the theory of complex multiplication from elliptic curves to abelian varieties, and they provided a partial solution to Hilbert's twelfth problem for F.

In one respect the result of Shimura and Taniyama falls short of generalizing the elliptic curve case: for an elliptic curve, the reflex field F is a complex quadratic extension of $\mathbb{Q}$; since one knows how complex conjugation acts on CM elliptic curves and their torsion points, the elliptic curve case provides a description of how the *full* group $\mathrm{Aut}(\mathbb{C}/\mathbb{Q})$ acts on CM elliptic curves and their torsion points. Shimura asked whether there was a similar result for abelian varieties, but concluded rather pessimistically that "In the higher-dimensional case, however, no such general answer seems possible" (Shimura 1977).

Grothendieck's theory of motives suggests the framework for an answer. The Hodge conjecture implies the existence of Tannakian category of CM-motives over $\mathbb{Q}$, whose motivic Galois group is an extension

$$1 \to S \to T \to \mathrm{Gal}(\mathbb{Q}^{\mathrm{al}}/\mathbb{Q}) \to 1$$

of $\mathrm{Gal}(\mathbb{Q}^{\mathrm{al}}/\mathbb{Q})$ (regarded as a pro-constant group scheme) by the Serre group S (a certain pro-torus). Étale cohomology defines a section λ of $T \to \mathrm{Gal}(\mathbb{Q}^{\mathrm{al}}/\mathbb{Q})$ over

[14]Rather, this is Serre and Tate's interpretation of what they prove; Shimura and Taniyama express their results in terms of ideals.

the finite adèles. The pair (T, λ) (tautologically) describes the action of $\mathrm{Aut}(\mathbb{C}/\mathbb{Q})$ on the CM abelian varieties and their torsion points. Deligne's theorem on Hodge classes on abelian varieties allows one to construct the pair (T, λ) without assuming the Hodge conjecture. To answer Shimura's question, it remains to give a direct explicit description of (T, λ).

Langlands's work on the zeta functions of Shimura varieties led him to define a certain explicit cocycle (Langlands 1979, §5), which Deligne recognized as conjecturally being that describing the pair (T, λ).

Tate was inspired by this to commence his own investigation of Shimura's question. He gave a simple direct construction of a map f that he conjectured describes how $\mathrm{Aut}(\mathbb{C}/\mathbb{Q})$ acts on the CM abelian varieties and their torsion points, and proved this up to signs. More precisely, he proved it up to a map e with values in an adèlic group such that $e^2 = 1$. See Tate (1981c).

It was soon checked that Langlands's and Tate's conjectural descriptions of how $\mathrm{Aut}(\mathbb{C}/\mathbb{Q})$ acts on the CM abelian varieties and their torsion points coincided, and a few months later Deligne (1982) proved that their conjectural descriptions are indeed correct.

3 Rigid Analytic Spaces

After Hensel introduced the p-adic number field $\mathbb{Q}_p$ in the 1890s, there were attempts to develop a theory of analytic functions over $\mathbb{Q}_p$, the most prominent being that of Krasner. The problem is that every disk D in $\mathbb{Q}_p$ can be written as a disjoint union of arbitrarily many open-closed smaller disks, and so there are too many functions on D that can be represented locally by power series. Outside a small group of mathematicians, p-adic analysis attracted little attention until the work of Dwork and Tate in late 1950s. In February, 1958, Tate sent Dwork a letter in which he stated a result concerning elliptic curves, and challenged Dwork to find a proof using p-adic analysis. In answering the letter, Dwork found "the first suggestion of a connection between p-adic analysis and the theory of zeta functions" (Dwork 1963).[15] By November, 1959, Dwork had found his famous proof of the rationality of the zeta function $Z(V, T)$ of an algebraic variety V over a finite field, a key point of which is to express $Z(V, T)$, which initially is a power series with integer coefficients, as a quotient of two p-adically entire functions (Dwork 1960).

In 1959 also, Tate discovered that, suitably normalized, certain classical formulas allow one to express many elliptic curves E over a nonarchimedean local field K as a quotient $E(K) = K^{\times}/q^{\mathbb{Z}}$. This persuaded him that there should exist a category in which E itself, not just its points, is a quotient; in other words, that there exists a category in which E, as an "analytic space", is the quotient of $K^{\times}$, as an "analytic space", by the discrete group $q^{\mathbb{Z}}$. Two years later, Tate constructed the correct category of "rigid analytic spaces", thereby founding a new subject in mathematics (with its own Math. Reviews number 14G22).

[15] Katz and Tate (1999, p. 343).

3.1 The Tate Curve

Let E be an elliptic curve over $\mathbb{C}$. The choice of a differential ω realizes $E(\mathbb{C})$ as the quotient $\mathbb{C}/\Lambda \simeq E(\mathbb{C})$ of $\mathbb{C}$ by the lattice of periods of ω. More precisely, it realizes the complex analytic manifold E^{an} as the quotient of the complex analytic manifold $\mathbb{C}$ by the action of the discrete group Λ.

For an elliptic curve E over a p-adic field K, there is no similar description of $E(K)$ because there are no nonzero discrete subgroups of K (if $\lambda \in K$, then $p^n\lambda \to 0$ as $n \to \infty$). However, there is an alternative uniformization of elliptic curves over $\mathbb{C}$. Let Λ be the lattice $\mathbb{Z}+\mathbb{Z}\tau$ in $\mathbb{C}$. Then the exponential map $\underline{e}\colon \mathbb{C} \to \mathbb{C}^\times$ sends $\mathbb{C}/\Lambda$ isomorphically onto $\mathbb{C}^\times/q^{\mathbb{Z}}$ where $q = \underline{e}(\tau)$, and so $\mathbb{C}^\times/q^{\mathbb{Z}} \simeq E^{\mathrm{an}}$ (as analytic spaces). If $\mathrm{Im}(\tau) > 0$, then $|q| < 1$, and the elliptic curve E_q is given by the equation

$$Y^2Z + XYZ = X^3 - b_2XZ^2 - b_3Z^3, \tag{22}$$

where

$$\begin{cases} b_2 = 5\displaystyle\sum_{n=1}^{\infty} \frac{n^3q^n}{1-q^n} = 5q + 45q^2 + 140q^3 + \cdots \\ b_3 = \displaystyle\sum_{n=1}^{\infty} \frac{7n^5+5n^3}{12}\,\frac{q^n}{1-q^n} = q + 23q^2 + 154q^3 + \cdots \end{cases} \tag{23}$$

are power series with integer coefficients. The discriminant and modular invariant of E_q are given by the usual formulas

$$\Delta = q\prod_{n\geq 1}(1-q^n)^{24} \tag{24}$$

$$j(E_q) = \frac{(1+48b_2)^3}{q\prod_{n\geq 1}(1-q^n)^{24}} = \frac{1}{q} + 744 + 196884q + \cdots. \tag{25}$$

Now let K be a field complete with respect to a nontrivial nonarchimedean valuation with residue field of characteristic $p \neq 0$, and let q be an element of $K^\times$ with $|q| < 1$. The series (23) converge in K, and Tate discovered[16] that (22) is an elliptic curve E_q such that $K'/q^{\mathbb{Z}} \simeq E_q(K')$ for all finite extension K' of K. It follows from certain power series identities, valid over $\mathbb{Z}$, that the discriminant and modular invariant of E_q are given by (24) and (25). Every $j \in K^\times$ with $|j| < 1$ arises from a q (determined by (25), which allows q to be expressed as a power series in $1/j$ with integer coefficients). The function field $K(E_q)$ of E_q consists of the quotients

[16] "I still remember the thrill and amazement I felt when it occurred to me that the classical formulas for such an isomorphism over $\mathbb{C}$ made sense p-adically when properly normalized." Tate (2008).

F/G of Laurent series

$$F = \sum_{-\infty}^{\infty} a_n z^n, \qquad G = \sum_{-\infty}^{\infty} b_n z^n, \quad a_n, b_n \in K,$$

converging for all nonzero z in $\mathbb{C}_p$, such that the F/G is invariant under $q^{\mathbb{Z}}$:

$$F(qz)/G(qz) = F(z)/G(z).$$

The elliptic curves E over K with $|j(E)| < 1$ that arise in this way are exactly those whose reduced curve has a node with tangents that are rational over the base field. They are now called Tate (elliptic) curves.

Tate's results were contained in a 1959 manuscript, which he did not publish until 1995, but there soon appeared several summaries of his results in the literature, and Roquette (1970) gave a very detailed account of the theory. The Tate curve has found many applications, for example, to Tate's isogeny conjecture (Serre 1968a; Tate 1995, p. 180) and to the study of elliptic modular curves near a cusp (Deligne and Rapoport 1973). Mumford (1972) generalized Tate's construction to curves of higher genus, and McCabe (1968) and Raynaud (1971) generalized it to abelian varieties of higher dimension.

3.2 Rigid Analytic Spaces

Tate's idea that his p-adic uniformization of elliptic curves indicated the existence of a general theory of p-adic analytic spaces was radically new. For example, Grothendieck was initially very negative.[17] However, when Tate began to work out his theory in the fall of 1961, Grothendieck, who was visiting Harvard at the time, became very optimistic,[18] and was very supportive.

Let K be a field complete with respect to a nontrivial nonarchimedean valuation, and let $\bar{K}$ be its algebraic closure. Tate began by introducing a new class of K-algebras. The Tate algebra $T_n = K\{X_1, \ldots, X_n\}$ consists of the formal power series in $K[[X_1, \ldots, X_n]]$ that are convergent on the unit ball,

$$B^n = \left\{(c_i)_{1 \le i \le n} \in \bar{K} \mid |c_i| \le 1\right\}.$$

[17] "Tate has written to me about his elliptic curve stuff, and has asked me if I had any ideas for a global definition of analytic varieties over complete valuation fields. I must admit that I have absolutely not understood why his results might suggest the existence of such a definition, and I remain skeptical. Nor do I have the impression of having understood his theorem at all; it does nothing more than exhibit, via brute formulas, a certain isomorphism of analytic groups." Grothendieck, letter to Serre, August 18, 1959.

[18] "Sooner or later it will be necessary to subsume ordinary analytic spaces, rigid analytic spaces, formal schemes, and maybe even schemes themselves into a single kind of structure for which all these usual theorems will hold." Grothendieck, letter to Serre, October 19, 1961.

Thus the elements of T_n are the power series

$$f = \sum a_{i_1 \cdots i_n} X_1^{i_1} \cdots X_n^{i_n}, \quad a_{i_1 \cdots i_n} \in K, \text{ such that } a_{i_1 \cdots i_n} \to 0 \text{ as } (i_1, \ldots, i_n) \to \infty.$$

Tate (1962c) shows that T_n is a Banach algebra for the norm $\|f\| = \sup |a_{i_1 \cdots i_n}|$, and that the ideals $\mathfrak{a}$ of T_n are closed and finitely generated. A quotient $T_n/\mathfrak{a}$ of T_n is a Banach algebra whose topology is independent of its presentation (because every homomorphism of such algebras is continuous). Such quotients are called affinoid (or Tate) K-algebras, and the category of affine rigid analytic spaces is the opposite of the category of affinoid K-algebras.

We need a geometric interpretation of this category. Tate showed that T_n is Jacobson (i.e., every prime ideal is an intersection of maximal ideals), and that the map $A \mapsto \max(A)$ sending an affinoid algebra to its set of maximal ideals is a functor: a homomorphism $\varphi\colon A \to B$ of affinoid algebras defines a map $\varphi^\circ\colon \max(B) \to \max(A)$. The set $\max(A)$ has the Zariski topology, which is very coarse, and a canonical topology induced from that of K. When K is algebraically closed, $\max(T_n) \simeq B^n$, and, by definition, $\max(A)$ can be realized as a closed subset of $\max(T_n)$ for some n.

Let $X = \max(A)$. One would like to define a sheaf $\mathcal{O}_X$ on X such that, for every open subset U isomorphic to B^n, $\mathcal{O}_X(U) \simeq T_n$. As noted at the start of this section, this is impossible. However, Tate's realized that it is possible to achieve something like this by allowing only certain "admissible" open subsets and certain "admissible" coverings. He defined an *affine subset* of X to be a subset Y such that the functor of affinoid K-algebras

$$B \rightsquigarrow \left\{\varphi\colon A \to B \mid \varphi^\circ\big(\max(B)\big) \subset Y\right\}$$

is representable (say, by $A \to A(Y)$). A subset Y of X is a *special affine subset* of X if there exist two finite families (f_i) and (g_j) of elements of A such that

$$Y = \left\{x \in X \mid \left|f_i(x)\right| \le 1,\ \left|g_j(x)\right| \ge 1,\ \text{all } i, j\right\}.$$

Every special affine subset is affine. Tate's acyclicity theorem (Tate 1962c, 8.2) says that, for every finite covering $(X_i)_{i \in I}$ of X by special affines, the Čech complex of the presheaf $Y \mapsto A(Y)$,

$$\begin{aligned} 0 \to A \to \prod_{i_0} A(X_{i_0}) \to \prod_{i_0 < i_1} A(X_{i_0} \cap X_{i_1}) \to \cdots \\ \to \prod_{i_0 < \cdots < i_p} A(X_{i_0} \cap \cdots \cap X_{i_p}) \to \cdots, \end{aligned}$$

is exact. In particular, $Y \mapsto A(Y)$ satisfies the sheaf condition on such coverings.

Using Tate's acyclicity theorem it is possible to define a collection of admissible open subsets of $X = \max(A)$ and admissible coverings of them for which there exists a functor $\mathcal{O}_X$ satisfying the sheaf conditions and such that $\mathcal{O}_X(Y) = A(Y)$

for any affine subset. Although the admissible open subsets and coverings don't form a topology in the usual sense, they satisfy the conditions necessary for them to support a sheaf theory—in fact, they form a Grothendieck topology. So, in this sense, Tate recovers analytic continuation.

For the final step, extending the category of affine rigid analytic spaces to a category of global rigid analytic spaces, Tate followed suggestions of Grothendieck. This step has since been clarified and simplified; see, for example, Bosch (2005), especially 1.12.

Tate reported on his work in a series of letters to Serre, who had them typed by IHES as the notes Tate (1962c). These notes were distributed to a number of mathematicians and libraries. They soon attracted the attention of the German school of complex analytic geometers, who were able to transfer many of their arguments and results to the new setting (e.g., Kiehl 1967). Already by 1984 to give a comprehensive account of the theory required a book of over 400 pages (Bosch et al. 1984). Tate did not publish his work, but eventually the editors of "Mir" published a Russian translation of his notes (Tate 1969a), and the editors of "Inventiones" published the original (Tate 1971).

There have been a number of extensions of Tate's theory. For example, following a suggestion of Grothendieck, Raynaud showed that it is possible to realize a rigid analytic space over a field K as the "generic fibre" of a formal scheme over the valuation ring of K. One problem with rigid analytic spaces is that, while they are adequate for the study of coherent sheaves, they have too few points for the study of locally constant sheaves—for example, there exist nonzero such sheaves whose stalks are all zero. Berkovich found a solution to this problem by enlarging the underlying set of a rigid analytic space without altering the sheaf of functions so that the spaces now support an étale cohomology theory (Berkovich 1990, 1993).

Rigid analytic spaces are now part of the landscape of arithmetic geometry: just as it is natural to regard the $\mathbb{R}$-points of a $\mathbb{Q}$-variety as a real analytic space, it has become natural to regard the $\mathbb{Q}_p$-points of the variety as a rigid analytic space. They have found numerous applications, for example, in the solution by Harbater and Raynaud of Abyhankar's conjecture on the étale fundamental groups of curves, and in the Langlands program (see Sect. 5.1).

4 The Tate Conjecture

This stuff is too beautiful not to be true

Tate[19]

The Hodge conjecture says that a rational cohomology class on a nonsingular projective variety over $\mathbb{C}$ is algebraic if it is of type (p, p). The Tate conjecture says

[19] As a thesis topic, Tate gave me the problem of proving a formula that he and Mike Artin had conjectured concerning algebraic surfaces over finite fields (Conjecture C below). One day he ran into me in the corridors of 2 Divinity Avenue and asked how it was going. "Not well" I said, "In

that an ℓ-adic cohomology class on a nonsingular projective variety over a finitely generated field k is in the span of the algebraic classes if it is fixed by the Galois group. (A field is finitely generated if it is finitely generated as a field over its prime field.)

4.1 Beginnings

In the last section of his talk at the 1962 International Congress, Tate states several conjectures.

4.1.1 *For every abelian variety A over a global field k and prime $\ell \neq \mathrm{char}(k)$, $Ш(A/k)(\ell)$ is finite.*

Let k be a global function field, and let k_0 be its finite field of constants, so that $k = k_0(C)$ for a complete nonsingular curve C over k_0. An elliptic curve A over k is the generic fibre of a map $X \to C$ with X a complete nonsingular surface over k_0, which may be taken to be a minimal. Tate showed that, in this case, 4.1.1 is

$$\begin{array}{ccc} X & \longleftarrow & A \\ \big\downarrow & & \big\downarrow \\ C & \longleftarrow & \mathrm{Spec}(k) \end{array}$$

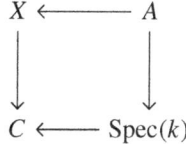

equivalent to the following conjecture.

4.1.2 *Let $q = |k_0|$. The $\mathbb{Z}_\ell$-submodule of $H^2_{\mathrm{et}}(X_{k_0^{\mathrm{al}}}, \mathbb{Z}_\ell)$ on which the Frobenius map acts as multiplication by q is exactly the submodule generated by the algebraic classes.*

---

one example, I computed the left hand side and got p^{13}; for the other side, I got p^{17}; 13 is not equal to 17, and so the conjecture is false." For a moment, Tate was taken aback, but then he broke into a grin and said "That's great! That's really great! Mike and I must have overlooked some small factor which you have discovered." He took me off to his office to show him. In writing it out in front of him, I discovered a mistake in my work, which in fact proved that the conjecture is correct in the example I considered. So I apologized to Tate for my carelessness. But Tate responded: "Your error was not that you made a mistake—we all make mistakes. Your error was not realizing that you must have made a mistake. This stuff is too beautiful not to be true." Benedict Gross tells of a similar experience, but as he writes: "John was so encouraging, saying that everyone made mistakes, and the key was to understand them and to keep thinking about the problem. I felt that one of his greatest talents as an advisor was to make his students feel like we were partners in a great enterprise, modern number theory."

As Tate notes, 4.1.2 makes sense for any complete nonsingular surface over k_0, and that, so generalized, it is equivalent to the following statement.[20]

4.1.3 *Let X be a complete nonsingular surface over a finite field. The order of the pole of $\zeta(X,s)$ at $s=1$ is equal to the number of algebraically independent divisors on X.*

Mumford pointed out that 4.1.3 implies that elliptic curves over a finite field are isogenous if and only if they have the same zeta function, and he proved this using results of Deuring (1941) on the lifting to characteristic 0 of the Frobenius automorphism.

In his talk at the 1964 Woods Hole conference, Tate vastly generalized these conjectures.

4.2 Statement of the Tate Conjecture

For a connected nonsingular projective variety V over a field k, we let $\mathcal{Z}^r(V)$ denote the $\mathbb{Q}$-vector space of algebraic cycles on V of codimension r, i.e., the $\mathbb{Q}$-vector space with basis the irreducible closed subsets of V of dimension $\dim V - r$. We let $H^r_{\text{et}}(V,\mathbb{Q}_\ell(s))$ denote the étale cohomology group of V with coefficients in the "Tate twist" $\mathbb{Q}_\ell(s)$ of $\mathbb{Q}_\ell$. There are cycle maps

$$c^r : \mathcal{Z}^r(V) \to H^{2r}_{\text{et}}\big(V,\mathbb{Q}_\ell(r)\big).$$

Assume that $\ell \neq \text{char}(k)$. Let $\bar{k}$ be an algebraically closed field containing k, and let $G(\bar{k}/k)$ be the group of automorphisms of $\bar{k}$ fixing k. Then $G(\bar{k}/k)$ acts on $H^{2r}_{\text{et}}(V_{\bar{k}},\mathbb{Q}_\ell(r))$, and the *Tate conjecture*[21] (Tate 1964a, Conjecture 1) is the following statement:[22]

> $T^r(V)$: When k is finitely generated, the $\mathbb{Q}_\ell$-space spanned by $c^r(\mathcal{Z}^r(V_{\bar{k}}))$ consists of the elements of $H^{2r}_{\text{et}}(V_{\bar{k}},\mathbb{Q}_\ell(r))$ fixed by some open subgroup of $G(\bar{k}/k)$.

[20] Assuming the Weil conjectures, which weren't proved until 1973.

[21] In the literature, a number of variants of $T^r(V)$, not obviously equivalent to it, are also called the Tate conjecture. It is not always easy to discern what an author means by the "Tate conjecture".

[22] Since Atiyah and Hirzebruch had already found their counterexample to an integral Hodge conjecture, Tate was not tempted to state his conjecture integrally.

Suppose for simplicity that $\bar{k}$ is an algebraic closure of k. For any finite extension k' of k in $\bar{k}$, there is a commutative diagram

$$\begin{array}{ccc} \mathcal{Z}^r(V_{\bar{k}}) & \xrightarrow{c^r} & H^{2r}_{\text{et}}(V_{\bar{k}}, \mathbb{Q}_\ell(r)) \\ \uparrow & & \uparrow \\ \mathcal{Z}^r(V_{k'}) & \xrightarrow{c^r} & H^{2r}_{\text{et}}(V_{k'}, \mathbb{Q}_\ell(r)), \end{array}$$

and the image of the right hand map is $H^{2r}_{\text{et}}(V_{\bar{k}}, \mathbb{Q}_\ell(r))^{G(\bar{k}/k')}$. As $\mathcal{Z}^r(V_{\bar{k}}) = \bigcup_{k'} \mathcal{Z}^r(V_{k'})$, we see that

$$c^r\big(\mathcal{Z}^r(V_{\bar{k}})\big) \subset \bigcup_{k'} H^{2r}_{\text{et}}\big(V_{\bar{k}}, \mathbb{Q}_\ell(r)\big)^{G(\bar{k}/k')}.$$

The content of the Tate conjecture is that the first set spans the space on the right. If an element of $\mathcal{Z}^r(V_{\bar{k}})$ is fixed by $G(\bar{k}/k')$, then it lies in $\mathcal{Z}^r(V_{k'})$, and so $T^r(V)$ implies that

$$c^r\big(\mathcal{Z}^r(V_{k'})\big)\mathbb{Q}_\ell = H^{2r}_{\text{et}}\big(V_{\bar{k}}, \mathbb{Q}_\ell(r)\big)^{G(\bar{k}/k')}; \qquad (26)$$

conversely, if (26) holds for all (sufficiently large) k', then $T^r(V)$ is true.

When asked about the origin of the Tate conjecture, Tate responded (Tate 2011):

> Early on I somehow had the idea that the special case about endomorphisms of abelian varieties over finite fields might be true. A bit later I realized that a generalization fit perfectly with the function field version of the Birch and Swinnerton-Dyer conjecture. Also it was true in various particular examples which I looked at and gave a heuristic reason for the Sato-Tate distribution. So it seemed a reasonable conjecture.

I discuss each of these motivations in turn.

4.3 Homomorphisms of Abelian Varieties

Let A be an abelian variety over a field k, let $\bar{k}$ be an algebraically closed field containing k, and let $G(\bar{k}/k)$ denote the group of automorphisms of $\bar{k}$ over k. For $\ell \neq \text{char}(k)$,

$$A \rightsquigarrow T_\ell A = \varprojlim A(\bar{k})_{\ell^n}$$

is a functor from abelian varieties over k to $\mathbb{Z}_\ell$-modules equipped with an action of $G(\bar{k}/k)$. The (Tate) *isogeny conjecture* is the following statement:

> $H(A, B)$: For abelian varieties A, B over a finitely generated field k, the canonical map
>
> $$\mathbb{Z}_\ell \otimes \text{Hom}(A, B) \to \text{Hom}(T_\ell A, T_\ell B)^{G(\bar{k}/k)}$$

is an isomorphism.

It follows from Weil's theory of correspondences and the interpretation of divisorial correspondences as homomorphisms, that, for varieties V and W,

$$\mathrm{NS}(V \times W) \simeq \mathrm{NS}(V) \oplus \mathrm{NS}(W) \oplus \mathrm{Hom}(A, B) \tag{27}$$

where A is the Albanese variety of V, B is the Picard variety of W, and NS denotes the Néron-Severi group. On comparing (27) with the decomposition of $H^2(V \times W, \mathbb{Q}(1))$ given by the Künneth formula, we find that, for varieties V and W over a finitely generated field k,

$$T^1(V \times W) \iff T^1(V) + T^1(W) + H(A, B). \tag{28}$$

When V is a curve, $T^1(V)$ is obviously true, and so, for elliptic curves E and E',

$$T^1(E \times E') \iff H(E, E').$$

At the time Tate made his conjecture, $H(E, E')$ was known for elliptic curves over a finite field as a consequence of work of Deuring (see above), and $H(E, E)$ was known for elliptic curves over number fields with at least one real prime (Serre 1964b).

Tate (1966b) proved $H(A, B)$ for all abelian varieties over finite fields (see below). As we discussed in (2.6), this has implication for the classification of abelian varieties over finite fields (and even cryptography).

Zarhin extended Tate's result to fields finitely generated over $\mathbb{F}_p$, and Faltings proved $H(A, B)$ for all abelian varieties over number fields in the same article in which he proved Mordell's conjecture. In fact, $H(A, B)$ has now been proved in all generality (Faltings and Wüstholz 1984).

Tate's theorem proves that T^1 is true for surfaces over finite fields that are a product of curves (by (28)). When Artin and Swinnerton-Dyer (1973) proved T^1 for elliptic $K3$ surfaces over finite fields, there was considerable optimism that T^1 would soon be proved for all surfaces over finite fields. However, there has been little progress in the years since then. By contrast, the Hodge conjecture is easily proved for divisors.

Tate's Proof of $H(A, B)$ over a Finite Field It suffices to prove the statement with $A = B$. As the map

$$\mathbb{Z}_\ell \otimes \mathrm{End}(A) \to \mathrm{End}(T_\ell A)^{G(\bar{k}/k)}$$

is injective, the problem is to construct enough endomorphisms of A. I briefly outline Tate's proof.

(a) If $H(A, A)$ is true for one prime $\ell \neq \mathrm{char}(k)$, then it is true for all. This allows Tate to choose an ℓ that is well adapted to his arguments.

(b) A polarization on A defines a skew-symmetric pairing $V_\ell A \times V_\ell A \to \mathbb{Q}_\ell$. Let W be a maximal isotropic subspace of $V_\ell A$ that is stable under $G(\bar{k}/k)$, and let

$$X_n = (T_\ell A \cap W) + \ell^n T_\ell A.$$

There is an infinite sequence of isogenies

$$\cdots \to B_n \to B_{n-1} \to \cdots \to B_1 \to B_0 = A$$

such that the image of $T_\ell B_n$ in $T_\ell A$ is X_n. Using a theorem of Weil, Tate shows that each B_n has a polarization *of the same degree* as the original polarization on A. As k is finite, this implies that the B_n fall into finitely many isomorphism classes. An isomorphism $B_n \to B_{n'}$, $n \neq n'$, gives a nontrivial isogeny $A \to A$.

(c) Having constructed one endomorphism of A not in $\mathbb{Z}$, Tate makes adroit use of the semisimplicity of the rings involved (and his choice of ℓ) to complete the proof.

4.4 Relation to the Conjectures of Birch and Swinnerton-Dyer

The original conjectures of Birch and Swinnerton-Dyer were stated for elliptic curves over $\mathbb{Q}$. Tate re-stated them more generally (see Sect. 2.3).

(A) For an abelian variety A over a global field K, the function $L(s,A)$ has a zero of order $r = \operatorname{rank} A(K)$ at $s = 1$.

(B) Moreover,

$$L^*(s,A) \sim \frac{|\text{Ш}(A)| \cdot |D|}{|A(K)_{\text{tors}}| \cdot |A'(K)_{\text{tors}}|}(s-1)^r \quad \text{as } s \to 1.$$

Let $f\colon V \to C$ be a proper map with fibres of dimension 1, where V (resp. C) is a nonsingular projective surface (resp. curve) over a finite field k. The generic fibre of f is a curve over the global field $k(C)$, and we let $A(f)$ denote its Jacobian variety (an abelian variety over $k(C)$). A comparison of the invariants of V with the invariants of $A(f)$ yields the following statement:

$$\text{Conjecture } T^1 \text{ holds for } V \iff \text{Conjecture (A) holds for } A(f).$$

In examining the situation further, Artin and Tate (Tate 1966e) were led to make the following (Artin-Tate) conjecture:

(C) For a projective smooth geometrically-connected surface V over a finite field k, the Brauer group $\operatorname{Br}(V)$ of V is finite, and

$$P_2(q^{-s}) \sim \frac{|\operatorname{Br}(V)| \cdot |D|}{q^{\alpha(X)}|\operatorname{NS}(V)_{\text{tors}}|^2}\left(1-q^{1-s}\right)^{\rho(V)} \quad \text{as } s \to 1$$

where $P_2(T)$ is the characteristic polynomial of the Frobenius automorphism acting on $H^2(V_{k^{\mathrm{al}}}, \mathbb{Q}_\ell)$, D is the discriminant of the intersection pairing on $\mathrm{NS}(V)$, $\rho(V)$ is the rank of $\mathrm{NS}(V)$, and $\alpha(V) = \chi(X, \mathcal{O}_X) - 1 + \dim(\mathrm{PicVar}(V))$.

Naturally, they also conjectured:

(d) Let $f\colon V \to C$ be a proper map, as above, and assume that f has connected geometric fibres and a smooth generic fibre. Then Conjecture (B) holds for $A(f)$ over $k(C)$ if and only if Conjecture (C) holds for V over k.

Tate explains that he gave this conjecture "only a small letter (d) as label, because it is of a more elementary nature than (B) and (C)", and indeed, it has been proved. Artin and Tate checked it directly when f is smooth and has a section, Gordon (1979) checked it when the generic fibre has a rational cycle of degree 1, and Milne (1982) checked it when this condition holds only locally. However, ultimately the proof of (d) came from a different direction, by combining the following two statements:

- Conjecture C holds for a surface V over a finite field if and only if $\mathrm{Br}(V)(\ell)$ is finite for some prime ℓ (Tate (1966e) ignoring the p part; Milne (1975) complete statement);
- Conjecture A holds for an abelian variety A over a global field of nonzero characteristic if and only if $\text{Ш}(A)(\ell)$ is finite for some ℓ (Kato and Trihan 2003).

In the situation of (d), $\mathrm{Br}(V)(\ell)$ is finite if and only if $\text{Ш}(A(f))(\ell)$ is finite (Liu et al. 2005).

The known cases of Conjecture B over function fields have proved useful in the construction of lattice packings.[23]

4.5 Poles of Zeta Functions

Throughout this subsection, V is a nonsingular projective variety over a field k. We regard algebraic cycles on V as elements of the $\mathbb{Q}$-vector spaces $\mathcal{Z}^r(V)$.

Algebraic cycles D and D' are said to be numerically equivalent if $D \cdot E = D' \cdot E$ for all algebraic cycles E on V of complementary dimension, and they are ℓ-homologically equivalent if they have the same class in $H^{2r}(V_{k^{\mathrm{al}}}, \mathbb{Q}_\ell(r))$. In his Woods Hole talk, Tate asked whether the following statement is always true:

$E^r(V)$: Numerical equivalence coincides with ℓ-homological equivalence for algebraic cycles on V of codimension r.

---

[23] "One of the most exciting developments has been Elkies' (sic) and Shioda's construction of lattice packings from Mordell-Weil groups of elliptic curves over function fields. Such lattices have a greater density than any previously known in dimensions from about 54 to 4096." Preface to Conway, J.H.; Sloane, N.J.A. Sphere packings, lattices and groups. Second edition. Springer-Verlag, New York, 1993.

This is now generally regarded as a folklore conjecture (it is also a consequence of Grothendieck's standard conjectures). Note that, like the Tate conjecture, $E^r(V)$ is an existence statement for algebraic cycles: for an algebraic cycle D, it says that there exists an algebraic cycle E of complementary dimension such that $D \cdot E \neq 0$ if there exists a cohomological cycle with this property.

Let $\mathcal{A}^r$ denote the image of $\mathcal{Z}^r(V)$ in $H^{2r}(V, \mathbb{Q}_\ell(r))$, and let $\mathcal{N}^r$ denote the subspace of classes numerically equivalent to zero. Thus, $\mathcal{A}^r$ (resp. $\mathcal{A}^r/\mathcal{N}^r$) is the $\mathbb{Q}$-space of algebraic classes of codimension r modulo homological equivalence (resp. modulo numerical equivalence). In particular, $\mathcal{A}^r/\mathcal{N}^r$ is independent of ℓ.

Now assume that k is finitely generated. We need to consider also the following statement:

> $S^r(V)$: The map $H^{2r}(V_{k^{\mathrm{al}}}, \mathbb{Q}_\ell(r))^{G(k^{\mathrm{al}}/k)} \to H^{2r}(V_{k^{\mathrm{al}}}, \mathbb{Q}_\ell(r))_{G(k^{\mathrm{al}}/k)}$ induced by the identity map is bijective.

When k is finite, this means that 1 occurs semisimply (if at all) as an eigenvalue of the Frobenius map acting on $H^{2r}(V_{k^{\mathrm{al}}}, \mathbb{Q}_\ell(r))$.

An elementary argument suffices to prove that the following three statements are equivalent (for a fixed variety V, integer r, and prime ℓ):

(a) $T^r + E^r$;
(b) $T^r + T^{\dim V - r} + S^r$;
(c) $\dim_{\mathbb{Q}}(\mathcal{A}^r/\mathcal{N}^r) = \dim_{\mathbb{Q}_l} H^{2r}(V_{k^{\mathrm{al}}}, \mathbb{Q}_\ell(r))^{G(k^{\mathrm{al}}/k)}$.

When k is finite, each statement is equivalent to:

(d) the order of the pole of $\zeta(V, s)$ at $s = r$ is equal to $\dim_{\mathbb{Q}}(\mathcal{A}^r/\mathcal{N}^r)$.

See Tate (1979, 2.9). Note that (d) is independent of ℓ.

Tate (1964a, Conjecture 2) conjectured the following general version of (d):

> $P^r(V)$: Let V be a nonsingular projective variety over a finitely generated field k. Let d be the transcendence degree of k over the prime field, augmented by 1 if the prime field is $\mathbb{Q}$. Then the $2r$th component $\zeta^{2r}(V, s)$ of the zeta function of V has a pole of order $\dim_{\mathbb{Q}}(\mathcal{A}^r(V))$ at the point $s = d + r$.

This is also known as the Tate conjecture. For a discussion of the known cases of $P^r(V)$, see Tate (1964a, 1994a).

When k is a global function field, statements (a), (b), (c) are independent of ℓ, and are equivalent to the statement that $\zeta_S^{2r}(V, s)$ has a pole of order $\dim_{\mathbb{Q}}(\mathcal{A}^r(V))$ at the point $s = d + r$; here $\zeta_S^{2r}(V, s)$ omits the factors at a suitably large finite set S of primes. This follows from Lafforgue's proof of the global Langlands correspondence and other results in Langlands program by an argument that will, in principle, also work over number fields.—see Lyons (2009).

In the presence of E^r, Conjecture $T^r(V)$ is equivalent to $P^r(V)$ if and only if the order of the pole of $\zeta^{2r}(V, s)$ at $s = r$ is $\dim_{\mathbb{Q}_l} H^{2r}(V_{k^{\mathrm{al}}}, \mathbb{Q}_\ell(r))^{G(k^{\mathrm{al}}/k)}$. This is known for some Shimura varieties.

The Sato-Tate Conjecture Let A be an elliptic curve over $\mathbb{Q}$. For a prime p of good reduction, the number N_p of points on A mod p can be written

$$N_p = p + 1 - a_p$$
$$a_p = 2\sqrt{p}\cos\theta_p, \quad 0 \leq \theta_p \leq \pi.$$

When A has complex multiplication over $\mathbb{C}$, it is easily proved that the θ_p are uniformly distributed in the interval $0 \leq \theta \leq \pi$ as $p \to \infty$. In the opposite case, Mikio Sato found computationally that the θ_p appeared to have a density distribution $\frac{2}{\pi}\sin^2\theta$.

Tate proved that, for a power of an elliptic curve, the $\mathbb{Q}$-algebra of algebraic cycles is generated modulo homological equivalence by divisor classes. Using this, he computed that, for an elliptic curve A over $\mathbb{Q}$ without complex multiplication,

$$\operatorname{rank}(\mathcal{A}^i(A^m)) = \binom{m}{i}^2 - \binom{m}{i-1}\binom{m}{i+1},$$

from which he deduced that Sato's distribution is the only symmetric density distribution for which the zeta functions of the powers of A have their zeros and poles in agreement with the Conjecture $P^r(V)$.

The conjecture that, for an elliptic curve over $\mathbb{Q}$ without complex multiplication, the θ_p are distributed as $\frac{2}{\pi}\sin^2\theta$ is known as the Sato-Tate conjecture. It has been proved only recently, as the fruit of a long collaboration (Richard Taylor, Michael Harris, Laurent Clozel, Nicholas Shepherd-Barron, Thomas Barnet-Lamb, David Geraghty). As did Tate, they approach the conjecture through the analytic properties of the zeta functions of the powers of A (Barnet-Lamb et al. 2011).[24]

Needless to say, the Sato-Tate conjecture has been generalized to motives. Langlands (2011) has pointed out that his functoriality conjecture contains a very general form of the Sato-Tate conjecture.

4.6 Relation to the Hodge Conjecture

For a variety V over $\mathbb{C}$, there is a well-defined cycle map

$$c^r : \mathcal{Z}^r(V) \to H^{2r}(V, \mathbb{Q})$$

(cohomology with respect to the complex topology). Hodge proved that there is a decomposition

$$H^{2r}(V, \mathbb{Q})_{\mathbb{C}} = \bigoplus_{p+q=2r} H^{p,q}, \quad \overline{H^{p,q}} = H^{q,p}.$$

[24]For expository accounts, see: Carayol (2008) and Clozel (2008).

In Hodge (1952), he observed that the image of c^r is contained in

$$H^{2r}(V, \mathbb{Q}) \cap V^{r,r}$$

and asked whether this $\mathbb{Q}$-module is exactly the image of c^r. This has become known as the Hodge conjecture.[25]

In his original article (Tate 1964), Tate wrote:

> I can see no direct logical connection between [the Tate conjecture] and Hodge's conjecture that a rational cohomology class of type (p, p) is algebraic.... However, the two conjectures have an air of compatibility.

Pohlmann (1968) proved that the Hodge and Tate conjectures are equivalent for CM abelian varieties, Piatetski-Shapiro (1971) proved that the Tate conjecture for abelian varieties in characteristic zero implies the Hodge conjecture for abelian varieties, and Milne (1999) proved that the Hodge conjecture for CM abelian varieties implies the Tate conjecture for abelian varieties over finite fields.

The relation between the two conjectures has been greatly clarified by the work of Deligne. He defines the notion of an absolute Hodge class on a (complete smooth) variety over a field of characteristic zero, and conjectures that every Hodge class on a variety over $\mathbb{C}$ is absolutely Hodge. The Tate conjecture for a variety implies that all absolute Hodge classes on the variety are algebraic. Therefore, in the presence of Deligne's conjecture, the Tate conjecture implies the Hodge conjecture. As Deligne has proved his conjecture for abelian varieties, this gives another proof of Piatetski-Shapiro's theorem.

The twin conjectures of Hodge and Tate have a status in algebraic and arithmetic geometry similar to that of the Riemann hypothesis in analytic number theory. A proof of either one for any significantly large class of varieties would be a major breakthrough. On the other hand, whether or not the Hodge conjecture is true, it is known that Hodge classes behave in many ways as if they were algebraic (Deligne 1982; Cattani et al. 1995). There is some fragmentary evidence that the same is true for Tate classes in nonzero characteristic (Milne 2009).

5 Lubin-Tate Theory and Barsotti-Tate Group Schemes

5.1 Formal Group Laws and Applications

Let R be a commutative ring. By a formal group law over R, we shall always mean a one-parameter commutative formal group law, i.e., a formal power series $F \in R[[X, Y]]$ such that

- $F(X, Y) = X + Y +$ terms of higher degree,

[25] Hodge actually asked the question with $\mathbb{Z}$-coefficients.

- $F(F(X,Y),Z) = F(X,F(Y,Z))$
- $F(X,Y) = F(Y,X)$.

These conditions imply that there exists a unique $i_F(X) \in X \cdot R[[X]]$ such that $F(X, i_F(X)) = 0$. A homomorphism $F \to G$ of formal group laws is a formal power series $f \in XR[[X]]$ such that $f(F(X,Y)) = G(f(X), f(Y))$.

The formal group laws form a $\mathbb{Z}$-linear category. Let $c(f)$ be the first-degree coefficient of an endomorphism f of F. If R is an integral domain of characteristic zero, then $f \mapsto c(f)$ is an injective homomorphism of rings $\mathrm{End}_R(F) \to R$. See Lubin (1964).

Let F be a formal group law over a field k of characteristic $p \neq 0$. A nonzero endomorphism f of F has the form

$$f = aX^{p^h} + \text{terms of higher degree}, \quad a \neq 0,$$

where h is a nonnegative integer, called the height of f. The height of the multiplication-by-p map is called the height of F.

Lubin-Tate Formal Group Laws and Local Class Field Theory Let K be a nonarchimedean local field, i.e., a finite extension of $\mathbb{Q}_p$ or $\mathbb{F}_p((t))$. Local class field theory provides us with a homomorphism (the local reciprocity map)

$$\mathrm{rec}_K : K^\times \to \mathrm{Gal}\big(K^{\mathrm{ab}}/K\big)$$

such that, for every finite abelian extension L of K in K^{ab}, rec_K induces an isomorphism

$$(-, L/K) : K^\times/\mathrm{Nm}\, L^\times \to \mathrm{Gal}(L/K);$$

moreover, every open subgroup $K^\times$ of finite index arises as the norm group of a (unique) finite abelian extension. This statement shows that the finite abelian extensions of K are classified by the open subgroups of $K^\times$ of finite index, but leaves open the following problem:

> Let L/K be the abelian extension corresponding to an open subgroup H of $K^\times$ of finite index; construct generators for L and describe how $K^\times/H$ acts on them.

Lubin and Tate (1965a) found an elegantly simple solution to this problem.

The choice of a prime element π determines a decomposition $K^\times = \mathcal{O}_K^\times \times \langle \pi \rangle$, and hence (by local class field theory) a decomposition $K^{\mathrm{ab}} = K_\pi \cdot K^{\mathrm{un}}$. Here K_π is a totally ramified extension of K with the property that π is a norm from every finite subextension. Since K^{un} is well understood, the problem then is to find generators for the subfields of K_π and to describe the isomorphism

$$(-, K_\pi/K) : \mathcal{O}_K^\times \to \mathrm{Gal}(K_\pi/K)$$

given by reciprocity map. Let $\mathcal{O} = \mathcal{O}_K$, let $\mathfrak{p} = (\pi)$ be the maximal ideal in $\mathcal{O}$, and let $q = (\mathcal{O} : \mathfrak{p})$.

Let $f \in \mathcal{O}[[T]]$ be a formal power series such that

$$\begin{cases} f(T) = \pi T + \text{ terms of higher degree} \\ f(T) \equiv T^q \text{ modulo } \pi\mathcal{O}[[T]], \end{cases}$$

for example, $f = \pi T + T^q$ is such a power series. An elementary argument shows that, for each $a \in \mathcal{O}$, there is a unique formal power series $[a]_f \in \mathcal{O}_K[[T]]$ such that

$$\begin{cases} [a]_f(T) = aT + \text{ terms of higher degree} \\ [a]_f \circ f = f \circ [a]_f. \end{cases}$$

Let

$$X_m = \{x \in K^{\text{al}} \mid |x| < 1, \ \overbrace{(f \circ \cdots \circ f)}^{m}(x) = 0\}.$$

Then Lubin and Tate (1965a) prove:

(a) the field $K[X_m]$ is the totally ramified abelian extension of K with norm group $U_m \times \langle \pi \rangle$ where $U_m = 1 + \mathfrak{p}_K^m$;
(b) the map

$$\begin{aligned} \mathcal{O}^\times / U_m &\to \operatorname{Gal}(K[X_m]/K) \\ u &\mapsto (x \mapsto [u^{-1}]_f(x)) \end{aligned}$$

is an isomorphism.

For example, if $K = \mathbb{Q}_p$, $\pi = p$, and $f(T) = (1+T)^p - 1$, then

$$X_m = \{\zeta - 1 \mid \zeta^{p^m} = 1\} \simeq \mu_{p^m}(K^{\text{al}})$$

and $[u]_f(\zeta - 1) = \zeta^{u^{-1}} - 1$.

Lubin and Tate (1965a) show that, for each f as above, there is a unique formal group law F_f admitting f as an endomorphism. Then X_m can be realized as a group of "torsion points" on F_f, which endows it with the structure of an $\mathcal{O}$-module for which it is isomorphic to $\mathcal{O}/\mathfrak{p}^m$. From this the statements follow in a straightforward way.

The proof of the above results does not use local class field theory. Using the Hasse-Arf theorem, one can show that $K_\pi \cdot K^{\text{un}} = K^{\text{ab}}$, and deduce local class field theory. Alternatively, using local class field theory, one can show that $K_\pi \cdot K^{\text{un}} = K^{\text{ab}}$, and deduce the Hasse-Arf theorem. In either case, one finds that the isomorphism in (b) is the local reciprocity map.

The F_f are called *Lubin-Tate formal group laws*, and the above theory is called *Lubin-Tate theory*.

Deformations of Formal Group Laws (Lubin-Tate Spaces) Let F be a formal group law of height h over a perfect field k of characteristic $p \neq 0$. We consider local artinian k-algebras A with residue field k. A deformation of F over such an A is a formal group law F_A over A such that $F_A \equiv F \bmod \mathfrak{m}_A$. An isomorphism of deformations is an isomorphism of $\varphi\colon F_A \to G_A$ of formal group laws such that $\varphi(T) \equiv T \bmod \mathfrak{m}_A$.

Let W denote the ring of Witt vectors with residue field k. Lubin and Tate (1966d) prove that there exists a formal group $\mathcal{F}(t_1, \ldots, t_{h-1})(X, Y)$ over $W[[t_1, \ldots, t_{h-1}]]$ with the following properties:

- $\mathcal{F}(0, \ldots, 0) \equiv F \bmod \mathfrak{m}_W$;
- for any deformation F_A of F, there is a unique homomorphism $W[[t_1, \ldots, t_{h-1}]] \to A$ sending $\mathcal{F}$ to F_A.

The results of Lubin and Tate are actually stronger, but this seems to be the form in which they are most commonly used.

In particular, the above result identifies the space of deformations of F with the formal scheme $\mathrm{Spf}(W[[t_1, \ldots, t_{h-1}]])$. These spaces are now called Lubin-Tate deformation spaces. Drinfeld showed that, by adding (Drinfeld) level structures, it is possible to construct towers of deformation spaces, called Lubin-Tate towers (Drinfeld 1974). These play an important role in the study of the moduli varieties of abelian varieties with PEL-structure and in the Langlands program. For example, it is known that both the Jacquet-Langlands correspondence and the local Langlands correspondence for GL_n can be realized in the étale cohomology of a Lubin-Tate tower (or, more precisely, in the étale cohomology of the Berkovich space that is attached to the rigid analytic space which is the generic fibre of the Lubin-Tate tower) (Carayol 1990; Boyer 1999; Harris and Taylor 2001).

5.2 Finite Flat Group Schemes

A group scheme G over a scheme S is finite and flat if $G = \mathrm{Spec}(A)$ with A locally free of finite rank as a sheaf of $\mathcal{O}_S$-modules. When A has constant rank r, G is said to have order r. A finite flat group scheme of prime order is necessarily commutative.

In his course on Formal Groups at Harvard in the fall of 1966, Tate discussed the following classification problem:

> let R be a local noetherian ring with residue field of characteristic $p \neq 0$; assume that R contains the $(p-1)$st roots of 1, i.e., that $R^\times$ contains a cyclic subgroup of order $p-1$; determine the finite flat group schemes of order p over R.

When R is complete, Tate found that such group schemes correspond to pairs of elements (a, c) of R such that $ac \in p \cdot R^\times$; two pairs (a, c) and (a_1, c_1) correspond to isomorphic groups if and only if $a_1 = u^{p-1}a$ and $c_1 = u^{1-p}c$ for some $u \in R^\times$.

These results were extended and completed in Tate and Oort (1970a). Let

$$\Lambda_p = \mathbb{Z}\left[\zeta, \frac{1}{p(p-1)}\right] \cap \mathbb{Z}_p$$

where ζ is a primitive $(p-1)$st root of 1 in $\mathbb{Z}_p$. Tate and Oort define a sequence $w_1 = 1, w_2, \ldots, w_p$ of elements of Λ_p in which $w_1, \ldots, w_{p-1}$ are units and $w_p = pw_{p-1}$. Then, given an invertible $\mathcal{O}_S$-module L and sections a of $L^{\otimes(p-1)}$ and b of $L^{\otimes(1-p)}$ such that $a \otimes b = w_p$, they show that there is a group scheme $G^L_{a,b}$ such that, for every S-scheme T,

$$G^L_{a,b}(T) = \left\{x \in \Gamma(T, L \otimes_{\mathcal{O}_S} \mathcal{O}_T) \mid x^{\otimes p} = a \otimes x\right\},$$

and the multiplication on $G^L_{a,b}(T)$ is given by

$$x_1 \cdot x_2 = x_1 + x_2 + \frac{b}{w_{p-1}} \otimes D_p(x_1 \otimes 1, 1 \otimes x_2),$$

where

$$D_p(X_1, X_2) = \frac{w_{p-1}}{1-p} \sum_{i=1}^{p-1} \frac{X_1^i}{w_i} \frac{X_2^{p-i}}{w_{p-i}} \in \Lambda_p[X_1, X_2].$$

Every finite flat group scheme of order p over S is of the form $G^L_{a,b}$ for some triple (L, a, b), and $G^L_{a,b}$ is isomorphic to $G^{L_1}_{a_1,b_1}$ if and only if there exists an isomorphism from L to L_1 carrying a to a_1 and b to b_1. The Cartier dual of $G^L_{a,b}$ is $G^{L^{-1}}_{b,a}$. The proofs of these statements make ingenious use of the action of $\mathbb{F}_p^\times$ on $\mathcal{O}_G$.

Tate and Oort apply their result to give a classification of finite flat group schemes of order p over the ring of integers in a number field in terms of idèle class characters. In particular, they show that the only such group schemes over $\mathbb{Z}$ are the constant group scheme $\mathbb{Z}/p\mathbb{Z}$ and its dual μ_p.

In the years since Tate and Oort wrote their article, the classification of finite flat commutative group schemes over various bases has been intensively studied, and some of the results were used in the proof of the modularity conjecture for elliptic curves (hence of Fermat's last theorem). Tate (1997a) has given a beautiful exposition of the basic theory of finite flat group schemes, including the results of Raynaud (1974) extending the above theory to group schemes of type $(p, \ldots, p)$.

5.3 Barsotti-Tate Groups (p-Divisible Groups)

Let A be an abelian variety over a field k. In his study of abelian varieties and their zeta functions, Weil used the ℓ-primary component $A(\ell)$ of the group $A(k^{\text{sep}})$ for ℓ a prime distinct from char(k). This is an ℓ-divisible group isomorphic to

$(\mathbb{Q}_\ell/\mathbb{Z}_\ell)^{2\dim A}$ equipped with an action of $\mathrm{Gal}(k^{\mathrm{sep}}/k)$. For $p = \mathrm{char}(k)$, it is natural to replace $A(\ell)$ with the direct system

$$A(p):\quad A_1 \hookrightarrow A_2 \hookrightarrow \cdots \hookrightarrow A_\nu \hookrightarrow A_{\nu+1} \hookrightarrow \cdots$$

where A_ν is the finite group *scheme* $\mathrm{Ker}(p^\nu\colon A \to A)$.

In the mid sixties, Serre and Tate[26] defined a *p-divisible group* of height h over a ring R to be a direct system $G = (G_\nu, i_\nu)_{\nu\in\mathbb{N}}$ where, for each $\nu \geq 0$, G_ν is a finite group scheme over R of order $p^{\nu h}$ and the sequence

$$0 \longrightarrow G_\nu \xrightarrow{i_\nu} G_{\nu+1} \xrightarrow{p^\nu} G_{\nu+1}$$

is exact. An abelian scheme A over R defines a p-divisible group $A(p)$ over R of height $2\dim(A)$.

The dual of a p-divisible group $G = (G_\nu, i_\nu)$ is the system $G' = (G'_\nu, i'_\nu)$ where G'_ν is the Cartier dual of G_ν and i'_ν is the Cartier dual of the map $G_{\nu+1} \xrightarrow{p} G_\nu$. It is again a p-divisible group.

Tate developed the basic theory p-divisible group in his article for the proceedings of the 1966 Driebergen conference (Tate 1967c) and in a series of ten lectures at the Collège de France in 1965–1966 (see Serre 1968b). He showed that p-divisible groups generalize formal Lie groups in the following sense: let R be a complete noetherian local ring with residue field k of characteristic $p > 0$; an n-dimensional commutative formal Lie group Γ over R can be defined to be a family $f(Y,Z) = (f_i(Y,Z))_{1\leq i\leq n}$ of n power series in $2n$ variables satisfying the conditions in the first paragraph of this section; if such a group Γ is divisible (i.e., $p\colon \Gamma \to \Gamma$ is an isogeny), then one can define the kernel G_ν of $p^\nu\colon \Gamma \to \Gamma$ as a group scheme over R; Tate shows that $\Gamma(p) = (G_\nu)_{\nu\geq 1}$ is a p-divisible group $\Gamma(p)$, and that the functor $\Gamma \rightsquigarrow \Gamma(p)$ is an equivalence from the category of divisible commutative formal Lie groups over R to the category of connected p-divisible groups over R.

The main theorem of Tate (1967c) states the following:

> Let R be an integrally closed, noetherian, integral domain whose field of fractions K is of characteristic zero, and let G and H be p-divisible groups over R. Then every homomorphism $G_K \to H_K$ of the generic fibres extends uniquely to a homomorphism $G \to H$.

In other words, the functor $G \rightsquigarrow G_K$ is fully faithful. This was extended to rings R of characteristic $p \neq 0$ by de Jong (1998).

Since their introduction, p-divisible groups have become an essential tool in the study of abelian schemes. We have already seen in (2.4) one application of p-divisible groups to the problem of lifting abelian varieties. Another application was to the proof of the Mordell conjecture (Faltings 1983).

---

[26]Usually this is credited to Tate alone, but Tate writes: "We were both contemplating them. I think it was probably Serre who first saw clearly the simple general definition and its relation to formal groups of finite height." The dual of a p-divisible group is often called the Serre dual.

In his talk at the 1970 International Congress, Grothendieck renamed p-divisible groups "Barsotti-Tate groups". Today, both terms are used.

5.4 Hodge-Tate Decompositions

Now let R be a complete discrete valuation ring of unequal characteristics, and let K be its field of fractions. Let K^{al} be an algebraic closure of K, and let C be the completion of K^{al}. As Serre noted (Serre 1968b, p. 322):

> One of the most surprising results of Tate's theory is the fact that *the properties of p-divisible groups are intimately related to the structure of C as a Galois module over* $\mathrm{Gal}(K^{\mathrm{al}}/K)$.

Let $\mathcal{G} = \mathrm{Gal}(K^{\mathrm{al}}/K)$, and let V be a C-vector space on which $\mathcal{G}$ acts semi-linearly. The Tate twist $V(i)$, $i \in \mathbb{Z}$, is V with $\mathcal{G}$ acting by

$$(\sigma, v) \mapsto \chi(\sigma)^i \cdot \sigma v, \quad \chi \text{ the } p\text{-adic cyclotomic character.}$$

Tate proved that $H^0(\mathcal{G}, C) = K$ and $H^1(\mathcal{G}, C) \approx K$, and that $H^q(\mathcal{G}, C(i)) = 0$ for $q = 0, 1$ and $i \neq 0$.[27] Using these statements, he proved that, for a p-divisible group G over R, there is a canonical isomorphism

$$\mathrm{Hom}(TG, C) \simeq t_{G'}(C) \oplus t_G^{\vee}(C)(-1), \tag{29}$$

where $TG = \varprojlim_{\nu} G_{\nu}(K^{\mathrm{al}})$ is the Tate module of G, t_G is the tangent space to G at zero, and G' is the dual p-divisible group. In particular TG determines the dimension of G, a fact that is used in the proof of the main theorem in the last subsection.

When G is the p-divisible group of an abelian scheme A over R, (29) can be written as:

$$H^1_{\mathrm{et}}(A_C, \mathbb{Q}_p) \otimes C \simeq H^1\left(A_C, \Omega^0_{A_C/C}\right) \oplus H^0\left(A_C, \Omega^1_{A_C/C}\right)(-1).$$

This result led Tate to make the following (Hodge-Tate) conjecture:[28]

> For every nonsingular projective variety X over K, there exists a canonical (Hodge-Tate) decomposition
>
> $$H^n_{\mathrm{et}}(X_C, \mathbb{Q}_p) \otimes_{\mathbb{Q}_p} C \simeq \bigoplus_{p+q=n} H^{p,q}(X_C)(-p) \tag{30}$$
>
> where $H^{p,q}(X_C) = H^q(X, \Omega^p_{X/K}) \otimes_K C$. This decomposition is compatible with the action of $\mathrm{Gal}(K^{\mathrm{al}}/K)$.

---

[27] Sen and Ax simplified and generalized Tate's proof that $C^{\mathcal{G}} = K$, and the result is now known as the Ax-Sen-Tate theorem.

[28] Tate 1967c, p. 180; see also Serre's summary of Tate's lectures (Serre 1968b, p. 324).

Tate's conjecture launched a new subject in mathematics, called p-adic Hodge theory. The isomorphism (30) can be regarded as a statement about the étale cohomology of X_C regarded as a module over $\mathrm{Gal}(K^{\mathrm{al}}/K)$. About 1980, Fontaine stated a series of successively stronger conjectures, beginning with the Hodge-Tate conjecture, that describe the structure of these Galois modules (Fontaine 1982).[29] Most of Fontaine's conjectures have now been proved. The Hodge-Tate conjecture itself was proved by Faltings (1988).

6 Elliptic Curves

Although elliptic curves are just abelian varieties of dimension one, their study is quite different. Throughout his career, Tate has returned to the study of elliptic curves.

6.1 Ranks of Elliptic Curves over Global Fields

Mordell proved that, for an elliptic curve E over $\mathbb{Q}$, the group $E(\mathbb{Q})$ is finite generated. At one time, it was widely conjectured that the rank of $E(\mathbb{Q})$ is bounded, but, as Cassels 1966 pointed out, this is implausible.[30] Tate and Shafarevich (1967d)[31] made it even less plausible by proving that, for elliptic curves over the global field $\mathbb{F}_p(t)$, the ranks are unbounded. Their examples are quadratic twists of a supersingular elliptic curve with coefficients in $\mathbb{F}_p$; in particular, they are isotrivial (i.e., have $j \in \mathbb{F}_p$). More recently, it has been shown that the ranks are unbounded even among the nonisotrivial elliptic curves over $\mathbb{F}_p(t)$ (Ulmer 2002). Meanwhile, the largest known rank for an elliptic curve over $\mathbb{Q}$ is 28.[32]

---

[29] See Fontaine (1982) and many other articles.

[30] "It has been widely conjectured that there is an upper bound for the rank depending only on the groundfield. This seems to me implausible because the theory makes it clear that an abelian variety can only have high rank if it is defined by equations with very large coefficients." Cassels (1966, p. 257).

[31] For a long time I was puzzled as to how this article came to be written, because I was not aware that Shafarevich had been allowed to travel to the West, but Tate writes: "sometime during the year 1965–1966, which I spent in Paris, Shafarevich appeared. There must have been a brief period when the Soviets relaxed their no-travel policy.... Shafarevich was in Paris for a month or so, and the paper grew out of some discussion we had. We both liked the idea of our having a joint paper, and I was happy to have it in Russian."

[32] Elkies, see http://web.math.hr/~duje/tors/tors.html.

6.2 Torsion Points on Elliptic Curves over $\mathbb{Q}$

Beppo Levi constructed elliptic curves E over $\mathbb{Q}$ having each of the groups

$$\begin{aligned} \mathbb{Z}/n\mathbb{Z} \quad & n = 1, 2, \ldots, 10, 12, \\ \mathbb{Z}/2\mathbb{Z} \times \mathbb{Z}/n\mathbb{Z} \quad & n = 2, 4, 6, 8, \end{aligned}$$

as the torsion subgroup of $E(\mathbb{Q})$, and he conjectured that this exhausted the list of possible such groups (Levi 1909).

Consistent with this, Mazur and Tate (1973c) show that there is no elliptic curve over $\mathbb{Q}$ with a rational point of order 13, or, equivalently, that the curve $X_1(13)$ that classifies the elliptic curves with a chosen point of order 13 has no rational points (except for its cusps). Ogg found a rational point of order 19 on the Jacobian J of $X_1(13)$, and Mazur and Tate show that J has exactly 19 rational points. They then deduce that $X_1(13)$ has no noncuspidal rational point by examining how it sits in its Jacobian.

The interest of their article is more in its methods than in the result itself.[33] The ring $\mathbb{Z}[\sqrt[3]{1}]$ acts on J, and Mazur and Tate perform a 19-descent by studying the flat cohomology of the exact sequence of group schemes

$$0 \to F \to J \xrightarrow{\pi} J \to 0$$

on $\operatorname{Spec} \mathbb{Z} \backslash (13)$, where π is one of the factors of 19 in $\mathbb{Z}[\sqrt[3]{1}]$. In a major work, Mazur developed these methods further, and completely proved Levi's conjecture (Mazur 1977).

The similar problem for an arbitrary number field K is probably beyond reach, but, following work of Kamienny, Merel (1996) proved that, for a fixed number field K, the order of the torsion subgroup of $E(K)$ for E an elliptic curve over K is bounded by a constant that depends only the degree of K over $\mathbb{Q}$.

6.3 Explicit Formulas and Algorithms

The usual Weierstrass form of the equation of an elliptic curve is valid only in characteristics $\neq 2, 3$. About 1965 Tate wrote out the complete form, valid in all characteristics, and even over $\mathbb{Z}$. For an elliptic curve over a nonarchimedean local field with perfect residue field, he wrote out an explicit algorithm (known as Tate's algorithm) for computing the minimal model of the curve and determining the Kodaira type of the special fibre. Ogg's formula then gives the conductor of the curve. The handwritten manuscript containing these formulas was invaluable to people working in the field. A copy, which had been sent to Cassels, was included, essentially verbatim, in the proceedings of the Antwerp conference (Tate 1975b).[34]

---

[33]About the same time, J. Blass found a more elementary proof of the same result.

[34]Tate writes: "Early in that summer [1965], Weil had told me of the idea that all elliptic curves over $\mathbb{Q}$ are modular [and that the conductor of the elliptic curve equals the conductor of the cor-

6.4 Analogues at p of the Conjecture of Birch and Swinnerton-Dyer

For an elliptic curve E over $\mathbb{Q}$, the conjecture of Birch and Swinnerton-Dyer states that

$$L(s,E) \sim \Omega \prod_{p \text{ bad}} c_p \frac{|\text{Ш}(E)| \cdot R}{|E(\mathbb{Q})_{\text{tors}}|^2}(s-1)^r \quad \text{as } s \to 1$$

where r is the rank of $E(\mathbb{Q})$, $\text{Ш}(E)$ is the Tate-Shafarevich group, R is the discriminant of the height pairing on $E(\mathbb{Q})$, Ω is the real period of E, and the $c_p = (E(\mathbb{Q}_p) \colon E^0(\mathbb{Q}_p))$. When E has good ordinary or multiplicative reduction at p, there is a p-adic zeta function $L_p(s,E)$, and Mazur, Tate, and Teitelbaum (1986) investigated whether the behaviour of $L_p(s,E)$ near $s=1$ is similarly related to the arithmetic invariants of E.[35] They found it is, but with one major surprise: there is an "exceptional" case in which $L_p(s,E)$ is related to an extended version of $E(\mathbb{Q})$ rather than $E(\mathbb{Q})$ itself. Supported by numerical evidence, they conjectured:

> BSD(p). When E has good ordinary or nonsplit multiplicative reduction at a prime p, the function $L_p(s,E)$ has a zero at $s=1$ of order at least $r = \operatorname{rank} E(\mathbb{Q})$, and $L_p^{(r)}(1,E)$ is equal to a certain expression involving $|\text{Ш}(E)|$ and a p-adic regulator $R_p(E)$. When E has split multiplicative reduction, it is necessary to replace r with $r+1$.

The L-function $L_p(s,E)$ is the p-adic Mellin transform of a p-adic measure obtained from modular symbols. The p-adic regulator is the discriminant of the canonical p-adic height pairing (augmented in the exceptional case). Much more is known about BSD(p) than the original conjecture of Birch and Swinnerton-Dyer. For example, Kato (2004) has proved the following statement:

> The function $L_p(s,E)$ has a zero at $s=1$ of order at least r (at least $r+1$ in the exceptional case). When the order of the zero equals its conjectured value, then the p-primary component of $\text{Ш}(E)$ is finite and $R_p(E) \neq 0$.

In the exceptional case, $E_{\mathbb{Q}_p}$ is a Tate elliptic curve and $L_p(1,E)=0$. On comparing their conjecture in this case with the original conjecture of Birch and Swinnerton-Dyer, the authors were led to the conjecture

$$L_p^{(1)}(1,E) = \frac{\log_p(q)}{\operatorname{ord}_p(q)} \frac{L(1,E)}{\Omega}$$

---

responding modular form]. That motivated Swinnerton-Dyer to make a big computer search for elliptic curves over $\mathbb{Q}$ with not too big discriminant, in order to test Weil's idea. But of course it was necessary to be able to compute the conductor to do that test. That was my main motivation."

[35] The authors assume that E is modular—at the time, it was not known that all elliptic curves over $\mathbb{Q}$ are modular.

where q is the period of the Tate curve $E_{\mathbb{Q}_p}$ and Ω is the real period of E. This became known as the Mazur-Tate-Teitelbaum conjecture. It was proved by Greenberg and Stevens (1993) for $p \neq 2, 3$, and by several authors in general.

Mazur and Tate (1987) state "refined" conjectures that avoid any mention of p-adic L-functions and, in particular, avoid the problem of constructing these functions. Let E be an elliptic curve over $\mathbb{Q}$. For a fixed integer $M > 0$, they use modular symbols to construct an element θ in the group algebra $\mathbb{Q}[(\mathbb{Z}/M\mathbb{Z})^\times/\{\pm 1\}]$. Let R be a subring of $\mathbb{Q}$ containing the coefficients of θ and such that the order the torsion subgroup of $E(\mathbb{Q})$ is invertible in R. The analogue of an L-function having a zero of order r at $s = 1$ is that θ lie in the rth power of the augmentation ideal I of the group algebra $R[(\mathbb{Z}/M\mathbb{Z})^\times/\{\pm 1\}]$. Assume that M is not divisible by p^2 for any prime p at which E has split multiplicative reduction. Then Mazur and Tate conjectured:

> Let r be the rank of $E(\mathbb{Q})$, and let r' be the number of primes dividing M at which E has split multiplicative reduction. Then $\theta \in I^{r+r'}$, and there is a formula (involving the order of Ш(E)) for the image of θ in $I^{r+r'}/I^{r+r'+1}$.

Again, the authors provide numerical evidence for their conjecture. Tate's student, Ki-Seng Tan, restated the Mazur-Tate conjecture for an elliptic curve over a global field, and he proved that part of the new conjecture is implied by the conjecture of Birch and Swinnerton-Dyer (Tan 1995).

In the first article discussed above, Mazur, Tate, Teitelbaum gave explicit formulas relating the canonical p-adic height pairings to a p-adic sigma function, and used the sigma function to study the height pairings. Mazur and Tate (1991), present a detailed construction of the p-adic sigma function for an elliptic curve E with good ordinary reduction over a p-adic field K, and they prove the properties used in the earlier article. In contrast to the classical sigma function, which is defined on the universal covering of E, the p-adic sigma function is defined on the formal group of E.

Finally, the article Mazur, Stein, and Tate (2006) studies the problem of efficiently computing of p-adic heights for an elliptic curve E over a global field K. This amounts to efficiently computing the sigma function, which in turn amounts to efficiently computing the p-adic modular form E_2.

6.5 Jacobians of Curves of Genus One

For a curve C of genus one over a field k, the Jacobian variety J of C is an elliptic curve over k. The problem is to compute a Weierstrass equation for J from an equation for C.

Weil (1954) showed that, when C is defined by an equation $Y^2 = f(X)$, $\deg f = 4$, then the Weierstrass equation of J can be computed using the invariant theory of the quartic of f, which goes back to Hermite.

An et al. (2001) showed how formulas from classical invariant theory give Weierstrass equations for J and for the map $C \to J$ when $\mathrm{char}(k) \neq 2, 3$ and C is a double cover of $\mathbb{P}^1$, a plane cubic, or a space quartic.

Tate and Rodrigues-Villegas found the Weierstrass equations over fields of characteristic 2 and 3, where the classical invariant theory no longer applies. Together with Artin they extended their result to an arbitrary base scheme S (Artin, Rodriguez-Villegas, Tate 2005). Specifically, let C be the family of curves over a scheme S defined, as a subscheme of $\mathbb{P}^2_S$, by a cubic $f \in \Gamma(S, \mathcal{O}_S)[X, Y, Z]$, and assume that the ten coefficients of f have no common zero. Then there is a Weierstrass equation

$$g\colon\quad Y^2Z + a_1XYZ + a_3YZ^2 = X^3 + a_2X^2Z + a_4XZ^2 + a_6Z^3, \quad a_i \in \Gamma(S, \mathcal{O}_S),$$

whose coefficients are given explicitly in terms of the coefficients of f, such that the functor $\mathrm{Pic}^0_{C/S}$ is represented by the smooth locus of the subscheme of $\mathbb{P}^2_S$ defined by g. A key ingredient of the proof is a characterization, over sufficiently good base schemes, of the group algebraic spaces that can be described by a Weierstrass equation.

6.6 Expositions

In 1961, Tate gave a series of lectures at Haverford College titled "Rational Points on Cubic Curves" intended for bright undergraduates in mathematics. Notes were taken of the lectures, and these were distributed in mimeographed form. The book, Silverman and Tate 1992, is a revision, and expansion, of the notes.

In the spring of 1960, the fall of 1967, 1975, ..., Tate gave courses on the arithmetic of elliptic curves, whose informal notes have influenced later expositions.

7 The K-Theory of Number Fields

7.1 K-Groups and Symbols

Grothendieck defined $K_0(X)$ for X a scheme in order to be able to state his generalization of the Riemann-Roch theorem. The topologists soon adapted Grothendieck's definition to topological spaces, and extended it to obtain groups K_n for all $n \in \mathbb{N}$.

For a commutative ring R, $K_0(R)$ is just the Grothendieck group of the category of finitely generated projective R-modules. In 1962, Bass and Schanuel defined $K_1(R)$, and in 1967, Milnor[36] defined $K_2(R)$. In the early 1970s, several authors suggested definitions for the higher K-groups, which largely coincided when this

[36] During a course at Princeton University; published as: Milnor (1971).

could be checked. Quillen's definition (Quillen 1973) was the most flexible, and it is his that has been adopted.

The Steinberg group ST(R) of a ring R is the group with generators

$$x_{ij}(r), \quad i, j = 1, 2, 3, \ldots, \; i \neq j, \; r \in R$$

and relations

$$x_{ij}(r)x_{ij}(s) = x_{ij}(r+s)$$
$$\left[x_{ij}(r), x_{kl}(s)\right] = \begin{cases} 1 & \text{if } i \neq l \text{ and } j \neq k \\ x_{il}(rs) & \text{if } i \neq l \text{ and } j = k. \end{cases}$$

The elementary matrices $E_{ij}(r) = I + re_{ij}$ in GL(R) satisfy these relations, and so there is a homomorphism $x_{ij}(r) \mapsto E_{ij}(r) \colon \mathrm{ST}(R) \to \mathrm{GL}(R)$. The groups $K_1(R)$ and $K_2(R)$ can be defined by the exact sequence

$$1 \to K_2(R) \to \mathrm{ST}(R) \to \mathrm{GL}(R) \to K_1(R) \to 1.$$

Let F be a field. A *symbol* on F with values in a commutative group C is defined to be a bimultiplicative map

$$(\,,\,)\colon F^{\times} \times F^{\times} \to C$$

such that $(a, 1-a) = 1$ whenever $a \neq 0, 1$. Matsumoto (1969) showed that the natural map

$$\{\,,\,\}\colon F^{\times} \times F^{\times} \to K_2F$$

is a universal symbol, i.e., that $K_2(F)$ is the free abelian group on the pairs $\{a, b\}$, $a, b \in F^{\times}$, subject only to the relations

$$\begin{aligned} \{aa', b\} &= \{a, b\}\{a', b\} && \text{all } a, a', b \in F^{\times} \\ \{a, bb'\} &= \{a, b\}\{a, b'\} && \text{all } a, b, b' \in F^{\times} \\ \{a, 1-a\} &= 1 && \text{all } a \neq 0, 1 \text{ in } F^{\times}. \end{aligned}$$

Examples of Symbols

(a) The tame (Hilbert) symbol. Let v be a discrete valuation of F, with residue field $\kappa(v)$. Then

$$(a, b)_v = (-1)^{v(a)v(b)} \frac{a^{v(b)}}{b^{v(a)}} \bmod \mathfrak{m}_v$$

is a symbol on F with values in $\kappa(v)^{\times}$.

(b) The Galois symbol (Tate 1970b, §1). For m not divisible by char(F), $H^1(F, \mu_m) \simeq F^{\times}/F^{\times m}$, and the cup-product pairing

$$H^1(F, \mu_m) \times H^1(F, \mu_m) \to H^2(F, \mu_m \otimes \mu_m)$$

gives a symbol on F with values in $H^2(F,\mu_m\otimes\mu_m)$. When F contains the mth roots of 1,

$$H^2(F,\mu_m\otimes\mu_m)\simeq H^2(F,\mu_m)\otimes\mu_m\simeq \mathrm{Br}(F)_m\otimes\mu_m$$

and the symbol was known classically.

(c) The differential symbol (Tate ibid.). For $p=\mathrm{char}(F)$,

$$f,g\mapsto\frac{df}{f}\wedge\frac{dg}{g}\colon F^\times\times F^\times\to\Omega^2_{F/\mathbb{F}_p}$$

is a symbol.

(d) On $\mathbb{C}$ there are no continuous symbols, but on $\mathbb{R}$ there is the symbol

$$(a,b)_\infty=\begin{cases}1 & \text{if } a>0 \text{ or } b>0\\ -1 & \text{otherwise.}\end{cases}$$

7.2 The Group K_2F for F a Global Field

Tate recognized that the study of the K_2 of a global field is related to classical objects in number fields, and sheds new light on them. He largely initiated the study of the K-groups of global fields and their rings of integers.

For a field F, $K_0F\simeq\mathbb{Z}$ is without particular interest. On the other hand, $K_1F\simeq F^\times$. For a number field, there is an exact sequence

$$0\to U_F\to F^\times\xrightarrow{(\mathrm{ord}_v)}\bigoplus_v\mathbb{Z}\to C_F\to 0$$

where v runs over the finite primes of F. Dirichlet proved that $U_F\approx\mu(F)\times\mathbb{Z}^{r_1+r_2-1}$, where r_1 and r_2 are the numbers of real and complex primes, and Dedekind proved that the class group C_F is finite. Thus understanding K_1F involves understanding the two basic objects in classical algebraic number theory.

Let F be a global field. For a noncomplex prime v of F, let $\mu_v=\mu(F_v)$ and let $m_v=|\mu_v|$. For a finite prime v of F, $\mathrm{Br}(F_v)\simeq\mathbb{Q}/\mathbb{Z}$, and so the Galois symbol with $m=m_v$ gives a homomorphism $\lambda_v\colon K_2F_v\to\mu_v$. Similarly, $(\ ,\ )_\infty$ gives a homomorphism $\lambda_v\colon K_2F_v\to\mu_v$ when v is real. The λ_v can be combined with the obvious maps $K_2F\to K_2F_v$ to give the homomorphism λ_F in the sequence

$$0\to\mathrm{Ker}(\lambda_F)\to K_2F\xrightarrow{\lambda_F}\bigoplus_v\mu_v\to\mu_F\to 0;$$

the direct sum is over the noncomplex primes of F, and the map from it sends $(x_v)_v$ to $\prod_v\frac{m_v}{m_F}x_v$ where $m_F=|\mu(F)|$. A product formula,

$$\prod(a,b)_v^{\frac{m_v}{m_F}}=1$$

shows that the sequence is a complex, and Moore (1968) showed that the cokernel of λ_F is μ_F. Thus, to compute K_2F, it remains to identify $\mathrm{Ker}(\lambda_F)$.

For $\mathbb{Q}$, Tate proved that $\mathrm{Ker}(\lambda_F)$ is trivial, and then observed that most of his argument was already contained in Gauss's first proof of the quadratic reciprocity law.

For a global field F, Bass and Tate proved that $\mathrm{Ker}(\lambda_F)$ is finitely generated, and that it is finite of order prime to the characteristic in the function field case. Garland (1971) proved that it is also finite in the number field case.

In the function field case, Tate proved that

$$\left|\mathrm{Ker}(\lambda_F)\right| = (q-1)\cdot\left(q^2-1\right)\cdot\zeta_F(-1). \tag{31}$$

For a number field, the Birch-Tate conjecture says that

$$\left|\mathrm{Ker}(\lambda_F)\right| = \pm w_2(F)\cdot\zeta_F(-1) \tag{32}$$

where $w_2(F)$ is the largest integer m such $\mathrm{Gal}(F^{\mathrm{al}}/F)$ acts trivially on $\mu_m(F^{\mathrm{al}})\otimes\mu_m(F^{\mathrm{al}})$ (Birch 1971; Tate 1970b). The odd part of this conjecture was proved by Wiles (1990).

When Quillen defined the higher K-groups, he proved that

$$K_2(\mathcal{O}_F) = \mathrm{Ker}\Bigl(K_2(F)\to\bigoplus_{v\text{ finite}}\mu_v\Bigr)$$

and so there is an exact sequence

$$0\to\mathrm{Ker}(\lambda_F)\to K_2(\mathcal{O}_F)\to\bigoplus_{v\text{ real}}\mu_v.$$

Thus the computation of $\mathrm{Ker}(\lambda_F)$ is closely related to that of $K_2(\mathcal{O}_F)$. Lichtenbaum (1972) generalized the Birch-Tate conjecture to the following statement: for all totally real number fields F[37]

$$\frac{|K_{4i-2}(\mathcal{O}_F)|}{|K_{4i-1}(\mathcal{O}_F)|} = \left|\zeta_F(1-2i)\right|,\quad \text{all } i\ge 1.$$

The Galois Symbol Tate proved (31) by using Galois symbols. For a global field F, he proved that the map

$$K_2F\to H^2\left(F,\mu_m^{\otimes 2}\right) \tag{33}$$

[37]The first test of the conjecture was for $F=\mathbb{Q}$ and $i=1$. Since $\zeta_{\mathbb{Q}}(-1)=-1/12$ and $K_2(\mathbb{Z})=\mathbb{Z}/2\mathbb{Z}$, the conjecture predicts that $|K_3(\mathbb{Z})|$ has 24 elements, but Lee and Szczarba showed that it has 48 elements. When a seminar speaker at Harvard mentioned this, and scornfully concluded that the conjecture was false, Tate responded from the audience "Only for 2". In fact, Lichtenbaum's conjecture is believed to be correct up to a power of 2.

defined by the Galois symbol induces an isomorphism

$$K_2F/(K_2F)^m \simeq H^2(F, \mu_m^{\otimes 2}) \tag{34}$$

when m is not divisible by char(F), and wrote "I don't know whether ... this holds for all fields" Tate (1970b, p. 208). Merkurjev and Suslin (1982) proved that it does hold for all fields.

Tate noted that the isomorphism (34) gives little information on Ker(λ_F) because $\bigcap_m (K_2F)^m$ is a subgroup of Ker(λ_F) of index at most 2, and Ker(λ_F) $\subset (K_2F)^m$ for all m not divisible by 8. He then defined more refined Galois symbols, which are faithful.

Fix a prime $\ell \neq$ char(F), and let $\mathbb{Z}_\ell(1) = \varprojlim_n \mu_{\ell^n}(F^{\mathrm{al}})$. This is a free $\mathbb{Z}_\ell$-module of rank 1 with an action of Gal(F^{al}/F), and we let $H^r(F, \mathbb{Z}_\ell(1)^{\otimes 2})$ denote the Galois cohomology group defined using continuous cocycles (natural topology on both Gal(F^{al}/F) and $\mathbb{Z}_\ell(1)^{\otimes 2}$). Tate proves that the maps (33) with $m = \ell^n$ lift to a map

$$K_2F \to H^2(F, \mathbb{Z}_\ell(1)^{\otimes 2}),$$

and that this map induces an isomorphism

$$K_2F(\ell) \to H^2(F, \mathbb{Z}_\ell(1)^{\otimes 2})_{\mathrm{tors}}.$$

As K_2F is torsion, with no char(F)-torsion, this gives a purely cohomological description of K_2F.

Notes The results of Tate in this subsection were announced in Tate (1970b), and proved in Tate (1973b, 1976b), or in Tate's appendix to Bass and Tate (1973a).

7.3 The Milnor K-Groups

Milnor (1969/1970) defines the (Milnor) K-groups of a field F as follows: regard $F^\times$ as a $\mathbb{Z}$-module; then $K_*^M F$ is the quotient of the tensor algebra of $F^\times$ by the ideal generated by the elements

$$a \otimes (1-a), \quad a \neq 0, 1.$$

This means that, for $n \geq 2$, $K_n^M F$ is the quotient of $(K_1F)^{\otimes n}$ by the subgroup generated by the elements

$$a_1 \otimes \cdots \otimes a_n, \quad a_i + a_{i+1} = 1 \text{ for some } i.$$

There is a canonical homomorphism $K_*F \to K_*^M F$ which induces isomorphisms $K_iF \to K_i^M F$ for $i \leq 2$. In the same article, Milnor defined for each discrete valuation v on F, a homomorphism

$$\partial_v \colon K_*F \to K_*\kappa(v)$$

of degree -1, where $\kappa(v)$ is the residue field.

Milnor (ibid.) quotes a theorem of Tate: for a global field F,

$$K_n^M F/2K_n^M F \simeq \bigoplus_v K_n^M F_v/2K_n^M F_v, \quad n \geq 3,$$

which implies that $K_n^M F/2K_n^M F \simeq (\mathbb{Z}/2\mathbb{Z})^{r_1}$ $(n \geq 3)$ where r_1 is the number of real primes of F. Bass and Tate (1973a) improve this statement by showing that

$$K_n^M F \simeq (\mathbb{Z}/2\mathbb{Z})^{r_1} \quad \text{for } n \geq 3.$$

The proof makes essential use of the "transfer maps"

$$\mathrm{Tr}\colon \quad K_* E \to K_* F,$$

defined whenever E is a finite field extension of F. Since these had only been defined for $n \leq 2$, a major part of the article is taken up with proving results on $K_* F$ for a general field, including the existence of the transfer maps.

The theorem of Bass and Tate completes the determination of the Milnor K-groups of a global field, except for K_1 and K_2.

7.4 *Other Results on K_2F*

Let F be a field containing a primitive mth root ζ of 1 for some $m > 1$. Tate (1976) showed that, when F is a global field, every element of K_2F killed by m can be represented as $\{\zeta, a\}$ for some $a \in F^\times$. Tate (1977b) examines whether this holds for other fields and obtains a number of positive results.

For a finite extension of fields $E \supset F$, there is a transfer (or trace) map $\mathrm{Tr}_{E/F}\colon K_2E \to K_2F$. As K_2E is generated by symbols $\{a, b\}$, in order to describe $\mathrm{Tr}_{E/F}$ it suffices to describe its action on each symbol. This Rosset and Tate (1983c) do.

8 The Stark Conjectures

In a series of four papers, Stark examined the behaviour of Artin L-series near $s = 0$ (equivalently, $s = 1$), and stated his now famous conjectures (Stark 1971). Tate gave a seminar at Harvard on Stark's conjectures in the spring of 1978, after Stark had given some talks on the subject in the number theory seminar in the fall of 1977. In 1980/81 Tate gave a course at the Université de Paris-Sud (Orsay) in which he clarified and extended Stark's work in important ways. The notes of Tate's course, when published in 1984, included most of the results known at that date, and became the basic reference for the Stark conjectures.

Let $\zeta_k(s)$ be the zeta function of a number field k. A celebrated theorem of Dedekind shows that

$$\zeta_k(s) \sim -\frac{R}{(e/h)} s^{r_1+r_2-1} \quad \text{as } s \to 0, \tag{35}$$

where h is the class number of k, R is its regulator, $e = |\mu(k)|$, and $r_1 + r_2 - 1$ is the rank of the group of units in k.

Let K be a finite Galois extension of k, with Galois group $G = G(K/k)$. Stark's insight was that the decomposition of $\zeta_K(s)$ into a product of Artin L-series indexed by the irreducible characters of G should induce an interesting decomposition of (35).

Stark's Main Conjecture Let $\chi: G \to \mathbb{C}$ be the character of a finite-dimensional complex representation $\rho: G \to \mathrm{GL}(V)$ of G. For a finite set S of primes of k containing the infinite primes, let

$$L(s,\chi) = \prod_{\mathfrak{p} \notin S} \frac{1}{\det(1 - \rho(\sigma_{\mathfrak{p}}) N\mathfrak{p}^{-s} | V^{I_{\mathfrak{P}}})}$$

(Artin L-function relative to S; cf. 1.1). Let S_K be the set of primes of K lying over a prime in S, let Y be the free $\mathbb{Z}$-module on S_K, and let X be the submodule of Y of elements $\sum n_w w$ such that $\sum n_w = 0$. Then $L(s,\chi)$ has a zero of multiplicity $r(\chi)$ at $s = 0$, where

$$r(\chi) = \dim_{\mathbb{C}} \mathrm{Hom}_G\big(V^\vee, X_{\mathbb{C}}\big).$$

Let U be the group of S_K-units in K. The unit theorem provides us with an isomorphism

$$\lambda\colon \quad U_{\mathbb{R}} \to X_{\mathbb{R}}, \qquad u \mapsto \sum_{w \in S_K} \log |u|_w w.$$

For each choice of an isomorphism of G-modules $f: X_{\mathbb{Q}} \to U_{\mathbb{Q}}$, Tate (1984, p. 26) defines the Stark regulator, $R(\chi, f)$, to be the determinant of the endomorphism of $\mathrm{Hom}_G(V^\vee, X_{\mathbb{C}})$ induced by $\lambda_{\mathbb{C}} \circ f_{\mathbb{C}}$. Then

$$L(s,\chi) \sim \frac{R(\chi,f)}{A(\chi,f)} s^{r(\chi)} \quad \text{as } s \to 0$$

for a complex number $A(\chi, f)$. The main conjecture of Stark, as formulated by Tate (1984, p. 27), says that

$$A(\chi,f)^\alpha = A\big(\chi^\alpha, f\big) \quad \text{for all automorphisms } \alpha \text{ of } \mathbb{C},$$

where $\chi^\alpha = \alpha \circ \chi$. In particular, $A(\chi, f)$ is an algebraic number, lying in the cyclotomic field $\mathbb{Q}(\chi)$. Tate proves that the validity of the conjecture is independent of both f and S, and that it suffices to prove it for irreducible characters χ of dimension 1 (application of Brauer's theorem p. 323).

Characters with Values in $\mathbb{Q}$ When the character χ takes its values in $\mathbb{Q}$, Stark's conjecture predicts that $A(\chi, f) \in \mathbb{Q}$ for all f. If, in addition, χ is a $\mathbb{Z}$-linear combination of characters induced from trivial characters, then the proof of the conjecture comes down to the case of a trivial character, where it follows from (35). Some multiple of χ has this form, and so this shows that some power of $A(\chi, f)$ is in $\mathbb{Q}$ (Stark 1975). Tate proves (1984, Chap. II) that $A(\chi, f)$ itself lies in $\mathbb{Q}$. His proof makes heavy use of the cohomology of number fields, including the theorems in Sect. 1.3.

The Case that $L(s, \chi)$ is Nonzero at $s = 0$ When $r(\chi) = 0$, the Stark regulator $R(\chi, f) = 1$, and Stark's conjecture becomes the statement:

$$L(0, \chi)^\alpha = L\big(0, \chi^\alpha\big) \quad \text{for all automorphisms } \alpha \text{ of } \mathbb{C}.$$

This special case of Stark's conjecture is also a special case of Deligne's conjecture on the critical values of motives (Deligne 1979, §6). Using a refinement of Brauer's theorem (cf. p. 323), Tate writes $L(s, \chi)$ as a sum of partial zeta functions:

$$L(s, \chi) = \sum_{\sigma \in G(K/k)} \chi(\sigma) \cdot \zeta(s, \sigma), \quad \zeta(s, \sigma) = \sum_{(\mathfrak{a}, K/k) = \sigma} N\mathfrak{a}^{-s}$$

(Tate 1984, III 1). According to an important theorem of Siegel (1970), $\zeta(0, \sigma) \in \mathbb{Q}$, which proves Stark's conjecture in this case.

The Case that $L(s, \chi)$ Has a First Order Zero at $s = 0$ By contrast, when $r(\chi) = 1$, the conjecture is still unknown, but it has remarkable consequences. Let $\mathbb{C}[G]$ be the group algebra of G, and let

$$e_\chi = \frac{\chi(1)}{|G|} \sum_{\sigma \in G} \chi\big(\sigma^{-1}\big) \cdot \sigma$$

be the idempotent in $\mathbb{C}[G]$ that projects every representation of G onto its χ-component. For an $a \in \mathbb{Q}(\chi)$ and a character χ with $r(\chi) = 1$, let

$$\pi(a, \chi) = \sum_{\alpha \in G(\mathbb{Q}(\chi)/\mathbb{Q})} a^\alpha \cdot L'\big(0, \chi^\alpha\big) \cdot e_{\bar{\chi}^\alpha} \in \mathbb{C}[G].$$

The character χ is realized on a $\mathbb{Q}[G]$-submodule U_W of $U_\mathbb{Q}$, and Stark's conjecture is true for χ if and only if

$$\pi(a, \chi) X_\mathbb{Q} = \lambda(U_W) \quad (\text{inside } X_\mathbb{C})$$

(Tate 1984, III 2.1). More explicitly, let Ψ be a set of irreducible characters $\chi \neq 1$ of G such that $r(\chi) = 1$, and assume that Ψ is stable under $\mathrm{Aut}(\mathbb{C})$. Let $(a_\chi)_{\chi \in \Psi}$ be a family of complex numbers such that $a_{\chi^\alpha} = (a_\chi)^\alpha$ for all $\alpha \in \mathrm{Aut}(\mathbb{C})$. If Stark's conjecture holds for the $\chi \in \Psi$, then for each prime v of S and extension of v to a prime w of K, there exists an integer $m > 0$ and an S-unit ε of K such that

$$\lambda(\varepsilon) = m \sum_{\chi \in \Psi} a_\chi \cdot L'(0, \chi) \cdot e_{\bar{\chi}} \cdot w; \tag{36}$$

once m has been fixed, ε is unique up to a root of 1 in K (ibid. III §3). The units ε arising (conjecturally) in this way are called Stark units. They are analogous to the cyclotomic units in cyclotomic fields.

Finer Conjectures when K/k is Abelian When K/k is abelian, (36) can be made into a more precise form of Stark's conjecture, which Tate denotes $\mathrm{St}(K/k, S)$ (Stark 1980; Tate 1984, IV 2). For a real prime w of K and certain hypotheses on S, $\mathrm{St}(K/k, S)$ predicts the existence of a unit $\varepsilon(K, S, w) \in U$ such that

$$\varepsilon(K, S, w)^\sigma = \exp\bigl(-2\zeta'(0, \sigma)\bigr), \quad \text{all } \sigma \in G.$$

When we use w to embed K in $\mathbb{R}$, the $\varepsilon(K, S, w)$ lie in the abelian closure of k in $\mathbb{R}$. In the case that k is totally real, Tate (1984, 3.8) determines the subfield they generate; for example, when $[K\colon \mathbb{Q}] = 2$, they generate the abelian closure of k in $\mathbb{R}$. This has implications for Hilbert's 12th problem. To paraphrase Tate (ibid. p. 95):

> If the conjecture $\mathrm{St}(K/k, S)$ is true in this situation, then the formula
>
> $$\varepsilon = \exp\bigl(-2\zeta'(0, 1)\bigr)$$
>
> gives generators of abelian extensions of k that are special values of transcendental functions. Finding generators of class fields of this shape is the vague form of Hilbert's 12th problem, and the Stark conjecture represents an important contribution to this problem. However, it is a totally unexpected contribution: Hilbert asked that we discover the functions that play, for an arbitrary number field, the same role as the exponential function for $\mathbb{Q}$ and the elliptic modular functions for a quadratic imaginary field. In contrast, Stark's conjecture, by using L-functions directly, bypasses the transcendental functions that Hilbert asked for. Perhaps a knowledge of these last functions will be necessary for the proof of Stark's conjecture.

Remarkably, $\mathrm{St}(K/k, S)$ is useful for the explicit computation of class fields, and has even been incorporated into the computer algebra system PARI/GP.

For an abelian extension K/k, Tate introduced another conjecture, combining ideas of Brumer and Stark, and which he calls the Brumer-Stark conjecture. Let S be a set of primes of k including a finite prime $\mathfrak{p}$ that splits completely in K, and let $T = S \backslash \{\mathfrak{p}\}$. Assume that T contains the infinite primes and the primes that ramify in K. Let

$$\theta_T(0) = \sum_{\chi \text{ irreducible}} L(0, \chi) e_{\bar{\chi}} \in \mathbb{C}[G].$$

Brumer conjectured that, for every ideal $\mathfrak{A}$ of K, $\mathfrak{A}^{e\theta_T(0)}$ is principal; the Brumer-Stark conjecture $\mathrm{BS}(K/k, T)$ says that $\mathfrak{A}^{e\theta_T(0)} = (\alpha)$ for an α satisfying certain conditions on the absolute values $|\alpha|_w$ ($w \in T$) and that $K(\alpha^{\frac{1}{e}})$ is an abelian extension of k (Tate 1984, 6.2).[38] Tate proved this conjecture for $k = \mathbb{Q}$ (ibid. 6.7) and for quadratic extensions K/k (Tate 1981b).

Function Fields All of the conjectures make sense for a global field k of characteristic $p \neq 0$. In this case, the Artin L-series are rational functions in q^{-s}, where q is the order of the field of constants, and Stark's main conjecture follows easily from the known properties of these functions. However, as Mazur pointed out, the Brumer-Stark conjecture is far from trivial for function fields. Tate gave a seminar in Paris in early fall 1980 in which he discussed the conjecture and some partial results he had obtained. Deligne attended the seminar, and later gave a proof of the conjecture using his one-motives. This proof is included in Chap. V of Tate (1984).

***p*-Adic Analogues** Tate's reformulation of Stark's conjecture helped inspire two p-adic analogues of his main conjecture, one for $s = 0$ (Gross) and one for $s = 1$ (Serre)—the absence of a functional equation for the p-adic L-series makes these distinct conjectures. In a 1997 letter, Tate proposed a refinement of Gross's conjecture. This letter was published, with additional comments, as Tate 2004.

There is much numerical evidence for the Stark conjectures, found by Stark and others. As Tate (1981a, p. 977) notes: "Taken all together, the evidence for the conjectures seems to me to be overwhelming".

9 Noncommutative Ring Theory

The Tate conjecture for divisors on a variety is related to the finiteness of the Brauer group of the variety, which is defined to be the set of the similarity classes of sheaves of (noncommutative) Azumaya algebras on the variety. This connection led M. Artin to an interest in noncommutative rings, which soon broadened beyond Azumaya algebras. Tate wrote a number of articles on noncommutative rings in collaboration with Artin and others.

9.1 Regular Algebras

A ring A is said to have *finite global dimension* if there exists an integer d such that every A-module has a projective resolution of length at most d. The smallest such d is then called the *global dimension* of A. Serre (1965) showed that a *commutative* ring is noetherian of finite global dimension if and only if it is regular.

[38] Recall that $e = |\mu(k)|$.

Let k be a field. Artin and Schelter (1987) defined a finitely generated k-algebra to be *regular* if it is of the form

$$A = k \oplus A_1 \oplus A_2 \oplus \cdots, \tag{37}$$

and

(a) A has finite global dimension (defined in terms of *graded* A-modules),
(b) A has polynomial growth (i.e., $\dim A_n$ is bounded by a polynomial function in n), and
(c) A is Gorenstein (i.e., the k-vector space $\mathrm{Ext}^i_A(k, A)$ has dimension 1 when i is the global dimension of A, and is zero otherwise).

The only commutative graded k-algebras satisfying these conditions and generated in degree 1 are the polynomial rings. It is expected that the regular algebras have many of the good properties of polynomial rings. For example, Artin and Schelter conjecture that they are noetherian domains. The dimension of a regular algebra is its global dimension.

In collaboration with Artin and others, Tate studied regular algebras, especially the classification of those of low dimension.

From now on, we require regular k-algebras to be generated in degree 1. Such an algebra is a quotient of a tensor algebra by a homogeneous ideal.

A regular k-algebra of dimension one is a polynomial ring, and one of dimension two is the quotient of the free associative algebra $k\langle X, Y\rangle$ by a single quadratic relation, which can be taken to be $XY - cYX$ $(c \neq 0)$ or $XY - YX - Y^2$. Thus, the first interesting dimension is three. Artin and Schelter (ibid.) showed that a regular k-algebra of dimension three either has three generators and three relations of degree two, or two generators and two relations of degree three. Moreover, they showed that the algebras fall into thirteen families. While the generic members of each family are regular, they were unable to show that all the algebras in the families are regular. Artin, Tate, Van den Bergh (1990a) overcame this problem, and consequently gave a complete classification of these algebras. Having found an explicit description of all the algebras, they were able to show that they are all noetherian.

These two articles introduced new geometric techniques into noncommutative ring theory. They showed that the regular algebras of dimension 3 correspond to certain triples $(E, \mathcal{L}, \sigma)$ where E is a one-dimensional scheme of arithmetic genus 1 which is embedded either as a cubic divisor in $\mathbb{P}^2$ or as a divisor of bidegree $(2, 2)$ in $\mathbb{P}^1 \times \mathbb{P}^1$, $\mathcal{L} = \mathcal{O}_E(1)$ is an invertible sheaf on E, and σ is an automorphism of E. The scheme E parametrizes the point modules for A, i.e., the graded cyclic right A-modules, generated in degree zero, such that $\dim_k(M_n) = 1$ for all $n \geq 0$. The geometry of (E, σ) is reflected in the structure of the point modules, and Artin, Tate, van den Bergh (1991a) exploit this relation to prove that the 3-dimensional regular algebra corresponding to a triple $(E, \mathcal{L}, \sigma)$ is finite over its centre if and only if the automorphism σ has finite order. They also show that noetherian regular k-algebras of dimension ≤ 4 are domains.

9.2 Quantum Groups

A bialgebra A over a field k is a k-module equipped with compatible structures of an associative algebra with identity and of a coassociative coalgebra with coidentity. A bialgebra is called a Hopf algebra if it admits an antipodal map (linear map $S\colon A \to A$ such that certain diagrams commute).

A bialgebra is said to be commutative if it is commutative as a k-algebra. The commutative bialgebras (resp. Hopf algebras) over k are exactly the coordinate rings of affine monoid schemes (resp. affine group schemes) over k.

Certain Hopf algebras (not necessarily commutative) are called quantum groups. For example, there is a standard one-parameter family $\mathcal{O}(\mathrm{GL}_n(q))$, $q \in k^\times$, of Hopf algebras that takes the value $\mathcal{O}(\mathrm{GL}_n)$ for $q = 1$. This can be regarded as a one-parameter deformation of $\mathcal{O}(\mathrm{GL}_n)$ by Hopf algebras, or of GL_n by quantum groups.

Artin, Schelter, and Tate (1991b) construct a family of deformations of $\mathcal{O}(\mathrm{GL}_n)$, depending on $1 + \binom{n}{2}$ parameters, which includes the family $\mathcal{O}(\mathrm{GL}_n(q))$. The algebras in the family are all twists of $\mathcal{O}(\mathrm{GL}_n(q))$ by 2-cocycles. They first construct a family of deformations of $\mathcal{O}(M_n)$ by bialgebras that are graded algebras generated in degree 1, have the same Hilbert series as the polynomial ring in n^2 variables, and are noetherian domains. The family of deformations of $\mathcal{O}(\mathrm{GL}_n)$ is then obtained by inverting the quantum determinant. The algebras in the family of deformations of $\mathcal{O}(M_n)$ are regular in the sense of (9.1), and so this gives a large class of regular algebras with the expected good properties.

9.3 Sklyanin Algebras

As noted in (9.1), regular algebras of degree 3 over a field k correspond to certain triples $(E, \mathcal{L}, \sigma)$ with E a curve, $\mathcal{L}$ an invertible sheaf of degree 3 on E, and σ an automorphism of E. When E is a nonsingular elliptic curve and σ is translation by a point P in $E(k)$, the algebra $A(E, \mathcal{L}, \sigma)$ is called a *Sklyanin algebra.* Let $U = \Gamma(E, \mathcal{L})$. This is a 3-dimensional k-vector space, and we can identify $U \otimes U$ with $\Gamma(E, \mathcal{L} \boxtimes \mathcal{L})$. The algebra $A(E, \mathcal{L}, \sigma)$ is the quotient of the tensor algebra $T(U)$ of U by $\{f \in U \otimes U \mid f(x, x + P) = 0\}$. It is essentially independent of $\mathcal{L}$, because any two invertible sheaves of degree 3 differ by a translation. More generally, there is a Sklyanin algebra $A(E, \mathcal{L}, \sigma)$ for every triple consisting of a nonsingular elliptic curve, an invertible sheaf $\mathcal{L}$ of degree d on E, and a translation by a point in $E(k)$. The algebra $A(E, \mathcal{L}, \sigma)$ has dimension d.

Artin, Schelter, and Tate (1994c) give a precise description of the centres of Sklyanin algebras of dimension three, and Smith and Tate (1994d) extend the description to those of dimension four.

Tate and van den Bergh (1996) prove that every d-dimensional Sklyanin algebra $A(E, \mathcal{L}, \sigma)$ is a noetherian domain, is Koszul, has the same Hilbert series as a polynomial ring in d variables, and is regular in the sense of (9.1); moreover, if σ has finite order, then $A(E, \mathcal{L}, \sigma)$ is finite over its centre.

10 Miscellaneous Articles

1951a Tate, John. On the relation between extremal points of convex sets and homomorphisms of algebras. Comm. Pure Appl. Math. 4, (1951). 31–32 Tate considers a convex set K of linear functionals on a commutative algebra A over $\mathbb{R}$. Under certain hypotheses on A and K, he proves that the extremal points of K are exactly the homomorphisms from A into $\mathbb{R}$.

1951b Artin, Emil; Tate, John T. A note on finite ring extensions. J. Math. Soc. Japan 3, (1951). 74–77 Artin and Tate prove that if S is a commutative finitely generated algebra over a noetherian ring R, and T is a subalgebra of S such that S is finitely generated as a T-module, then T is also finitely generated over R. This statement generalizes a lemma of Zariski, and is now known as the Artin-Tate lemma. There are various generalizations of it to noncommutative rings.

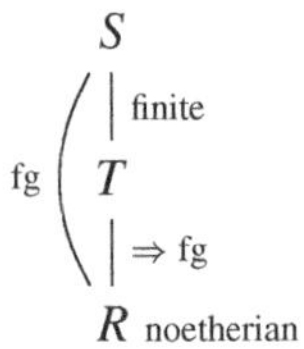

1952a Tate, John. Genus change in inseparable extensions of function fields. Proc. Amer. Math. Soc. 3, (1952). 400–406 Let C be a complete normal geometrically integral curve over a field k of characteristic p, and let C' be the curve obtained from C by an extension of the base field $k \to k'$. If k' is inseparable over k, then C' need not be normal, and its normalization $\tilde{C}'$ may have genus $g(\tilde{C}')$ less than the genus $g(C)$ of C. However, Tate proves that

$$\frac{p-1}{2} \text{ divides } g(C) - g\big(\tilde{C}'\big). \tag{38}$$

In particular, the genus of C can't change if $g(C) < (p-1)/2$ (which implies that C is smooth in this case).

Statement (38) is widely used. Tate derives it from a "Riemann-Hurwitz formula" for purely inseparable coverings, which he proves using the methods of the day (function fields and repartitions). A modern proof has been given of (38) (Schröer 2009), but not, as far as I know, of the more general formula.

1952b Lang, Serge; Tate, John. On Chevalley's proof of Luroth's theorem. Proc. Amer. Math. Soc. 3, (1952). 621–624 Chevalley (1951, p. 106) proved Lüroth's theorem in the following form: let k_0 be a field, and let $k = k_0(X)$ be the field of rational functions in the symbol X (i.e., k is the field of fractions of the polynomial ring $k_0[X]$); then every intermediate field k', $k_0 \subsetneq k' \subset k$, is of the form $k_0(f)$ for some $f \in k$.

A classical proof of Lüroth's theorem uses the Riemann-Hurwitz formula. Let k be a function field in one variable over a field k_0, and let k' be an intermediate field; the Riemann-Hurwitz formula shows that, if k/k' is separable, then

$$g(k') \leq g(k).$$

Therefore, if k has genus zero, so also does k'; if, in addition, k has a prime of degree 1, so also does k', and so k' is a rational field (by a well-known criterion).

However, if k/k' is not separable, it may happen that $g(k') > g(k)$. Chevalley proved Lüroth's theorem in nonzero characteristic by showing directly that, when $k = k_0(X)$, every intermediate field k has genus zero. Lang and Tate generalized Chevalley's argument to prove:

> Let k be a function field in one variable over a field k_0, and let k' be an intermediate field; if k is separably generated over k_0, then $g(k') \leq g(k)$.

In other words, they showed that Chevalley's argument doesn't require that $k = k_0(X)$ but only that it be separably generated over k_0. They also prove a converse statement:

> A field of genus zero that is not separably generated over its field of constants contains subfields of arbitrarily high genus.

Finally, to complete their results, they exhibit a field of genus zero, not separably generated over its field of constants.

1955b Brauer, Richard; Tate, John. On the characters of finite groups. Ann. of Math. (2) 62, (1955). 1–7 Recall (p. 261) that Brauer's theorem says that every character χ of a finite group G can be expressed in the form

$$\chi = \sum_i n_i \operatorname{Ind} \chi_i, \quad n_i \in \mathbb{Z},$$

with the χ_i *one-dimensional* characters on subgroups of G (as conjectured by Artin). Brauer and Tate found what is probably the simplest known proof of Brauer's theorem. Recall that a group is said to be elementary if it can be expressed as the product of a cyclic group with a p-group for some prime p. An elementary group is nilpotent, and so every irreducible character of it is induced from a one-dimensional character on a subgroup. Let G be a finite group, and let $\mathcal{H}$ be a set of subgroups of G. Consider the following three $\mathbb{Z}$-submodules of the space of complex-valued functions on G:

$$X(G) = \operatorname{span}\{\text{irreducible characters of } G\} \quad \text{(module of virtual characters)}$$

$$Y = \operatorname{span}\{\text{characters of } G \text{ induced from an irreducible character of an } H \text{ in } \mathcal{H}\}$$

$$U = \left\{\text{class functions } \chi \text{ on } G \text{ such that } \chi|H \in X(H) \text{ for all } H \text{ in } \mathcal{H}\right\}.$$

Brauer and Tate show that

$$U \supset X(G) \supset Y,$$

that U is a ring, and that Y is an ideal of U. Using this, they show that if $\mathcal{H}$ consists of the elementary subgroups of G, then $U = Y$, thereby elegantly proving not only Artin's conjecture (the equality $X(G) = Y$), but also the main theorem of Brauer (1953) (the equality $U = X(G)$).

1957 Tate, John. Homology of Noetherian rings and local rings. Illinois J. Math. 1 (1957), 14–27 Tate makes systematic use of the skew-commutative graded differential algebras over a noetherian commutative ring R to obtain results concerning R and its quotient rings. The differential of such an R-algebra allows it to be regarded as a complex, and Tate proves that every quotient $R/\mathfrak{a}$ of R has a free resolution that is an R-algebra (in the above sense). Such resolutions are now called Tate resolutions.

Let R be a local noetherian ring with maximal ideal $\mathfrak{m}$. The Betti series of R is defined to be the formal power series $\mathcal{R} = \sum_{r\geq 0} b_r Z^r$ with b_r equal to the length of $\mathrm{Tor}_r^R(R/\mathfrak{m}, R/\mathfrak{m})$. Serre (1956) showed that R is regular if and only if $\mathcal{R}$ is a polynomial, in which case $\mathcal{R} = (1+Z)^d$ with $d = \dim(R)$. Tate showed that $\mathcal{R} = (1+Z)^d/(1-Z^2)^{b_1-d}$ if R is a complete intersection. In general, he showed that the natural homomorphism

$$\bigwedge^{*} \mathrm{Tor}_1^R(R/\mathfrak{m}, R/\mathfrak{m}) \to \mathrm{Tor}^R(R/\mathfrak{m}, R/\mathfrak{m})$$

is injective and realizes $\mathrm{Tor}^R(R/\mathfrak{m}, R/\mathfrak{m})$ as a free module over $\bigwedge^* \mathrm{Tor}_1^R(R/\mathfrak{m}, R/\mathfrak{m})$ with a homogeneous basis. If R is regular, then the homomorphism is an isomorphism; conversely, if the homomorphism is an isomorphism on one homogeneous component of degree ≥ 2, then R is regular.

1962a Fröhlich, A.; Serre, J.-P.; Tate, J. A different with an odd class. J. Reine Angew. Math. 209 (1962) 6–7 Let A be a Dedekind domain with field of fractions K, and let B be the integral closure of A in a finite separable extension of K. The different $\mathfrak{D}$ of B/A is an ideal in B, and its norm $\mathfrak{d}$ is the discriminant ideal of B/A. The ideal class of $\mathfrak{d}$ is always a square, and Hecke (1954, §63) proved that the ideal class of $\mathfrak{D}$ is a square when K is a number field, but the authors show that it need not be a square otherwise. Specifically, they construct examples of affine curves over perfect fields whose coordinate rings A have extensions B for which the ideal class of the different is not a square.[39]

This is not a major result. However, Martin Taylor (2006) writes:

> [This article and Fröhlich's earlier work on discriminants] seems to have marked the start of [his] interest in parity questions. He would go on to be interested in whether conductors of real-valued characters were squares; this in turn led to questions about the signs of Artin root numbers—an issue that lay right at the heart of his work on Galois modules.

---

[39] Hecke's theorem can be proved for global fields of characteristic $p \neq 0$ by methods similar to those of Hecke (Armitage 1967).

Fröhlich's work on Artin root numbers and Galois module structures was his most important.

1963 Sen, Shankar; Tate, John. Ramification groups of local fields. J. Indian Math. Soc. (N.S.) 27 1963 197–202 (1964) Let F be a field, complete with respect to a discrete valuation, and let K be a finite Galois extension of F. Assume initially that the residue field is finite, and let W be the Weil group of K/F (extension of $G(K/F)$ by $K^\times$ determined by the fundamental class of K/F). Shafarevich showed that there is a homomorphism s making the following diagram commute

$$\begin{array}{ccccccccc} 1 & \longrightarrow & K^\times & \longrightarrow & W & \longrightarrow & G(K/F) & \longrightarrow & 1 \\ & & \downarrow{\scriptstyle r} & & \downarrow{\scriptstyle s} & & \| & & \\ 1 & \longrightarrow & G(K^{\mathrm{ab}}/K) & \longrightarrow & G(K^{\mathrm{ab}}/F) & \longrightarrow & G(K/F) & \longrightarrow & 1, \end{array}$$

where r is the reciprocity map. For a real $t > 0$, let $G(K^{\mathrm{ab}}/K)^t$ denote the tth ramification subgroup of $G(K^{\mathrm{ab}}/K)$. Then

$$r^{-1}\big(G\big(K^{\mathrm{ab}}/K\big)^t\big) = U_K^t \stackrel{\mathrm{def}}{=} \begin{cases} \{u \in K^\times \mid \mathrm{ord}_K(u) = 0\} \text{ if } t = 0 \\ \{u \in K^\times \mid \mathrm{ord}_K(u-1) \geq t\} \text{ if } t > 0. \end{cases} \tag{39}$$

Artin and Tate (1961) proved the existence of the Shafarevich map s for a general class formation. When the residue field of F is algebraically closed, the groups $\pi_1(U_K)$ (fundamental group of U_K regarded as a pro-algebraic group) form a class formation, and so the above diagram exists with $K^\times$ replaced by $\pi_1(U_K)$. In this case,

$$r^{-1}\big(G\big(K^{\mathrm{ab}}/K\big)^t\big) = \pi_1\big(U_K^t\big). \tag{40}$$

In both cases, Sen and Tate give a description of the subgroups $s^{-1}(G(K^{\mathrm{ab}}/F)^t)$ of W generalizing those in (39) and (40), which can be considered the case $K = F$. Specifically, let $G(K/F)_x$, $x \geq 0$, denote with ramification groups of K/F with the lower numbering, and let

$$\varphi(x) = \int_0^x \frac{du}{(G(K/F)_0 \colon G(K/F)_u)} \quad \text{for } x \geq 0.$$

For $u \in W$, let $m(u) > 0$ be the smallest integer such that $u^{m(u)} \in K^\times$ (resp. $\pi_1(U_K)$). Then

$$W^{\varphi(x)} = \big\{u \in W \mid u^{m(u)} \in U_K^{m(u)\cdot x}\big(\text{resp. } \pi_1\big(U_K^{m(u)\cdot x}\big)\big)\big\}.$$

1964c Tate, John. Nilpotent quotient groups. Topology 3 1964 suppl. 1 109–111 For a finite group G, subgroup S, and positive integer p, there are restriction maps r and transfer maps t,

$$H^i(G, \mathbb{Z}/p\mathbb{Z}) \xrightarrow{r^i} H^i(S, \mathbb{Z}/p\mathbb{Z}) \xrightarrow{t^i} H^i(G, \mathbb{Z}/p\mathbb{Z}), \quad i \geq 0,$$

whose composite is multiplication by $(G\colon S)$.

Let S be Sylow p-subgroup of G (so p is prime). If S has a normal p-complement in G, then the restriction maps are isomorphisms, and Atiyah asked whether the converse is true. Thompson pointed out that the answer is yes, and that results of his and Huppert show that one need only require that r^1 is an isomorphism. Tate gives a very short cohomological proof of a somewhat stronger result.

Specifically, for a finite group G, define a descending sequence of normal subgroups of G as follows:

$$G_0 = G, \qquad G_{n+1} = (G_n)^p[G, G_n] \quad \text{for } n \geq 0, \qquad G_\infty = \bigcap_{n=0}^{\infty} G_n$$

(p not necessarily prime). Thus, G/G_1 (resp. G/G_∞) is the largest quotient group of G that is abelian of exponent p (resp. nilpotent and p-primary). Let S be a subgroup of G of index prime to p. The following three conditions are (obviously) equivalent,

- the restriction map $r^1\colon H^1(G, \mathbb{Z}/p\mathbb{Z}) \to H^1(S, \mathbb{Z}/p\mathbb{Z})$ is an isomorphism,
- the map $S/S_1 \to G/G_1$ is an isomorphism,
- $S \cap G_1 = S_1$,

and Tate proves that they imply

- $S \cap G_n = S_n$ for all $1 \leq n \leq \infty$.

When S is a Sylow p-subgroup of G, $S \cap G_\infty = 1$, and so the conditions imply that G_∞ is a normal p-complement of S in G (thereby recovering the Huppert-Thompson theorem).

1968b Tate, John. Residues of differentials on curves. Ann. Sci. École Norm. Sup. (4) 1 (1968) 149–159 Tate defines the residues of differentials on curves as the traces of certain "finite potent" linear maps. From his definition, all the standard theorems on residues follow naturally and easily. In particular, the residue formula

$$\sum_{P \in C} \operatorname{res}_p(\omega) = 0 \quad (C \text{ a complete curve})$$

follows directly, without computation, from the finite dimensionality of the cohomology groups $H^0(C, \mathcal{O}_C)$ and $H^1(C, \mathcal{O}_C)$ "almost as though one had an abstract Stokes's Theorem available".

A linear map $\theta\colon V \to V$ is finite potent if $\theta^n V$ is finite dimensional for some n. The trace $\operatorname{Tr}_V(\theta)$ of such a map can be defined to be its trace on any finite dimensional subspace W of V such that $\theta W \subset W$ and $\theta^n V \subset W$ for some n. Many of the properties of the usual trace continue to hold, but not all. For example, linearity fails even for two finite potent operators on an infinite-dimensional vector space (Gonzalez and Romo 2012). Tate defines the residue of a differential $f\,dg$ at a closed point p of a curve C to be the trace of the commutator $[f_p, g_p]$, where f_p, g_p are representatives of f, g in a certain subspace of $\operatorname{End}(k(C)_p)$.

Tate's approach to residues has found its way into the text books (e.g., Iwasawa 1993). Elzein (1971) used Tate's ideas to give a definition of the residue that recaptures both Leray's in the case of a complex algebraic variety and Grothendieck's in the case of a smooth integral morphisms of relative dimension n.

Others have adapted his proof of the residue formula to other situations; for example, Arbarello et al. (1989) use it to prove an "abstract reciprocity law" for tame symbols on a curve over an algebraically closed field, and Beilinson et al. (2002) use it to prove a product formula for ε-factors in the de Rham setting.

In reading Tate's article, Beilinson recognized that a certain linear algebra construction there can be reformulated as the construction of a canonical central extension of Lie algebras. This led to the notion of a Tate extension in various settings; see Beĭlinson and Schechtman (1988), and Beilinson and Drinfeld (2004, 2.7).

1978a Cartier, P.; Tate, J. A simple proof of the main theorem of elimination theory in algebraic geometry. Enseign. Math. (2) 24 (1978), no. 3-4, 311–317 The authors give an elementary one-page proof of the homogeneous form of Hilbert's theorem of zeros:

> let $\mathfrak{a}$ be a graded ideal in a polynomial ring $k[X_0, \ldots, X_n]$ over a field k; either the radical of $\mathfrak{a}$ contains the ideal $(X_0, \ldots, X_n)$, or $\mathfrak{a}$ has a nontrivial zero in an algebraic closure of k.

From this, they quickly deduce the main theorem of elimination theory, both in its classical form and in its modern form:

> let $A = \bigoplus_{d \geq 0} A_d$ be a graded commutative algebra such that A is generated as an A_0-algebra by A_1 and each A_0-module A_d is finitely generated; then the map of topological spaces $\mathrm{proj}(A) \to \mathrm{spec}(A_0)$ is closed.

1989 Gross, B.; Tate, J. Commentary on algebra. A century of mathematics in America, Part II, 335–336, Hist. Math., 2, Amer. Math. Soc., Providence, RI (1989) For the bicentenary of Princeton University in 1946, there was a three-day conference in which various distinguished mathematicians discussed Problems in Mathematics. Artin, Brauer, and others contributed to the discussion on algebra, and in 1989 Gross and Tate wrote a commentary on their remarks. For example:

> Artin's belief that "whatever can be said about non-Abelian class field theory follows from what we know now," and that "our difficulty is not in the proofs, but in learning what to prove," seems overly optimistic.

1994b Tate, John. The non-existence of certain Galois extensions of $\mathbb{Q}$ unramified outside 2. Arithmetic geometry (Tempe, AZ, 1993), 153–156, Contemp. Math., 174, Amer. Math. Soc., Providence, RI (1994) In a 1973 letter to Tate, Serre suggested that certain two-dimensional mod p representations of $\mathrm{Gal}(\mathbb{Q}^{\mathrm{al}}/\mathbb{Q})$ should be modular. In response, Tate verified this for $p = 2$ by showing that every two-dimensional mod 2 representation unramified outside 2 has zero trace. The article is based on his letter.

Serre's suggestion became Serre's conjecture on the modularity of two-dimensional mod p representations, which attracted much attention because of its relation to the modularity conjecture for elliptic curves over $\mathbb{Q}$ and Fermat's last theorem. Serre's conjecture was recently proved by an inductive argument that uses Tate's result as one of the base cases (Khare and Wintenberger 2009).

1996a Tate, John; Voloch, José Felipe. Linear forms in p-adic roots of unity. Internat. Math. Res. Notices (1996), no. 12, 589–601 The authors make the following conjecture: for a semi-abelian variety A over $\mathbb{C}_p$ and a closed subvariety X, there exists a lower bound $c > 0$ for the p-adic distance of torsion points of A, not in X, to X. Here, as usual, $\mathbb{C}_p$ is the completion of an algebraic closure of $\mathbb{Q}_p$. They prove the conjecture for the torus

$$A = \operatorname{Spec} \mathbb{C}_p\big[T_1, T_1^{-1}, \ldots, T_n, T_n^{-1}\big].$$

This comes down to proving the following explicit statement: for every hyperplane

$$a_1 T_1 + \cdots + a_n T_n = 0$$

in $\mathbb{C}_p^n$, there exists a constant $c > 0$, depending on $(a_1, \ldots, a_n)$, such that, for any n-tuple $\zeta_1, \ldots, \zeta_n$ of roots of 1 in $\mathbb{C}_p$, either $a_1\zeta_1 + \cdots + a_1\zeta_n = 0$ or $|a_1\zeta_1 + \cdots + a_1\zeta_n| \geq c$.

2002a Tate, John. On a conjecture of Finotti. Bull. Braz. Math. Soc. (N.S.) 33 (2002), no. 2, 225–229 In his study of the Teichmüller points in canonical lifts of elliptic curves, Finotti was led to a conjecture on remainders of division by polynomials (Finotti 2004). He checked it by computer for all primes $p \leq 877$, and Tate proved it in general. The statement is:

> Let k be a field of characteristic $p = 2m + 1 \geq 5$. Let $F \in k[X]$ be a monic cubic polynomial, and let A be the coefficient of X^{p-1} in F^m. Let $G \in k[X]$ be a polynomial of degree $3m + 1$ such that $G' = F^m - AX^{p-1}$. Then the remainder in the division of G^2 by $X^p F^{m+1}$ has degree $\leq 5m + 2 = \frac{5p-1}{2}$.

Acknowledgements I thank B. Gross for help with dates, J.-P. Serre for correcting a misstatement, and J. Tate for answering my queries and pointing out some mistakes.

Appendix: Bibliography of Tate's Articles

1950s

1950 Tate, John, Fourier Analysis in Number Fields and Hecke's Zeta Functions, Ph.D. thesis, Princeton University. Published as 1967b.

1951a Tate, John. On the relation between extremal points of convex sets and homomorphisms of algebras. Comm. Pure Appl. Math. 4 (1951) 31–32.

1951b Artin, Emil; Tate, John T. A note on finite ring extensions. J. Math. Soc. Japan 3 (1951) 74–77.

1952a Tate, John. Genus change in inseparable extensions of function fields. Proc. Amer. Math. Soc. 3 (1952) 400–406.

1952b Lang, Serge; Tate, John. On Chevalley's proof of Luroth's theorem. Proc. Amer. Math. Soc. 3 (1952) 621–624.

1952c Tate, John. The higher dimensional cohomology groups of class field theory. Ann. of Math. (2) 56 (1952) 294–297.

1954 Tate, John, The Cohomology Groups of Algebraic Number Fields, pp. 66-67 in Proceedings of the International Congress of Mathematicians, Amsterdam (1954). Vol. 2. Erven P. Noordhoff N. V., Groningen; North-Holland Publishing Co., Amsterdam (1954). iv+440 pp. 19

1955a Kawada, Y.; Tate, J. On the Galois cohomology of unramified extensions of function fields in one variable. Amer. J. Math. 77 (1955) 197–217.

1955b Brauer, Richard; Tate, John. On the characters of finite groups. Ann. of Math. (2) 62 (1955) 1–7.

1957 Tate, John. Homology of Noetherian rings and local rings. Illinois J. Math. 1 (1957) 14–27.

1958a Mattuck, Arthur; Tate, John. On the inequality of Castelnuovo-Severi. Abh. Math. Sem. Univ. Hamburg 22 (1958) 295–299.

1958b Tate, J. WC-groups over $\mathfrak{p}$-adic fields. Séminaire Bourbaki; 10e année: 1957/1958. Textes des conférences; Exposés 152 à 168; 2e éd. corrigée, Exposé 156, 13 pp. Secrétariat mathématique, Paris (1958) 189 pp (mimeographed).

1958c Lang, Serge; Tate, John. Principal homogeneous spaces over abelian varieties. Amer. J. Math. 80 (1958) 659–684.

1958d Tate, John, Groups of Galois Type (published as Chap. VII of Lang 1967; reprinted as Lang 1996).[40]

1959a Tate, John. Rational points on elliptic curves over complete fields, manuscript 1959. Published as part of 1995.

1959b. Tate, John. Applications of Galois cohomology in algebraic geometry. (Written by S. Lang based on letters of Tate 1958–1959). Chap. X of: Lang, Serge. Topics in cohomology of groups. Translated from the 1967 French original by the author. Lecture Notes in Mathematics, 1625. Springer-Verlag, Berlin (1996). vi+226 pp.

1960s

1961 Artin, E., and Tate., J. Class Field Theory, Harvard University, Department of Mathematics, 1961.[41] Notes from the Artin-Tate seminar on class field theory

[40]Lang calls his Chap. VII an "unpublished article of Tate", but gives no date. In his MR review, Shatz writes that "It appears here in almost the same form the reviewer remembers from the original seminar of Tate in 1958."

[41]The original notes don't give a date or a publisher. I copied this information from the footnote p. 162 of 1967a. The volume was prepared by the staff of the Institute of Advanced Study, but it was distributed by the Harvard University Mathematics Department.

given a Princeton University 1951–1952. Reprinted as 1968c, 1990b; second edition 2009.
1962a Fröhlich, A.; Serre, J.-P.; Tate, J. A different with an odd class. J. Reine Angew. Math. 209 (1962) 6–7.
1962b Tate, John. Principal homogeneous spaces for Abelian varieties. J. Reine Angew. Math. 209 (1962) 98–99.
1962c Tate, John, Rigid analytic spaces. Private notes, reproduced with(out) his permission by I.H.E.S (1962). Published as 1971b; Russian translation 1969a.
1962d Tate, John. Duality theorems in Galois cohomology over number fields. (1963) Proc. Internat. Congr. Mathematicians (Stockholm, 1962) pp. 288–295 Inst. Mittag-Leffler, Djursholm
1963 Sen, Shankar; Tate, John. Ramification groups of local fields. J. Indian Math. Soc. (N.S.) 27 (1963) 197–202 (1964).
1964a Tate, John. Algebraic cohomology classes, Woods Hole 1964, 25 pages. In: Lecture notes prepared in connection with seminars held at the Summer Institute on Algebraic Geometry, Whitney Estate, Woods Hole, MA, July 6–July 31 (1964). Published as 1965b; Russian translation 1965c.
1964b Tate, John (with Lubin and Serre). Elliptic curves and formal groups, 8 pages. In: Lecture notes prepared in connection with seminars held at the Summer Institute on Algebraic Geometry, Whitney Estate, Woods Hole, MA, July 6–July 31 (1964).
1964c Tate, John. Nilpotent quotient groups. Topology 3 (1964) suppl. 1 109–111.
1965a Lubin, Jonathan; Tate, John. Formal complex multiplication in local fields. Ann. of Math. (2) 81 (1965) 380–387.
1965b Tate, John T. Algebraic cycles and poles of zeta functions. (1965) Arithmetical Algebraic Geometry (Proc. Conf. Purdue Univ., 1963) pp. 93–110 Harper & Row, New York
1965c Tate, John. Algebraic cohomology classes. (Russian) Uspehi Mat. Nauk 20 (1965) no. 6 (126) 27–40.
1965d Tate, John. Letter to Cassels on elliptic curve formulas. Published as 1975b.
1966a Tate, John. Multiplication complexe formelle dans les corps locaux. 1966 Les Tendances Géom. en Algèbre et Théorie des Nombres pp. 257–258 Éditions du Centre National de la Recherche Scientifique, Paris.
1966b Tate, John. Endomorphisms of abelian varieties over finite fields. Invent. Math. 2 (1966) 134–144.
1966c Tate, J. The cohomology groups of tori in finite Galois extensions of number fields. Nagoya Math. J. 27 (1966) 709–719.
1966d Lubin, Jonathan; Tate, John. Formal moduli for one-parameter formal Lie groups. Bull. Soc. Math. France 94 (1966) 49–59.
1966e Tate, John T. On the conjectures of Birch and Swinnerton-Dyer and a geometric analog. 1966. Séminaire Bourbaki: Vol. 1965/66, Expose 306.
1966f Tate, John. Letter to Springer, January 13 (1966). (Contains proofs of some of the theorems announced in 1962d.)
1967a Tate, J. T. Global class field theory. (1967) Algebraic Number Theory (Proc. Instructional Conf., Brighton, 1965) pp. 162–203 Thompson, Washington, D.C.

1967b Tate, J. T. Fourier analysis in number fields and Hecke's zeta-functions. (1967) Algebraic Number Theory (Proc. Instructional Conf., Brighton, 1965) pp. 305–347 Thompson, Washington, D.C.

1967c Tate, J. T. p-divisible groups. (1967) Proc. Conf. Local Fields (Driebergen, 1966) pp. 158–183 Springer, Berlin.

1967d Tate, John. Shafarevich, I. R. The rank of elliptic curves. (Russian) Dokl. Akad. Nauk SSSR 175 (1967) 770–773.

1968a Serre, Jean-Pierre; Tate, John. Good reduction of abelian varieties. Ann. of Math. (2) 88 (1968) 492–517.

1968b Tate, John. Residues of differentials on curves. Ann. Sci. École Norm. Sup. (4) 1 (1968) 149–159.

1968c Artin, E.; Tate, J. Class field theory. W. A. Benjamin, Inc., New York-Amsterdam (1968) xxvi+259 pp.

1969a Tate, John; Rigid analytic spaces. (Russian) Mathematics: periodical collection of translations of foreign articles, Vol. 13, No. 3 (Russian), pp. 3–37. Izdat. "Mir", Moscow (1969).

1969b Tate, John, Classes d'isogénie des variétés abéliennes sur un corps fini (d'après T. Honda) Séminaire Bourbaki 352 (1968/1969).

1969c Tate, John, K_2 of global fields, AMS Taped Lecture (Cambridge, Masss., Oct. 1969).

1970s

1970a Tate, John; Oort, Frans. Group schemes of prime order. Ann. Sci. École Norm. Sup. (4) 3 (1970) 1–21.

1970b Tate, John. Symbols in arithmetic. Actes du Congrès International des Mathèmaticiens (Nice, 1970), Tome 1, pp. 201–211. Gauthier-Villars, Paris (1971).

1971 Tate, John. Rigid analytic spaces. Invent. Math. 12 (1971) 257–289.

1973a Bass, H.; Tate, J. The Milnor ring of a global field. Algebraic K-theory, II: "Classical" algebraic K-theory and connections with arithmetic (Proc. Conf., Seattle, Wash., Battelle Memorial Inst., 1972), pp. 349–446. Lecture Notes in Math., Vol. 342, Springer, Berlin (1973).

1973b Tate, J. Letter from Tate to Iwasawa on a relation between K_2 and Galois cohomology. Algebraic K-theory, II: "Classical" algebraic K-theory and connections with arithmetic (Proc. Conf., Seattle Res. Center, Battelle Memorial Inst., 1972), pp. 524–527. Lecture Notes in Math., Vol. 342, Springer, Berlin (1973).

1973c Mazur, B.; Tate, J. Points of order 13 on elliptic curves. Invent. Math. 22 (1973/74), 41–49.

1974a Tate, John T. The arithmetic of elliptic curves. Invent. Math. 23 (1974), 179–206.

1974b Tate, J. The 1974 Fields medals. I. An algebraic geometer. Science 186 (1974), no. 4158, 39–40.

1975a Tate, J. The work of David Mumford. Proceedings of the International Congress of Mathematicians (Vancouver, B. C., 1974), Vol. 1, pp. 11–15. Canad. Math. Congress, Montreal, Que., 1975.

1975b Tate, J. Algorithm for determining the type of a singular fiber in an elliptic pencil. Modular functions of one variable, IV (Proc. Internat. Summer School, Univ. Antwerp, Antwerp, 1972), pp. 33–52. Lecture Notes in Math., Vol. 476, Springer, Berlin (1975).

1976a Tate, J. Problem 9: The general reciprocity law. Mathematical developments arising from Hilbert problems (Proc. Sympos. Pure Math., Northern Illinois Univ., De Kalb, Ill., 1974), pp. 311–322. Proc. Sympos. Pure Math., Vol. XXVIII, Amer. Math. Soc., Providence, R.I. (1976).

1976b Tate, John. Relations between K_2 and Galois cohomology. Invent. Math. 36 (1976), 257–274.

1977a Tate, J. On the torsion in K_2 of fields. Algebraic number theory (Kyoto Internat. Sympos., Res. Inst. Math. Sci., Univ. Kyoto, Kyoto, 1976), pp. 243–261. Japan Soc. Promotion Sci., Tokyo (1977).

1977b Tate, J. T. Local constants. Prepared in collaboration with C. J. Bushnell and M. J. Taylor. Algebraic number fields: L-functions and Galois properties (Proc. Sympos., Univ. Durham, Durham, 1975), pp. 89–131. Academic Press, London (1977).

1978a Cartier, P.; Tate, J. A simple proof of the main theorem of elimination theory in algebraic geometry. Enseign. Math. (2) 24 (1978), no. 3-4, 311–317.

1978b Tate, John. Fields medals. IV. Mumford, David; An instinct for the key idea. Science 202 (1978), no. 4369, 737–739.

1979 Tate, J. Number theoretic background. Automorphic forms, representations and L-functions (Proc. Sympos. Pure Math., Oregon State Univ., Corvallis, Ore., 1977), Part 2, pp. 3–26, Proc. Sympos. Pure Math., XXXIII, Amer. Math. Soc., Providence, R.I. (1979).

1980s

1981a Tate, John. On Stark's conjectures on the behavior of $L(s,\chi)$ at $s=0$. J. Fac. Sci. Univ. Tokyo Sect. IA Math. 28 (1981), no. 3, 963–978 (1982).

1981b Tate, John. Brumer-Stark-Stickelberger. Seminar on Number Theory, 1980–1981 (Talence, 1980–1981), Exp. No. 24, 16 pp., Univ. Bordeaux I, Talence (1981).

1981c Tate, John. On conjugation of abelian varieties of CM type. Handwritten notes (1981).

1983a Tate, J. Variation of the canonical height of a point depending on a parameter. Amer. J. Math. 105 (1983), no. 1, 287–294.

1983b Mazur, B.; Tate, J. Canonical height pairings via biextensions. Arithmetic and geometry, Vol. I, 195–237, Progr. Math., 35, Birkhäuser Boston, Boston, MA (1983).

1983c Rosset, Shmuel; Tate, John. A reciprocity law for K_2-traces. Comment. Math. Helv. 58 (1983), no. 1, 38–47.

1984 Tate, John. Les conjectures de Stark sur les fonctions L d'Artin en $s=0$. Notes of a course at Orsay written by Dominique Bernardi and Norbert Schappacher. Progress in Mathematics, 47. Birkhäuser Boston, Inc., Boston, MA (1984).

1986 Mazur, B.; Tate, J.; Teitelbaum, J. On p-adic analogues of the conjectures of Birch and Swinnerton-Dyer. Invent. Math. 84 (1986), no. 1, 1–48.

1987 Mazur, B.; Tate, J. Refined conjectures of the "Birch and Swinnerton-Dyer type". Duke Math. J. 54 (1987), no. 2, 711–750.

1989 Gross, B.; Tate, J. Commentary on algebra. A century of mathematics in America, Part II, 335–336, Hist. Math., 2, Amer. Math. Soc., Providence, RI (1989).

1990s

1990a Artin, M.; Tate, J.; Van den Bergh, M. Some algebras associated to automorphisms of elliptic curves. The Grothendieck Festschrift, Vol. I, 33–85, Progr. Math., 86, Birkhäuser Boston, Boston, MA (1990).

1990b Artin, Emil; Tate, John. Class field theory. Second edition. Advanced Book Classics. Addison-Wesley Publishing Company, Advanced Book Program, Redwood City, CA (1990). xxxviii+259 pp. ISBN: 0-201-51011-1

1991a Artin, M.; Tate, J.; Van den Bergh, M. Modules over regular algebras of dimension 3. Invent. Math. 106 (1991), no. 2, 335–388.

1991b Artin, Michael; Schelter, William; Tate, John. Quantum deformations of GL_n. Comm. Pure Appl. Math. 44 (1991), no. 8-9, 879–895.

1991c Mazur, B.; Tate, J. The p-adic sigma function. Duke Math. J. 62 (1991), no. 3, 663–688.

1992 Silverman, Joseph H.; Tate, John. Rational points on elliptic curves. Undergraduate Texts in Mathematics. Springer-Verlag, New York (1992). x+281 pp.

1994a Tate, John. Conjectures on algebraic cycles in l-adic cohomology. Motives (Seattle, WA, 1991), 71–83, Proc. Sympos. Pure Math., 55, Part 1, Amer. Math. Soc., Providence, RI (1994).

1994b Tate, John. The non-existence of certain Galois extensions of $\mathbb{Q}$ unramified outside 2. Arithmetic geometry (Tempe, AZ, 1993), 153–156, Contemp. Math., 174, Amer. Math. Soc., Providence, RI (1994).

1994c Artin, Michael; Schelter, William; Tate, John. The centers of 3-dimensional Sklyanin algebras. Barsotti Symposium in Algebraic Geometry (Abano Terme, 1991), 1–10, Perspect. Math., 15, Academic Press, San Diego, CA (1994).

1994d Smith, S. P.; Tate, J. The center of the 3-dimensional and 4-dimensional Sklyanin algebras. Proceedings of Conference on Algebraic Geometry and Ring Theory in honor of Michael Artin, Part I (Antwerp, 1992). K-Theory 8 (1994), no. 1, 19–63.

1995 Tate, John. A review of non-Archimedean elliptic functions. Elliptic curves, modular forms, & Fermat's last theorem (Hong Kong, 1993), 162–184, Ser. Number Theory, I, Int. Press, Cambridge, MA (1995).

1996a Tate, John; Voloch, José Felipe. Linear forms in p-adic roots of unity. Internat. Math. Res. Notices (1996), no. 12, 589–601.

1996b Tate, John; van den Bergh, Michel. Homological properties of Sklyanin algebras. Invent. Math. 124 (1996), no. 1-3, 619–647.

1997a Tate, John. Finite flat group schemes. Modular forms and Fermat's last theorem (Boston, MA, 1995), 121–154, Springer, New York (1997).

1997b Tate, J. The work of David Mumford. Fields Medallists' lectures, 219–223, World Sci. Ser. 20th Century Math., 5, World Sci. Publ., River Edge, NJ (1997).

1999 Katz, Nicholas M.; Tate, John. Bernard Dwork (1923–1998). Notices Amer. Math. Soc. 46 (1999), no. 3, 338–343.

2000s

2000 Tate, John. The millennium prize problems I, Lecture by John Tate at the Millenium Meeting of the Clay Mathematical Institute, May 2000, Paris. Video available from the CMI website.

2001 Tate, John. Galois cohomology. Arithmetic algebraic geometry (Park City, UT, 1999), 465–479, IAS/Park City Math. Ser., 9, Amer. Math. Soc., Providence, RI (2001).

2002 Tate, John. On a conjecture of Finotti. Bull. Braz. Math. Soc. (N.S.) 33 (2002), no. 2, 225–229.

2004 Tate, John. Refining Gross's conjecture on the values of abelian L-functions. Stark's conjectures: recent work and new directions, 189–192, Contemp. Math., 358, Amer. Math. Soc., Providence, RI (2004).

2005 Artin, Michael; Rodriguez-Villegas, Fernando; Tate, John. On the Jacobians of plane cubics. Adv. Math. 198 (2005), no. 1, 366–382.

2006 Mazur, Barry; Stein, William; Tate, John. Computation of p-adic heights and log convergence. Doc. Math. (2006), Extra Vol., 577–614 (electronic).

2008 Tate, John. Foreword to p-adic geometry, 9–63, Univ. Lecture Ser., 45, Amer. Math. Soc., Providence, RI (2008).

2009 Artin, Emil; Tate, John. Class field theory. New edition. TeXed and slightly revised from the original 1961 version. AMS Chelsea Publishing, Providence, RI (2009). viii+194 pp.

2010s

2011 Raussen, Martin; Skau, Christian. Interview with Abel Laureate John Tate. Notices Amer. Math. Soc. 58 (2011), no. 3, 444–452.

2011 Tate, John. Stark's basic conjecture. Arithmetic of L-functions, 7–31, IAS/Park City Math. Ser., 18, Amer. Math. Soc., Providence, RI.

Added September 2012 I should have mentioned the work of Tate on liftings of Galois representations, as included in Part II of: Serre, J.-P. Modular forms of weight one and Galois representations. Algebraic number fields: L-functions and Galois properties (Proc. Sympos., Univ. Durham, Durham, 1975), pp. 193–268. Academic Press, London (1977). See also: Variations on a theorem of Tate. Stefan Patrikis, arXiv:1207.6724.

Also, "An oft cited (1979) letter from Tate to Serre on computing local heights on elliptic curves." was posted on the arXiv by Silverman (arXiv:1207.5765).

The collected works of Tate, which will include other unpublished letters, is in preparation.

References

An, S.Y., Kim, S.Y., Marshall, D.C., Marshall, S.H., McCallum, W.G., Perlis, A.R.: Jacobians of genus one curves. J. Number Theory **90**(2), 304–315 (2001)

Arbarello, E., De Concini, C., Kac, V.G.: The infinite wedge representation and the reciprocity law for algebraic curves. In: Theta Functions—Bowdoin 1987, Part 1, Brunswick, ME, 1987. Proc. Sympos. Pure Math., 49, Part 1, pp. 171–190. Am. Math. Soc., Providence (1989)

Armitage, J.V.: On a theorem of Hecke in number fields and function fields. Invent. Math. **2**, 238–246 (1967)

Artin, E.: Über die Zetafunktionen gewisser algebraischer Zahlkörper. Math. Ann. **89**(1–2), 147–156 (1923)
Artin, E.: Beweis des allgemeinen Reziprozitätsgesetzes. Abhandlungen Hamburg **5**, 353–363 (1927)
Artin, M., Schelter, W.F.: Graded algebras of global dimension 3. Adv. Math. **66**(2), 171–216 (1987)
Artin, M., Swinnerton-Dyer, H.P.F.: The Shafarevich–Tate conjecture for pencils of elliptic curves on K3 surfaces. Invent. Math. **20**, 249–266 (1973)
Barnet-Lamb, T., Geraghty, D., Harris, M., Taylor, R.: A family of Calabi–Yau varieties and potential automorphy II. P.R.I.M.S. **47**, 29–98 (2011)
Bass, H., Schanuel, S.: The homotopy theory of projective modules. Bull. Am. Math. Soc. **68**, 425–428 (1962)
Beilinson, A., Bloch, S., Esnault, H.: ε-factors for Gauss–Manin determinants. Dedicated to Yuri I. Manin on the occasion of his 65th birthday. Mosc. Math. J. **2**(3), 477–532 (2002)
Beilinson, A., Drinfeld, V.: Chiral Algebras. American Mathematical Society Colloquium Publications, vol. 51. Am. Math. Soc., Providence (2004)
Beĭlinson, A.A., Schechtman, V.V.: Determinant bundles and Virasoro algebras. Commun. Math. Phys. **118**(4), 651–701 (1988)
Berkovich, V.G.: Spectral Theory and Analytic Geometry over Non-Archimedean Fields. Mathematical Surveys and Monographs, vol. 33. Am. Math. Soc., Providence (1990)
Berkovich, V.G.: Etale cohomology for non-Archimedean analytic spaces. Inst. Hautes Études Sci. Publ. Math. **78**, 5–161 (1993)
Birch, B.J.: K_2 of global fields. 1969 number theory institute. In: Proc. Sympos. Pure Math., Vol. XX, State Univ. New York, Stony Brook, NY, 1969, pp. 87–95. Am. Math. Soc., Providence (1971)
Bosch, S.: Lectures on Formal and Rigid Geometry, Preprint 378 of the SFB Geometrische Strukturen in der Mathematik, Münster (2005)
Bosch, S., Güntzer, U., Remmert, R.: Non-Archimedean analysis. In: A Systematic Approach to Rigid Analytic Geometry. Grundlehren der Mathematischen Wissenschaften, vol. 261. Springer, Berlin (1984)
Boyer, P.: Mauvaise réduction des variétés de Drinfeld et correspondance de Langlands locale. Invent. Math. **138**(3), 573–629 (1999)
Brauer, R.: On Artin's L-series with general group characters. Ann. of Math. (2) **48**, 502–514 (1947)
Brauer, R.: A characterization of the characters of groups of finite order. Ann. of Math. (2) **57**, 357–377 (1953)
Carayol, H.: Nonabelian Lubin–Tate theory. In: Automorphic Forms, Shimura Varieties, and L-Functions, Vol. II, Ann Arbor, MI, 1988. Perspect. Math., vol. 11, pp. 15–39. Academic Press, Boston (1990)
Carayol, H.: La conjecture de Sato–Tate (d'après Clozel, Harris, Shepherd-Barron, Taylor). Séminaire Bourbaki. Vol. 2006/2007. Astérisque No. 317 (2008), Exp. No. 977, ix, 345–391
Cartan, H., Eilenberg, S.: Homological Algebra. Princeton University Press, Princeton (1956)
Cassels, J.W.S.: Arithmetic on curves of genus 1. IV. Proof of the Hauptvermutung. J. Reine Angew. Math. **211**, 95–112 (1962)
Cassels, J.W.S.: Arithmetic on curves of genus 1. VII. The dual exact sequence. J. Reine Angew. Math. **216**, 150–158 (1964)
Cassels, J.W.S.: Arithmetic on curves of genus 1. VIII. On conjectures of Birch and Swinnerton-Dyer. J. Reine Angew. Math. **217**, 180–199 (1965)
Cassels, J.W.S.: Diophantine equations with special reference to elliptic curves. J. Lond. Math. Soc. **41**, 193–291 (1966)
Cassels, J.W.S., Fröhlich, A. (eds.): Algebraic Number Theory. Proceedings of an Instructional Conference Organized by the London Mathematical Society. Academic Press, London (1967). Thompson Book, Washington, DC

Cattani, E., Deligne, P., Kaplan, A.: On the locus of Hodge classes. J. Am. Math. Soc. **8**(2), 483–506 (1995)
Chevalley, C.: Introduction to the Theory of Algebraic Functions of One Variable. Mathematical Surveys, vol. VI. Am. Math. Soc., New York (1951)
Chevalley, C.: Class Field Theory. Nagoya University, Nagoya (1954)
Clozel, L.: The Sato–Tate Conjecture. Current Developments in Mathematics, 2006, pp. 1–34. International Press, Somerville (2008)
de Jong, A.J.: Homomorphisms of Barsotti–Tate groups and crystals in positive characteristic. Invent. Math. **134**(2), 301–333 (1998). Erratum: ibid. **138**(1), 225 (1999)
Deligne, P.: Variétés abéliennes ordinaires sur un corps fini. Invent. Math. **8**, 238–243 (1969)
Deligne, P.: Les constantes des équations fonctionnelles des fonctions L. In: Modular Functions of One Variable, II, Proc. Internat. Summer School, Univ. Antwerp, Antwerp, 1972. Lecture Notes in Math., vol. 349, pp. 501–597. Springer, Berlin (1973)
Deligne, P.: Valeurs de fonctions L et périodes d'intégrales. In: Proc. Sympos. Pure Math., XXXIII, Oregon State Univ., Corvallis, OR, 1977. Automorphic Forms, Representations and L-Functions, pp. 313–346. Am. Math. Soc., Providence (1979). Part 2
Deligne, P.: Motifs et groupe de Taniyama. In: Hodge Cycles, Motives, and Shimura Varieties. Lecture Notes in Mathematics, vol. 900, pp. 261–279. Springer, Berlin (1982)
Deligne, P.: Hodge cycles on Abelian varieties (notes by J.S. Milne). In: Hodge Cycles, Motives, and Shimura Varieties. Lecture Notes in Math., vol. 900, pp. 9–100. Springer, Berlin (1982)
Deligne, P., Rapoport, M.: Les schémas de modules de courbes elliptiques. In: Modular Functions of One Variable, II, Proc. Internat. Summer School, Univ. Antwerp, Antwerp, 1972. Lecture Notes in Math., vol. 349, pp. 143–316. Springer, Berlin (1973)
Deuring, M.: Die Typen der Multiplikatorenringe elliptischer Funktionenkörper. Abh. Math. Sem. Hansischen Univ. **14**, 197–272 (1941)
Deuring, M.: Die Zetafunktion einer algebraischen Kurve vom Geschlechte Eins. Nachr. Akad. Wiss. Göttingen, pp. 85–94 (1953)
Douady, A.: Cohomologie des groupes compacts totalement discontinus (d'après des notes de Serge Lang sur un article non publie de Tate). Séminaire Bourbaki, Vol. 5, Exp. No. 189, pp. 287–298. Soc. Math. France, Paris (1959)
Drinfeld, V.G.: Elliptic modules. Mat. Sb. (N.S.) **94**(136), 594–627 (1974), 656 (in Russian)
Dwork, B.: On the Artin root number. Am. J. Math. **78**, 444–472 (1956)
Dwork, B.: On the rationality of the zeta function of an algebraic variety. Am. J. Math. **82**, 631–648 (1960)
Dwork, B.: A deformation theory for the zeta function of a hypersurface. In: Proc. Internat. Congr. Mathematicians, Stockholm, 1962, pp. 247–259. Inst. Mittag-Leffler, Djursholm (1963)
Elzein, F.: Résidus en géométrie algébrique. C. R. Acad. Sci. Paris Sér. A-B **272**, A878–A881 (1971)
Faltings, G.: Endlichkeitssätze für abelsche Varietäten über Zahlkörpern. Invent. Math. **73**(3), 349–366 (1983). Erratum: Ibid. no. 2, 381 (1984)
Faltings, G.: Arithmetische Kompaktifizierung des Modulraums der abelschen Varietäten. In: Workshop Bonn 1984, Bonn, 1984. Lecture Notes in Math., vol. 1111, pp. 321–383. Springer, Berlin (1985)
Faltings, G.: p-Adic Hodge theory. J. Am. Math. Soc. **1**(1), 255–299 (1988)
Faltings, G., Wüstholz, G. (eds.): Rational Points. Papers from the Seminar Held at the Max-Planck-Institut für Mathematik, 1983/1984. Aspects of Mathematics, E6. Vieweg, Braunschweig (1984)
Farrell, F.T.: An extension of Tate cohomology to a class of infinite groups. J. Pure Appl. Algebra **10**(2), 153–161 (1977/78)
Finotti, L.R.A.: Canonical and minimal degree liftings of curves. J. Math. Sci. Univ. Tokyo **11**(1), 1–47 (2004)
Fontaine, J.-M.: Sur certains types de représentations p-adiques du groupe de Galois d'un corps local; construction d'un anneau de Barsotti–Tate. Ann. of Math. (2) **115**(3), 529–577 (1982)
Garland, H.: A finiteness theorem for K_2 of a number field. Ann. of Math. (2) **94**, 534–548 (1971)

Gonzalez, J.R., Romo, F.P.: A Negative Answer to the Question Of the Linearity of Tate's Trace for the Sum of Two Endomorphisms (preprint 2012)
Gordon, W.J.: Linking the conjectures of Artin–Tate and Birch–Swinnerton-Dyer. Compos. Math. **38**(2), 163–199 (1979)
Green, W.: Heights in families of Abelian varieties. Duke Math. J. **58**(3), 617–632 (1989)
Green, M., Griffiths, P., Kerr, M.: Mumford–Tate domains. Boll. Unione Mat. Ital. (9) **3**(2), 281–307 (2010)
Greenberg, R., Stevens, G.: p-Adic L-functions and p-adic periods of modular forms. Invent. Math. **111**(2), 407–447 (1993)
Grothendieck, A.: Sur une note de Mattuck–Tate. J. Reine Angew. Math. **200**, 208–215 (1958)
Grothendieck, A.: Standard conjectures on algebraic cycles. In: Algebraic Geometry. Internat. Colloq., Tata Inst. Fund. Res., Bombay, 1968, pp. 193–199. Oxford Univ. Press, London (1969)
Harris, M., Taylor, R.: The Geometry and Cohomology of Some Simple Shimura Varieties. Annals of Mathematics Studies, vol. 151. Princeton University Press, Princeton (2001). With an appendix by Vladimir G. Berkovich
Hecke, E.: Eine neue Art von Zetafunktionen und ihre Beziehungen zur Verteilung der Primzahlen. I. Math. Z. **1**, 357–376 (1918)
Hecke, E.: Eine neue Art von Zetafunktionen und ihre Beziehungen zur Verteilung der Primzahlen. II. Math. Z. **6**, 11–51 (1920)
Hecke, E.: Vorlesungen über die Theorie der algebraischen Zahlen, 2. Aufl. Akademische Verlagsgesellschaft, Leipzig (1954) (German)
Hodge, W.V.D.: The topological invariants of algebraic varieties. In: Proceedings of the International Congress of Mathematicians, Cambridge, MA, 1950, vol. 1, pp. 182–192. Am. Math. Soc., Providence (1952)
Honda, T.: Isogeny classes of Abelian varieties over finite fields. J. Math. Soc. Jpn. **20**, 83–95 (1968)
Iwasawa, K.: A note on functions. In: Proceedings of the International Congress of Mathematicians, Cambridge, MA, 1950, vol. 1, p. 322. Am. Math. Soc., Providence (1952)
Iwasawa, K.: Letter to J. Dieudonné. In: Zeta Functions in Geometry, April 8, 1952. Adv. Stud. Pure Math., vol. 21, pp. 445–450. Kinokuniya, Tokyo (1992)
Iwasawa, K.: Algebraic Functions. Translations of Mathematical Monographs, vol. 118. Am. Math. Soc., Providence (1993). Translated from the 1973 Japanese edition by Goro Kato
Kato, K.: p-Adic Hodge theory and values of zeta functions of modular forms. In: Cohomologies p-adiques et applications arithmétiques. III. Astérisque, vol. 295, pp. 117–290 (2004)
Kato, K., Trihan, F.: On the conjectures of Birch and Swinnerton-Dyer in characteristic $p > 0$. Invent. Math. **153**(3), 537–592 (2003)
Khare, C., Wintenberger, J.-P.: Serre's modularity conjecture. I. Invent. Math. **178**(3), 485–504 (2009). II. Ibid. 505–586
Kiehl, R.: Der Endlichkeitssatz für eigentliche Abbildungen in der nichtarchimedischen Funktionentheorie. Invent. Math. **2**, 191–214 (1967). Theorem A und Theorem B in der nichtarchimedischen Funktionentheorie. Ibid. 256–273
Krull, W.: Galoissche Theorie der unendlichen algebraischen Erweiterungen. Math. Ann. **100**, 687–698 (1928)
Kudla, S.: In: Bernstein, J., Gelbart, S. (eds.) An Introduction to the Langlands Program, p. 133. Birkhäuser, Boston (2003)
Lang, S.: Diophantine approximations on toruses. Am. J. Math. **86**, 521–533 (1964)
Lang, S.: Rapport sur la cohomologie des groupes. Benjamin, New York (1967)
Langlands, R.P.: Automorphic representations, Shimura varieties, and motives. In: Ein Märchen. Automorphic Forms, Representations and L-Functions, Proc. Sympos. Pure Math., Oregon State Univ., Corvallis, OR, 1977. Proc. Sympos. Pure Math., vol. XXXIII, pp. 205–246. Am. Math. Soc., Providence (1979). Part 2
Langlands, R.P.: Reflexions on receiving the Shaw prize. In: On Certain L-Functions. Clay Math. Proc., vol. 13, pp. 297–308. Am. Math. Soc., Providence (2011)
Levi, B.: Sull'equazione indeterminata del 3° ordine. Rom. 4. Math. Kongr. **2**, 173–177 (1909)

Lichtenbaum, S.: On the values of zeta and L-functions. I. Ann. of Math. (2) **96**, 338–360 (1972)
Liu, Q., Lorenzini, D., Raynaud, M.: On the Brauer group of a surface. Invent. Math. **159**(3), 673–676 (2005)
Lubin, J.: One-parameter formal Lie groups over p-adic integer rings. Ann. of Math. (2) **80**, 464–484 (1964)
Lyons, C.: A rank inequality for the Tate conjecture over global function fields. Expo. Math. **27**(2), 93–108 (2009)
Manin, J.I.: The Tate height of points on an Abelian variety, its variants and applications. Izv. Akad. Nauk SSSR, Ser. Mat. **28**, 1363–1390 (1964) (Russian)
Matsumoto, H.: Sur les sous-groupes arithmétiques des groupes semi-simples déployés. Ann. Sci. Éc. Norm. Super. **2**, 1–62 (1969)
Mazur, B.: Modular curves and the Eisenstein ideal. Inst. Hautes Études Sci. Publ. Math. **47**, 33–186 (1977)
McCabe, J.: p-Adic theta functions. Ph.D. thesis, Harvard (1968), 222 pp.
Merel, L.: Bornes pour la torsion des courbes elliptiques sur les corps de nombres. Invent. Math. **124**(1–3), 437–449 (1996)
Merkurjev, A.S., Suslin, A.A.: K-cohomology of Severi-Brauer varieties and the norm residue homomorphism. Izv. Akad. Nauk SSSR, Ser. Mat. **46**(5), 1011–1046 (1982), 1135–1136 (Russian)
Messing, W.: The Crystals Associated to Barsotti–Tate Groups: With Applications to Abelian Schemes. Lecture Notes in Mathematics, vol. 264. Springer, Berlin (1972)
Milne, J.S.: Weil–Châtelet groups over local fields. Ann. Sci. Éc. Norm. Super. **3**, 273–284 (1970); ibid. **5**, 261–264 (1972)
Milne, J.S.: On a conjecture of Artin and Tate. Ann. of Math. (2) **102**(3), 517–533 (1975)
Milne, J.S.: Comparison of the Brauer group with the Tate–Shafarevich group. J. Fac. Sci. Univ. Tokyo (Shintani Memorial Volume) IA **28**, 735–743 (1982)
Milne, J.S.: Arithmetic Duality Theorems. Perspectives in Mathematics, vol. 1. Academic Press, Boston (1986)
Milne, J.S.: Lefschetz motives and the Tate conjecture. Compos. Math. **117**(1), 45–76 (1999)
Milne, J.S.: Rational Tate classes. Mosc. Math. J. **9**(1), 111–141 (2009)
Milnor, J.: Algebraic K-theory and quadratic forms. Invent. Math. **9**, 318–344 (1969/1970)
Milnor, J.: Introduction to Algebraic K-Theory. Annals of Mathematics Studies, vol. 72. Princeton University Press, Princeton (1971)
Moore, C.C.: Group extensions of p-adic and adelic linear groups. Inst. Hautes Études Sci. Publ. Math. **35**, 157–222 (1968)
Mumford, D.: Families of Abelian varieties. In: Algebraic Groups and Discontinuous Subgroups. Proc. Sympos. Pure Math., Boulder, Colo, 1965, pp. 347–351. Am. Math. Soc., Providence (1966)
Mumford, D.: An analytic construction of degenerating curves over complete local rings. Compos. Math. **24**, 129–174 (1972)
Nakayama, T.: Cohomology of class field theory and tensor product modules I. Ann. of Math. **65**, 255–267 (1957)
Néron, A.: Modèles minimaux des variétés abéliennes sur les corps locaux et globaux. Inst. Hautes Études Sci. Publ. Math. No. 21 (1964), 128 pp.
Néron, A.: Quasi-fonctions et hauteurs sur les variétés abéliennes. Ann. of Math. (2) **82**, 249–331 (1965)
Piatetski-Shapiro, I.I.: Interrelations between the Tate and Hodge hypotheses for Abelian varieties. Mat. Sb. (N.S.) **85**(127), 610–620 (1971) (Russian)
Pohlmann, H.: Algebraic cycles on Abelian varieties of complex multiplication type. Ann. of Math. (2) **88**, 161–180 (1968)
Quillen, D.: Higher algebraic K-theory. I. Algebraic K-theory, I: Higher K-theories. In: Proc. Conf., Battelle Memorial Inst., Seattle, Wash., 1972. Lecture Notes in Math., vol. 341, pp. 85–147. Springer, Berlin (1973)

Raynaud, M.: Variétés abéliennes et géométrie rigide. Actes du Congrès International des Mathématiciens, Nice, 1970, pp. 473–477. Gauthier-Villars, Paris (1971), Tome 1

Raynaud, M.: Schémas en groupes de type (p, ..., p). Bull. Soc. Math. Fr. **102**, 241–280 (1974)

Roquette, P.: Arithmetischer Beweis der Riemannschen Vermutung in Kongruenzfunktionenkörpern beliebigen Geschlechts. J. Reine Angew. Math. **191**, 199–252 (1953)

Roquette, P.: Analytic Theory of Elliptic Functions over Local Fields. Hamburger Mathematische Einzelschriften (N.F.), vol. 1. Vandenhoeck & Ruprecht, Göttingen (1970)

Schappacher, N.: The Bourbaki Congress at El Escorial and other mathematical (non)events of 1936. In: The Mathematical Intelligencer, Special issue International Congress of Mathematicians, Madrid, August 2006, pp. 8–15 (2006)

Schröer, S.: On genus change in algebraic curves over imperfect fields. Proc. Am. Math. Soc. **137**(4), 1239–1243 (2009)

Serre, J.-P.: Cohomologie et arithmétique, Séminaire Bourbaki 1952/1953, no. 77

Serre, J.-P.: Sur la dimension homologique des anneaux et des modules noethériens. In: Proceedings of the International Symposium on Algebraic Number Theory, Tokyo & Nikko, 1955, pp. 175–189. Science Council of Japan, Tokyo (1956)

Serre, J.-P.: Cohomologie Galoisienne. In: Cours au Collège de France (1962–1963), 2nd edn. Lecture Notes in Mathematics, vol. 5. Springer, Berlin (1964a)

Serre, J.-P.: Groupes de Lie l-adiques attachés aux courbes elliptiques. In: Colloque de Clermond-Ferrand. Les Tendances Géom. en Algébre et Théorie des Nombres, pp. 239–256 (1964b). Éditions du Centre National de la Recherche Scientifique, Paris (1966)

Serre, J.-P.: Algèbre locale. In: Multiplicités, 2nd edn. Lecture Notes in Mathematics, vol. 11. Springer, Berlin (1965)

Serre, J.-P.: Abelian l-Adic Representations and Elliptic Curves. Benjamin, New York (1968a)

Serre, J.-P.: Œuvres, Vol. II. Springer, Berlin (1968b)

Shimura, G.: Reduction of algebraic varieties with respect to a discrete valuation of the basic field. Am. J. Math. **77**, 134–176 (1955)

Shimura, G.: On Abelian varieties with complex multiplication. Proc. London Math. Soc. (3) **34**(1), 65–86 (1977)

Shimura, G., Taniyama, Y.: Complex Multiplication of Abelian Varieties and Its Applications to Number Theory. Publications of the Mathematical Society of Japan, vol. 6. Mathematical Society of Japan, Tokyo (1961)

Siegel, C.L.: Über die Fourierschen Koeffizienten von Modulformen. Nachr. Akad. Wiss. Göttingen Math.-Phys. Kl. **II**, 15–56 (1970)

Stark, H.M.: Values of L-functions at $s = 1$. I. L-Functions for quadratic forms. Adv. Math. **7**, 301–343 (1971)

Stark, H.M.: Values of L-functions at $s = 1$. II. Artin L-functions with rational characters. Adv. Math. **17**(1), 60–92 (1975)

Stark, H.M.: Values of L-functions at $s = 1$. III. Totally real fields and Hilbert's twelfth problem. Adv. Math. **22**(1), 64–84; (1976)

Stark, H.M.: Values of L-functions at $s = 1$. IV. First derivatives at $s = 0$. Adv. Math. **35**(3), 197–235 (1980)

Tan, K.-S.: Refined theorems of the Birch and Swinnerton-Dyer type. Ann. Inst. Fourier (Grenoble) **45**(2), 317–374 (1995)

Taylor, M.J.: Obituary: Albrecht Fröhlich, 1916–2001. Bull. Lond. Math. Soc. **38**(2), 329–350 (2006)

Ulmer, D.: Elliptic curves with large rank over function fields. Ann. of Math. (2) **155**(1), 295–315 (2002)

Weil, A.: Remarques sur des résultats recents de C. Chevalley. C. R. Acad. Sci., Paris **203**, 1208–1210 (1936)

Weil, A.: Sur les courbes algébriques et les variétés qui s'en déduisent. Publ. Inst. Math. Univ. Strasbourg 7 (1945)

Weil, A.: Variétés abéliennes et courbes algébriques. Actualités Sci. Ind., no. 1064 = Publ. Inst. Math. Univ. Strasbourg 8 (1946). Hermann & Cie., Paris (1948)

Weil, A.: Sur la théorie du corps de classes. J. Math. Soc. Jpn. **3**, 1–35 (1951)
Weil, A.: Remarques sur un mémoire d'Hermite. Arch. Math. (Basel) **5**, 197–202 (1954)
Wiles, A.: The Iwasawa conjecture for totally real fields. Ann. of Math. (2) **131**(3), 493–540 (1990)
Zariski, O.: Complete linear systems on normal varieties and a generalization of a lemma of Enriques-Severi. Ann. of Math. (2) **55**, 552–592 (1952)
Zariski, O.: Scientific report on the second summer institute, III algebraic sheaf theory. Bull. Am. Math. Soc. **62**, 117–141 (1956)

List of Publications for John Torrence Tate

1950

[1] *Fourier Analysis in Number Fields and Hecke's Zeta Functions*. PhD thesis, Harvard University. Published in *Algebraic Number Theory (Proc. Instructional Conf., Brighton 1965)*, pages 305–347. Thompson, Washington, DC, 1967.

1951

[2] On the relation between extremal points of convex sets and homomorphisms of algebras. *Comm. Pure Appl. Math.*, 4:31–32.

[3] (with E. Artin). A note on finite ring extensions. *J. Math. Soc. Japan*, 3:74–77.

1952

[4] Genus change in inseparable extensions of function fields. *Proc. Amer. Math. Soc.*, 3:400–406.

[5] (with S. Lang). On Chevalley's proof of Luroth's theorem. *Proc. Amer. Math. Soc.*, 3:621–624.

[6] The higher dimensional cohomology groups of class field theory. *Ann. of Math. (2)*, 56:294–297.

1955

[7] (with Y. Kawada). On the Galois cohomology of unramified extensions of function fields in one variable. *Amer. J. Math.*, 77:197–217.

[8] (with R. Brauer). On the characters of finite groups. *Ann. of Math. (2)*, 62:1–7.

1957

[9] Homology of Noetherian rings and local rings. *Illinois J. Math.*, 1:14–27.

1958

[10] (with A. Mattuck). On the inequality of Castelnuovo–Severi. *Abh. Math. Sem. Univ. Hamburg*, 22:295–299.

H. Holden, R. Piene (eds.), *The Abel Prize 2008-2012*,
DOI 10.1007/978-3-642-39449-2_16,

[11] *WC-groups over $\mathfrak{p}$-adic fields*, volume 13 of *Séminaire Bourbaki; 10e année: 1957/1958. Textes des conférences; Exposés 152 à 168; 2e éd. corrigée, Exposé 156*. Secrétariat mathématique, Paris.

[12] (with S. Lang). Principal homogeneous spaces over Abelian varieties. *Amer. J. Math.*, 80:659–684.

1962

[13] (with A. Fröhlich and J.-P. Serre). A different with an odd class. *J. Reine Angew. Math.*, 209:6–7.

[14] Principal homogeneous spaces for Abelian varieties. *J. Reine Angew. Math.*, 209:98–99.

1963

[15] Duality theorems in Galois cohomology over number fields. In *Proc. Internat. Congr. Mathematicians (Stockholm, 1962)*, pages 288–295. Inst. Mittag-Leffler, Djursholm.

[16] (with S. Sen). Ramification groups of local fields. *J. Indian Math. Soc. (N.S.)*, 27:197–202 (1964).

1964

[17] Nilpotent quotient groups. *Topology*, 3(suppl. 1):109–111.

1965

[18] (with J. Lubin). Formal complex multiplication in local fields. *Ann. of Math. (2)*, 81:380–387.

[19] Algebraic classes of cohomologies. *Uspehi Mat. Nauk*, 20(6 (126)):27–40.

[20] Algebraic cycles and poles of zeta functions. In *Arithmetical Algebraic Geometry (Proc. Conf. Purdue Univ., 1963)*, pages 93–110. Harper & Row, New York, 1965.

1966

[21] Multiplication complexe formelle dans les corps locaux. In *Les Tendances Géom. en Algèbre et Théorie des Nombres*, pages 257–258. Éditions du Centre National de la Recherche Scientifique, Paris.

[22] Endomorphisms of Abelian varieties over finite fields. *Invent. Math.*, 2:134–144.

[23] The cohomology groups of tori in finite Galois extensions of number fields. *Nagoya Math. J.*, 27:709–719.

[24] (with J. Lubin). Formal moduli for one-parameter formal Lie groups. *Bull. Soc. Math. France*, 94:49–59.

[25] On the conjectures of Birch and Swinnerton-Dyer and a geometric analog. In *Séminaire Bourbaki, 1965/1966*. Exposé 306, 415–440.

1967

[26] Global class field theory. In *Algebraic Number Theory (Proc. Instructional Conf., Brighton, 1965)*, pages 162–203. Thompson, Washington, DC.

[27] p-divisible groups. In *Proc. Conf. Local Fields (Driebergen, 1966)*, pages 158–183. Springer, Berlin.
[28] (with I. Shafarevich). The rank of elliptic curves. *Dokl. Akad. Nauk SSSR*, 175:770–773.

1968

[29] (with J.-P. Serre). Good reduction of Abelian varieties. *Ann. of Math. (2)*, 88:492–517.
[30] Residues of differentials on curves. *Ann. Sci. École Norm. Sup. (4)*, 1:149–159.
[31] (with E. Artin). *Class field theory*. W. A. Benjamin, Inc., New York-Amsterdam. Second edition, Advanced Book Classics, Addison-Wesley Publishing Company, Redwood City, CA, 1990. Reprinted by AMS Chelsea Publishing, Providence, RI, 2009.

1970

[32] (with F. Oort). Group schemes of prime order. *Ann. Sci. École Norm. Sup. (4)*, 3:1–21.

1971

[33] Symbols in arithmetic. In *Actes du Congrès International des Mathématiciens (Nice, 1970), Tome 1*, pages 201–211. Gauthier-Villars, Paris.
[34] Rigid analytic spaces. *Invent. Math.*, 12:257–289. Also available in *Mathematics: periodical collection of translations of foreign articles*, 13(3):3–37, Izdat, "Mir", Moscow, 1969 (in Russian).
[35] Classes d'isogénie des variétés abéliennes sur un corps fini (d'après Z. Honda). In *Séminaire Bourbaki 1968/69*. Exposé 352, 95–110.

1973

[36] (with H. Bass). The Milnor ring of a global field. In *Algebraic K-theory, II: "Classical" algebraic K-theory and connections with arithmetic (Proc. Conf., Seattle, Wash., Battelle Memorial Inst., 1972)*, pages 349–446. Lecture Notes in Math., Vol. 342. Springer, Berlin.
[37] Letter from Tate to Iwasawa on a relation between K_2 and Galois cohomology. In *Algebraic K-theory, II: "Classical" algebraic K-theory and connections with arithmetic (Proc. Conf., Seattle Res. Center, Battelle Memorial Inst., 1972)*, pages 524–527. Lecture Notes in Math., Vol. 342. Springer, Berlin.
[38] (with B. Mazur). Points of order 13 on elliptic curves. *Invent. Math.*, 22:41–49.

1974

[39] The arithmetic of elliptic curves. *Invent. Math.*, 23:179–206.
[40] The 1974 Fields medals. I. An algebraic geometer. *Science*, 186(4158):39–40. Also available in *Fiz.-Mat. Spis. B″lgar. Akad. Nauk.*, 18(51)(1):68–72 (in Bulgarian).

1975

[41] The work of David Mumford. In *Proceedings of the International Congress of Mathematicians (Vancouver, BC, 1974), Vol. 1*, pages 11–15. Canad. Math. Congress, Montreal, Quebec.

[42] Algorithm for determining the type of a singular fiber in an elliptic pencil. In *Modular functions of one variable, IV (Proc. Internat. Summer School, Univ. Antwerp, Antwerp, 1972)*, pages 33–52. Lecture Notes in Math., Vol. 476. Springer, Berlin.

1976

[43] Problem 9: The general reciprocity law. In *Mathematical developments arising from Hilbert problems (Proc. Sympos. Pure Math., Northern Illinois Univ., De Kalb, Ill., 1974)*, pages 311–322. Proc. Sympos. Pure Math., Vol. XXVIII. Amer. Math. Soc., Providence, RI.

[44] Relations between K_2 and Galois cohomology. *Invent. Math.*, 36:257–274.

1977

[45] On the torsion in K_2 of fields. In *Algebraic number theory (Kyoto Internat. Sympos., Res. Inst. Math. Sci., Univ. Kyoto, Kyoto, 1976)*, pages 243–261. Japan Soc. Promotion Sci., Tokyo.

[46] Local constants. In *Algebraic number fields: L-functions and Galois properties (Proc. Sympos., Univ. Durham, Durham, 1975)*, pages 89–131. Academic Press, London. Prepared in collaboration with C.J. Bushnell and M.J. Taylor.

1978

[47] (with P. Cartier). A simple proof of the main theorem of elimination theory in algebraic geometry. *Enseign. Math. (2)*, 24(3–4):311–317.

[48] (with D. Mumford). Fields medals. IV. An instinct for the key idea. *Science*, 202(4369):737–739.

1979

[49] Number theoretic background. In *Automorphic forms, representations and L-functions (Proc. Sympos. Pure Math., Oregon State Univ., Corvallis, Ore., 1977), Part* 2, Proc. Sympos. Pure Math., XXXIII, pages 3–26. Amer. Math. Soc., Providence, RI.

1981

[50] On Stark's conjectures on the behavior of $L(s, \chi)$ at $s = 0$. *J. Fac. Sci. Univ. Tokyo Sect. IA Math.*, 28(3):963–978 (1982).

[51] Brumer–Stark–Stickelberger. In *Seminar on Number Theory, 1980–1981 (Talence, 1980–1981)*, pages 16, Exp. No. 24. Univ. Bordeaux I, Talence.

1983

[52] Variation of the canonical height of a point depending on a parameter. *Amer. J. Math.*, 105(1):287–294.

[53] (with B. Mazur). Canonical height pairings via biextensions. In *Arithmetic and geometry, Vol. I*, volume 35 of *Progr. Math.*, pages 195–237. Birkhäuser Boston, Boston, MA.

[54] (with S. Rosset). A reciprocity law for K_2-traces. *Comment. Math. Helv.*, 58(1):38–47.

1984

[55] *Les conjectures de Stark sur les fonctions L d'Artin en* $s = 0$, volume 47 of *Progress in Mathematics*. Birkhäuser Boston Inc., Boston, MA. Lecture notes edited by Dominique Bernardi and Norbert Schappacher.

1986

[56] (with B. Mazur and J. Teitelbaum). On p-adic analogues of the conjectures of Birch and Swinnerton-Dyer. *Invent. Math.*, 84(1):1–48.

1987

[57] (with B. Mazur). Refined conjectures of the "Birch and Swinnerton-Dyer type". *Duke Math. J.*, 54(2):711–750.

1989

[58] (with B. Gross). Commentary on algebra. In *A century of mathematics in America, Part II*, volume 2 of *Hist. Math.*, pages 335–336. Amer. Math. Soc., Providence, RI.

1990

[59] (with M. Artin and M. Van den Bergh). Some algebras associated to automorphisms of elliptic curves. In *The Grothendieck Festschrift, Vol. I*, volume 86 of *Progr. Math.*, pages 33–85. Birkhäuser Boston, Boston, MA.

1991

[60] (with M. Artin and M. Van den Bergh). Modules over regular algebras of dimension 3. *Invent. Math.*, 106(2):335–388.

[61] (with M. Artin and W. Schelter). Quantum deformations of GL_n. *Comm. Pure Appl. Math.*, 44(8-9):879–895.

[62] (with B. Mazur). The p-adic sigma function. *Duke Math. J.*, 62(3):663–688.

1992

[63] (with J. Silverman). *Rational points on elliptic curves*. Undergraduate Texts in Mathematics. Springer-Verlag, New York.

1994

[64] (with M. Artin and W. Schelter). The centers of 3-dimensional Sklyanin algebras. In *Barsotti Symposium in Algebraic Geometry (Abano Terme, 1991)*, volume 15 of *Perspect. Math.*, pages 1–10. Academic Press, San Diego, CA.

[65] The non-existence of certain Galois extensions of **Q** unramified outside 2. In *Arithmetic geometry (Tempe, AZ, 1993)*, volume 174 of *Contemp. Math.*, pages 153–156. Amer. Math. Soc., Providence, RI.
[66] (with S. Smith). The center of the 3-dimensional and 4-dimensional Sklyanin algebras. In *Proceedings of Conference on Algebraic Geometry and Ring Theory in honor of Michael Artin, Part I (Antwerp, 1992)*, volume 8, pages 19–63.
[67] Conjectures on algebraic cycles in l-adic cohomology. In *Motives (Seattle, WA, 1991)*, volume 55 of *Proc. Sympos. Pure Math.*, pages 71–83. Amer. Math. Soc., Providence, RI.

1995

[68] WC-groups over $\mathfrak{p}$-adic fields. In *Séminaire Bourbaki, Vol. 4*, pages 265–277, Exp. No. 156. Soc. Math. France, Paris.
[69] A review of non-Archimedean elliptic functions. In *Elliptic curves, modular forms, & Fermat's last theorem (Hong Kong, 1993)*, Ser. Number Theory, I, pages 162–184. Int. Press, Cambridge, MA.

1996

[70] (with J. Voloch). Linear forms in p-adic roots of unity. *Internat. Math. Res. Notices*, (12):589–601.
[71] (with M. van den Bergh). Homological properties of Sklyanin algebras. *Invent. Math.*, 124(1-3):619–647.

1997

[72] Finite flat group schemes. In *Modular forms and Fermat's last theorem (Boston, MA, 1995)*, pages 121–154. Springer, New York.
[73] The work of David Mumford. In *Fields Medallists' lectures*, volume 5 of *World Sci. Ser. 20th Century Math.*, pages 219–223. World Sci. Publ., River Edge, NJ.

1999

[74] (with N. Katz). Bernard Dwork (1923–1998). *Notices Amer. Math. Soc.*, 46(3):338–343.

2001

[75] Galois cohomology. In *Arithmetic algebraic geometry (Park City, UT, 1999)*, volume 9 of *IAS/Park City Math. Ser.*, pages 465–479. Amer. Math. Soc., Providence, RI.

2002

[76] *The millennium prize problems*. Springer VideoMATH. Springer-Verlag, Berlin. A lecture by John Tate, CMI Millennium Meeting Collection.
[77] On a conjecture of Finotti. *Bull. Braz. Math. Soc. (N.S.)*, 33(2):225–229.

2004

[78] Refining Gross's conjecture on the values of Abelian L-functions. In *Stark's conjectures: recent work and new directions*, volume 358 of *Contemp. Math.*, pages 189–192. Amer. Math. Soc., Providence, RI.

2005

[79] (with M. Artin and F. Rodriguez-Villegas). On the Jacobians of plane cubics. *Adv. Math.*, 198(1):366–382.

2006

[80] (with B. Mazur and W. Stein). Computation of p-adic heights and log convergence. *Doc. Math.*, (Extra Vol.):577–614 (electronic).

2008

[81] Foreword. In *p-adic geometry*, volume 45 of *University Lecture Series*, page ix. American Mathematical Society, Providence, RI, 2008. Lectures from the 10th Arizona Winter School held at the University of Arizona, Tucson, AZ, March 10–14, 2007.

2011

[82] Stark's basic conjecture. In *Arithmetic of L-functions*, volume 18 of *IAS/Park City Math. Ser.*, pages 7–31. Amer. Math. Soc., Providence, RI, 2011.

2012

[83] Introduction [Brief biography of Serge Lang]. In *Number theory, analysis and geometry*, pages xv–xx. Springer, New York, 2012.

Curriculum Vitae for John Torrence Tate Jr.

Born:	March 13, 1925 in Minneapolis, USA
Degrees/education:	Bachelor of Arts, Harvard University, 1946
	PhD, Princeton University, 1950
Positions:	Instructor, Princeton University, 1950–1953
	Assistant Professor, Harvard University, 1954–1959
	Professor, Harvard University, 1959–1966
	Perkins Professor, Harvard University, 1966–1990
	Sid W. Richardson Chair in Mathematics, University of Texas at Austin, 1990–2009
Visiting positions:	Columbia University, 1953–1954
	Institute for Advanced Study, Princeton, 1959
	University of California at Berkeley, 1963
	Institut des Hautes Études Scientifiques at Bures-sur-Yvette, 1965–1966 and 1968–1969
	Université de Paris at Orsay, 1980–1981
	Princeton University, 1992
	École Normale Supérieure, Paris, 1992
Memberships:	National Academy of Sciences, USA, 1969
	Académie des sciences, France, 1992
	London Mathematical Society (Honorary member), 1999
	Norwegian Academy of Science and Letters, 2010
Awards and prizes:	Frank Nelson Cole Prize for Number Theory, 1956
	Leroy P. Steele Prize for Lifetime Achievement, 1995
	Wolf Prize, 2002
	Abel Prize, 2010

H. Holden, R. Piene (eds.), *The Abel Prize 2008-2012*,
DOI 10.1007/978-3-642-39449-2_17, © Springer-Verlag Berlin Heidelberg 2014

2011

John W. Milnor

"for pioneering discoveries in topology, geometry and algebra"

ABEL
PRISEN

Autobiography

John Milnor

I grew up in Maplewood, New Jersey, a suburban community where at least half of the adult male population took the train into New York City every weekday morning. My father, an electrical engineer with Western Union, was no exception. He worked particularly on undersea cable engineering,[1] and obtained many patents for devices which helped to optimize telegraphic transmission. My mother was an enthusiastic amateur artist. During the depression, she organized a Toy Lending Library to help parents who couldn't afford toys.

I was painfully shy, and socially backward as a youngster. It didn't help that my parents bought a farm an hours drive to the west when I was four years old. After that time, every summer and every weekend was spent on the farm. I certainly enjoyed the rolling countryside, and the animals. But it meant that I was isolated from anyone my own age much of the time.

My father and my brother Bob, who is seven years older, were both adept with tools, and were always happy building things. (There was a hydroponic garden in the barn, and a ten inch telescope, permanently mounted under a sliding shed out in the field.) Bob built an elaborate model railway system. I was fascinated by the relay switching circuits used to control it; but wasn't much help in actually constructing anything. With World War II looming, he took an accelerated degree in aeronautical engineering at the University of Michigan, and spent the rest of the war in the army, working on aircraft maintenance.

[1] See J. Willard Milnor, *Submarine cable telegraphy*, Transactions of the American Institute of Electrical Engineers **41** (1922) 20–38.

Electronic supplementary material Supplementary material is available in the online version of this chapter at http://dx.doi.org/10.1007/978-3-642-39449-2_18. Videos can also be accessed at http://www.springerimages.com/videos/978-3-642-39449-2.

J. Milnor (✉)
Institute for Mathematical Sciences, Stony Brook University, Stony Brook, NY 11794-3600, USA
e-mail: jack@math.sunysb.edu

H. Holden, R. Piene (eds.), *The Abel Prize 2008-2012*,
DOI 10.1007/978-3-642-39449-2_18, © Springer-Verlag Berlin Heidelberg 2014

My father J. Willard Milnor and mother Emily Cox Milnor in 1928, a few years before I was born

We had a few cows and chickens, at least during the war years. I even learned how to milk a cow

With my brother Bob in the late 1930s

Thus I spent a great deal of time by myself, reading everything I could get my hands on. I loved Bertrand Russell's "History of Western Philosophy", and was quite intrigued by the few mathematics books that my father owned. These included a calculus text for engineers, and a translation of a very brief German text on complex function theory, which was fascinating but very mysterious. However, I certainly never thought of mathematics as a career at that point.

One memorable event was a cross-country road trip which Bob and I took in 1948. In particular, we both took an exciting lesson in rock climbing in Wyoming. We have both had a love of the mountains since that time.[2]

At age seventeen I enrolled in Princeton, and was almost immediately captivated by the mathematical world. It was not that I wasn't interested in other subjects; but I found everything else much harder. Furthermore, the friendly ambience in the mathematics department felt wonderful. There was a cosmopolitan atmosphere, created by many distinguished refugees from Nazi Germany and elsewhere in Europe.

One particularly memorable course was taught by Ralph Fox. The subject was point set topology in the manner of R.L. Moore. This meant that Fox provided the definitions and theorems, while we were required to find the proofs, without any help from books. Later Fox introduced me to 3-dimensional topology, which many people in the department then seemed to think of as a boring backwater, although the field has really come into its own in recent years. I wrote both my senior thesis and my doctoral thesis, under Fox's direction, on the theory of higher order linking invariants.

Another memorable course, completely opposite in style, was a totally polished presentation of algebraic number theory, by Emil Artin. Although the course was very enjoyable, I was chagrined to realize some time later that I had no idea what an "algebraic number" is. As far as I can remember, that particular topic had never been mentioned. Nevertheless, that course, and also contact with junior faculty members such as Serge Lang and John Tate, served me well in later years.

A third memorable class, on elementary differential geometry, was taught by Albert Tucker. In particular, Tucker introduced us to a problem of Karol Borsuk on the total curvature of knots, which I was happily able to solve.

One particularly enjoyable feature of the department was the commons room, which was open at all times. There was often some game such as Go or Kriegspiel in progress, usually surrounded by a crowd of kibitzers. In fact there was an active group in Princeton studying Game Theory, headed by Tucker, but also including younger people such as David Gale, Harold Kuhn, and John Nash. I became actively involved for a few years, and spent several summers at The Rand Corporation in California working in Game Theory. (They were particularly interested in this field because of its possible military applications.) However, I eventually lost enthusiasm for the subject, since it seemed to me that mathematics could play only a limited role. Any really important application would also involve questions of politics, sociology, and psychology, which were completely foreign to me.

---

[2]Some of my adventures (and misadventures) in the mountains are described in the dedication pages of my "Collected Papers III".

With a hyperbolic blackboard, Princeton 1963. Alfred Eisenstaedt/Time & Life Pictures/Getty Images. Reprinted with permission

During my time as a graduate student in Princeton, I took a year off to study geometry under Heinz Hopf in Zürich. The time in Switzerland, was useful and enjoyable, but the difference in style was amazing to me. When I acted like a Princeton student, and interrupted the lecture to ask a question, everyone turned around to stare at me as if I were totally crazy.

I returned to Princeton with a Swiss wife, Brigitte. Over the next ten years, we had three children. Stefan, the oldest, is a computer hardware engineer, who shares my love of the mountains. After many treks in Nepal, he and his wife Lisa adopted two Nepali children. Daniel runs a successful business as a commercial artist in Switzerland, and Gabrielle, the youngest, lives with her blacksmith partner and two horses in the California hills.

During the many years I spent at Princeton,[3] my primary focus was on the topology of manifolds, and the tools from algebraic topology needed to understand them. Here I benefited very much from the presence of Norman Steenrod and John Moore. This was a golden time in topology. The work of Jean-Pierre Serre had made homotopy groups accessible, and the work of René Thom had provided an unexpected and surprising relationship between homotopy theory and the study of smooth manifolds.[4] Furthermore Raoul Bott's work had provided an amazingly simple description for the stable homotopy groups of classical groups. The confluence of these new ideas, together with the well established techniques of cohomology theory, obstruction theory, fiber bundle theory, and characteristic classes led to solutions for many problems which had seemed completely intractable.

[3]For more about my mathematical life in Princeton, see: *Growing Up in the Old Fine Hall*, in "Prospects in Mathematics", edited by H. Rossi, AMS, 1998.

[4]One conversation with Thom, in which he described how to "kill" a homotopy class by a surgery construction, was particularly important to me.

With grandchildren Kavi and Deepa, in Provincetown

The power of these new methods was brought home to me when I discovered what I first thought was a contradiction in mathematics. I had been studying the classification problem for closed $2n$-dimensional manifolds which are $(n-1)$-connected. The homotopy theory of such a manifold is relatively easy to describe, since it has homology only in dimensions 0, n, and $2n$. In dimension eight, many such examples can be constructed by starting with a fiber bundle

$$D^4 \xrightarrow{\subset} E^8 \twoheadrightarrow S^4$$

over S^4 with the closed disk D^4 as fiber. Whenever the boundary manifold $M^7 = \partial E^8$ is a topological sphere, one can paste a copy of the disk D^8 onto E^8 to obtain the required closed 8-dimensional manifold. In fact, for many examples of smooth D^4-bundles over S^4, one can easily check that the boundary is a homotopy 7-sphere. Assuming the Generalized Poincaré Hypothesis, this seemed to lead to a contradiction. It was easy to compute the characteristic classes for the 8-manifold constructed in this way, but in many cases the results contradicted the restrictions on characteristic classes for smooth closed manifolds which followed from the work of Thom and Hirzebruch. My first thought was that the boundary was only a homotopy sphere, not a topological sphere, so that I had found a counterexample to the 7-dimensional Poincaré Hypothesis. However, a little careful analysis showed that the boundary was indeed a topological sphere. Thus I had actually constructed examples of smooth 7-dimensional manifolds which are homeomorphic, but not diffeomorphic, to the standard 7-sphere. In effect, I had answered a question which, as far as I know, no one had ever asked.

A year or so later, Michel Kervaire and I discovered that we were working on very similar ideas, and decided to combine forces, leading to our work on "Groups of Homotopy Spheres."

I was particularly lucky during these years to have a wonderful group of graduate students and junior faculty, who helped me convert some of my lectures into print. Thus my lectures on Morse Theory, inspired by Bott, were put into book form by

In Princeton, perhaps in the 1970s

Michael Spivak and Robert Wells. Similarly, my lectures on Stephen Smale's h-Cobordism Theorem were converted into published form by Larry Siebenmann and Jonathan Sondow, and my Characteristic Classes lectures by Jim Stasheff.

Contacts between American and European mathematicians always play a very important role. I was very grateful for the opportunity to visit the IHES near Paris many times during this period. The annual Arbeitstagung organized by Fritz Hirzebruch in Bonn also provided a wonderful opportunity for keeping up with the latest developments.

Unfortunately, my marriage to Brigitte fell apart. In 1967, I left Princeton, spending a year at UCLA and two years at MIT. Then in 1970, I returned to Princeton, but at the Institute for Advanced Study, where I spent 20 happy years. During this time, work in algebraic topology inevitably led to related problems in pure algebra. Jean-Pierre Serre's "Cours d'Arithmétique" provided a marvelously readable introduction to quadratic forms, which form an indispensable tool in studying the homotopy theory of manifolds. Michael Atiyah taught me the importance of K-theory, and I learned the related subject of Algebraic K-theory through the work of Hyman Bass. The group K_0A of a ring A can be thought of as a simplified description of the class of finitely generated projective modules over A, while K_1A is closely related to J.H.C. Whitehead's theory of the "simple homotopy type" of a finite simplicial complex. I was happy to find a useful definition of the group K_2A of a ring A, and made a completely ad hoc definition of what are now called the "Milnor K-groups" $K_n^M F$ of a field F for higher values of n. These groups are of interest because of their close relation with the theory of quadratic forms and with Galois cohomology, as proved later in the work of Kazuya Kato and Vladimir Voevodsky.[5] (A few years later, Daniel Quillen constructed a more generally useful theory of higher K groups.)

[5]Compare: *On the Milnor Conjectures, history, influence, applications*, by Albrecht Pfister, Jahresber. Deutsch. Math.-Verein, **102** (2000) 15–41.

With Dusa in the California High Sierra in 1980, and in Arizona, many years later

Another development that I was particularly happy about in this period was the book "Singular Points of Complex Hypersurfaces," which made a useful contribution to elementary algebraic geometry, even though I was completely untrained in that field.

During the late seventies, my attention drifted towards a different area. Under the influence of Bill Thurston, and later of Adrien Douady, I became very much interested in the theory of dynamical systems. In particular, the abundance of fascinating problems which can be directly visualized seemed very attractive.

The theory of dynamics in one complex variable had received a powerful start in the early 20-th century through the work of Pierre Fatou and Gaston Julia. But there was a long hiatus until their work was brought to new life in the late 20-th century through the work of many mathematicians such as Adrien Douady, John Hubbard, Bill Thurston, and Dennis Sullivan. The advent of computers which could bring the abstract mathematics to life with vivid illustrations, played an important role. One particularly interesting feature is that dynamics over the complex numbers, where all of the powerful classical machinery is available, often provides an essential tool in understanding problems in real dynamics which seem a priori much simpler.

After a second failed marriage, I met and married Dusa McDuff. We have one child Thomas who is now a graduate student in Vancouver. (She also has one child Anna, by a previous marriage.) A high point of every year is the family get together, which takes place at Daniel's vacation house in Provincetown.

In 1989 I left the Institute for Advanced Study, and moved to Stony Brook University. In part this was because I missed the regular contact with students. However, the fact Dusa had been in Stony Brook for some years was the deciding factor.

At Stony Brook, I organized a small "Institute for Mathematical Sciences", with the help of Misha Lyubich, who has now taken over as its director. Over the years, many young (and not so young) mathematicians have spent time at our Institute, many of them working in the field of dynamics.

In front of the Abel monument, Oslo 2011

Although I am now formally working only half-time at Stony Brook, I still feel that happiness consists of thinking about mathematical problems.

Milnor's Work in Algebra and Its Ramifications

Hyman Bass

1 Introduction

While John Milnor was preeminently a topologist and geometer, he saw algebra as not only a resource but also, at times, an area of deep and engaged interest in its own right. He attributes some of his algebraic sensibilities to his exposure to Serge Lang, John Tate, and, especially, Emil Artin at Princeton. He wrote in [27] that, "attempts to solve topological problems led directly to serious questions in algebra, which of course had an addictive fascination of their own." Indeed, during the period 1965–1973, Milnor made transformative contributions to algebra, with substantial and lasting impacts on algebraic groups, algebraic K-theory, and quadratic forms. Though his own algebraic interventions were somewhat bounded in time, it is striking how, in several cases, his work helped set important research agendas for years to follow. Indeed, work subsequent to Milnor's in algebraic K-theory earned two Fields Medals (Quillen and Voevodsky). These ramifications of Milnor's work are what I shall try to highlight here. Other discussions of Milnor's work in algebra can be found in the references [7] and [27].

Part II is a brief discussion of Milnor's work with John Moore [19] on Hopf algebras, work that followed Milnor's important paper [18] on the Steenrod algebra and its dual. Part III discusses Milnor's introduction of ideas about growth of finitely generated groups, and relations of curvature properties of a compact manifold to the growth of its fundamental group. This precipitated the question: Which groups have polynomial growth? This question was finally resolved in a landmark paper of M. Gromov [10] that laid one of the foundations of geometric group theory. Part IV de-

Electronic supplementary material Supplementary material is available in the online version of this chapter at http://dx.doi.org/10.1007/978-3-642-39449-2_19. Videos can also be accessed at http://www.springerimages.com/videos/978-3-642-39449-2.

H. Bass (✉)
Department of Mathematics, University of Michigan Ann Arbor, 3864 East Hall,
525 E. University, Ann Arbor, MI 48109-1109, USA
e-mail: hybass@umich.edu

H. Holden, R. Piene (eds.), *The Abel Prize 2008-2012*,
DOI 10.1007/978-3-642-39449-2_19,

scribes the work of Milnor (with Bass and Serre) on the Congruence Subgroup Property (CSP) for SL_n $(n \geq 3)$ and Sp_{2n} $(n \geq 2)$ [3]. This revealed deep connections with number theory, and established initial methods for investigating the CSP for other algebraic groups over global fields, a large program of research that continues to this day. See Prasad and Rapinchuk [31] for survey of the current state of the field. The final Part V presents some highlights of Milnor's work on algebraic K-theory and quadratic forms. While the congruence subgroup theorem had some direct connections with the functor K_1, it also revealed links with work of C. Moore [28], H. Matsumoto [14] and R. Steinberg [39] about central extensions that Milnor used to develop his definition of the functor K_2. This is all exposed in Milnor's excellent introduction to algebraic K-theory [26]. For a field $\mathbf{F}$, Milnor obtains a presentation of $K_2(\mathbf{F})$ generated by "Steinberg symbols," $\{x, y\}$ with $x, y \in \mathbf{F}^{\bullet} = K_1(\mathbf{F})$. These are bi-multiplicative in (x, y) and are trivial when $x + y = 1$. In his milestone 1970 paper, Algebraic K-theory and quadratic forms [24], Milnor uses this presentation to make $K_n(\mathbf{F})$ $(n = 0, 1, 2)$ the low degree terms of a graded ring, now denoted $K_*^M(\mathbf{F})$ (to distinguish it from the K-groups soon after introduced by Quillen [32]), and called the *Milnor ring* of $\mathbf{F}$. Assume now that $\mathrm{char}(\mathbf{F}) \neq 2$, and put $k_*^M(\mathbf{F}) = K_*^M(\mathbf{F})/2K_*^M(\mathbf{F})$. Then Milnor constructs two homomorphisms:

$$s_n^{\mathbf{F}}: k_n^M(\mathbf{F}) \to \mathrm{GW}_n(\mathbf{F}) \quad \text{and} \quad h_n^{\mathbf{F}}: k_n^M(\mathbf{F}) \to H^n(\mathbf{F}, \mu_2)$$

where $\mathrm{GW}_n(\mathbf{F})$ is the degree n component of the associated graded ring of the Witt–Grothendieck ring of quadratic forms over $\mathbf{F}$, and $H_n(\mathbf{F}, \mu_2)$ is the Galois cohomology with coefficient in $\mu_2 = \{\pm 1\}$. What came to be known as the *Milnor Conjectures* were his questions about whether $s_n^{\mathbf{F}}$ and $h_n^{\mathbf{F}}$ are isomorphisms. That $s_n^{\mathbf{F}}$ is always an isomorphism was proved in 2007 by Orlov, Vishik, and Voevodsky [29]. (And Kato [11] proved a version of this when $\mathrm{char}(\mathbf{F}) = 2$.) The proof that $h_n^{\mathbf{F}}$ is always an isomorphism was proved in a monumental and original work for which Voevodsky received the Fields Medal. Interestingly, Voevodsky's methods, using motivic cohomology, return to algebraic topology ideas, introducing a motivic analogue of singular cohomology, and make strategic use of an analogue of the Steenrod algebra, wherein he uses ideas from Milnor's 1958 paper [18] cited at the beginning.

2 Hopf Algebras

The theory of Hopf algebras emerged from Hopf's work on the homology of Lie groups, work later elaborated by John Moore. At about the same time, pursuant to a suggestion of Steenrod, Milnor [18] showed that the Steenrod algebra had a natural a Hopf algebra structure. This confluence of interest led Milnor and Moore [19] to collaborate on a project to study the general structure of graded Hopf algebras. The topological relevance comes from the fact that, if G is a pathwise connected homotopy associative H-space with unit, then the Hurewicz homomorphism $\lambda: \pi_*(G, K) \to H_*(G, K)$ of graded Lie algebras induces an isomorphism of Hopf algebras $U(\pi_*(G, K)) \to H_*(G, K)$.

The main results of [19] describe the structure of a graded connected primitively generated Hopf algebra A over a field K of characteristic zero. The primitive elements $P(A)$ form a graded Lie algebra, and for any graded Lie algebra L, the universal enveloping algebra $U(L)$ has the structure of a graded connected primitively generated Hopf algebra. A fundamental structure theorem asserts that the composite functors, $P \circ U$ and $U \circ P$, are isomorphic to the respective identity functors. In particular, A is canonically isomorphic, as Hopf algebra, to $U(P(A))$. In the course of proving this, Milnor and Moore establish a sharpened version of the Poincaré–Birkhoff–Witt Theorem: The associated graded algebra of $U(L)$ is Hopf-algebra isomorphic to the symmetric algebra of the underlying graded vector space of L.

When A has commutative multiplication (and K is only required to be perfect), Milnor and Moore retrieve and strengthen classical theorems of Borel, Leray, and Samelson. For example, when A is further of finite type as a graded vector space, then A is a tensor product of single generator Hopf algebras.

3 Growth of Groups

In [20] Milnor introduced the growth function $g(r)$ of a group G with respect to a finite set S of generators of G; $g(r)$ denotes the number of elements of G of length less than or equal to r as a group word in S. The group G is said to have polynomial growth of degree less than or equal to d if $g(r) \le ar^d$ for some $a > 0$, and exponential growth if $g(r) \ge a^r$ for some $a > 1$. Such properties depend only on G, not S, and are shared by subgroups of finite index.

Let M be a complete Riemannian n-manifold, with $G = \pi_1(M)$ acting (freely) on its universal cover $\tilde{M}$. Choose $x \in \tilde{M}$, and let $B(r)$ denote the ball of radius r in $\tilde{M}$ centered at x. Then it is shown in [21] that, for the fundamental group G, the function $g(r)$ above has the same asymptotic growth properties as the functions $\#(G \cdot x \cap B(r))$ and $\mathrm{Vol}(B(r))$. The latter are related to curvature properties of M, and thus Milnor proved results relating curvature to the growth of $\pi_1(M)$.

Results of this genre were further pursued by Wolf [45]. Suppose that G has a nilpotent subgroup H of finite index, with descending central series $H = H_1 \ge H_2 \ge H_3 \ge \cdots$ Let $r_i = \mathrm{rank}(H_i/H_{i+1})$ and put

$$d(G) = \sum_{i \ge 1} i r_i \le d'(G) = \sum_{i \ge 1} 2^{i-1} r_i.$$

(These depend only on G, not H.) Wolf in [45] shows then that G has polynomial growth of degree between $d(G)$ and $d'(G)$. Bass later showed, in [5], that G has polynomial growth of degree exactly $d(G)$. Wolf further shows that if G is virtually polycyclic, then either G is virtually nilpotent, and hence has polynomial growth as above, or else it has exponential growth. Milnor's note [22] shows in fact that any finitely generated solvable group is either virtually polycyclic or else of exponential growth.

These results together imply that: If G is virtually solvable, then either G is virtually nilpotent, and hence of polynomial growth of degree $d(G)$, or else G is of exponential growth. Two questions naturally emerged from this: Question I: Must the growth of any finitely generated group be either polynomial or exponential? Conjecture II: Every finitely generated group of polynomial growth is virtually nilpotent. Question I was actually posed by Milnor as an Advanced Problem in the MAA Monthly [23]. It took 15 years before Grigorchuk [9] found a counterexample, a group of "intermediate growth."

As for Conjecture II, the results described above show that it holds for virtually solvable groups. It then follows that it is true for linear groups, using the "Tits alternative" [41]: A finitely generated linear group G is either virtually solvable or else G contains a non-Abelian free group (and is hence of exponential growth). Using this, Conjecture II was finally proved in a landmark paper of Gromov [10] that has since made the ideas of metric geometry a fundamental tool of combinatorial group theory. Gromov's idea was, roughly speaking, to map the group G of polynomial growth into a Lie group L, and hence in a linear group, and so that the kernel has smaller growth degree, thus setting up an induction. Relative to a finite set of generators of G, word length defines a kind of norm on G, and hence also a metric. Gromov obtains the sought after Lie group L as a "limit" of a suitable sequence of micro-scalings of the metric space G, in the same way that we can picture $\mathbf{R}^n$ as the limit of the sequence $(1/N!) \cdot \mathbf{Z}^n$ (as $N \to \infty$).

4 The Congruence Subgroup Problem

Let $Z = Z_S$ be the ring of S-integers in a global field k, where S is a finite set of places of k containing all Archimedean places. For readers not so familiar with this context, the main gist of what follows can be understood by taking $Z = \mathbf{Z}$, the integers, $k = \mathbf{Q}$, the rational numbers, and $S = \{\infty\}$. Let $G \subseteq \mathrm{GL}_n$ be a linear algebraic group defined over k, and let $\Gamma = G(Z) = G(k) \cap \mathrm{GL}_n(Z)$. For any ideal $J \neq 0$ of Z we have the principal congruence subgroup of level J,

$$\Gamma_J = \mathrm{Ker}\big(\Gamma = G(Z) \to G(Z/J)\big).$$

Since the ring Z/J is finite, Γ_J has finite index in Γ. The Congruence Subgroup Property (CSP) for G and Z (or for G, k, and S) asserts, conversely, that every finite index subgroup of Γ is a congruence subgroup, i.e. contains Γ_J for some $J \neq 0$. The CSP does not depend on the linear embedding $G \subseteq \mathrm{GL}_n$.

Some Examples

1. When G is a unipotent group and k is a number field, the CSP is easily shown to hold.
2. For $\Gamma = \mathrm{SL}_2(\mathbf{Z})$ it was already known by Fricke and Klein in the 19th century that the CSP fails; in fact Γ is virtually a free group, and so has an abundance

of non-congruence subgroups of finite index. Positive results for $\mathrm{SL}_n(\mathbf{Z})$ with $n \geq 3$, were obtained independently by Bass–Lazard–Serre [2] and Mennicke [15].
3. For $G = \mathrm{GL}_1$, and, more generally, for G an algebraic torus, the CSP was proved by Chevalley [8].
4. When G is an Abelian variety (no longer a linear algebraic group), $G(k)$ is a finitely generated Abelian group (Mordell–Weil), and Serre [36] defines an S-congruence subgroup of $G(k)$ to be the intersection of $G(k)$ with an open subgroup of $\prod_{v \in S} G(k_v)$ (in which $G(k)$ is diagonally embedded). With this interpretation, he proves the CSP in [36] and [38].

The paper [3] of Bass–Milnor–Serre, offers an exact determination of whether, and even to what extent, the CSP holds for $\mathrm{SL}_n(Z)$ ($n \geq 3$) and $\mathrm{Sp}_{2n}(Z)$ ($n \geq 2$). This section briefly recounts the content of [3] as well as the ensuing research that it precipitated. In this I draw largely on the excellent survey [31] by Prasad and Rapinchuk. The case $G = \mathrm{SL}_n$ ($n \geq 3$) is treated in [3] as follows. (The case of Sp_{2n} ($n \geq 2$) is similar.) Let E be the subgroup of Γ generated by elementary matrices, and, for each ideal $J \neq 0$ of Z, let E_J be the normal subgroup of E generated by elementary matrices in Γ_J. Then the following results are proved in [3]:

1. E_J is a normal subgroup of finite index in Γ, and every finite index subgroup of Γ contains E_J for some $J \neq 0$. Moreover, $C_J = \Gamma_J / E_J$ is central in Γ / E_J.
2. Each element c of C_J is represented by an element s of $\Gamma_J \cap \mathrm{SL}_2(Z)$. If (a, b) is the first row of s, then c depends only on (a, b) ($\equiv (1, 0) \bmod J$), so we can write $c = \left[\begin{smallmatrix} b \\ a \end{smallmatrix}\right]$. (This notation was first introduced and exploited by Milnor, and became a primary tool in later computations of K_1 groups in algebraic K-theory.)
3. The symbols $\left[\begin{smallmatrix} b \\ a \end{smallmatrix}\right]$ satisfy the following properties:
   a. They are bi-multiplicative as a function of $(a, b) \in W_J = \{(a, b) \equiv (1, 0) \bmod J \text{ in } Z^2\}$.
   b. $\left[\begin{smallmatrix} b \\ a \end{smallmatrix}\right]$ is unchanged if we add tb to a for any $t \in Z$, or if we add ta to b for any $t \in J$.

Maps from W_J to an Abelian group satisfying these properties are called *Mennicke symbols* (of level J) in [3], and it is shown in [3] that the above $\left[\begin{smallmatrix} b \\ a \end{smallmatrix}\right]$ is a *universal* Mennicke symbol. In particular, it is independent of $n \geq 3$. (For these results, Z could be any Dedekind domain.)

When Z is the ring of S-integers of a global field k, as above, Chap. I of [3] determines all Mennicke symbols of level J for each $J \neq 0$ in Z. If S is not totally imaginary, they are all trivial, and hence $C_J = \{1\}$ for all J. If S is totally imaginary, then there is an $r = r(J) \geq 1$ such that k contains μ_r, the group of rth roots of unity, and the map

$$W_J \to \mu_r, \qquad (a, b) \mapsto \left(\frac{b}{a}\right),$$

given by the rth power Legendre symbol, is a universal Mennicke symbol, and hence $C_J \cong \mu_r$. If μ_m is the group of all roots of unity in k, and if J is divisible by $m \cdot \prod_{p|m} p^{1/(p-1)}$ then $r(J) = m$, and so $\varprojlim C_J \cong \mu_m$.

Chapter IV of [3] provides a measure of the failure of the CSP, by an Abelian group $C^S = C^S(G(k))$, the so called *congruence kernel*, defined below. In the case $G = \mathrm{SL}_n$ discussed above,

$$C^S = \varprojlim C_J = \varprojlim \Gamma_J/E_J.$$

It is this formulation that frames subsequent research on the CSP.

The *S-congruence topology* on $G(k)$ is that for which the congruence subgroups of Γ form a base for neighborhoods of 1. It is the topology induced by the diagonal embedding of $G(k)$ in $G(A_S)$, where A_S is the ring of S-adèles of k. Thus, the S-congruence completion, $\overline{G(k)}$, is just the closure of $G(k)$ in $G(A_S)$. Kneser's Strong Approximation Theorem [12], valid under rather general conditions on G, asserts that $\overline{G(k)} = G(A_S)$. The closure $\overline{\Gamma}$ of Γ in $\overline{G(k)}$, the congruence completion of Γ, is a profinite open subgroup of $\overline{G(k)}$.

The *S-arithmetic topology* on $G(k)$ is that for which the S-arithmetic groups, i.e. subgroups of $G(k)$ commensurable with $G(Z)$, form a base for neighborhoods of 1. Let $\widehat{G(k)}$ denote the S-arithmetic completion of $G(k)$. The closure $\hat{\Gamma}$ of Γ in $G(k)$ is just the profinite completion of Γ, and it is a compact open subgroup of $\widehat{G(k)}$.

Since the S-arithmetic topology refines the S-congruence topology, we have an exact sequence $E(G, S)$

$$1 \to C^S \to \widehat{G(k)} \to \overline{G(k)} \to 1$$

that restricts to an exact sequence $E(\Gamma, S)$

$$1 \to C^S \to \hat{\Gamma} \to \overline{\Gamma} \to 1.$$

The CSP is just the assertion that the S-congruence and S-arithmetic topologies coincide, and so:

The CSP is equivalent to the condition: $C^S = \{1\}$.

With this formulation, we can now state the main result of [3]. We shall say that *S is totally imaginary* if every place in S is complex, i.e. $k_v = \mathbf{C}$ for all v in S. This means that Z_S is just the ring of algebraic integers in a totally imaginary number field.

Congruence Subgroup Theorem (14.1 in [3]) *For* $G = \mathrm{SL}_n$ $(n \geq 3)$ *or* Sp_{2n} $(n \geq 2)$, $C^S \cong \mu(k)$, *the roots of unity in* k, *if* S *is totally imaginary and* $C^S = \{1\}$ *otherwise.*

In particular, though the CSP fails in the totally imaginary case, its failure is precisely measured by the finite group $\mu(k)$. In fact, the isomorphism with $\mu(k)$ is intimately connected with power residue symbols in class field theory, as described above.

The results of [3] were shown to bear a close relation with the work of C. Moore [28] on "metaplectic covering groups." In fact, when S is totally imaginary, $E(G, S)$

is exactly the metaplectic covering group defined by Moore. It was also shown in [3] that the finiteness of C^S implies certain rigidity properties of Γ, namely that any linear representation of Γ is virtually the restriction of an algebraic representation of G. The precise calculations in [3] also yielded some calculations of certain algebraic K-groups, $SK_1(\mathbf{Z}\pi)$, with π a finite Abelian group.

The Congruence Subgroup paper [3] naturally drew interest in the CSP for other algebraic groups G, and a substantial body of work (e.g. [33]) has been done in this direction, for which the survey [31] of Prasad and Rapinchuk is an excellent reference. First, Matsumoto [14] extended the above results to all Chevalley groups of rank ≥ 2. The subsequent results are due largely to Serre, Raghunathan, Prasad, and Rapinchuk. As Serre pointed out early, the CSP fails badly (C^S is infinite if G is not simply connected).

The discussion in [31] explains how the investigation of the CSP quickly focuses on the case when G is absolutely simple, and simply connected (SSC), an assumption we now make. Define $\text{rk}_S(G) = \sum_{v \in S} \text{rk}_v(G)$, where $\text{rk}_v(G)$ is the rank of the algebraic group G over the completion k_v. Let V_∞ denote the set of arcimedian places of k. assumed to be contained in S. Then [31] formulates the following *Congruence Subgroup Conjecture*, slightly corrected from the original version posed by Serre:

- *Higher rank:* If $\text{rk}_S(G) \geq 2$, then C^S is finite if and only if $\text{rk}_v(G) > 0$ for all $v \in S \setminus V_\infty$.
- *Rank 1:* If $\text{rk}_S(G) = 1$, then C^S is infinite.

Serre formulated (essentially) this conjecture after treating the rank 1 case when $G = \text{SL}_2$ [37]. In the positive direction, Theorem 2 of [31] asserts: Assuming that $\text{rk}_S(G) > 0$, and $\text{rk}_v(G) > 0$ for all $v \in S \setminus V_\infty$, then C^S is finite if and only if it is central in $\widehat{G(k)}$.

In particular, the higher rank congruence subgroup conjecture is equivalent to the centrality of C^S. The centrality of C^S has been proved in many cases, for example when G is k-isotropic (Raghunathan [34], and [31]) but many anisotropic cases remain open, and seem to require new methods.

Some of the results in [31] require the following Margulis–Platonov Conjecture (MP), known to be true in many cases: Let T be the (finite) set of places $v \notin V_\infty$ such that $\text{rk}_v(G) = 0$. If N is a non-central normal subgroup of $G(k)$, then there is an open normal subgroup W of $G_T = \prod_{v \in T} G(k_v)$ such that $N = W \cap G(k)$.

This is used in linking the CSP to the notion of *bounded generation*. A discrete group Γ is said to be boundedly generated if there exists $s_1, s_2, \ldots, s_t$ in Γ such that $\Gamma = \langle s_1 \rangle \langle s_2 \rangle \cdots \langle s_t \rangle$ (a finite product of cyclic groups). A similar notion is defined for profinite groups, with $\langle s_i \rangle$ replaced by the closure of the cyclic group generated by s_i. In this connection we have the following result.

Theorem (Platonov–Rapinchuk [30], Lubotzky [13]) *Assuming (SSC) and (MP), C^S is central if and only if $\hat{\Gamma}$ is boundedly generated.*

5 Algebraic K-Theory and Quadratic Forms

This exposition draws much from the excellent survey of Merkurjev [17].

Milnor's $K_2(R)$ An early precursor of algebraic K-theory (specifically K_1) was J.H.C. Whitehead's work on the torsion $\tau(f)$ of a homotopy equivalence $f: X \to Y$ of finite complexes, an invariant that detects whether f is a *simple homotopy equivalence*. It lives in the "Whitehead group," that we can now describe as $\mathrm{Wh}(\pi) = K_1(\mathbf{Z}\pi)/(\pm\pi)$, where π is the (common) fundamental group of X and Y. Milnor was interested in Whitehead torsion (see [20]), as well as in other algebraic invariants, such as C.T.C. Wall's obstruction to finiteness of a CW complex X dominated by a finite complex, an obstruction that lives in a quotient of $K_0(\mathbf{Z}\pi)$, with $\pi = \pi_1(X)$. While the general philosophy of algebraic topology was to reduce hard topology to easier (usually homological) algebra, these algebraic K-groups proved to be difficult to calculate. The computational tools offered by algebraic K-theory no doubt provided some of the early motivation for Milnor's interest in the subject. In fact, the study of K_1 for rings of arithmetic type, for example group rings of finite groups, led quite naturally to the congruence subgroup problem, discussed above.

Both algebraic and topological K-theory were inspired by Grothendieck's introduction of the "Grothendieck group," now denoted K_0 in algebra (or K^0 in topology) in the proof of his general version of the Riemann–Roch Theorem in algebraic geometry. Topological K-groups were defined by applying K^0 to suspensions. But, lacking a good notion of suspension, the algebraic theory succeeded at first only in defining serviceable notions of K_0 and K_1. Nonetheless, enough interesting mathematical connections and applications could be made, even with only these two functors, to stimulate interest in a possible higher algebraic K-theory [4]. When Milnor entered the subject he made the first major expansion of the horizon, in his Princeton Studies monograph, Introduction to Algebraic K-theory [26]. In addition to providing a beautifully exposed and accessible introduction to the subject, Milnor there introduces his functor K_2, he shows that it fits naturally with K_0 and K_1, and he provides some first substantial calculations of it.

Milnor's definition of K_2 was inspired by work of Robert Steinberg [39] on generators and relations for Chevalley groups, and related universal covering groups. These ideas surfaced in the work of C. Moore [28], in work related to the Congruence Subgroup Theorem, and in Matsumoto's proof of the CSP for Chevalley groups of rank ≥ 2 [11]. For a ring R, Milnor used the Steinberg relations on the elementary matrices, that generate the subgroup $E_n(R)$ of $\mathrm{GL}_n(R)$, to define the covering group $\mathrm{St}_n(R) \to E_n(R)$. Taking the inductive limit as $n \to \infty$, this gives a universal central extension

$$1 \to K_2(R) \to \mathrm{St}(R) \to E(R) \to 1$$

in which $K_2(R) \cong H_2(E(R), \mathbf{Z})$. From this definition he showed directly that $K_2(\mathbf{Z})$ has order 2. When R is commutative and $u, v \in R^\bullet$, the unit group of R, then there is defined an element $\{u, v\}$ in $K_2(R)$, called a *Steinberg symbol*. It is bi-multiplicative in (u, v) and is trivial if $v = 1-u$. When R is a field, the Steinberg symbols generate $K_2(R)$, and Milnor presents Matsumoto's work to show that the Steinberg symbols and the above relations furnish a presentation of $K_2(R)$. He further presents Tate's computation of $K_2(\mathbf{Q})$, based on one of Gauss's proofs of quadratic reciprocity.

The Milnor Ring $K_*^M(\mathbf{F})$ Milnor's monograph [26] appeared in 1971. At the Battelle Conference on Algebraic K-theory, in the summer of 1972, Quillen appeared with his definitive definitions of the higher algebraic K-functors [32], compatible with the previously defined K_i for $i \le 2$.

The next major phase of Milnor's work related to K-theory was the seminal 1970 paper [24], *Algebraic K-theory and Quadratic Forms*. Let $\mathbf{F}$ be a field, and let $u \mapsto \{u\}$ denote the isomorphism $\mathbf{F}^\bullet \to K_1(\mathbf{F})$ (written additively). Let $T(K_1(\mathbf{F}))$ denote the tensor algebra over $\mathbf{Z}$ of $K_1(\mathbf{F})$. Then Milnor defines the graded ring of K-groups,

$$K_*^M(\mathbf{F}) = T\big(K_1(\mathbf{F})\big)/J = \bigoplus_{n\ge 0} K_n^M(\mathbf{F})$$

where J is the ideal generated by the degree two elements $\{u\} \otimes \{1-u\}$ for all $u \ne 0, 1$ in $\mathbf{F}$. It follows from this that $K_0^M(\mathbf{F}) = \mathbf{Z} = K_0(\mathbf{F})$, $K_1^M(\mathbf{F}) \cong \mathbf{F}^\bullet \cong K_1(\mathbf{F})$, and

$$\begin{aligned} K_2^M(\mathbf{F}) &= \frac{K_1(\mathbf{F}) \otimes K_1(\mathbf{F})}{(\text{the subgroup generated by all } \{u\} \otimes \{1-u\},\ u \ne 0, 1)} \\ &= K_2(\mathbf{F}), \end{aligned}$$

the latter being essentially the presentation of $K_2(\mathbf{F})$ described above in terms of Steinberg symbols $\{u, v\}$, which we can now interpret as the product $\{u\}\{v\}$ in $K_*^M(\mathbf{F})$. More generally, for $u_1, u_2, \ldots, u_n$ in $\mathbf{F}^\bullet$, we write

$$\{u_1, u_2, \ldots, u_n\} = \text{the product } \{u_1\} \cdot \{u_2\} \cdot \cdots \cdot \{u_n\} \text{ in } K_*^M(\mathbf{F}).$$

This Milnor K-ring, $K_*^M(\mathbf{F})$, is skew commutative, in the sense that $xy = (-1)^{nm} yx$ for x in $K_n^M(\mathbf{F})$ and y in $K_m^M(\mathbf{F})$.

Let v be a discrete valuation of $\mathbf{F}$ with valuation ring R and residue field $\mathbf{F}(v)$. Denote the homomorphism $R \to \mathbf{F}(v)$ by $a \mapsto \bar{a}$. The homomorphism

$$v\colon \mathbf{F}^\bullet = K_1(\mathbf{F}) \to \mathbf{Z} = K_0\big(\mathbf{F}(v)\big)$$

extends to a degree -1 homomorphism,

$$\begin{aligned} &\partial_v\colon K_*^M(\mathbf{F}) \to K_*^M\big(\mathbf{F}(v)\big), \quad \text{such that} \\ &\partial_v\big(\{a_1, a_2, \ldots, a_n\}\big) = v(a_1)\{\bar{a}_2, \ldots, \bar{a}_n\} \quad \text{for } a_1 \in \mathbf{F}^\bullet \text{ and } a_2, \ldots, a_n \in R^\bullet. \end{aligned}$$

Milnor uses these maps to calculate the Milnor K-ring of a rational function field $\mathbf{F}(t)$. Each monic irreducible polynomial $p = p(t)$ defines a discrete valuation on

$\mathbf{F}(t)$ with residue field $\mathbf{F}(p) = \mathbf{F}[t]/p\mathbf{F}[t]$, so, from above, we have homomorphisms, for $n \geq 1$, $\partial_p : K_n^M(\mathbf{F}(t)) \to K_{n-1}^M(\mathbf{F}(p))$. Milnor shows [24, Theorem 2.3], using methods that Tate used for $n = 2$, that these assemble into a split exact sequence,

$$0 \to K_n^M(\mathbf{F}) \to K_n^M(\mathbf{F}(t)) \to \bigoplus_p K_{n-1}^M(\mathbf{F}(p)) \to 0$$

where the direct sum is over all monic irreducible polynomials p.

The Homomorphism $s_n^{\mathbf{F}} : k_n^M(\mathbf{F}) \to \mathbf{GW}_n(\mathbf{F})$ Milnor next relates the ring

$$k_*^M(\mathbf{F}) = K_*^M(\mathbf{F})/2K_*^M(\mathbf{F}) = \bigoplus_{n \geq 0} k_n^M(\mathbf{F})$$

to the Witt ring $W(\mathbf{F})$ of quadratic forms over $\mathbf{F}$. For this he assumes that $\mathrm{char}(\mathbf{F}) \neq 2$, but as Merkurjev [17] points out, this is unnecessary if one uses instead the Witt ring of symmetric bilinear forms. (In a separate paper [25], for fields $\mathbf{F}$ of characteristic 2, Milnor obtains an elegant description of the structure of the Witt ring of quadratic forms over $\mathbf{F}$.)

Consider the category of pairs (V, b) where V is a finite dimensional $\mathbf{F}$-module on which b is a non-degenerate symmetric bilinear form. The Witt–Grothendieck ring $\hat{W}(\mathbf{F})$ is the Grothendieck ring of this category, with respect to direct sum and tensor product. (V, b) is said to be *metabolic* if there is a self-orthogonal subspace W of V. The Witt ring $W(\mathbf{F})$ is the quotient of $\hat{W}(\mathbf{F})$ by the ideal generated by metabolic forms. We write $[b]$ for the class of (V, b) in $W(\mathbf{F})$. If b and b' are anisotropic and $[b] = [b']$ then (V, b) and (V', b') are isomorphic. The map, $(V, b) \mapsto (\dim(V) \bmod 2)$, defines a ring homomorphism $W(\mathbf{F}) \to \mathbf{Z}/2\mathbf{Z}$, whose kernel $I(\mathbf{F})$ is called the *fundamental ideal* of $W(\mathbf{F})$. Its powers define the associated graded ring of $W(\mathbf{F})$:

$$\mathrm{GW}_*(\mathbf{F}) = \bigoplus_{n \geq 0} I^n(\mathbf{F})/I^{n+1}(\mathbf{F}) = \bigoplus_{n \geq 0} \mathrm{GW}_n(\mathbf{F}).$$

Milnor asked [24, (4.4)] if the intersection of the $I^n(\mathbf{F})$ is zero; this was later proved by Aronson and Pfister [1].

For $a, a_1, a_2, \ldots, a_n$ in $\mathbf{F}$ write $\langle a_1, a_2, \ldots, a_n \rangle$ for the bilinear form on $\mathbf{F}^n$ with diagonal matrix $\mathrm{diag}(a_1, a_2, \ldots, a_n)$, $\langle\langle a \rangle\rangle = \langle 1, -a \rangle$, and

$$\langle\langle a_1, a_2, \ldots, a_n \rangle\rangle = \langle\langle a_1 \rangle\rangle \otimes \langle\langle a_2 \rangle\rangle \otimes \cdots \otimes \langle\langle a_n \rangle\rangle.$$

The latter is called an *n-fold Pfister form*. Their classes additively generate $I^n(\mathbf{F})$.

The map $\mathbf{F}^\bullet \to I(\mathbf{F})$ sending a to $[\langle\langle a \rangle\rangle] \bmod I^2(\mathbf{F})$ defines a homomorphism

$$s_1^{\mathbf{F}} : k_1^M(\mathbf{F}) = \mathbf{F}^\bullet/\mathbf{F}^{\bullet 2} \to \mathrm{GW}_1(\mathbf{F}).$$

If $a \neq 0, 1$ then $\langle\langle a, 1-a\rangle\rangle$ can be shown to be metabolic, so $[\langle\langle a, 1-a\rangle\rangle] = 0$, and so $s_1^{\mathbf{F}}$ extends to a graded ring homomorphism

$$s_*^{\mathbf{F}}\colon k_*^M(\mathbf{F}) \to \mathrm{GW}_*(\mathbf{F}).$$

What had come to be known as a Milnor Conjecture ($\mathrm{MQ}^{\mathbf{F}}(n)$):

$$s_n^{\mathbf{F}} \text{ is an isomorphism}$$

was in fact only a question posed by Milnor [24, (4.3)]. Milnor [24, (4.1)] proves $\mathrm{MQ}^{\mathbf{F}}(n)$ for $n \leq 2$ when $\mathrm{char}(\mathbf{F}) \neq 2$. Moreover, Milnor obtains a calculation of $W(\mathbf{F}(t))$ analogous to that above of $K_*^M(\mathbf{F}(t))$ [24, (5.3)]. When $\mathrm{char}(\mathbf{F}) = 2$, $\mathrm{MQ}^{\mathbf{F}}(n)$ was proved by Kato [11], in 1982. For $\mathrm{char}(\mathbf{F}) \neq 2$, $\mathrm{MQ}^{\mathbf{F}}(n)$ was finally proved in general in 2005 by Orlov, Vishik, and Voevodsky [29, Sect. 4].

The Homomorphism $h_n^{\mathbf{F}}\colon k_n^M(\mathbf{F}) \to H^n(\mathbf{F})$ Let $\mathbf{F}$ be a field of characteristic $\neq 2$, with separable closure $\mathbf{F}_s$, and Galois group $G = G_F = \mathrm{Gal}(\mathbf{F}_s/\mathbf{F})$. Let μ_n denote the group of nth roots of unity. The map $q\colon x \mapsto x^2$ on $\mathbf{F}_s^\bullet$ yields an exact sequence of G-modules,

$$1 \to \mu_2 \to \mathbf{F}_s^\bullet \xrightarrow{q} \mathbf{F}_s^\bullet \to 1$$

whence an exact cohomology sequence,

$$H^0(G, \mathbf{F}_s^\bullet) \xrightarrow{2} H^0(G, \mathbf{F}_s^\bullet) \to H^1(G, \mu_2) \to H^1(G, \mathbf{F}_s^\bullet).$$

By Hilbert's Theorem 90, $H^1(G, \mathbf{F}_s^\bullet) = 0$, and so the latter becomes an exact sequence

$$\mathbf{F}^\bullet \xrightarrow{q} \mathbf{F}^\bullet \to H^1(G, \mu_2) \to 0.$$

Whence an isomorphism ($\mathrm{MH}^{\mathbf{F}}(1)$):[1]

$$h_1^{\mathbf{F}}\colon k_1^M(\mathbf{F}) = \mathbf{F}^\bullet/\mathbf{F}^{\bullet 2} \to H^1(G, \mu_2).$$

It was shown in Bass–Tate [6] that, for $u \neq 0, 1$ in $\mathbf{F}^\bullet$, $h_1^{\mathbf{F}}(u) \cap h_1^{\mathbf{F}}(1-u) = 0$ in $H^2(G, \mu_2)$. Thus $h_1^{\mathbf{F}}$ extends to a graded ring homomorphism

$$h_*^{\mathbf{F}}\colon k_*^M(\mathbf{F}) \to H^*(G, \mu_2).$$

Milnor states [24, p. 340], "I don't know of any examples for which the homomorphism $h_*^{\mathbf{F}}$ fails to be bijective." This led to what came to be known as a conjecture of Milnor ($\mathrm{MH}^{\mathbf{F}}(n)$):

$$h_n^{\mathbf{F}}\colon k_n^M(\mathbf{F}) \to H^n(G, \mu_2) \text{ is an isomorphism.}$$

---

[1] When $\mathrm{char}(\mathbf{F}) = 2$ there are natural analogues of $h_n^{\mathbf{F}}$ and $\mathrm{MH}^{\mathbf{F}}(n)$, but with $H^n(G, \mu_2)$ replaced by groups $H^n(\mathbf{F})$ defined in terms of differentials. (See [17, p. 5].)

Milnor proved this when the field $\mathbf{F}$ is finite, local, global, or real closed. He also showed that $\mathrm{MH}^{\mathbf{F}}$ implies $\mathrm{MH}^{\mathbf{F}((t))}$.

$\mathrm{MH}^{\mathbf{F}}(2)$ was proved by Merkurjev [16], using work of Suslin on a k_*^M-generalization of Hilbert's Theorem 90, as well as computations of Quillen of the K-theory of quadrics. These methods were extended by Rost [35] to prove $\mathrm{MH}^{\mathbf{F}}(3)$. However these methods relied on a close connection of the lower K^M-groups with Quillen K-theory, a connection no longer available for larger n.

The proof of $\mathrm{MH}^{\mathbf{F}}(n)$ in general was achieved in groundbreaking work of Voevodsky [43]. First, using joint work with Suslin [40], he was able to reinterpret the Milnor groups $K_*^M(\mathbf{F})$ as motivic cohomology groups. In this context he showed that $\mathrm{MH}^{\mathbf{F}}(n)$ follows from a motivic generalization of Hilbert's Theorem 90. At this point Voevodsky introduced motivic methods analogous to those in algebraic topology [42]. For each smooth scheme X over $\mathbf{F}$, he introduced a Cech simplicial scheme, from which he defined motivic cohomology groups of X, analogous to singular cohomology for CW complexes. On these Voevodsky defined a motivic Steenrod algebra [44], and established results analogous to those from Milnor's famous paper [18] on the Steenrod algebra and its dual. This impressive arsenal of tools was finally mobilized to prove the motivic version of Hilbert's Theorem 90, and hence $\mathrm{MH}^M(n)$ in general.

References

1. Arason, J., Pfister, A.: Beweis des Krullschen Durchschnittsatzes für den Wittring. Invent. Math. **12**, 173–176 (1971)
2. Bass, H., Lazard, M., Serre, J.-P.: Sous-groupes d'indices finis dans $\mathrm{SL}(n, Z)$. Bull. Am. Math. Soc. **70**, 385–392 (1964)
3. Bass, H., Milnor, J., Serre, J.-P.: Solution of the congruence subgroup problem for SL_n $(n \geq 3)$ and Sp_{2n} $(n \geq 2)$. Publ. Math. IHES **33**, 59–137 (1967) (Erratum: On a functorial property of power residue symbols. Publ. Math. IHES **44** (1974), 241–244)
4. Bass, H.: Algebraic K-Theory. Benjamin, New York (1968)
5. Bass, H.: On the degree of growth of a finitely generated nilpotent group. Proc. Lond. Math. Soc. **25**, 603–614 (1972)
6. Bass, H., Tate, J.: The Milnor ring of a global field. Springer LNM, vol. 342, pp. 349–446 (1973)
7. Bass, H.: John Milnor the algebraist. In: Topological Methods in Modern Mathematics. A Symposium in Honor of John Milnor's Sixtieth Birthday, pp. 45–83 (1993). Publish or Perish
8. Chevalley, C.: Deux théoèmes d'arithmétique. J. Math. Soc. Jpn. **3**, 36–44 (1951)
9. Grigorchuk, R.I.: On the Milnor problem of group growth. Dokl. Akad. Nauk SSSR **271**, 30–33 (1983)
10. Gromov, M.: Groups of polynomial growth and expanding maps. Publ. Math. IHES **53**, 53–73 (1981)
11. Kato, K.: Symmetric bilinear forms, quadratic forms and Milnor K-theory in characteristic two. Invent. Math. **66**(3), 493–510 (1982)
12. Kneser, M.: Strong approximation, I, II. Algebraic groups and discontinuous subgroups. In: Proc. Symp. Pure Math. Amer Math. Soc., vol. IX, pp. 187–196 (1966)
13. Lubotzky, A.: Subgroup growth and congruence subgroups. Invent. Math. **119**, 267–295 (1995)

14. Matsumoto, H.: Sur les sous-groupes arithmétiques des groupes semi-simples déployés. Ann. Sci. Éc. Norm. Super. **2**, 1–62 (1969)
15. Mennicke, J.: Finite factor groups of the unimodular group. Ann. Math. **81** (1965)
16. Merkurjev, A.S.: On the norm residue symbol of degree 2. Dokl. Akad. Nauk SSSR **261**(3), 542–547 (1981)
17. Merkurjev, A.S.: Developments in algebraic K-theory and quadratic forms after the work of Milnor. In: Collected Papers of John Milnor, V. Algebra, pp. 399–417. Am. Math. Soc., Providence (2010)
18. Milnor, J.: The Steenrod algebra and its dual. Ann. of Math. (2) **67**, 150–171 (1958)
19. Milnor, J., Moore, J.C.: On the structure of Hopf algebras. Ann. Math. **81**, 211–264 (1965)
20. Milnor, J.: Whitehead torsion. Bull. Amer. Math. Soc. (N. S.) **72**, 358–426 (1966)
21. Milnor, J.: A note on curvature and the fundamental group. J. Differ. Geom. **2**, 1–7 (1968)
22. Milnor, J.: Growth of finitely generated solvable groups. J. Differ. Geom. **2**, 447–449 (1968)
23. Milnor, J.: Advanced problem 5603. MAA Monthly **75**, 685–686 (1968)
24. Milnor, J.: Algebraic K-theory and quadratic forms. Invent. Math. **9**, 318–344 (1969/1970)
25. Milnor, J.: Symmetric inner products in characteristic 2. Prospects in mathematics. In: Proc. Sympos., Princeton Univ., Princeton, NJ, 1970. Ann. of Math. Studies, vol. 70, pp. 59–75. Princeton Univ. Press, Princeton (1971)
26. Milnor, J.: Introduction to Algebraic K-Theory. Annals of Mathematics Studies, vol. 72. Princeton University Press, Princeton (1971)
27. Milnor, J.: Collected papers of John Milnor, v. In: Bass, H., Lam, T.-Y. (eds.) Algebra. Am. Math. Soc., Providence (2010)
28. Moore, C.: Group extensions of p-adic and adèlic groups. Publ. Math. IHES **35**, 5–70 (1968)
29. Orlov, D., Vishik, A., Voevodsky, V.: An exact sequence for $K_*^M/2$ with applications to quadratic forms. Ann. of Math. (2) **165**(1), 1–13 (2007)
30. Platonov, V.P., Rapinchuk, A.S.: Abstract properties of S-arithmetic groups and the congruence subgroup problem. Russian Acad. Sci. Izv. Math. **40**, 455–476 (1993)
31. Prasad, G., Rapinchuk, A.S.: Developments of the congruence subgroup problem after the work of Bass, Milnor, and Serre. In: Bass, H., Lam, T.-Y. (eds.) Collected Papers of John Milnor, V. Algebra. Am. Math. Soc., Providence (2010)
32. Quillen, D.: In: Higher Algebraic K-Theory I. Lecture Notes in Math., vol. 341, pp. 85–147. Springer, Berlin (1973)
33. Raghunathan, M.S.: On the congruence subgroup problem. Publ. Math. IHES **46**, 107–161 (1976)
34. Raghunathan, M.S.: On the congruence subgroup problem II. Invent. Math. **85**, 73–117 (1986)
35. Rost, M.: On Hilbert Satz 90 for K_3 for degree-two extensions, available as http://www.mathematik.uni-bielefeld.de/~rost/K3-86.html (1986)
36. Serre, J.-P.: Sur les groupes de congruence des variétés abéliennes. Izv. Akad. Nauk SSSR, Ser. Mat. **28**, 3–18 (1964)
37. Serre, J.-P.: Le probème des groupes de congruence pour SL_2. Ann. Math. **92**, 489–527 (1970)
38. Serre, J.-P.: Sur les groupes de congruence des variétés abéliennes II. Izv. Akad. Nauk SSSR, Ser. Mat. **35**, 731–735 (1971)
39. Steinberg, R.: Générateurs, relations et revêtements de groupes algébriques. In: Colloq. Théorie des Groupes Algébriques, Bruxelles, Librairie Universitaire, Louvain, 1962, pp. 113–127. Gauthier-Villars, Paris (1962)
40. Suslin, A., Voevodsky, V.: Bloch–Kato conjecture and motivic cohomology with finite coefficients. In: The Arithmetic and Geometry of Algebraic Cycles, Banff, AB, 1998. NATO Sci. Ser. C Math. Phys. Sci., vol. 548, pp. 117–189. Kluwer Academic, Dordrecht (2000)
41. Tits, J.: Free subgroups of linear groups. J. Algebra **20**, 250–270 (1972)
42. Voevodsky, V.: Triangulated categories of motives over a field. In: Cycles, Transfers, and Motivic Homology Theories. Ann. of Math. Stud., vol. 143, pp. 188–238. Princeton Univ. Press, Princeton (2000)
43. Voevodsky, V.: Motivic cohomology with $\mathbf{Z}/2$-coefficients. IHES Math. Publ. **98**, 59–104 (2003)

44. Voevodsky, V.: Reduced power operations in motivic cohomology. Publ. Math. IHES **98**, 1–57 (2003)
45. Wolf, J.A.: Growth of finitely generated solvable groups, and curvature of Riemannian manifolds. J. Differ. Geom. **2**, 421–446 (1968)

John Milnor's Work in Dynamics

Mikhail Lyubich

1 Preface

John Milnor fell in love with One-Dimensional Dynamics, real and complex, in the mid 1970's. His first work, joint with Thurston, developed a combinatorial one-dimensional theory called Kneading Theory. It sparked a firework of activity that completely changed the face of the field. Besides the development of Kneading Theory, Milnor has contributed by clarifying important concepts, proposing inspiring conjectures, carefully developing foundational themes, exploring new interesting dynamical families, writing an introductory book [40] and expository articles from which three generations of students have learned the subject, creating a gallery of beautiful dynamical objects, and last but not least: by bringing to the area a wonderful atmosphere of excitement and dedication to research.

This paper can be viewed as an update of my article in the volume dedicated to Milnor's 60th birthday [28]. Though overlap is inevitable, I have tried to reduce it to minimum. For a more systematic story of real and complex one-dimensional dynamics (focused on the unimodal case) the reader can take a look at my recent survey [34].

Electronic supplementary material Supplementary material is available in the online version of this chapter at http://dx.doi.org/10.1007/978-3-642-39449-2_20. Videos can also be accessed at http://www.springerimages.com/videos/978-3-642-39449-2.

M. Lyubich (✉)
Institute for Mathematical Sciences, Stony Brook University, Stony Brook, NY 11794-3660, USA
e-mail: mlyubich@math.sunysb.edu

H. Holden, R. Piene (eds.), *The Abel Prize 2008-2012*,
DOI 10.1007/978-3-642-39449-2_20, © Springer-Verlag Berlin Heidelberg 2014

2 Selected Themes

2.1 *Kneading Theory*

Let us consider the class $\mathscr{S}$ of continuous piecewise monotone interval maps $f : I \to I$. The Milnor-Thurston Kneading Theory[1] provides us with a combinatorial classification of the corresponding dynamical systems. To such a map one can associate a sequence called the kneading invariant that determines a symbolic model for the map. This model is nicely semi-conjugate to the original map.

More precisely, let $I_0, \dots, I_l$ be the tiling of I into intervals of monotonicity (*laps*) of f. (Assume for definiteness that f is increasing on I_0.) Let $c_k = I_k \cap I_{k+1}$ be the extrema of f. A map $f \in \mathscr{S}$ with l extrema is called *l-modal*, and we let $\mathscr{S}_l$ be the space of such maps.

To each point $x \in I$ we can associate a symbolic sequence $\bar{\epsilon}(x) = (\epsilon_0, \epsilon_1, \dots)$, where the symbols ϵ_k assume the values I_j and c_j according to whether $f^k x \in \operatorname{int} I_j$ or $f^k x = c_j$ (where $\operatorname{int} I_j$ is understood in the relative topology of I). Moreover if $\epsilon_k = c_j$ for some k (so that x is a precritical point), we stop, so the sequence $\bar{\epsilon}$ is finite in this case.

The space Σ_f of all admissible sequences with the shift transformation σ acting on it is called the *symbolic model* of f. Two maps with the same symbolic model are called *combinatorially equivalent.*

The symbolic sequences $\bar{\epsilon}(fc_j)$ of the critical values form the *kneading invariant* $\kappa(f)$ of f. Milnor and Thurston showed that the whole combinatorial model is determined by the kneading invariant, and described all admissible kneading sequences. In this way Kneading Theory provides us with a full combinatorial classification of the one-dimensional dynamics in question.

An important special case of the theory covers the dynamics of *unimodal maps* $f \in \mathscr{S}_1$, i.e., the maps with one extremum c. A remarkable conclusion of Kneading Theory is that the real quadratic family $f_a : x \mapsto ax(1-x)$, $a \in [1,4]$, is *full* in the space of unimodal maps, in the sense that any map $f \in \mathscr{S}_1$ is combinatorially equivalent to some f_a. As the latter can be studied with powerful methods of holomorphic dynamics, this put polynomials into a very special position in the dynamical world.

A beautiful problem was raised in a preliminary version of [48]: Does the kneading invariant $\kappa(f_a)$ depend monotonically on a (with respect to a natural "twisted lexicographic order" on the space of kneading sequences)? The problem was resolved affirmatively in the final version of the paper. The proof is based on ideas of *holomorphic dynamics*, more precisely, on *Thurston's Rigidity Theorem* (see [17]) asserting that a superattracting parameter $a \in [1,4]$ (i.e., such that the critical point is periodic) is determined by its kneading invariant. It was the first deep application of ideas of Holomorphic Dynamics to real dynamics.

A more general *Rigidity Conjecture* that naturally emerged from the above theory is that every non-periodic kneading invariant is realized by a *single* quadratic

[1]A preliminary version of the theory had been developed in Metropolis-Stein-Stein [38].

map f_a, $a \in [1,4]$. (Note that periodic kneading sequences correspond to *hyperbolic* maps f_a, i.e. maps having an attracting cycle; they are realized on intervals of parameters.) This conjecture is equivalent to the density of hyperbolic maps in the real quadratic family. It was proved in [30, 22]. Methods of holomorphic dynamics play a crucial role in the proof, and until now no purely real argument has been found.

2.2 Milnor's Attractors

The notion of a "strange" attractor played an inspiring role in the 1970–1980's. Any invariant set with somewhat complicated topology that attracts "many" points was regarded to be a strange attractor. Examples included the Smale solenoid, Lorenz, Hénon, and Feigenbaum attractors. The notion itself was coined by Ruelle and Takens who proposed it as a mathematical foundation for the turbulence phenomenon.

However, mathematically the situation deteriorated fast because of the lack of agreement what exactly these creatures are? Should the notion be topological or measure-theoretical? Can it be so broad that attractors govern the behavior of "most" orbits for "most" systems? (and what is the exact meaning of "most", anyway?) In a conceptual article [41], Milnor gave an overview of this unsatisfactory situation and proposed a general notion of attractor.

Let $f : M \to M$ be a smooth dynamical system. A closed invariant set $A \subset M$ is called a *(measure-theoretic) attractor* if

- Its basin of attraction $\mathscr{B}(A) = \{x \in M : f^n x \to A\}$ has positive Lebesgue measure;
- Any proper closed invariant subset $A' \subset A$ has a smaller basin: $l(\mathscr{B}(A')) < l(\mathscr{B}(A))$ (where l stands for the Lebesgue measure).

An important feature of this definition is that it does not require that A is a *trapped attractor*. The latter would have an invariant neighborhood where all orbits are attracted to A. This requirement would exclude some very interesting examples like the Feigenbaum attractor.

Milnor showed that any smooth dynamical system has a unique *global attractor* A_f that attracts *almost all* orbits. He posed a problem as to whether this attractor can be decomposed into finitely many *minimal* attractors, and if so what is their structure.

This problem was particularly emphasized in dimension one, and stimulated a lot of activity in this direction. In a series of papers by Alexander Blokh and the author, it was confirmed that a finite decomposition into minimal attractors for one-dimensional maps does exist, see [5]. Moreover, these attractors can be of *four different types: attracting cycles*, (topologically) *transitive cycles of intervals*, *Feigenbaum attractors*, and *"wild" attractors*. The latter is a Cantor attractor contained in a transitive cycle of intervals. We could not resolve the problem of whether such wild attractors may or may not exist, and it remained open for a few more years.

A breakthrough came in a joint work of Milnor with the author [35] where it was proved that the key example, the *Fibonacci quadratic map*, does not have a wild attractor. (This map can be characterized by the property that the closest returns of the critical point occur at the Fibonacci moments.) In [29], this was generalized to all quadratic polynomials. These developments were largely based upon complex methods. A purely real argument was eventually found by Weixiao Shen [54].

On the other hand, Bruin, Keller, Nowicki and van Strien proved that the Fibonacci map $z \mapsto z^d + c$ of sufficiently high degree has a wild attractor [11].

More recently, it was shown that a complex quadratic map $f_c : z \mapsto z^2 + c$ may have a wild attractor as well. Namely, there exist quadratic maps f_c whose Julia set $J_c = J(f_c)$ has positive area, see Buff–Cheritat [13] and Avila–Lyubich [2]. The global measure-theoretic attractor for such a map is the union of $\{\infty\}$ and a nowhere dense subset A_c of J_c that attracts almost all points of J_c. Moreover, in examples constructed in [2], A_c is a minimal Cantor attractor.

These results give a clear picture in dimension one. But what about higher dimensions? Of course, one cannot anticipate anymore that the global attractor can always be decomposed into finitely many minimal ones as the *Newhouse phenomenon* gives an obstruction for this: there are maps that have infinitely many attracting cycles. Moreover, these maps can densely fill some parameter domain.

It is still conceivable that from probabilistic point of view, the Newhouse phenomenon is neglectable. The appropriate probabilistic notion (in infinitely dimensional space of systems) goes back to Kolmogorov: some property is considered to be *typical* if it is satisfied for almost all parameters in a generic one-parameter family of systems. The corresponding conjectures in this spirit were articulated by Jacob Palis in [50].

Milnor–Palis Conjecture For a typical smooth dynamical system $f : M \to M$, the global attractor A_f is decomposed into finitely many minimal attractors A_i. Moreover, for almost every point $x \in M$, the ω-limit set $\omega(x)$ is equal to one of the A_i.

In fact, Palis put forward a stronger conjecture asserting that typically each minimal attractor supports a unique *SRB measure*[2] μ that governs behavior of Lebesgue almost all points $x \in M$. The latter means that

$$\frac{1}{n}\sum_{k=0}^{n-1} \phi(f^k x) \to \int \phi \, d\mu \quad \text{as } n \to \infty$$

for any continuous function $\phi \in C(M)$.

For real analytic one-dimensional unimodal maps the strong Palis Conjecture was proven in [32, 33, 3]. However, already in dimension two the above conjectures are wide open.

[2]This abbreviation stands for Sinai-Ruelle-Bowen.

2.3 *Self-similarity and Hairiness of the Mandelbrot Set*

2.3.1 Pinched Model for Julia Sets

Let us now provide some brief background in holomorphic dynamics.

Consider a polynomial $f : \mathbb{C} \to \mathbb{C}$ of degree $d \geq 2$ normalized so that its leading coefficient is equal to 1. The *basin of infinity* $\mathscr{B}_f(\infty)$ is the set of points z whose orbits $\{f^n z\}_{n=0}^{\infty}$ escape to infinity. The complementary set of non-escaping points is called the *filled Julia set* $K(f)$. The *Julia set* $J(f)$ is the boundary of $K(f)$ (and of $\mathscr{B}_f(\infty)$ as well).

The (filled) Julia set is connected if and only if none of the critical points c_i of f escape to ∞. In this case, the basin of infinity can be uniformized by the complement of the unit disk, $\Phi : \mathbb{C} \setminus \bar{\mathbb{D}} \to \mathscr{B}_f(\infty)$ conjugating f to the pure power $z \mapsto z^d$. If additionally $J(f)$ is locally connected, then by the classical *Carathéodory Theorem*, the Riemann mapping Φ extends continuously to the unit circle $\mathbb{T} = \partial\mathbb{D}$, so it semi-conjugates the map $z \mapsto z^d$ on $\mathbb{T}$ to the dynamics on the Julia set $J(f)$.

This leads to the *pinched model* for the filled Julia set that can be obtained by pinching the unit disk $\bar{\mathbb{D}}$ along some geodesic lamination (see [14, 56]). More precisely, for any point $\zeta \in J(f)$, let us consider its full preimage $\Phi^{-1}(\zeta) \subset \mathbb{T}$, and let $C(\zeta) \subset \mathbb{D}$ be its convex hull in the unit disk $\mathbb{D}$ viewed as the hyperbolic plane. If $\bar{C}(\zeta)$ is a hyperbolic geodesic or a finite sided hyperbolic polygon, let us pinch it to a point. It turns out that the quotient set obtained this way is naturally homeomorphic to $K(f)$ (whenewer $K(f)$ is locally connected).

2.3.2 Hyperbolic and Superattracting Polynomials

Let α be a periodic point of period p, and let $\boldsymbol{\alpha} = \{f^k \alpha\}_{k=0}^{p-1}$ be the corresponding cycle. The derivative $\lambda = (f^p)'(\alpha)$ is called the *multiplier* of α (and its cycle). The periodic point and its cycle are called *attracting* if $|\lambda| < 1$. They are called *superattracting* if $\lambda = 0$ (note that such a cycle contains a critical point).

The basin $\mathscr{B}(\boldsymbol{\alpha}) = \mathscr{B}_f(\boldsymbol{\alpha})$ of an attracting cycle is the set of points whose orbits converge to $\boldsymbol{\alpha}$. The *immediate basin* $\mathscr{B}^*(f^k\alpha)$ is the component of $\mathscr{B}(\boldsymbol{\alpha})$ containing $f^k\alpha$. The union

$$\mathscr{B}^*(\boldsymbol{\alpha}) = \bigcup_{k=0}^{p-1} \mathscr{B}^*(f^k\alpha)$$

is called the immediate basin of the cycle $\boldsymbol{\alpha}$.

A classical Fatou–Julia Theorem asserts that $\mathscr{B}(\boldsymbol{\alpha})$ contains a critical point. It follows that a polynomial of degree d can have at most $d - 1$ attracting cycles.

A polynomial is called *hyperbolic* if all its critical orbits converge to attracting cycles. It is called *superattracting* if all these cycles are superattracting.

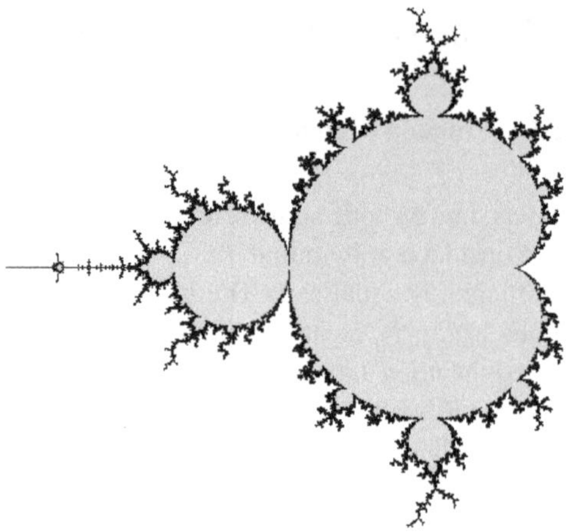

Fig. 1 Mandelbrot set. It encodes in one picture all beauty and subtlety of the complex quadratic family. Its pieces attached to the main cardioid are called the limbs

2.3.3 Little Mandelbrot Copies and Tuning

Let us now consider the quadratic family $f_c : z \mapsto z^2 + c$. The *Mandelbrot set* M is the set of parameters c for which the Julia set $J(f_c)$ is connected. One of the prominent features of the Mandelbrot set is that it contains many copies of itself that look identical to the main set (Fig. 1). Douady and Hubbard showed that these copies can be produced by a surgery called *tuning*.

Let us take some superattracting quadratic polynomial $f_s : z \mapsto z^2 + s$ with period $p > 1$. For this map, the critical point 0 is periodic with period p, and its cycle $\boldsymbol{\alpha}$ is superattracting. Consider the immediate basin $\mathscr{B}_s^*(0)$. Then the return map $g = f^p : \mathscr{B}_s^*(0) \to \mathscr{B}_s^*(0)$ is a branched covering of degree two. Moreover, the uniformization of $\mathscr{B}_s^*(0)$ by the unit disk, $\phi : (\mathbb{D}, 0) \to (\mathscr{B}_s^*(0), 0)$, brings g to the simplest possible form, $\phi^{-1} \circ g \circ \phi = f_0 : z \mapsto z^2$. In fact, one can show that $\mathscr{B}_s^*(0)$ is a Jordan disk, so ϕ extends to a homeomorphism between $\bar{\mathbb{D}}$ and $\bar{\mathscr{B}}_s^*(0)$.

Take now another quadratic polynomial $f_c : z \mapsto z^2 + c$ with connected Julia set, i.e., $c \in M$. If its Julia set is also locally connected then it can be obtained from the map f_0 by pinching the unit disk along some geodesic lamination. By means of the Riemann mapping ϕ, this pinching can be carried to the immediate basin $\mathscr{B}_s^*(0)$. By means of the dynamics, it can be then executed consistently on all components of the basin $\mathscr{B}_s(\boldsymbol{\alpha})$. It will produce a map that gives a topological model for some quadratic polynomial called the *tuning* of f_s by f_c (or just "tuning s by c").

Douady and Hubbard [16] justified this procedure indirectly using the inverse operation called *renormalization*. In fact, they proved existence of the "tuning" $s \star c$ for an arbitrary $c \in M$, even when the Julia set $J(f_c)$ is not locally connected.

2.3.4 Milnor's Observations

Thus, to any superattracting parameter s of period $p > 1$ and any other parameter $c \in M$ corresponds the tuning $s \star c$, which gives us a map $\sigma_s : M \to M_s$. The image

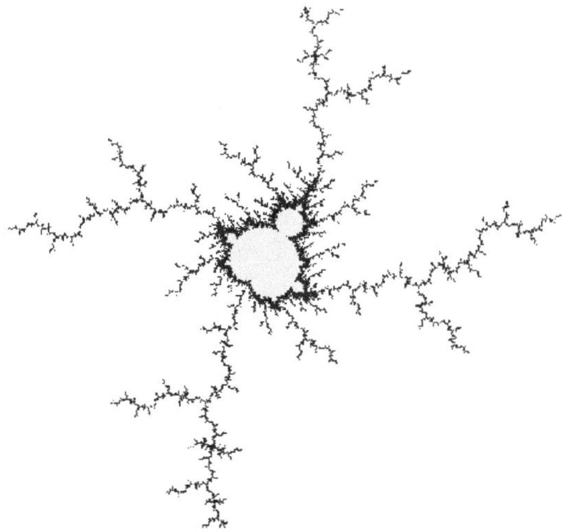

Fig. 2 A baby Mandelbrot copy

of this map is a little Mandelbrot copy $M' = s \star M$. Iterating this map, we obtain the nest of copies, $M \subset M' \supset M^n \supset \cdots$ centered at the points $s^{\star n}$.

In [42], Milnor conjectured that *the intersection of these Mandelbrot sets is a single point* $s^{\star\infty}$, *and the tuning map is conformal at this point.* This would immediately imply the *self-similarity* of M at $c^{\star\infty}$, which is clearly observed on the computer pictures. This Conjecture is a clean generalization to the complex setting of the *Universality Conjecture* for period doublings put forward by Feigenbaum, Coullet and Tresser in 1970's. (For the complex period triplings, this conjecture was formulated by Gol'berg, Sinai and Khanin [55].)

Milnor also made a surprising observation that the Mandelbrot set possesses a *hairiness* property at $c^{\star\infty}$: magnifications of M near $c^{\star\infty}$ appear to fill densely the complex plane. So, the little copies M^n are small pieces of M (almost identical to M itself) but decorated with a bundle of hairs that almost completely fill a neighborhood of $c^{\star\infty}$. On the pictures, the local magnifications of M rapidly start to look homogeneously black (see Fig. 2).

It turns out that Milnor's Self-Similarity and Hairiness conjectures are intimately related. In [36], McMullen proved the dynamical version of the Hairiness Conjecture and derived from it the dynamical version of the Self-Similarity Conjecture. (Here "dynamical" means the version concerning the Julia sets, rather than the "parameter" Mandelbrot set.) In [31], the original parameter conjectures were confirmed.

2.4 Beyond the Quadratic Family

At the time when the world was rotating around the quadratic family $f_c : z \mapsto z^2 + c$, Milnor kept reminding people that there are plenty of very interesting research areas beyond it, concerning polynomial dynamics of higher degree and rational dynamics. As the quadratic family was eventually getting exhausted, with few outstanding very difficult problems remaining, this Milnor's work assumed greater resonance becoming a foundation for further exploration.

2.4.1 Combinatorics of External Rays

Combinatorial issues were always central in Milnor's work. An important problem Milnor addressed in his joint work with Lisa Goldberg [21] is combinatorics of rays landing at the fixed points of polynomials with connected Julia sets.

In the quadratic case, the situation is as follows [15]. There are two fixed points, α and β, that play very different dynamical role. Namely, β is the landing point of the 0-external ray $\mathcal{R}_0$. It is either repelling or parabolic with multiplier 1. The other point, α, is either non-repelling, or is the landing point of finitely many (more than 1) rays $\mathcal{R}_{\theta_i}$ that are cyclically permuted by the dynamics, with some rotation number p/q. Thus, the external angles θ_i form a *rotation set* under the doubling dynamics $\theta \mapsto 2\theta \bmod \mathbb{Z}$ on the circle $\mathbb{R}/\mathbb{Z}$. The rotation number p/q is called the *combinatorial rotation number* of f.

The set of parameters c for which the quadratic polynomial f_c has combinatorial rotation number p/q form a *p/q-limb* $\mathcal{L}_{p/q}$ of the Mandelbrot set M. It can also be characterized as the part of M "attached" to the parabolic parameter $c_{p/q}$ on the main cardioid where the α-fixed point is parabolic with rotation number p/q (i.e., with multiplier $\lambda = e^{2\pi i p/q}$). See Fig. 3.

This theory is extremely important as it allows us to penetrate into the combinatorial structure of dynamical and parameter objects. In particular, it is a beginning of the *Yoccoz puzzle* construction (closely related to the Branner-Hubbard tableaux [10]) which gives a key to many crucial problems in one-dimensional dynamics, real and complex, see survey [34]. (Yoccoz has never published his results, so most people have learned them from expository articles by Hubbard [23] and Milnor [45].)

Goldberg and Milnor carried the above theory further to the higher degree case. They showed that the set of the external angles $\theta_i \in \mathbb{R}/\mathbb{Z}$ of the rays $\mathcal{R}_{\theta_i}$ landing at fixed points of f is the union of sets $T_j \subset T$ satisfying the following properties (see Fig. 4):

P1. Each T_j is the rotation set under the dynamics of $z \mapsto z^d$ with some rotation number p_j/q_j.
P2. The T_j are disjoint and pairwise unlinked.
P3. The union of all T_j with zero rotation number is the set of fixed angles $k/(d-1), k = 0, 1, \dots, d-2$.
P4. Each pair of sets $T_i \neq T_j$ with non-zero rotation number is separated with a set T_k with zero rotation number. (This property is particularly interesting as it was not visible in the quadratic case.)

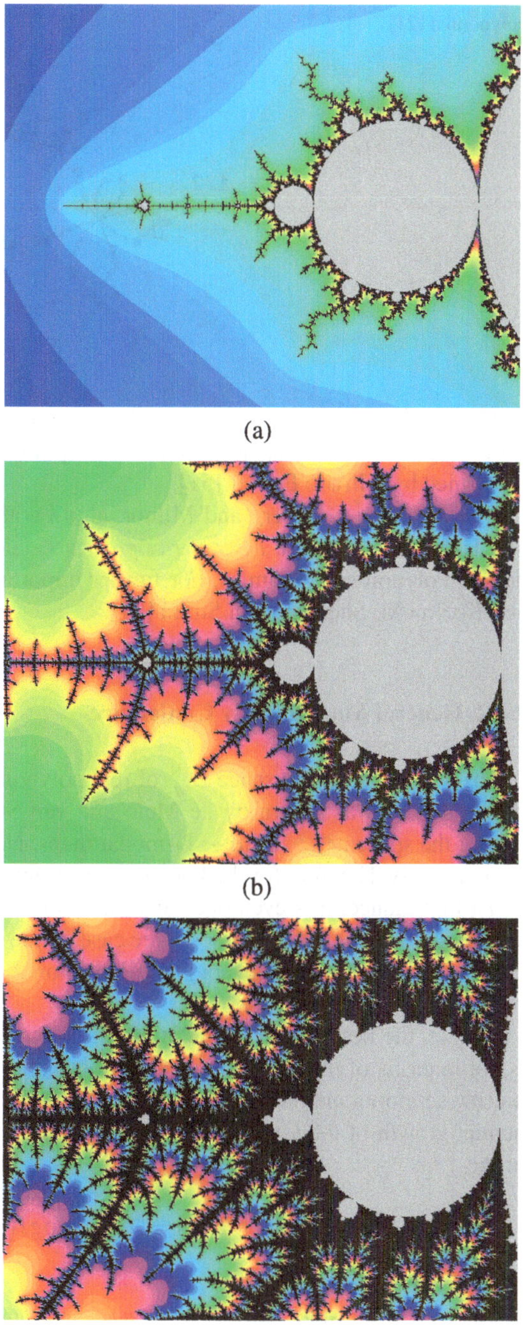

Fig. 3 Two consecutive magnifications of the Mandelbrot set near the Feigenbaum point illustrating the hairiness phenomenon [42]

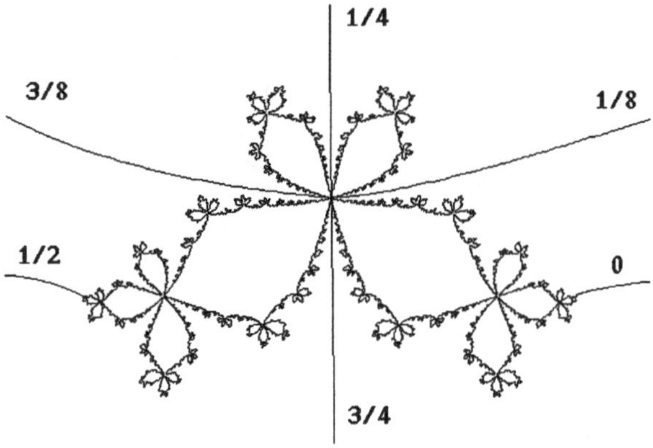

Fig. 4 Fixed point ray portrait for a cubic polynomial [21]

Goldberg and Milnor conjectured that these properties are also sufficient for the set of angles to be realizable by some fixed point ray portrait. This conjecture was confirmed by Alfredo Poirier [51].

The paper by Goldberg and Milnor, along with the paper by Bielifield, Fisher and Hubbard [4] on the *critical ray portraits*, laid down a foundation for higher degree Polynomial Dynamics. For further important advances, see e.g., Kiwi [26] and Kozlovski, Shen and van Strien [27].

2.4.2 General Monotonicity Conjecture

Let us now consider the space $\mathscr{P}_d^{\mathbb{R}}$ of real polynomials of degree d with all critical points real. It was shown by de Melo and van Strien [39] that this is a *full* section of the space $\mathscr{S}_{d-1}$ of $(d-1)$-modal maps. Next, one can wonder whether the Monotonicity Theorem for the kneading invariant can be extended to this space?

At first glance, it is even unclear what is the meaning of "monotonicity" of the kneading invariant in the higher degree case, as the space $\mathscr{P}_d^{\mathbb{R}}$ of real polynomials of degree d (up to affine conjugacy) has dimension $d-1$. A natural way was proposed by Milnor in [43].

In fact, the best known formulation of the monotonicity of the quadratic family is not in terms of the kneading invariant but rather in terms of *entropy* $h(f)$. For a piecewise monotone interval map f, the latter can be defined as the rate of exponential growth of the *lap number* $l(f^n)$ (i.e., the number of monotonicity intervals of the f^n):

$$h(f) = \lim_{n\to\infty} \frac{1}{n} \log l\big(f^n\big)$$

(see Misiurewicz-Szlenk [37]).

Let us now consider entropy $h(f)$ as a function on the space $\mathscr{P}_d^{\mathbb{R}}$. *Isentropes* are level sets of this function. *Connectivity of all isentropes* gives a good sense to the entropy monotonicity conjecture in the higher degree case.

In the cubic case (when the parameter space is two-dimensional), the problem was analyzed and resolved by Milnor and Tresser [43, 49]. In fact, this work provides a precise combinatorial description of the isentropes in the cubic family.

The general Monotonicity Conjecture in all spaces $\mathscr{P}_d^{\mathbb{R}}$ was recently established by Henk Bruin and Sebastian van Strien [12]. It required the general version of the Density of Hyperbolicity for all degrees that had been proved by Kozlovski, Shen and van Strien [27].

2.4.3 Rational Dynamics

General rational maps have many new features compared with polynomials, and their dynamics is much less explored. In this direction Milnor has written a foundational paper [44] with a discussion of the moduli space $\mathscr{M}_2$ of degree two rational maps f, i.e., the space of these maps up to Möbius conjugacy. He showed that $\mathscr{M}_2$ is an orbifold with the underlying space biholomorphically equivalent to $\mathbb{C}^2$, naturally parametrized by two symmetric functions of the multipliers μ_1, μ_2, μ_3 of the fixed points.

Milnor went on to describe a natural compactification of $\mathscr{M}_2$, to explore one-dimensional dynamical slices $\mathrm{Per}_n(\mu)$ comprising maps that have a cycle of period n with multiplier μ, to give a rough classification of hyperbolic components (following an important work by Mary Rees [53]), and to take a look at the real slice of $\mathscr{M}_2$.

Since then, essentially any discussion of the rational dynamics begins with description of Milnor's results, making use of the terminology and notation introduced in [44]. See Epstein [19], DeMarco [18], Petersen and Uhre [52].

2.5 Two-Dimensional Dynamics

Milnor's work sparked an intense interest to yet another direction of research: Higher Dimensional Complex Dynamics. An article by Friedland and Milnor [20] was probably the first paper since Fatou's time dedicated to global complex dynamics in several variables. It particularly emphasized the role of (generalized) complex Hénon maps

$$H : \mathbb{C}^2 \to \mathbb{C}^2, \qquad (x, y) \mapsto \big(p(x) - \delta y, x\big)$$

(where p is a polynomial of degree > 1) by deriving from a classical Jung's Theorem that any non-elementary polynomial automorphism of $\mathbb{C}^2$ can be decomposed into Hénon maps. (Note that the *real* Hénon family originally appeared as a simple model exhibiting strange attractors, see [24].)

Then the authors made first steps in understanding the global dynamics of complex Hénon maps, their periodic points, and topological entropy. This work was followed up by Hubbard and Oberste-Vorth, Bedford and Smillie, Fornaess and Sibony,

Fig. 5 For the real Desboves map with parameters $(\frac{1}{3}, 0, -\frac{1}{3})$, the Fermat curve seems to be the global attractor [7]

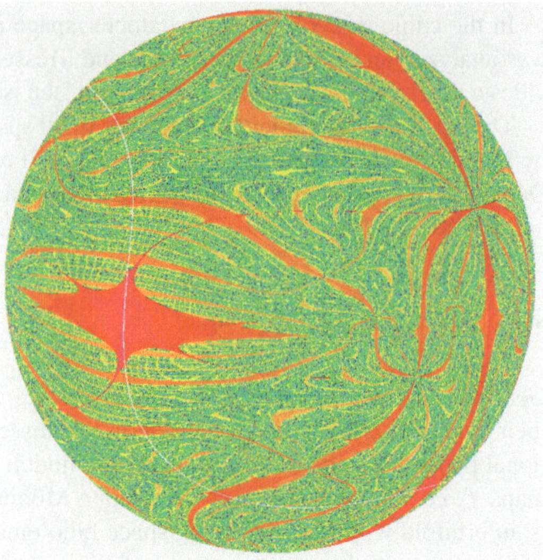

Ueda, among many others, and by now has developed into a deep and flourishing area on the borderline of dynamics and complex geometry.

Milnor himself left the subject until early 2000's when he started collaboration with Araceli Bonifant and Marius Dabija exploring the family of generalized *Desboves maps* [7]. This is a 3-parameter family of rational maps $\mathbb{P}^2 \to \mathbb{P}^2$,

$$F = F_{a,b,c} : [x : y : z] \mapsto \big[x\big(y^3 - z^3\big) : y\big(z^3 - x^3\big) : z\big(x^3 - y^3\big) \\ + (a : b : c)\big(x^3 + y^3 + z^3\big)\big],$$

that preserves the Fermat elliptic curve $A = \{x^3 + y^3 + z^3 = 0\}$.

Revisiting Milnor's favorite theme, the authors asked themselves whether A can be a measure-theoretic attractor for F. They found out numerically that for some values of parameters (a, b, c), F has a negative transverse Lyapunov exponent on A, which makes A a measure-theoretic attractor indeed (and it appears to be the global one in some case, see Fig. 5). However, the authors showed that a non-singular elliptic curve is never a *trapped attractor* (i.e., it never attracts a full neighborhood of itself)—one more confirmation of good sense for the measure-theoretic approach.

On the other hand, for some other parameter values, the experiment indicated that A may lose its attractiveness giving it to a cycle of two Herman rings that serve as an attractor.

The authors went on to study *elementary* parameters for which the map F preserves a pencil of lines through some point p. This point can be made attracting, in which case its basin $\mathscr{B}(p)$ is the whole Fatou set. At the same time, the Lyapunov exponent of the Fermat curve A can be made negative, so for these parameters, A becomes a measure-theoretic attractor. Thus, in this case the Julia set $J(F)$ is a nowhere dense set of positive volume! Even better, this Julia can contain an invariant line L that can also have a negative Lyapunov exponent. In this case the Julia

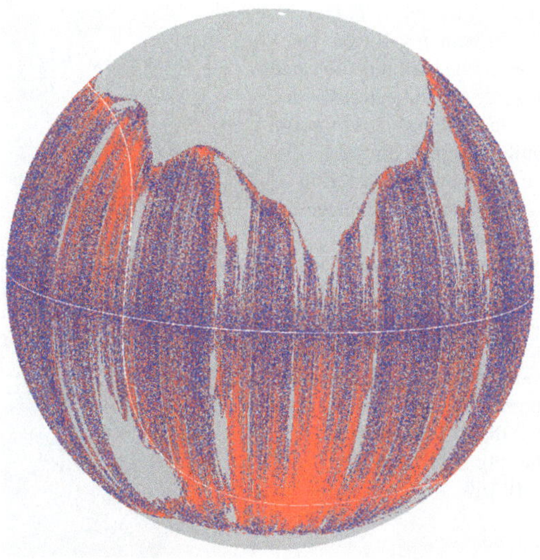

Fig. 6 Plot for the elementary map with parameters $(a, b, c) = (-1, \frac{1}{3}, 1)$. In this case, the pencil of great circles through the north pole is invariant under the dynamics. There are three attractors: the Fermat curve, the equator, and the north pole, each marked in *white*. The corresponding attracting basins are colored *red*, *blue*, and *grey* respectively. (However, the closely intermingled *blue* and *red* yield a *purple* effect.) See [7]

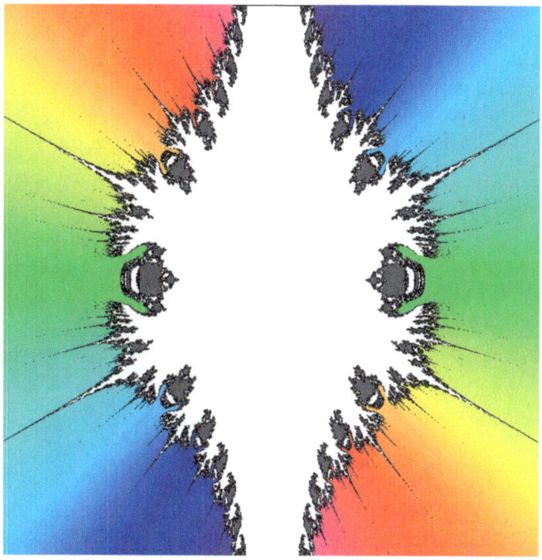

Fig. 7 This picture represents a family of rational maps $f_q : z \mapsto z^2(z-q)/(1+\bar{q}z)$ commuting with the antipodal involution of the sphere. A surprising virtue of this family is abundance of maps that have an Herman ring. Each map in the colored region has a well defined rotation number, coded by its color. Irrational rotation numbers correspond to Herman rings [6]

set contains two measure-theoretic attractors, A and L, with *intermingled basins*: both basins are dense in the Julia set (see Fig. 6). It is the first appearance in the rational dynamics of a very interesting phenomenon discovered earlier in [1, 25]. (For instance, in the latter work, Ittai Kan constructed a cylinder map fibered over the circle with the property that the top and the bottom of the cylinder are attractors with dense basins.)

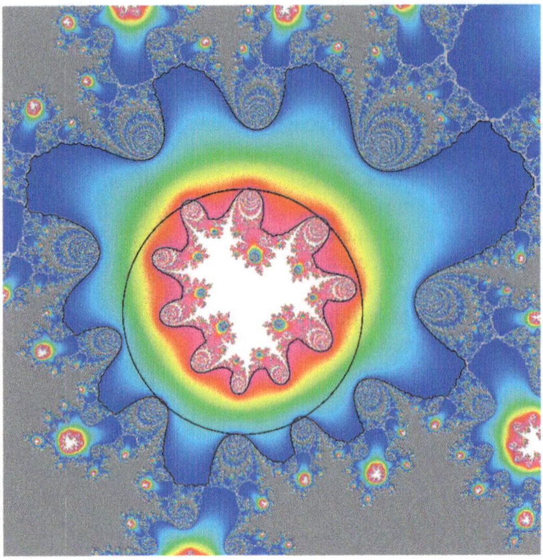

Fig. 8 Julia set of f_q (see Fig. 7) with $q = 4 + 3i$. The inner *white* region is the basin of zero; and the outer *darker* region is the basin of infinity; while the colored region consists of an Herman ring together with its preimages. Note that the picture is symmetric with respect to the unit circle. One interesting feature is the pair of repelling period ten orbits, one inside and one outside, which crowd the ring. This reflects the fact that the rotation number of the ring is very close to 7/10 [6]

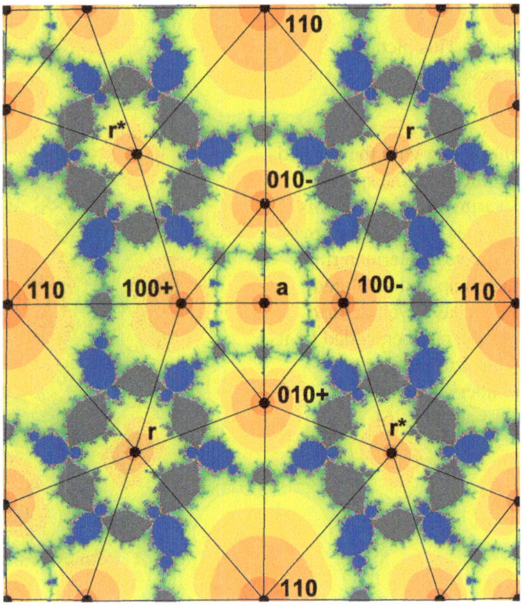

Fig. 9 This picture shows the Universal covering of the space of cubic polynomials with one periodic critical point of period 3 [8]

2.6 Art Gallery

Let us conclude with an exhibition of a few dynamics pictures from various Milnor's papers: see Fig. 7–10.

We stop here. Of course, our brief survey is far from being complete. The reader is encouraged to go directly to Milnor's papers in dynamics (whose collection up

Fig. 10 This is a Markov tiling for the torus covering of a "Lattès map" f [46]. The Julia set of f is the whole sphere that can be naturally realised as an invariant Peano curve known as the Heighway Dragon. The depicted fractal tiling gives a dynamical approximation to the Dragon. This example inspired a deep work by Bonk and Mayer [9]

to 2000 has just appeared as a separate volume [47]) and to learn more about this beautiful world.

Acknowledgement I thank Araceli Bonifant and John Milnor for helpful comments and for assistance with pictures.

References

1. Alexander, J.C., Kan, I., Yorke, J., You, Z.: Int. J. Bifurc. Chaos **2**, 795–813 (1992)
2. Avila, A., Lyubich, M.: Feigenbaum Julia Sets of Positive Area (2011). Manuscript
3. Avila, A., Lyubich, M., de Melo, W.: Regular or stochastic dynamics in real analytic families of unimodal maps. Invent. Math. **154**, 451–550 (2003)
4. Bielifield, B., Fisher, Y., Hubbard, J.: The classification of critically preperiodic polynomials as dynamical systems. J. Am. Math. Soc. **5**, 721–762 (1992)
5. Blokh, A., Lyubich, M.: Measurable dynamics of S-unimodal maps of the interval. Ann. Sci. Éc. Norm. Super. **24**, 545–573 (1991)
6. Bonifant, A., Buff, X., Milnor, J.: On antipode preserving cubic maps, in preparation
7. Bonifant, A., Dabija, M., Milnor, J.: Elliptic curves as attractors in $\mathbb{P}^2$. I Dynamics. Exp. Math. **16**, 385–420 (2007)
8. Bonifant, A., Kiwi, J., Milnor, J.: Cubic polynomial maps with periodic critical orbit. II. Escape regions. Conform. Geom. Dyn. **14**, 68–112 (2010), 190–193. (See also Errata: Cubic polynomial maps with periodic critical orbit. II. Escape regions.)
9. Bonk, M., Mayer, D.: Expanding Thurston maps (2010). arXiv:1009.3647
10. Branner, B., Hubbard, J.: The iteration of cubic polynomials, Part II. Acta Math. **169**, 229–325 (1992)
11. Bruin, H., Keller, G., Nowicki, T., van Strien, S.: Wild Cantor attractors exist. Ann. Math. **143**, 97–130 (1996)
12. Bruin, H., van Strien, S.: Monotonicity of entropy for real multimodal maps (2009). arXiv:0905.3377

13. Buff, X., Cheritat, A.: Examples of Julia sets with positive area. Ann. Math. **176**, 673–746 (2012)
14. Douady, A.: Description of compact sets in $\mathbb{C}$. In: Topological Methods in Modern Mathematics. A Symposium in Honor of John Milnor's 60th Birthday (1993). Publish or Perish
15. Douady, A., Hubbard, J.H.: Étude dynamique des polynômes complexes. Parties I et II. Publications Mathématiques d'Orsay, 84-2 & 85-4
16. Douady, A., Hubbard, J.H.: On the dynamics of polynomial-like maps. Ann. Sci. Éc. Norm. Super. **18**, 287–343 (1985)
17. Douady, A., Hubbard, J.: A proof of Thurston's topological characterization of rational functions. Acta Math. **171**, 263–297 (1993)
18. DeMarco, L.: The moduli space of quadratic rational maps. J. Appl. Math. Simul. **20**, 321–355 (2007)
19. Epstein, A.: Bounded hyperbolic components of quadratic rational maps. Ergod. Theory Dyn. Syst. **20**, 727–748 (2000)
20. Friedlend, S., Milnor, J.: Dynamical properties of plane polynomial automorphisms. Ergod. Theory Dyn. Syst. **9**, 67–99 (1989)
21. Goldberg, L., Milnor, J.: Fixed points of polynomial maps. Part II. Fixed point portraits. Ann. Sci. Éc. Norm. Super. **26**, 51–98 (1993)
22. Graczyk, J., Swiatek, G.: Generic hyperbolicity in the logistic family. Ann. Math. **146**, 1–52 (1997)
23. Hubbard, J.H.: Local connectivity of Julia sets and bifurcation loci: three theorems of J.-C. Yoccoz. In: Topological Methods in Modern Mathematics. A Symposium in Honor of John Milnor's 60th Birthday (1993). Publish or Perish
24. Hénon, M.: A two-dimensional mapping with a strange attractor. Commun. Math. Phys. **50**, 69–77 (1976)
25. Kan, I.: Open set of diffeomorphisms having two attractors, each with an everywhere dense basin. Bull. Am. Meteorol. Soc. **31**, 68–74 (1994)
26. Kiwi, J.: Rational rays and critical portraits of complex polynomials. Thesis, Stony Brook (1997)
27. Kozlovsky, O., Shen, W., van Strien, S.: Rigidity for real polynomials. Ann. Math. **165**, 749–841 (2007)
28. Lyubich, M.: Back to the origin: Milnor's program in dynamics. In: Topological Methods in Modern Mathematics. A Symposium in Honor of John Milnor's 60th Birthday (1993). Publish or Perish
29. Lyubich, M.: Combinatorics, geometry and attractors of quasi-quadratic maps. Ann. Math. **140**, 347–404 (1994)
30. Lyubich, M.: Dynamics of quadratic polynomials, I–II. Acta Math. **178**, 185–297 (1997)
31. Lyubich, M.: Feigenbaum-Coullet-Tresser universality and Milnor's hairiness conjecture. Ann. Math. **149**, 319–420 (1999)
32. Lyubich, M.: Dynamics of quadratic polynomials, III. Parapuzzle and SBR measure. In: Géométrie complexe et systémes dynamiques. Asterisque Volume in Honor of Douady's 60th Birthday, vol. 261, pp. 173–200 (2000)
33. Lyubich, M.: Almost every real quadratic map is either regular or stochastic. Ann. Math. **156**, 1–78 (2002)
34. Lyubich, M.: Fourty years of unimodal dynamics: on the occasion of Artur Avila winning the Brin prize. J. Mod. Dyn. **6**, 183–204 (2012)
35. Lyubich, M., Milnor, J.: The Fibonacci unimodal map. J. Am. Math. Soc. **6**, 425–457 (1993)
36. McMullen, C.: Renormalization and Three Manifolds which Fiber over the Circle. Princeton University Press, Princeton (1996)
37. Misiurewicz, M., Szlenk, H.: Entropy for piecewise monotone mappings. Studia Math. **67**, 45–63 (1980)
38. Metropolis, M., Stein, M., Stein, P.: On finite limit sets for transformations of the unit interval. J. Comb. Theory **15**, 25–44 (1973)
39. de Melo, W., van Strien, S.: One-Dimensional Dynamics. Springer, Berlin (1993)

40. Milnor, J.: Dynamics in One Complex Variable. Annals of Math. Studies, vol. 160. Princeton University Press, Princeton (2006)
41. Milnor, J.: On the concept of attractor. Commun. Math. Phys. **99**, 177–195 (1985)
42. Milnor, J.: Self-similarity and hairiness in the Mandelbrot set. In: Computers in Geometry and Topology. Lect. Notes in Pure Appl. Math., vol. 114, pp. 211–257 (1989)
43. Milnor, J.: Remarks on iterated cubic maps. Exp. Math. **67**, 5–24 (1992)
44. Milnor, J.: Rational maps with two critical points. Exp. Math. **9**, 481–522 (2000)
45. Milnor, J.: Local connectivity of Julia sets: Expository Lectures. In: Lei, T. (ed.) The Mandelbrot set, Theme and Variations. London Math. Soc. Lecture Note Series, vol. 274, pp. 67–116. Cambridge Univ Press, Cambridge
46. Milnor, J.: Pasting together Julia sets: a worked out example of mating. Exp. Math. **13**, 55–92 (2004)
47. Collected papers of John Milnor VI. Dynamical Systems (1953–2000) Amer. Math. Soc. 2012
48. Milnor, J., Thurston, W.: On iterated maps of the interval. In: Alexander, J. (eds.) Dynamical Systems, Proc. U. Md., 1986–1987. Lect. Notes Math., vol. 1342, pp. 465–563 (1988)
49. Milnor, J., Tresser, C.: On entropy and monotonicity for real cubic maps. Commun. Math. Phys. **209**, 123–178 (2000)
50. Palis, J.: A global view of dynamics and a conjecture of the denseness of finitude of attractors. Astérisque **261**, 335–348 (2000)
51. Poirier, A.: On the realization of the fixed point portraits (1992). arXiv:math/9201296
52. Petersen, C., Uhre, E.: The structure of parabolic slices $\mathrm{Per}_1(e^{2\pi i p/q})$ in the moduli space of quadratic rational maps, in preparation
53. Rees, M.: Components of degree two rational maps. Invent. Math. **100**, 357–382 (1990)
54. Shen, W.: Decay of geometry for unimodal maps: an elementary proof. Ann. Math. **163**, 383–404 (2006)
55. Gol'berg, A.I., Sinai, Ya.G., Khanin, K.M.: Universal properties of sequences of period-tripling bifurcations. Russ. Math. Surv. **38**, 187–188 (1983)
56. Thurston, W.: On the geometry and dynamics of iterated rational maps. In: Schleicher, D., Peters, A.K. (eds.) Complex Dynamics: Friends and Families, pp. 3–110

John W. Milnor's Work on the Classification of Differentiable Manifolds

L.C. Siebenmann

John Willard Milnor (nicknamed Jack) was born in New Jersey in 1931. Ever since the mid 1950's, he has been among the world's leading mathematicians. One of the greatest of his many contributions to mathematics was 'surgery on manifolds', which is still the dominant method for the classification of various sorts of manifold of dimension ≥ 5. I will first devote several sections to his 1956 discovery of exotic smooth manifold structures that revealed unsuspected subtlety lurking since the 19th century in the concept of differentiable manifold. Then I will go on to describe how his surgery method germinated and grew up in the subsequent classification of exotic smooth structures on spheres. His contributions to surgery, made in the decade 1955–1965, remain fundamental to the highly evolved surgery theory of today. Surgery was at the same time a unifying motivation for his important parallel work in algebraic topology.

1 Some Preliminaries

I adopt common terminology concerning topology and manifolds. Topological spaces will all be metrizable, and all manifolds will be of finite dimension. All maps will be at least continuous unless there is some contrary indication. A *closed manifold* is by definition a compact manifold with empty boundary.

The term *smooth* will normally be synonymous with infinitely differentiable, i.e. of class $\mathscr{C}^\infty$. A *diffeomorphism* is a one-to-one smooth map of smooth manifolds whose inverse map is also smooth. Degree $+1$ diffeomorphism of oriented smooth

Electronic supplementary material Supplementary material is available in the online version of this chapter at http://dx.doi.org/10.1007/978-3-642-39449-2_21. Videos can also be accessed at http://www.springerimages.com/videos/978-3-642-39449-2.

L.C. Siebenmann (✉)
Laboratoire de Mathématique, Université de Paris-Sud, 91405 Orsay, France
e-mail: lcs98@free.fr

H. Holden, R. Piene (eds.), *The Abel Prize 2008-2012*,
DOI 10.1007/978-3-642-39449-2_21, © Springer-Verlag Berlin Heidelberg 2014

manifolds will be manifolds is denoted $\approx$. Where clarifications concerning smooth manifolds and differential topology are wanted, I suggest the reader first consult Milnor's justly famous 1965 primer [M65e] of 70 pages.

A *smooth triangulation* of a smooth manifold M is a homeomorphism $K \to M$ from a simplicial complex K onto M that is smooth and nonsingular on every closed simplex of K. It endows M with a PL (= *piecewise linear*) manifold structure; see [M56d, App. 2]. It endows K with what is called a *compatible smooth structure*. For a bootstrapping development of PL topology, see [RoS72] or the references there. For theorems of existence of such triangulations and also of 'near uniqueness', see [WhdJ40] or [Mu63]. This 'near uniqueness', first proved by J.H.C. Whitehead, is a strong technical condition implying, in particular, that two such PL structures on M are related by a PL homeomorphism that is homotopic to the identity map of M. Most of this article can be understood without the notions of this paragraph.

Homology with integer coefficients $\mathbb{Z}$ will be used, unless the contrary is indicated. A (continuous) map of topological spaces $h\colon X \to X'$ is said to be a *homotopy equivalence* if there exists a map $h'\colon X' \to X$ such that the compositions $h'h\colon X \to X$ and $hh'\colon X' \to X'$ are both homotopic to the identity. One then writes $X \simeq X'$. A well known theorem of J.H.C. Whitehead asserts that every map $g\colon X \to X'$ of simply connected manifolds (or ANRs) that induces an isomorphism of homology is a homotopy equivalence. Every closed oriented n-manifold M clearly admits a degree $+1$ map $f\colon M \to \mathbb{S}^n$; thus a simply connected closed manifold with the same homology groups as $\mathbb{S}^n$ is homotopy equivalent to $\mathbb{S}^n$. It is called a *homotopy n-sphere*. A smooth homotopy n-sphere that is known to be *not* diffeomorphic to $\mathbb{S}^n$ will already qualify for the epithet *exotic*.

The old-fashioned abbreviation $\mathbb{Z}_n$ will be used for the integers $\mathbb{Z}$ modulo an integer $n > 0$. Mathematical entities with a standard meaning usually appear in roman type (i.e. not italic). For example, the tangent vector bundle of a smooth manifold X will be denoted $\mathrm{T}X$.

Articles of Milnor are identified by a tag of the form [Mxyz], where two digits xy identify its year, and z is a lower case letter that further locates it in any one the several complete lists of Milnor's publications; see [Milnor]. For example, [M59d] refers to the fourth distinct article in the list of Milnor's papers for the year 1959. Such tags serve to pinpoint the article in volumes of Milnor's collected papers [M07a], [M09a], etc. where it is supported by related material.

The body of this article has four parts, and each part (beyond this first one) is divided into several sections. Thus, Sect. 2.13 refers to the thirteenth section in Part 2.

In a survey of this length I am sure to leave some errors and omissions that beg for comment. Readers are invited to communicate any such to the author.

For assisting me by answering queries, or by offering comments on a preliminary version, I owe thanks to several mathematicians: G. Brumfiel, J.F. Davis, C. Escher, T. Kragh, J. Lannes, A. Ranicki, D. Zagier, and particularly to Jack Milnor himself.

2 The Discovery of Exotic 7-Spheres

2.1 Synopsis

As a young postdoctoral researcher of 25 years of age, with however over 5 years of experience in research, Milnor was the first to discover exotic smooth manifold structures. In [M56d] of 1956, for every oriented smooth homotopy 7-sphere, he defined, in $\mathbb{Z}_7$, a new degree $+1$ diffeomorphism invariant (called "lambda"). It is clearly zero for the standard 7-sphere $\mathbb{S}^7$ and he calculated that it assumes four distinct nonzero values on a certain known family of oriented homotopy 7-spheres, namely those occurring as oriented orthogonal $\mathbb{S}^3$-bundles over $\mathbb{S}^4$ having Euler class ± 1. In summary, he could exhibit four exotic oriented homotopy 7-spheres in the chosen family, which his lambda invariant showed to be pairwise *not* degree $+1$ diffeomorphic.

On the other hand, Milnor constructed, on each member of that family, an explicit smooth real-valued function with exactly two critical points, both nondegenerate, and deduced that each is homeomorphic to $\mathbb{S}^7$.

In concluding [M56d], Milnor constructed, from each of his newfound exotic 7-spheres, a closed topological 8-manifold, and gave reason to suspect that none of them admits a smooth manifold structure—as was indeed proved later on.

Part 3 of this article will describe Milnor's discovery of surgery and the subsequent calculation of groups of exotic spheres in high dimensions. It is convenient at this point to insert a limited synopsis of it inasmuch as it is a direct sequel to the discovery of exotic 7-spheres.

In the period 1958–1962, two essentially disjoint research projects respectively of Milnor (see [M59c], [M59d]) and of Steven Smale (see ([Sm60], [Sm61], [Sm62]) combined to establish that the degree $+1$ diffeomorphism classes of all oriented smooth manifolds homeomorphic to $\mathbb{S}^7$ form an Abelian group, here denoted $\mathscr{S}_7$, that is isomorphic to the cyclic group $\mathbb{Z}_{28}$. Milnor's lambda invariant is a surjective homomorphism from $\mathscr{S}_7$ onto $\mathbb{Z}_7$. In $\mathscr{S}_7$, addition corresponds to the well-known geometric *connected sum* operation, and algebraic change of sign corresponds to reversal of manifold orientation. Connected sum of oriented connected smooth manifolds is well defined in all dimensions (up to degree $+1$ diffeomorphism); in dimension 2, it is the operation one sees when two soap bubbles touch and coalesce to form one! The fact that connected sum preserves the property of being homeomorphic to a sphere was established by B. Mazur's topological Schoenflies Theorem [Maz59]. Thus the group $\mathscr{S}_n$ can be defined for all dimensions n; see [M11a].

Milnor's role was to use his freshly discovered surgery techniques to show in [M59d] that a similarly defined group, Θ_7, formed from all smooth oriented homotopy 7-spheres viewed up to a seemingly looser equivalence called *degree* $+1$ *h-cobordism* is isomorphic to $\mathbb{Z}_{28}$.

Smale's role was to give an autonomous proof of his famous *h-cobordism theorem*, which is discussed in Sect. 2.16. It easily implies that the 'forgetting' map $\mathscr{S}_n \to \Theta_n$, is a group isomorphism for all dimensions $n \geq 6$. It likewise implies that a forgetting map $\Gamma_n \to \Theta_n$ is an isomorphism for $n \geq 6$ where Γ_n is the group of oriented smooth *twisted n-spheres* to be defined in Sect. 2.13.

The generalization of Milnor's calculation of Θ_n for $n = 7$, to an almost complete calculation for all dimensions $n \geq 5$, was achieved by extension of the new surgery techniques of [M59d] in a famous collaboration of Milnor with the French topologist Michel Kervaire. The article [KM63] to be discussed in Part 3 was the culmination.

2.2 1956: Why the Surprise? Some History

Milnor's 1956 discovery of exotic spheres was quite a surprise to all mathematicians and scientists who work with manifolds. In recalling his discovery, Milnor expressed this mildly as follows in 2011, see [RS11].

> "This was something which hadn't been expected, and I am not aware that anybody had explicitly asked the question; we just assumed the answer was obvious."

But why the surprise? Ever since the pioneering period 1892–1905 when H. Poincaré made a wide-ranging study of smooth manifolds, their smooth triangulations,[1] their homology, and their fundamental group, all evidence had concurred with the following common but tacit and somewhat vague assumption:

(A) *Given a manifold M of any reasonable sort, one can endow it with a smooth manifold structure. And given between two smooth manifolds a homeomorphism $h\colon M \to M'$ of any reasonable sort, one can find a diffeomorphism $f\colon M \to M'$ approximating h.*[2]

Motivating (A) was perhaps the belief that differential analysis as developed since the time of Newton and Leibniz should and will be available to study the topology of all manifolds. In Poincaré's time, the use of the so called 'uniformization theorem' of Klein, Poincaré, and Koebe to help classify 2-manifolds, was a tempting paradigm.

It is appropriate to add that evidence comforting this assumption (A) had been piling up:

---

[1]Poincaré used triangulations as an essential tool in defining homology, and in establishing Poincaré duality for closed oriented manifolds. Anticipating [WhdJ40] (as Poincaré seemingly did) one can quickly deduce invariance of his homology under diffeomorphisms and also under homeomorphisms that are PL with respect to smooth triangulations; indeed, exercising hindsight, and using the simple bisection operation of J.W. Alexander [Alr30] and K. Reidemeister [Reid38], this deduction is easy, see [Sieb80].

[2]Ever since F. Hausdorff's 1914 monograph [Hau14], "homeomorphism" has consistently meant a one-to-one continuous map between topological spaces such that the inverse map is also continuous. However, for manifolds, Poincaré used the term "homéomorphe" (in French) to mean sometimes diffeomorphic and sometimes PL homeomorphic. In dynamics and in discussing dimension he did indeed sometimes use "homéomorphe" in its modern meaning!

- In 1885, K. Weierstrass proved [Weier85] that an arbitrary real continuous function on the unit interval [0, 1] can be uniformly approximated by polynomials; his method was very general and has enjoyed wide application to this day.
- Among D. Hilbert's famous 23 problems of 1900, was this question (in his 5th Problem): *Does a topological group that is known to be a topological manifold necessarily admit a smooth manifold structure making the group multiplication smooth?* The answer was yes, but the proof was difficult. For compact groups G, J. von Neumann proved this in 1927 [vonN27] using the Peter–Weyl embedding [PW27] of G as subgroup of a classical matrix group. Then many famous mathematicians contributed to a proof in general. It was complete by 1952; see [Pal09] for an excellent overview.
- In the 1930s, H. Whitney [Why36] showed that, for $k \geq 1$, given a class $\mathscr{C}^k$-smooth manifold, i.e. one with an atlas $\mathscr{A}$ of mutually $\mathscr{C}^k$-compatible charts, one can find for M an atlas $\mathscr{A}'$ of $\mathscr{C}^\infty$-compatible charts (or even analytically compatible charts) that is $\mathscr{C}^k$-compatible with the given atlas $\mathscr{A}$.
- In 1940, J.H.C. Whitehead established an essentially unique PL manifold structure for every smooth manifold M [WhdJ40]; it can be specified by a homeomorphism $f\colon K \to M$ where K is a simplicial complex and f is smooth and nonsingular on each closed simplex of K.
- In the 1940s, S. Cairns [Cai44] (cf. [WhdJ61]) made progress in the opposite direction by compatibly smoothing all PL manifolds of dimension ≤ 4, and established uniqueness up to diffeomorphism of such smoothings in dimensions ≤ 3. (It was perhaps no accident that Whitney and Cairns, who worked on smoothing problems, had been students at Harvard University of George Birkhoff, who in turn was an eminent disciple of Poincaré.)
- Only the question whether or not all merely topological manifolds are triangulable as simplicial complex (the *Triangulation Conjecture*), and the question whether or not such triangulations are unique up to some sort of subdivision (the *Hauptvermutung*) had been *explicitly* raised (see H. Tietze 1908 [Ttz08], E. Steinitz 1908 [Stz08], H. Kneser 1925 [Kn25], and J.W. Alexander 1932 [Alr32]). In 1952, to these two known conjectures, E. Moise [Moi52] provided a positive answer in dimensions ≤ 3.

In 1905, Poincaré had formulated the *Classical Poincaré Conjecture* that a closed smooth 3-manifold with trivial fundamental group is diffeomorphic to the 3-sphere. It was recently established by Grigory Perelman. Incidentally, Milnor boldly and accurately heralded that solution [M03a], [M06a], converting many a skeptic into an admirer. I believe that, before Milnor's 1956 breakthrough, the similar *Generalized Poincaré Conjecture* had not been explicitly formulated. Soon thereafter, it became (usually!) the conjecture that every closed smooth n-manifold having the same homotopy type as the standard sphere $\mathbb{S}^n$ is homeomorphic to $\mathbb{S}^n$.

To close this historical introduction, I quote Friedrich Hirzebruch's 1954 comment [Hirz54, p. 214] indicating that, for one expert, assumption (A) had, become a recognized problem, one perhaps too hard to attack frontally.

> "No manifold is known which can carry two different differentiable structures. But it is not unlikely that it will be easier to solve Problem 1 [*topolog-*

ical invariance of Pontrjagin classes] than to prove (or disprove?) that every manifold carries at most one differentiable structure."

Two years later, Hirzebruch's own work was to contribute to Milnor's disproof! A decade after that, Hirzebruch was first to discover exotic spheres in classical algebraic geometry; they arose as links of isolated Brieskorn singularities of complex hypersurfaces; see [Brie66], [Hirz66a].

2.3 Milnor's Incendiary 1956 Article Appears

Milnor's incendiary 1956 [M56d] article is not easy to understand on its own. It was submitted to the prestigious journal "Annals of Mathematics" on June 14th 1956, and published a very few months later in the September issue; clearly it was considered exceptionally urgent.

Milnor is rightly considered one of the finest expositors of mathematical research active since the 1950s, as his three subsequent Amer. Math. Society Steele prizes testify. He soon compensated his initial terseness with a mimeographed article of November 1956 that beautifully filled in much geometric background material. (That exposition was formally published for the first time in 2007 in volume III of Milnor's collected papers, along with further helpful introductory remarks.) A few months later, in spring 1957, Milnor elucidated much of the necessary algebraic topology in a lecture course on characteristic cohomology classes for vector bundles, and R. Thom's cobordism theory.[3] Written up by J. Stasheff, those notes were cherished by generations of students of differential topology. The 1974 monograph by Milnor and Stasheff [MSt74] is a vastly expanded version that is also meant for experts.

On the other hand, Milnor's admirable zeal and talent for streamlining and improving existing mathematics has mostly been devoted to theories of others upon which he has wished to build, and very much less to theories he has himself built. Thus the terse style adopted for the urgent discoveries in the article [M56d] has remained an obstacle to its being fully understood and appreciated by nonexperts and students of today. This partly explains why my discussion of [M56d] will be more didactic than for surgery theory. Additionally, the article [M56d] was a starting point, whilst surgery theory has had a long trajectory during which it has been repeatedly renewed and reexpounded by successive generations, see, for example, [Brow72], [Wall70], [Cap00], [Cap01].

2.4 From Thom's Cobordism to Diffeomorphism?

At this point I set the scene for readers wanting to understand Milnor's key 1956 article [M56d]. Hopefully my historical and technical observations will help make all

[3] The mimeographed notes for it are still extant at http://www.maths.ed.ac.uk/%7Eaar/surgery/.

aspects of [M56d] accessible to non-experts and students. Other resources and commentaries for understanding [M56d] can be found in volume III [M07a] of Milnor's collected papers, or in [M00b], [M09c] and [M11a].

A quick explanation of oriented cobordism is needed; see [MSt74] for more detail. Let W be a smooth, compact and oriented $(n+1)$-manifold whose *boundary* is the closed n-manifold V with the naturally induced orientation. It will often be convenient to say that W is an (oriented) *coboundary* for V. Suppose that $\partial W = V$ is the disjoint union $V' \sqcup V''$ of two closed oriented n-manifolds. Then W is said to give an *oriented cobordism of* $-V'$ *to* V'', or a *oriented null-cobordism of* $V = V' \sqcup V''$; here $-V$ denotes V with its orientation reversed. Quite different cobordism equivalence classifications arise when different sorts of manifolds are used, for example unoriented manifolds, or oriented manifolds with a framing.

In 1954, R. Thom had published [Thom54] an elegant and complete *unoriented cobordism* classification of closed smooth manifolds. In principle he also obtained an *oriented cobordism* classification but his calculations were complete only in dimensions ≤ 7. In higher dimensions they were complete 'rationally'; rational cobordism equivalence allows V equivalent to V' if, for some non-zero positive integer s, there is an oriented cobordism, from the s-fold disjoint sum $sV := V \sqcup \cdots \sqcup V$, to sV' similarly defined. Thom's 'rational' calculation was enough to establish the Hirzebruch–Thom signature formula in all dimensions; it expresses in terms of Pontrjagin characteristic classes (see [Hirz56] and [MSt74]) the signature of any smooth oriented closed manifold of dimension divisible by 4, say $4k$. Signature here refers to the usual algebraic signature of the nonsingular real intersection form on $2k$-dimensional real homology. When this bilinear form is expressed by a diagonal matrix, its signature is the number of positive diagonal entries minus the number of negative diagonal entries.

By 1956, Milnor was perhaps wondering whether related methods might help to advance from Thom's 1954 oriented cobordism classification of smooth closed manifolds into the finer diffeomorphism classification about which little was then known in dimensions greater than 3.

2.5 Milnor's Test Manifolds

As part of a simplest prototype classification problem, Milnor was examining the infinite family of smooth compact 8-manifolds W (along with their nonempty closed 7-dimensional boundaries) that occur as total space of the well known smooth oriented orthogonal 4-disk bundles ξ over the standard 4-sphere $\mathbb{S}^4$. This total space W is the union of all the 4-disk fibers of ξ, and its boundary ∂W is the union of all the 3-sphere fiber boundaries. Classic references for bundles are [Strd51] and [Hus66]; we will use (often implicitly) the most basic concepts of bundle theory such as structure group, principal bundle, and associated bundle with given fiber, including the usual conventions concerning left and right actions of structure group.

The bundle structure group of ξ is the so called special orthogonal group SO(4) consisting of the degree $+1$ isometries of the standard unit 4-ball $\mathbb{B}^4$ in $\mathbb{R}^4$. Up to smooth bundle equivalence fixing the base $\mathbb{S}^4$, these bundles are neatly classified by homotopy classes of smooth maps $\mathbb{S}^3 \to \mathrm{SO}(4)$, or equivalently, by the elements of the homotopy group $\pi_3(\mathrm{SO}(4))$. But SO(4) is smoothly the Cartesian product $\mathrm{SO}(3) \times \mathbb{S}^3$ where $\mathbb{S}^3$ is the 3-sphere of unit norm quaternions, which we often denote by **Q**; this splitting of SO(4) is determined by the standard section of the projection $\mathrm{SO}(4) \to \mathbb{S}^3$ induced by the right action of **Q** on $\mathbb{B}^4$. (Beware: As Lie group, SO(4) is not this product.) One deduces natural isomorphisms

$$\pi_3\big(\mathrm{SO}(4)\big) = \pi_3\big(\mathrm{SO}(3)\big) \oplus \pi_3\big(\mathbb{S}^3\big) = \mathbb{Z} \oplus \mathbb{Z}.$$

Viewed up to bundle isomorphism fixing base, these bundles ξ thus form a doubly indexed family $\xi(a,b)$, where (a,b) is an element of $\mathbb{Z} \oplus \mathbb{Z}$.

The canonical generator of $\pi_3(\mathrm{SO}(3))$ is the degree $+2$ covering map $\mathbf{Q} = \mathbb{S}^3 \to \mathbb{P}^3_{\mathbb{R}} = \mathrm{SO}(3)$ of real projective 3-space realized by W.R. Hamilton's orthogonal action $\mathbf{Q} \times \mathbb{R}^3 \ni (u,v) \mapsto u\,v\,u^{-1} \in \mathbb{R}^3$.

In [Strd51], the smooth structures mentioned above are absent; but Weierstrass' smooth approximation method lets one introduce them with no resulting change to the bundle classification.

This classification is a case of a 1939 result of J. Feldbau, which features prominently in N. Steenrod's 1951 monograph [Strd51] on fiber bundles. It is helpful to observe that the Abelian addition in $\pi_3(\mathrm{SO}(4))$ corresponds to a geometric 'base-connected-sum' operation on the isomorphism classes of these bundles with base $\mathbb{S}^4$. The classifying element in $\pi_3(\mathrm{SO}(4))$ of such a bundle ξ coincides with the primary obstruction to sectioning the so-called principal SO(4) bundle of ξ. Similar statements apply to all $\mathrm{SO}(n)$ bundles over all spheres.

Let $W(a,b)$ denote the total space of $\xi(a,b)$, viewed as a smooth compact 8-manifold, and endowed with the orientation for which the zero section has intersection number $+1$ with each oriented fiber. Clearly, $W(a,b)$ retracts by deformation onto the bundle zero-section; thus it has the homotopy type of $\mathbb{S}^4$; it's boundary $\partial W(a,b)$ is a closed oriented and simply connected 7-manifold to be denoted $M(a,b)$. Milnor found his first exotic 7-spheres among these 7-manifolds $M(a,b)$.

2.6 *Towards an Easy 'Endoscopic' Classification of these 8-Manifolds*

The degree $+1$ diffeomorphism classification of these 8-manifolds $W(a,b)$ is so close to the above SO(4) bundle classification of $\xi(a,b)$ that is easy to grasp, by examining their internal structure. The adjective 'endoscopic' derives from the ancient Greek roots 'endon' = *within*, and 'skopein' = *to see*. Robert Langlands has often employed 'endoscopic' in mathematics.

A *weak* $\mathrm{O}(2n)$ *bundle equivalence* $\xi \to \xi'$ of $\mathrm{O}(2n)$ bundles over a sphere $\mathbb{S}^{2n}$ is a $\mathrm{O}(2n)$ bundle map that induces on the common base $\mathbb{S}^{2n}$ either the identity or the (degree -1) antipodal map. It is thus a diffeomorphism $E(\xi) \to E(\xi')$ of the total spaces. We require bundle equivalences *not* specified as weak to induce the identity on the common base.

Preliminary Classification Theorem *Suppose that* $f: W(a,b) \to W(a',b')$ *is a diffeomorphism. Then there exists a smooth isotopy of* f *to a diffeomorphism* g *that coincides on a neighborhood of the zero-section with a weak* O(4) *bundle equivalence* $g_0: \xi(a,b) \to \xi(a',b')$.

There is a direct geometric proof of this based on a 1961 embedding uniqueness theorem of A. Haefliger, namely [Haef61b, Theorem 5.1, p. 76]. In the present context, it is a natural generalization of Whitney's famous 1944 embedding uniqueness theorem [Why44]. Haefliger's theorem provides a smooth deformation of f through diffeomorphisms to a diffeomorphism f' such that f' identifies the zero-section of $\xi(a,b)$ to that of $\xi(a',b')$ either by the antipodal map (the zero sections being copies of $\mathbb{S}^4$) or by the identity map. Then an application of the well known tubular neighborhood uniqueness theorem (see [Lang62]) completes the construction of g_0. □

Corollary 1 *With the same data, suppose that* f *is of degree* $+1$, *and that it induces the identity map* $\mathbb{Z} = \mathrm{H}_4(W(a,b)) \to \mathrm{H}_4(W(a',b')) = \mathbb{Z}$ *where the identifications to* $\mathbb{Z}$ *come from the zero sections. Then* $(a,b) = (a',b')$.

Proof The diffeomorphism g_0 of the theorem is then an SO(4) bundle equivalence, not just a weak O(4) bundle equivalence. □

Corollary 2 *The degree* ± 1 *diffeomorphism classification of the manifolds* $W(a,b)$ *coincides with the classification of the bundles* $\xi(a,b)$ *up to weak* O(4) *bundle equivalence.*

It is known that this weak classification of the bundles $\xi(a,b)$ corresponds to dividing the plane of parameters (a,b) by a certain linear reflection group of order 4. See Sect. 2.12 for details.

2.7 Towards a Classification of the 7-Manifolds $M(a,b)$

Given the above conceptually simple diffeomorphism classification of the 8-manifolds $W(a,b)$, it is tempting to believe that similar considerations would let one classify up to diffeomorphism the closed 7-manifold boundaries $M(a,b) := \partial W(a,b)$ of these classified 8-manifolds. Each boundary $M(a,b)$ inherits from $\xi(a,b)$ a fibration $\partial\xi(a,b)$ by 3-spheres that again has structure group SO(4). Their classification as smooth SO(4)-bundles with fiber $\mathbb{S}^3$ is easily seen to be the same as for the bundles $\xi(a,b)$. But the *diffeomorphism* classification of $M(a,b)$ turned out to be very difficult, and also very different from anything contemplated before, and thus very exciting!

Remark Recently, these fibrations $\partial\xi(a,b)$ of $M(a,b)$ have been *visually* described by Niles Johnson [John12] using quaternion notations of Milnor to be introduced presently. It would be nice to have some visual clue to the appearance of exotic spheres among these $M(a,b)$. As already mentioned, in 1959, Milnor was to determine that there are exactly 28 degree $+1$ diffeomorphism classes of smooth homotopy 7-spheres. Then, in [EK62] (see also the end of Sect. 2.11), it was further determined that 16 of these 28 classes occur among the manifolds $M(a,b)$, each of the 16 occurring for infinitely many values of the pair (a,b). Hence there are numerous diffeomorphism relations beyond known 'fibered' relations (to be further explained and exploited below; see Sect. 2.12) that correspond to reversal of fiber and/or base orientations. Have any of them been described in concrete terms? See Milnor's query on page 403 line 3 in [M56d]. See also the relatively recent article of D. Crowley and C. Escher [CrE03]; it answered many outstanding questions about the 7-manifolds $M(a,b)$ and includes an extensive survey of what was known about them in 2003.

The well understood Euler class $\mathrm{e}(\xi(a,b))$ in $\mathrm{H}^4(\mathbb{S}^4)=\mathbb{Z}$ is the primary obstruction to sectioning the $\mathbb{S}^3$ bundle $\partial\xi(a,b)$ with base $\mathbb{S}^4$, and one observes that $\mathrm{e}(\xi(a,b))=b$. Indeed, since $\partial\xi(a,0)$ has a canonical section, $\mathrm{e}(\partial\xi(a,b))=0+\mathrm{e}(\partial\xi(0,b))$; and one can pursue definitions to see that $\mathrm{e}(\partial\xi(0,b))=b$.

One readily shows that b is the self intersection number of the zero section of $\xi(a,b)$ in the 8-manifold $W(a,b)$. It follows that its (simply connected!) boundary $\partial W(a,b)=M(a,b)$ has exactly the homology of $\mathbb{S}^7$ if and only if $b=\pm 1$. Thus the manifolds $M(a,\pm 1)$ are exactly the homotopy 7-spheres among the manifolds $M(a,b)$.

Remark In the last two decades differential geometers have discovered, on each of Milnor's *exotic* homotopy 7-spheres $M(a,\pm 1)$, Riemannian metrics with all their sectional curvatures non-negative; see [GrZ00] and [JoW08] for many related results. Whether *any* exotic smooth homotopy sphere has a metric with strictly positive sectional curvatures remains a wide open question.

2.8 *Milnor's* SO(4) *Bundle Notations*

For the calculation of the needed first Pontrjagin class of the above bundles, Milnor made a nicely optimized choice of notations as follows. Identify $\mathbb{S}^3$ with the unit norm quaternions $\mathbf{Q}$. Define $f_{h,j}\colon \mathbb{S}^3\to \mathrm{SO}(4)$ by $f_{h,j}(u)\colon v\mapsto u^h v u^j$. Note that $f_{1,-1}$ is, up to sign, W.R. Hamilton's covering map $\mathbf{Q}\to\mathbb{P}^3=\mathrm{SO}(3)$ identifying antipodes.

Denote by $\xi_{h,j}$ the 4-disk bundle over $\mathbb{S}^4$ determined by $f_{h,j}$. Since $u^h\,v\,u^j=u^h\,v\,u^{-h}\,u^{h+j}$, it is clear that $\xi_{h,j}=\xi(h,h+j)$ or inversely $\xi(a,b)=\xi_{a,b-a}$. Thus, we also have identifications $W_{h,j}=W(h,h+j)$ of their respective naturally oriented total spaces. And likewise of their oriented boundary 7-manifolds $M_{h,j}=M(h,h+j)$.

Remark Although I find Milnor's notations and conventions relating to the bundles $\xi_{h,j}$ to be clear and optimal, they are not universally followed. It might be helpful to specify their relationship to others, notably those of [JW55] and [CrE03].

2.9 The First Pontrjagin Class

Milnor's calculation in [M56d] of the Pontrjagin class $\mathrm{p}_1(\xi_{h,j}) \in \mathrm{H}^4(\mathbb{S}^4)$ assumes the rather technical fact that fiber orientation reversal in the bundle $\xi_{h,j}$ transforms it into the bundle $\xi_{-j,-h}$. I will prove this 'ab initio' using quaternions. Milnor's calculation also assumes the fact that fiber orientation reversal leaves $\mathrm{p}_1(\xi_{h,j})$ unaltered, I will deduce this from the property (needed elsewhere) that p_1 depends only on the stabilized bundle with group SO(5) (or SO). Thus, this section intended for readers who want direct and elementary proofs of these two known facts.

Since the Pontrjagin class of an SO(4) bundle over $\mathbb{S}^4$ depends only on the stabilization of the bundle to group SO(5), let us examine the surjective stabilization map s in the following segment:

$$\mathbb{Z} \cong \pi_4(\mathbb{S}^4) \to \pi_3(\mathrm{SO}(4)) \xrightarrow{s} \pi_3(\mathrm{SO}(5)) = \pi_3(\mathrm{SO}) \qquad (*)$$

of the exact sequence of the fibration $\mathrm{SO}(4) \to \mathrm{SO}(5) \to \mathbb{S}^4$.

Consider the two maps λ, and ρ from $\mathbb{S}^3 = \mathbf{Q}$ to SO(4) defined for $v \in \mathbb{S}^3 \subset \mathbb{R}^4$ by, respectively, left and right quaternion multiplication:

$$\lambda(u)\colon v \mapsto uv, \quad \text{and} \quad \rho(u)\colon v \mapsto vu.$$

Their homotopy classes $[\lambda]$ and $[\rho]$ are free generators of the rank 2 free Abelian group $\pi_3(\mathrm{SO}(4))$; indeed, in $\pi_3(\mathrm{SO}(4))$, one clearly has $[\xi_{h,j}] = h[\lambda] + j[\rho]$. On the other hand:

Theorem 1 (Steenrod–Whitehead) $s[\lambda] = -s[\rho]$ *in* $\pi_3(\mathrm{SO}(5))$.

This theorem is an immediate consequence of [Strd51, §23.6], which is based on G.W. Whitehead's article [WhdG42] of 1942 written under N. Steenrod's direction. I will presently explain an abbreviated proof of this theorem suggested by Milnor's use of fiber-orientation reversal on the bundles $\xi_{h,j}$. But first some corollaries.

Corollary A $\pi_3(\mathrm{SO}(5)) \cong \mathbb{Z}$.

Proof By the theorem, the element $[\lambda] + [\rho]$ in $\pi_3(\mathrm{SO}(4)) = \mathbb{Z}^2$ lies in the kernel of s. Since this element is primitive (indivisible) in $\mathbb{Z}^2$, it generates a summand K of $\pi_3(\mathrm{SO}(4))$ with infinite cyclic quotient. Thus, the exactness at $\pi_3(\mathrm{SO}(4))$ of the above displayed sequence shows that K is the whole kernel of s. □

Corollary B *The two generators of* $\pi_3(\mathrm{SO}(5))$ *(differing only by sign) are* $s[\lambda]$ *and* $s[\rho]$.

Corollary C *With respect to the basis* $[\lambda]$, $[\rho]$ *of* $\pi_3(\mathrm{SO}(4)) \cong \mathbb{Z}^2$, *and the basis* $s[\lambda]$ *of* $\pi_3(\mathrm{SO}(5)) \cong \mathbb{Z}$, *the map* s *assigns* $(h, j) \mapsto h - j$.

We now explain a proof of the above theorem; it is just an assemblage of known observations.

Any homotopy group $\pi_k(\mathrm{SO}(n))$, $n \geq 1$, can be shown, see [Strd51, §16], to be the set of free homotopy classes of maps $\mathbb{S}^k \to \mathrm{SO}(n)$, the addition being induced by either the standard coproduct $\mathbb{S}^k \to \mathbb{S}^k \vee \mathbb{S}^k$ or by composition of the isometries in $\mathrm{SO}(n)$.

The group $\pi_0(\mathrm{O}(n)) = \mathbb{Z}_2$ determines an involutive homotopy class α of maps $\mathrm{SO}(n) \to \mathrm{SO}(n)$ that is well defined, as follows. Let β and β' be any degree -1 isometries of $\mathbb{S}^n$. The map $\mathrm{SO}(n) \to \mathrm{SO}(n)$:

$$\mathrm{SO}(n) \ni x \quad \mapsto \quad \beta' x \beta \in \mathrm{SO}(n)$$

has by definition homotopy class α. This α is well defined; indeed, any two choices of β are joined by a path in $\mathrm{SO}(n)$, and similarly for β'. It is easily verified that $\alpha\alpha$ is the class of the identity map. Given any homotopy class $[g]$ of maps $X \to \mathrm{SO}(n)$, we can define $\alpha([g])$ to be the homotopy class of the following map $X \to \mathrm{SO}(n)$:

$$X \ni x \quad \mapsto \quad \beta' g(x) \beta \in \mathrm{SO}(n).$$

The inclusion-induced maps $s\colon \pi_* \mathrm{SO}(n) \to \pi_* \mathrm{SO}(n+1)$ are easily seen to commute with the involutions α on the graded groups $\pi_* \mathrm{SO}(n)$; thus $s\,\alpha = \alpha\, s$ when defined.

For n odd, the involution α on $\pi_* \mathrm{SO}(n)$ is the identity; indeed one can choose β and β' to both be the antipodal involution of $\mathbb{S}^{n-1}$, which has degree -1 and commutes with all of $\mathrm{SO}(n)$. The situation is quite different in even dimensions; for example, on $\pi_1(\mathrm{SO}(2)) \cong \mathbb{Z}$, the involution α clearly reverses sign.

It follows from the last two observations that α is the identity on the stable groups $\pi_k(\mathrm{SO}) \cong \pi_k(\mathrm{SO}(n))$, $n \geq k+2$.

Proposition 1 *Let* ξ *be a bundle over* $\mathbb{S}^k$ *with fiber* $\mathbb{S}^{n-1}$, *with group* $\mathrm{SO}(n)$, *and classified by* $[\xi] \in \pi_{k-1}(\mathrm{SO}(n))$. *Then the element* $\alpha[\xi] \in \pi_{k-1}(\mathrm{SO}(n))$ *classifies the bundle* ξ^α *derived from* ξ *by reversal of the orientation of each fiber.*

Proof The proof lies in the definitions. □

Consider quaternion inversion $\beta : \mathbf{Q} \to \mathbf{Q}$ mapping $u \mapsto u^{-1}$, and note that $\beta : uv \mapsto v^{-1}u^{-1}$. This β is the degree -1 isometry of $\mathbb{S}^3 = \mathbf{Q}$ that fixes the center $\{-1, 1\}$ of $\mathbf{Q}$ and is the antipodal involution on the orthogonal 2-sphere consisting of all square roots of -1, in particular W.R. Hamilton's six operators $\pm\mathbf{i}$, $\pm\mathbf{j}$, $\pm\mathbf{k}$.

Proposition 2 *The involution α of $\pi_3(\mathrm{SO}(4)) \cong \mathbb{Z}^2$ exchanges the pairs* $([\lambda], [\rho])$ *and* $(-[\rho], -[\lambda])$. *In other words* $\xi^{\alpha}_{h,j} \cong \xi_{-j,-h}$.

Proof It will suffice to prove that $\alpha([\rho]) = -[\lambda]$ since applying α to both sides of this last equality then yields $[\rho] = -\alpha([\lambda])$. Now, $\alpha([\rho])$ is by definition the homotopy class of the map that to $u \in \mathbf{Q}$ associates the composed isometry $\beta\rho\beta(u) \in \mathrm{SO}(4)$. This isometry acts as follows on any point $x \in \mathbb{S}^3$:

$$x \overset{\beta}{\longmapsto} x^{-1} \overset{\rho}{\longmapsto} x^{-1}u \overset{\beta}{\longmapsto} u^{-1}x.$$

Thus $\beta\rho\beta(u)$ is the isometry $\lambda(u)^{-1}$. But $u \mapsto \lambda(u)^{-1}$ represents $-[\lambda]$ in $\pi_3(\mathrm{SO}(4))$. □

Assertion *Proposition* 2 *implies the Steenrod–Whitehead Theorem* 1 *above.*

Proof Apply to $\pi_3(\mathrm{SO}(5))$ the above commutation relation $s\alpha = \alpha s$. On the element $[\lambda]$, it tells us that $s\alpha[\lambda] = -s[\rho]$ is equal to $\alpha s[\lambda] = s[\lambda]$. Hence $s[\lambda] = -s[\rho]$ as required. □

Theorem 2 $\mathrm{p}_1(\xi_{h,j}) = \pm 2(h - j) \in \mathrm{H}^4(\mathbb{S}^4) = \mathbb{Z}$.

Remark The sign is constant (independent of h and j). Because he exploits only p_1^2, Milnor makes no effort in [M56d] to determine which sign is the correct one.

The proof and what follows involve some *tangent bundles* and related conventions. Given X a smooth oriented n-dimensional manifold, $\mathrm{T}X$ will denote its oriented tangent vector bundle with group $\mathrm{SL}(n)$. Its reduction to group $\mathrm{SO}(n)$ is (up to bundle isomorphism) a bijective operation, so the result is also denoted $\mathrm{T}X$. The associated *stable tangent bundle* of X will be denoted τX; it corresponds bijectively to the associated $\mathrm{SO}(n+k)$ bundle for $k \geq 1$.

When q is any stable characteristic class such as a Pontrjagin class, and X is a manifold, the notation $q(X)$ will often abbreviate $q(\tau X) = q(\mathrm{T}X)$.

Proof of Theorem 2 in outline By a Whitney sum formula (see [MSt74]), the first Pontrjagin class $\mathrm{p}_1(\xi) \in \mathrm{H}^4(\mathbb{S}^4) = \mathbb{Z}$ of an SO bundle over $\mathbb{S}^4$ is an additive function of the classifying element $[\xi] \in \pi_3(\mathrm{SO}) = \mathbb{Z}$. Thus, by Corollaries A, B, and C above, there exists an integer constant c such that $p_1(\xi_{h,j}) = c(h - j) \in \mathbb{Z} = \mathrm{H}^4(\mathbb{S}^4)$.

Appealing to Hirzebruch's proof [Hirz53] that the first Pontrjagin class $\mathrm{p}_1(\tau\mathbb{P}^8)$ of the quaternion projective plane $\mathbb{P}^8$ is $\pm 2 \in \mathbb{Z} = \mathrm{H}^4(\mathbb{S}^4) = \mathrm{H}^4(\mathbb{P}^8)$, and a geometric (stable) identification of $\xi_{1,0}$ or $\xi_{0,1}$ with the restriction of the tangent bundle $\mathrm{T}\mathbb{P}^8$ to the quaternion projective line $\mathbb{S}^4$ in $\mathbb{P}^8$, Milnor concludes that $c = \pm 2$. See [M56d, §3] for more details. □

2.10 Exotic Homotopy 7-Spheres Appear

Applied to any closed smooth oriented 8-manifold X, the Thom–Hirzebruch signature theorem expresses the signature $\sigma(X) \in \mathbb{Z}$ in terms of the 8- and 4-dimensional integral Pontrjagin classes $\mathrm{p}_2(X)$ and $\mathrm{p}_1(X)$ of the tangent bundle of X, as follows:

$$45\,\sigma(X) = 7\,\mathrm{p}_2(X) \cdot [X] - \mathrm{p}_1^2(X) \cdot [X] \quad \text{in } \mathbb{Z}.$$

Here, $\cdot[X]$ indicates Kronecker product of the immediately preceding cohomology class in $\mathrm{H}^8(X)$, with the orientation class $[X]$ generating $\mathrm{H}_8(X) \cong \mathbb{Z}$. In situations where we know nothing about $\mathrm{p}_2(X)$ we can get rid of its term by passing to residue classes modulo 7 thus:

$$3\,\sigma(X) = 0 + 6\,\mathrm{p}_1^2(X) \cdot [X] \bmod 7.$$

Then multiplying by 2 in the field $\mathbb{Z}_7$ and transposing gives:

$$2\,\mathrm{p}_1^2(X) \cdot [X] - \sigma(X) = 0 \bmod 7 \qquad (*)$$

The equation $(*)$ can serve as a test for the existence of X.

Let the left hand side of equation $(*)$ above be denoted $\Lambda(X)$, viewed as an expression with value in $\mathbb{Z}_7$.

We can use $(*)$ to quickly prove by 'reductio ad absurdum' that most of the homotopy 7-spheres among the 7-manifolds $M_{h,j}$ are *exotic*, i.e. not diffeomorphic to $\mathbb{S}^7$. Recall that $M_{h,j}$ is a homotopy 7-sphere precisely if $h + j = \mathrm{e}(\xi_{h,j}) = \pm 1$ in $\mathrm{H}^4(\mathbb{S}^4) = \mathbb{Z}$.

Initially, we fix $h + j = +1$ so that $M_{h,j} = M_{h,1-h}$ and $\mathrm{p}_1(W_{h,j})$ is $\pm 2(h - j) = \pm 2(2h - 1) \in \mathrm{H}^4(W) \cong \mathbb{Z}$. Aiming for a contradiction, we suppose $M_{h,1-h} \approx \mathbb{S}^7$ and close up $W_{h,1-h}$ by gluing on a copy of $\mathbb{B}^8$ to form a closed oriented smooth 8-manifold X^8. Its signature is $+1$, the self-intersection number of the zero-section of $\xi_{h,1-h}$ in $W_{h,1-h}$. Thus, substituting into the expression $\Lambda(X)$ above the first Pontrjagin class $\pm 2(2h - 1)$ just determined, and then simplifying, one gets $\Lambda(X) = (2h - 1)^2 - 1 \in \mathbb{Z}_7$. As $h \in \mathbb{Z}$ runs through the values $\ldots, 1, 2, 3, 4, 5, 6, 7, \ldots$, the expression $(2h - 1)$ takes the values $\ldots, 1, 3, 5, 0, 2, 4, 6, \ldots$ in $\mathbb{Z}_7$ and the expression $\Lambda(X) \in \mathbb{Z}_7$ takes the values $\ldots, 0, 1, 3, 6, 3, 1, 0, \ldots$ Since the resulting sequence of values of $\Lambda(X)$ is clearly periodic with period 7, we conclude that 5 out of 7 of the oriented homotopy 7-spheres in the sequence $M_{h,1-h}$, $h \in \mathbb{Z}$, yield $\Lambda(X) \neq 0$ and hence are exotic.

We will soon see (in the next section) that, up to orientation reversal, all the exotic 7-spheres that one can similarly detect for $h + j = -1$, have already been detected above for $h + j = +1$.

Remark Recalculating $\Lambda(X)$ as above but *without* Hirzebruch's result that, in the expression $\mathrm{p}_1(\xi_{h,j}) = c(h - j)$, the integer constant c is ± 2, one gets $\Lambda(X) = 2c^2(2h - 1)^2 \bmod 7$ with c still unknown. Thus, the same $M_{h,j}$ would be detected to be exotic homotopy 7-spheres *unless* c is divisible by 7; but if c were so divisible none would be detected!

2.11 Milnor's Invariant $\boldsymbol{\lambda}$ and Its Refinement μ

The expression $\Lambda(X)$ can also be defined for any oriented smooth compact 8-manifold X whose boundary is a homotopy sphere; of course, the orientation class $[X]$ then lies in $\mathrm{H}_8(X, \partial X)$ rather than in $\mathrm{H}_8(X)$. In order to prove that there exist several oriented smooth homotopy 7-spheres *no two of which are degree* $+1$ *diffeomorphic*, Milnor showed, using equation $(*)$ for closed smooth manifolds, and some homology calculations, that in this context, the value of $\Lambda(X) \in \mathbb{Z}_7$ is an invariant of the degree $+1$ diffeomorphism type of the oriented homotopy sphere ∂X. Thus, for ∂X the expression $\Lambda(X) \in \mathbb{Z}_7$ defines Milnor's $\boldsymbol{\lambda}$-*invariant* $\boldsymbol{\lambda}(\partial X) := \Lambda(X)$. In particular $\boldsymbol{\lambda}(M_{h,j}) = \Lambda(W_{h,j}) \in \mathbb{Z}_7$ whenever $h + j = \pm 1$. For details and a somewhat broader definition of the $\boldsymbol{\lambda}$-invariant, see [M56d, §1]. (This bold $\boldsymbol{\lambda}$ must not be confused with λ indicating left multiplication by quaternions.)

Thom proved in [Thom54] that every smooth oriented closed 7-manifold M has an oriented coboundary W^8. Thus Milnor's $\boldsymbol{\lambda}$-invariant in $\mathbb{Z}_7$ is defined for every oriented smooth homotopy 7-sphere M.

Since $\Lambda(-W) = -\Lambda(W)$, one has $\boldsymbol{\lambda}(-M) = -\boldsymbol{\lambda}(M)$. Since 7 is odd, this reveals that $\boldsymbol{\lambda}(M) \neq 0$, implies M is *chiral*—meaning that $M \not\approx -M$.

Milnor's invariant $\boldsymbol{\lambda}$ *is additive for connected sum of oriented smooth homotopy 7-spheres.* This is an important and easily proved property left unmentioned

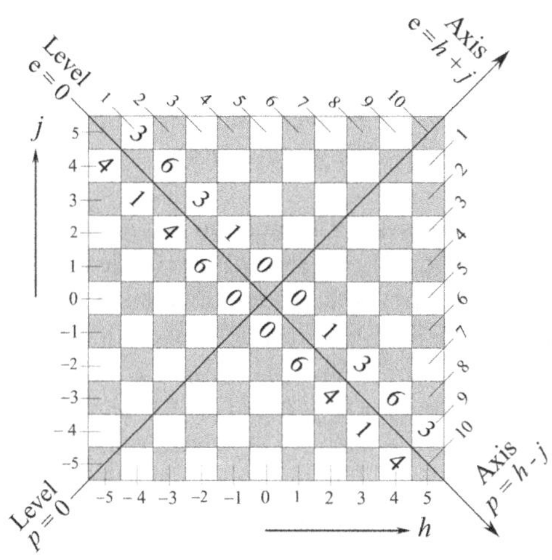

Fig. 1 $\boldsymbol{\lambda}$-*invariants of homotopy 7-spheres in the* h, j-*plane*. Each square of the checkerboard is centered on an integer point $(h, j) \in \mathbb{Z}^2 \subset \mathbb{R}^2$ to which is associated an oriented closed 7-manifold $M_{h,j}$ that is the total space of a smooth fibration $\partial\xi_{h,j}$ with fiber $\mathbb{S}^3$ group SO(4) and base $\mathbb{S}^4$. The origin (0, 0) is the intersection of the two diagonals drawn. The white squares are those for which $h + j \equiv 1 \bmod 2$. The homotopy 7-spheres among the $M_{h,j}$ are characterized by $h + j = \pm 1$ and so correspond to the white squares adjacent to the antidiagonal $x + y = 0$; on each is marked the value of Milnor's invariant $\boldsymbol{\lambda}(M_{h,j})$. A second coordinate system on $\mathbb{R}^2$ is given by the values of the functions $\mathrm{e} = x + y$ and $p = x - y$. At any integer point (h, j), the value of $\mathrm{e} = h + j$ gives the Euler class $\mathrm{e}(\partial\xi_{h,j}) \in \mathrm{H}^4(\mathbb{S}^4) = \mathbb{Z}$, and that of $p = h - j$ is up to sign $\frac{1}{2}$ of the value of the Pontrjagin class $\mathrm{p}_1(\partial\xi_{h,j}) \in \mathrm{H}^4(\mathbb{S}^4) = \mathbb{Z}$

in [M56d], but used in followup articles, for example [M59c] of 1959.[4] To prove it, let W and W' be oriented coboundaries for the smooth homotopy 7-spheres M and M' respectively. Then a boundary connected sum of W with W' is a suitable coboundary to show that $\lambda(M \sharp M') = \lambda(M) + \lambda(M')$.

Consequently, if $\lambda(M) \neq 0$ for an oriented smooth homotopy 7-sphere M, then M must generate, by iterated connected sum, at least seven oriented smooth homotopy 7-spheres no two of which are degree $+1$ diffeomorphic; indeed their λ-invariants assume all values in $\mathbb{Z}_7$. Incidentally, each of these of these generated homotopy spheres can be a *twisted sphere* (see next section) since the connected sum operation can clearly preserve this property.

Above we have calculated $\lambda(M_{h,j})$ when $h + j = +1$, while leaving aside those $M_{h,j}$ for which $h + j = -1$. This is legitimate because $M_{-h,-j} \approx -M_{h,j}$ by an obvious fibered map that reverses base orientation but preserves fiber orientation. (See the first lemma of the next section for more details.)

The values assumed by λ on the homotopy spheres $M_{h,j}$ are indicated in Fig. 1. Notice that they correspond to the white squares adjacent to the anti-diagonal, with Euler class $+1$ above the antidiagonal and Euler class -1 below. Above the antidiagonal, the sequence of values in $\mathbb{Z}_7$ assumed by the invariant λ is 7-periodic with repeating pattern $0, 1, 3, 6, 3, 1, 0$. Crossing the antidiagonal merely changes sign in $\mathbb{Z}_7$.

Thus the set of values assumed by Milnor's invariant λ on all the fibered and oriented smooth homotopy 7-spheres among the manifolds $M_{h,j}$ are precisely these five values: 0, 1, 6, 3, 4 mod 7.

Corollary *No smooth homotopy 7-sphere M with λ-invariant 2 or 5 in $\mathbb{Z}_7$ can admit a fibration $\mathbb{S}^3 \to M \to \mathbb{S}^4$ with group* SO(4).

In many situations, Milnor's 1959 evaluation of Θ_{4k+3}, $k \geq 1$, and notably of $\Theta_7 \cong \mathbb{Z}_{28}$ (see Sect. 2.1, Sect. 3.5 and [M59d]), provided no convenient means to decide when two given homotopy spheres are degree $+1$ diffeomorphic. A couple of years later (see [EK61], [EK62]), James Eells and Nicolas Kuiper constructed a powerful refinement μ of Milnor's lambda invariant λ that does the job *completely* for Milnor's homotopy 7-spheres $M_{h,j}$, $h + j = \pm 1$, and many others. It provides an Abelian group isomorphism $\mu : \Theta \to \mathbb{Z}_{28}$ whose reduction modulo 7 coincides with Milnor's λ. In addition to the Thom-Hirzebruch signature theorem, their construction uses an integrality property of the Atiyah-Hirzebruch $\hat{A}$-genus [AH59], [Hirz66a] (one already used in [M59d]).

For dimension 7, there is a succinct development of μ with its basic properties (quite analogous to those of λ) which one can find on the first page of Sect. 2 in [M65c]. For the calculations to follow, we need this property: If W is any smooth

[4]Why was this additivity left unmentioned in 1956? Perhaps because connected sum of two oriented connected n-manifolds without boundary had not yet been proved to be well-defined up to degree $+1$ diffeomorphism. However this does not invalidate the additivity as asserted and used here. The tubular neighborhood uniqueness theorem, see [Lang62], for the special case of point submanifolds easily implies well-definition of connected sum.

compact oriented 8-manifold with boundary $\partial W = M \simeq \mathbb{S}^7$, and $\mathrm{H}_*(W) = \mathrm{H}_*(\mathbb{S}^4)$, then

$$\boldsymbol{\mu}(M) = \big(p_1^2(W) \cdot [W, M] - 4\sigma(W)\big)/32 \mod 28.$$

Note that, since $\mathbb{Z}_{28}$ decomposes naturally as $\mathbb{Z}_7 \oplus \mathbb{Z}_4$, the value of $\hat{\boldsymbol{\lambda}} := \boldsymbol{\mu} \bmod 4$, together with that of $\boldsymbol{\lambda}$ in $\mathbb{Z}_7$, determines the value of $\boldsymbol{\mu}$ in $\mathbb{Z}_{28}$.

Recalling that the Pontrjagin class p_1 of $W_{h,1-h}$ is $\pm 2(h-j) \in \mathrm{H}^4(W) = \mathbb{Z}$, the value of $\boldsymbol{\mu}$ on $M_{h,1-h}$ simplifies to $h(h-1)/2 \bmod 28$, while that of $\boldsymbol{\lambda}$ (already calculated) is $4(2h-1)^2 = h(h-1)/2 \bmod 7$.

One can now quickly calculates that, for $h = 1, 2, \ldots, 8$, the values assumed by $\hat{\boldsymbol{\lambda}}$ are 0, 1, 3, 2, 2, 3, 1, 0; and they continue for $h > 8$ by repeating with period eight. This periodic sequence corresponds to the points $(h, 1-h)$ on the (infinite) checkerboard of Fig. 1 starting at the point $(1, 0)$ and stepping southeastwards on the squares of the white bishop's corridor just above the antidiagonal. Similarly (or consequently), the values of $\boldsymbol{\mu}(M_{h,1-h}) \in \mathbb{Z}_{28}$ for $h = 1, \ldots, 56 = 7 \times 8$ are 0, 1, 3, 6, 10, 15, 21, 0, 8, 17, 27, 10, 22, 7, 21, 8, 24, 13, 3, 24, 13, 3, 22, 14, 7, 1, 24, 20, 17, 15, 14; 14, ..., 2, 1. The last 28 values 14, ..., 3, 2, 1 (those after the semicolon) are the same as the first 28 values except for reversed order. In view of the weak SO(4) bundle isomorphisms described in the next section, this completes the degree +1 diffeomorphism classification of Milnor's spheres $M_{h,j}$, $h + j = \pm 1$.

From these $\boldsymbol{\mu}$ values, one can, for example, conclude that, of the 15 degree ± 1 diffeomorphism classes represented in $\Theta_7 \cong \mathbb{Z}_{28}$, those *not* represented by one of Milnor's $M_{h,j}$, $h + j = \pm 1$, are the three classes whose $\pm\boldsymbol{\mu}$ invariant is one of: ± 5, ± 9, ± 12.

2.12 Weak Equivalences Among the SO(4) *Disk Bundles*

There are three interesting families of fiber preserving diffeomorphisms among the 8-manifolds $W_{h,j}$ (and 7-manifolds $M_{h,j} = \partial W_{h,j}$). They are specified in Lemmas 1, 2, 3 below. Each arises from a weak O(4) bundle equivalence between the corresponding bundles $\xi_{h,j}$ that are *not* an SO(4) bundle equivalence. They have been helpful in studying Milnor's exotic 7-spheres and explain much of what meets the eye in Fig. 1. This section is not essential to what follows.

Recall that a weak $\mathrm{O}(2n)$ bundle equivalence $\xi \to \xi'$ of $\mathrm{O}(2n)$ bundles over a sphere $\mathbb{S}^{2n}$ is a $\mathrm{O}(2n)$ bundle map that induces on the common base $\mathbb{S}^{2n}$ either the identity or the (degree -1) antipodal map.

Recall that the *total space* $W_{h,j}$ of the SO(4) bundle $\xi_{h,j}$ over $\mathbb{S}^4$ is oriented by the convention that the intersection number (geometric or homological) of oriented *base* and oriented *fiber* be $+1$; it is thus -1 in the manifold $-W_{h,j}$. Each of these three orientations (of base, fiber and total space) can be construed as a preferred generator of a homology group isomorphic to $\mathbb{Z}$; for example; the fiber orientation amounts to a preferred generator of $\mathrm{H}_4(W, \partial W) \cong \mathbb{Z} \cong \mathrm{H}^4(W, \partial W)$. It follows that each diffeomorphism $f: W \to W'$ between two of the oriented manifolds $\pm W_{h,j}$ has a well defined degree ± 1 on each of base, fiber, and total space. The same is true for every homotopy equivalence $f: (W, \partial W) \to (W', \partial W')$.

Lemma 1 (Base orientation reversal) *For every pair (h, j) in $\mathbb{Z}^2$, the two fiber bundles $\xi_{h,j}$ and $\xi_{-h,-j}$ have Euler class and Pontrjagin class differing by sign only. Their naturally oriented total spaces $W_{h,j}$ and $W_{-h,-j}$ are degree -1 diffeomorphic by an* O(4) *bundle map $\xi_{h,j} \to \xi_{-h,-j}$ that reverses base orientation, but preserves fiber orientations. In particular, $W_{h,j} \approx -W_{-h,-j}$ and $M_{h,j} \approx -M_{-h,-j}$.*

Recall that $\approx$ indicates degree $+1$ diffeomorphism.

Proof Note that (h, j) and $(-h, -j)$ are related by the *central inversion* of $\mathbb{R}^2 \supset \mathbb{Z}^2$, i.e. rotation of angle 180° about the origin; it changes the sign of both $h + j$ and $h - j$. The weak O(4) bundle equivalence $\xi_{h,j} \to \xi_{-h,-j}$ can be constructed as follows. Pull back the bundle $\xi_{h,j}$ by a the antipodal (degree -1) involution r of $\mathbb{S}^4$; then the canonical map $r^*\xi_{-h,-j} \to \xi_{-h,-j}$ is essentially the required bundle map; indeed, by the Feldbau classification, $r^*\xi_{-h,-j}$ is isomorphic as SO(4) bundle to $\xi_{h,j}$. □

Lemma 2 (Fiber orientation reversal) *For every pair (h, j) in $\mathbb{Z}^2$, the two fiber bundles $\xi_{h,j}$ and $\xi_{-j,-h}$ have Euler class differing by sign, but Pontrjagin class unchanged. Their naturally oriented total spaces $W_{h,j}$ and $W_{-j,-h}$ are degree -1 diffeomorphic by a weak* O(4) *bundle equivalence $\xi_{h,j} \to \xi_{-j,-h}$ that reverses fiber orientation, but preserves base orientation. In particular, $W_{h,j} \approx -W_{-j,-h}$ and $M_{h,j} \approx -M_{-j,-h}$.*

Proof Note that (h, j) and $(-j, -h)$ are related by reflection in the anti-diagonal $h = -j$ of $\mathbb{R}^2$; it changes the sign of $h + j$ while leaving $h - j$ unchanged. This lemma is less trivial, but it follows directly from Propositions 1 and 2 in our calculation of Pontrjagin class in Sect. 2.9. □

Lemma 3 (Base and fiber orientation reversal) *For every pair (h, j) in $\mathbb{Z}^2$, the two fiber bundles $\xi_{h,j}$ and $\xi_{j,h}$ have the same Euler class, but Pontrjagin classes differing by sign. Their naturally oriented total spaces $W_{h,j}$ and $W_{j,h}$ are degree $+1$ diffeomorphic by a weak* O(4) *bundle equivalence $\xi_{h,j} \to \xi_{j,h}$ that reverses both fiber and base orientations. In particular, $W_{h,j} \approx W_{j,h}$ and $M_{h,j} \approx M_{j,h}$.*

Proof Note that (h, j) and (j, h) are related by reflection in the diagonal $h = j$ of $\mathbb{R}^2$; it changes the sign of $h - j$ while leaving $h + j$ unchanged. This lemma follows immediately from the preceding two. □

Theorem *Let $f: \xi_{h,j} \to \xi_{h',j'}$ be a weak* O(4) *bundle equivalence. Then either f respects all three orientations (of base, fiber, and total space) in which case f is an* SO(4) *bundle equivalence, or else f belongs to the unique one of the above 3* *classes that reverses the same orientations as f or equivalently induces the same sign changes on the Euler and Pontrjagin classes.*

Proof The case where f preserves all three orientations is trivial. If it does not, observe that there exists a weak O(4) bundle equivalence $g\colon \xi_{h,j} \to \xi_{h'',j''}$ that belongs to one of the three classes and reverses exactly the same orientations as does f. Then the composed weak O(4) bundle equivalence $h = gf^{-1} : \xi_{h',j'} \to \xi_{h'',j''}$ preserves all three orientations and hence is an SO(4) bundle map. From this it follows by composition that $f = hg$ is a weak O(4) bundle equivalence of the same class as g. □

Combining the above theorem with the classification Sect. 2.3 of the diffeomorphisms $W_{h,j} \to W_{h',j'}$ in terms of weak O(4) bundle equivalences $\xi_{h,j} \to W_{h',j'}$ we get:

Corollary 1 *The following conditions are equivalent*:

(i) $W_{h,j}$ *is degree* $+1$ *diffeomorphic to* $W_{h',j'}$.
(ii) (h, j) *and* $(h' j')$ *both lie in the same orbit of the orthogonal reflection of* $\mathbb{R}^2$ *in the main diagonal* $x = y$.
(iii) $(h', j') = (h, j)$ *or* $(h', j') = (j, h)$.

Corollary 2 *The following conditions are equivalent*:

(i) $W_{h,j}$ *is degree* ± 1 *diffeomorphic to* $W_{h',j'}$.
(ii) (h, j) *and* $(h' j')$ *of lie in the same orbit of the order* 4 *group generated by the orthogonal reflection in the diagonal* $x = y$ *and the orthogonal reflection in the antidiagonal* $x = -y$.
(iii) $|h + j| = |h' + j'|$ *and* $|h - j| = |h' - j'|$.

2.13 Twisted Spheres Appear

To complement his proof of exoticity, Milnor established a homeomorphism from every $M(a, 1) = M_{a,1-a}$ to $\mathbb{S}^7$ by a surprising 'deus ex machina'. He presented explicit (and amazingly simple!) smooth functions $h_a\colon M_{a,1-a} \to \mathbb{R}$ (for all $a \in \mathbb{Z}$), leaving the reader to check that each has just two critical points, both nondegenerate; they are necessarily the two points where the maximum and minimum values are attained. It had been known, from M. Morse [Mors25] of 1925 or G. Reeb [Reeb49] of 1949 (by different proofs!), that the existence of such a function h_a is equivalent to $M_{a,1-a}$ being a *twisted sphere*; in other words, $M_{a,1-a}$ is then of the form $M = B_+ \cup B_-$ where B_+ and B_- are diffeomorphic to the standard ball, while $B_+ \cap B_-$ is the common boundary sphere of B_+ and B_- in M. Every twisted sphere (in any dimension) is trivially homeomorphic (by coning) to a standard sphere. Better, the uniqueness clause of Whitehead's smooth triangulation theory [WhdJ40] of 1940 lets one show that, with any smooth triangulation, every smooth twisted sphere is PL homeomorphic to a standard PL sphere; this is an observation of R. Thom; see [M56e] for details.

The oriented twisted n-spheres of any dimension n, viewed up to degree $+1$ diffeomorphism, form an Abelian group whose addition arises from connected sum; it is denoted Γ_n. These groups appear as the obstruction groups for the theory of compatible smoothings of PL manifolds; see [Hir63] and earlier articles of J. Munkres cited there.

2.14 Conjecturally Nonsmoothable Manifolds Appear

In his 1956 colophon to [M56d], Milnor raised an important two-fold question:

(i) Is the closed PL 8-manifold $\widehat{W}_i$, obtained by first smoothly triangulating W_i and then adding the simplicial cone on its boundary, always (or sometimes) *non-smoothable* when $M_i = \partial W_i$ is one of Milnor's exotic 7-spheres?

By November of 1956, R. Thom had proved this to be *always* so, by applying his cobordism methods; see the reference in [M56e] to a 1956 preprint for [Thom58]. For all PL manifolds, he defined Pontrjagin classes in rational (not integral!) cohomology. These rational Pontrjagin classes are shown to agree with the rational images of the classical integral Pontrjagin classes whenever the PL manifold is smoothable, and at the same time, the usual Hirzebruch–Thom signature formula is established for PL manifolds. This discovery of Thom encouraged Milnor to reformulate his $\boldsymbol{\lambda}$-invariant and its later generalizations to take rational values modulo 1, beginning in [M56e] of November 1956.

(ii) The same question as (i) *modified* by allowing *arbitrary* smoothings of $\widehat{W}_i$ viewed now as a merely topological manifold.

Milnor observed that homeomorphism invariance of p_1 for smoothings of $W_i - \partial W_i$ would make such smoothings absurd. Indeed, his $\boldsymbol{\lambda}$-invariant would be $\neq 0$ for the boundary of a small smooth 8-disk in the smoothed $\widehat{W}_i$.

Although Milnor's examples above were the first non-smoothable topological manifolds to be *exhibited*, they were not the first closed topological manifolds to be *proved* non-smoothable. The first example was established in dimension 10 by Michel Kervaire in 1960 [Kerv60]. Shortly thereafter, in [Sm60], [Sm62], Steven Smale noticed similar but easier examples in infinitely many dimensions $12, 28, 44, \ldots$ based on the 1959 "plumbing construction" of Milnor [M59d] (see Sect. 3.3). Kervaire's also proved that his 10-manifold does not have the homotopy type of any closed smooth manifold. We will return to this 'homotopy non-smoothability' phenomenon in discussing surgery; see Sect. 3.3.

Several years later, in 1965, topological invariance of all rational Pontrjagin classes of smooth and PL manifolds was established by S. Novikov [Nov65]. Still later, rational Pontrjagin classes for all topological manifolds could be constructed by Thom's method assuring continued validity of the signature formula (see [Kahn72] of 1972); but that construction had to await a topological transversality theorem established by R. Kirby and this author, see [KirS71], [Sieb71] and [KirS77].

2.15 Comments on Motivation and Strategy

Milnor's article [M56d] that unveiled exotic spheres could well leave the reader with the impression that Milnor's original motivations included the ambition to test for exotic smooth structures. Milnor has recently made it very clear that exactly the contrary was true:[5]

> ... I was trying to study 3-connected 8-manifolds. The case $H_4 = 0$ seemed too hard. For $H_4 = \mathbb{Z}$, one can assume that the 4-skeleton is a 4-sphere. To build an 8-manifold, one can try to fatten it up by taking a tubular normal bundle neighborhood, and then adjoin an 8-cell. This worked so beautifully that I came up with many manifolds which couldn't possibly exist ...

> ... at the time I was unaware that there could be any difference between smooth and topological manifolds.

Thus his 1956 discovery of exoticity was the result of analyzing entirely unexpected paradoxes. This contrasts with F. Hirzebruch's keen awareness, already in 1954, of the possibility of exoticity, noted at the end of the historical Sect. 2.2.

Milnor's overall 1956 strategy for proving exoticity turned out to be fundamental in his soon to germinate surgery technique for manifold classification. Indeed, he had extracted subtle diffeomorphism invariants of closed manifolds M from rather basic characteristic class invariants of compact oriented manifolds W with boundary M. Following usage of Eells and Kuiper [EK62] (and others), I often call W a *coboundary* for M. This coboundary strategy of Milnor has remained a feature of most diffeomorphism classifications of closed manifolds in dimensions ≥ 5. As a doctoral student of Prof. Ralph Fox in Princeton, Milnor had been a 3-dimensional knot theorist, and a somewhat analogous strategy had been well known there for decades; namely, the use of a suitable surface whose boundary is the knot, as a tool to extract knot invariants such as J.W. Alexander's knot polynomial, or knot signature (see H. Seifert [Seif36] of 1936). Admittedly, in homology theory, the linking of cycles in manifolds of dimensions ≥ 3 had already been studied using chains of which those cycles are the boundary. On the other hand, treating an abstract closed manifold as if it were a cycle was distinctly new in the art of classifying manifolds up to diffeomorphism.

---

[5]The following two quotes are taken with permission from Milnor's correspondence with this author in October 2013. Compare the historical comments by Milnor in the introduction to Part I of [M07a].

Gradually, an explanation emerged of the need for Milnor's coboundary approach in [M56d]. It was proved that, for any homotopy 7-sphere Σ, the tangent vector bundle $T\Sigma$ is trivial (not just stably!); see the proof in Sect. 3.3. The bundle $T\mathbb{S}^7$ was long known to be trivial because $\mathbb{S}^7$ is the 7-sphere of Cayley numbers of unit norm; indeed any 7-frame at the north pole of $\mathbb{S}^7$ can be translated by right (or left) Cayley multiplication to produce a field of 7-frames trivializing $T\mathbb{S}^7$. Since $T\Sigma$ is also trivial, it offers no useful bundle invariants!

2.16 Smale's Dramatic Explanation of Milnor's 'deus ex machina'

Milnor's nondegenerate Morse functions $h_a\colon M(a,1) \to \mathbb{R}$ of 1956 with just two critical points begged to be explained; they had cost him some experimentation to get rid of parasitic critical points and were chronologically the last aspect of exoticity he established. But of that effort no wisdom or conjecture appeared in [M56d].[6] In 1960, Steven Smale gave dramatically general explanations using his high dimensional handlebody theory; see his articles [Sm60], [Sm61] and [Sm62]; his most widely used result (in [Sm62]) is the following h-Cobordism Theorem.

Data *Consider a compact smooth n-manifold W whose boundary is the disjoint union of closed* $(n-1)$*-manifolds V and* V'*. Suppose that the inclusions* $V \to W \leftarrow V'$ *are homotopy equivalences.*

h-Cobordism theorem ([Sm62], cf. Milnor's 1965 lectures [M65d]) *If W is simply connected and of dimension* ≥ 6*, then W is diffeomorphic to* $[0,1] \times V$.

Given the above data, the 'triad' $(W;\, V, V')$ is called an *h-cobordism*, and V, V' are said to be (unoriented) *h-cobordant*. The theorem (when applicable) clearly implies that V and V' are diffeomorphic by a diffeomorphism in the homotopy class $V \to V'$ forming a homotopy commutative triangle with the above inclusions $V \to W \leftarrow V'$.

When, furthermore, W is oriented (and its boundary components have the induced orientation) the induced homotopy equivalence $-V \to V'$ has degree $+1$. Thus one says that the triad $(W;\, V, V')$ gives an *oriented h-cobordism from* $-V$ *to* V', and one also says that $-V$ and V' are *degree* $+1$ *h-cobordant*. Smale's h-cobordism theorem, whenever applicable, establishes that $-V \approx V'$ by a diffeomorphism in the induced homotopy class.

It is an easy consequence of the above h-cobordism theorem that, *for* $n \geq 6$, *every smooth homotopy n-sphere is a twisted n-sphere* (as already defined). Thus,

[6] On the other hand, Milnor was able to explicitly construct such Morse functions for further examples of exotic spheres in [M59b]. And, M. Kervaire did as much for his 9-dimensional exotic sphere in [Kerv60].

Smale established in dimensions ≥ 6 a strong version of the Generalized Poincaré Conjecture.

The h-cobordism relation had been introduced by R. Thom in [Thom58] of 1958—under the less satisfactory name *J-equivalence*. In [M59d] of 1959, Milnor introduced the Abelian groups Θ_n of oriented smooth homotopy n-spheres up to oriented h-cobordism (then still called J-equivalence!), the addition corresponding to connected sum.

Of particular relevance here is Milnor's use of surgery in Theorem 5.15 of [M59d] to show that $\Theta_5 = 0$. This means that every smooth homotopy 5-sphere V bounds a compact contractible 6-manifold, say M. Then deletion from M of the interior of a smooth compact 6-disk in $\operatorname{Int} M$ yields a 6-dimensional h-cobordism from V to a standard 5-sphere $\mathbb{S}^5$, and application of Smale's h-cobordism theorem in dimension 6 then completes the proof that $M \approx \mathbb{S}^5$ establishing a strongest possible smooth Poincaré conjecture in dimension 5.

In Sect. 2.1 we have mentioned natural Abelian group isomorphisms $\Gamma_n \cong \mathscr{S}_n \cong \Theta_n$ valid for $n \geq 5$; they follow from the proved Poincaré conjectures mentioned above.

In [KM63], Kervaire and Milnor used surgery to prove that every smooth homotopy 4-sphere Σ^4 bounds a smooth compact contractible 5-manifold; thus $\Theta_4 = 0$. But the question whether Σ^4 is diffeomorphic to $\mathbb{S}^4$ is wide open; nothing (beyond countability) is known about the Abelian group $\mathscr{S}_4$ mentioned in Sect. 2.1. Moreover, in the mid 1980s, Simon Donaldson showed that the 5-dimensional (smooth) h-cobordism theorem is false in general, even when applied to complex surfaces; see [Don87].

For more information about low dimensions see [M11a].

Smale proved his h-cobordism theorem by geometrically manipulating and simplifying structures *internal* to W, namely 'handle decompositions' of W on V (or dually of W on V'). Re-expressed in the essentially equivalent language of Morse functions, this means he manipulated and simplified the critical point structure and Thom gradient structure of Morse functions $f\colon (W; V, V') \to ([0,1]; 0, 1)$ that have only generic critical points, all of them lying in the interior of W. Smale's theory is thus 'internal', or 'endoscopic'. In contrast, Milnor's coboundary approach to revealing and measuring the exoticity of homotopy spheres is 'external'.

There is a parallel piecewise linear (= PL) version of Smale's h-cobordism theorem. It has been expounded by several authors; see [RoS72] and references there. Also, there is a topological (= TOP) one; see [KirS77].

3 The Early Achievements of Surgery

3.1 A Rough Description of Surgery

In 1959, shortly before Smale's handlebody theory was announced, Milnor reported in [M59c] on a new *external* methodology called *surgery* initially conceived to systematicly classify smooth homotopy spheres of dimensions ≥ 5, up to degree $+1$

h-cobordism; it conspicuously orchestrated three specialties: geometric differential topology, homotopy theory, and algebra of quadratic forms.

Surgery is, roughly speaking, a procedure for finding an 'optimal' manifold in some sort of a cobordism class of closed manifolds of dimension $n \geq 5$. The manifolds and cobordisms involved usually have some extra structure, i.e., they are equipped with some auxiliary data that is to be suitably maintained at each step, for example an orientation and a normal framing. 'Optimal' initially meant simply connected and having least homology. In a preliminary phase of Milnor's surgery, it is necessary to assure simple connectedness or more. Milnor explained such preliminary steps in a pleasant expository article [M61a] of 1961. Each individual step in the surgery procedure involves an *elementary cobordism*, i.e., a triad $(W; V, V')$ of dimension $n+1$ with a Morse function on it having a single critical point, of index $\lambda+1$ say. The two boundary components of V and V' of W are thus non-critical levels of the Morse function, and are related by what is called a *spherical modification*; this alters the n-manifold V by extracting from V a copy of $\mathbb{S}^\lambda$ and replacing it, in a very precise way, by a copy of $\mathbb{S}^{n-\lambda-1}$. This optimization procedure usually encounters a single obstruction to complete success, one lying in a group here denoted $\mathbb{L}_n$, which depends only modulo 4 on the dimension $n \geq 5$ of the manifold undergoing spherical modifications. For surgery on simply connected manifolds, the obstruction group is normally one of these four:

$$\mathbb{L}_{4k} = \mathbb{Z}, \qquad \mathbb{L}_{4k+1} = 0, \qquad \mathbb{L}_{4k+2} = \mathbb{Z}_2, \qquad \mathbb{L}_{4k+3} = 0.$$

- the obstruction in $\mathbb{L}_{4k} = \mathbb{Z}$ arises as $1/8$ of the signature of a nonsingular *even* and symmetric intersection form in the dimension $2k$ integral homology of the $4k$-manifold undergoing surgery.
- the obstruction in $\mathbb{L}_{4k+2} = \mathbb{Z}_2$ is the Arf invariant that, along with rank, classifies a certain nonsingular quadratic form over $\mathbb{Z}_2$ that arises in the dimension $2k+1$ homology (coefficients in $\mathbb{Z}_2$) of the $(4k+2)$-manifold being surgered. See [Arf41], [LorR11] and [Brow72].

3.2 The Springtime of Surgery

Already early in 1959, in a widely circulated mimeographed article [M59d], Milnor developed surgery in a context general enough to make progress on the problem of determining in any given dimension $n \geq 5$ the Abelian group Θ_n classifying oriented smooth homotopy n-spheres up to degree $+1$ h-cobordism (or equivalently degree $+1$ diffeomorphism if we anticipate Smale's h-cobordism theorem); the group addition arises from connected sum. He made great progress in the dimensions $n = 4k+3$ by use of signature obstructions in $\mathbb{L}_{4k}$. For $n = 4k+1$, a preliminary result was announced involving Arf invariants in $\mathbb{L}_{4k+2}$. But these results were by no means definitive; presumably for that reason, publication was deferred to the almost definitive 1963 article [KM63] written in collaboration with Michael Kervaire.

Milnor thanks Thom for communicating, in an example, the central idea of surgery. He also mentions early surgery done independently by A. Wallace [Waa60] and attributes to him the term *spherical modification*. In [Waa60] Wallace had studied a medly of surgery and handlebody theory [Waa60] before Smale announced his solution of the Generalized Poincaré Conjecture (smooth version). In [Waa61] he was able to give a quite readable proof of that conjecture in dimensions ≥ 6, based on Smale's somewhat sketchy announcement [Sm60]. His subsequent efforts to prove the h-cobordism theorem fell short.

3.3 The First Flowering of Surgery

In the Kervaire–Milnor article [KM63] of 1963, the surgery obstruction groups featured *triumphantly* in an almost complete calculation of the groups Θ_n of smooth homotopy spheres, for $n \geq 5$. The construction of a surgery obstruction theory as sketched above constitutes the major part of their article. The fact that surgery with these obstruction groups $\mathbb{L}_n$ occurs similarly and decisively in several classification problems of manifold theory became apparent only later.

The first formally published announcement of results obtained using surgery obstructions appeared in [M59c] of 1959 where Milnor succinctly stated two revolutionary results:

(a) The group Θ_n of smooth homotopy n-spheres is finite for all $n \neq 3$. (I will sketch the surgical proof presented by Kervaire and Milnor in [KM63] of 1963.)
(b) For $n = 4k - 1 > 3$, the group Θ_n contains a cyclic subgroup, (namely bP_n) of a precisely specified order > 1 (see below for more details).

In reading about the calculations of Θ_n, one cannot fail to notice a pervading dependence on information from R. Bott's famous periodicity theorems of 1957–1959 for the stable homotopy groups $\pi_{n-1}(\mathrm{SO}) := \pi_{n-1}(\mathrm{SO}(N))$, $N \geq n+1$. For $n = 2, 3, 4, \ldots$ these classify stable oriented vector bundles over $\mathbb{S}^n$. This sequence of groups depends on n only modulo 8 and for $n \equiv 1, 2, 3, 4, 5, 6, 7, 8$, the group is:

$$\mathbb{Z}_2, \quad \mathbb{Z}_2, \quad 0, \quad \mathbb{Z}, \quad 0, \quad 0, \quad 0, \quad \mathbb{Z}.$$

This and other Bott periodicities are explained *ab initio* in Milnor's lectures on Morse theory [M63d] of 1963 that were beautifully written up with his students M. Spivak and R. Wells.

Also indispensible (although less conspicuous) are results from Frank Adams proof [Ad58], [Ad60] of the 'Hopf Invariant 1' conjecture.

Here is a another prerequisite stated in a widely useful form that can be rather simply proved geometrically.

Theorem 1 *If M and M' are closed smooth manifolds and $f\colon M \to M'$ is a homotopy equivalence, then f induces a well defined fiber homotopy equivalence of their stable tangent bundles.*

The first proof [At61] of this theorem was by M. Atiyah in 1961 using Spanier–Whitehead duality. A year earlier, Milnor and Spanier [MSp60] had proved the case of homotopy spheres (needed in [KM63]), and had conjectured the general case.

A naive geometric proof of this was implicit in the proof by B. Mazur [Maz61] in 1961 of his 'stable diffeomorphism theorem' for such manifolds; and a neat variant was implicit in Milnor's 1961 article [M61b] giving counterexamples to the 'Hauptvermutung' for simplicial complexes.

Remark The geometric proof above can easily be generalized to compact manifolds with boundary, provided one supposes f to be a pair homotopy equivalence of pairs $f : (M, \partial M) \to (M', \partial M')$.

A widely used proof is even simpler to explain, as follows. Let W' be a tubular neighborhood of an embedding $i\colon M' \to \mathbb{R}^n$, $n > 2 \dim M'$. Approximate the composition $i \circ f$ by an embedding $j\colon M \to \mathbb{R}^n$ whose image lies in $\operatorname{Int} W'$ and choose a tubular neighborhood W of $j(M)$ lying in $\operatorname{Int} W'$. Then the triad $(W' - \operatorname{Int} W; \partial W, \partial W')$, can be shown to be an h-cobordism from the normal sphere bundle of M to the normal sphere bundle of M'. It induces a fiber-homotopy equivalence of these normal bundles. From it, the wanted fiber homotopy equivalence of stable tangent bundles can be rather formally deduced.

This argument can be adapted to define for any finite simply connected complex enjoying Poincaré duality a stable normal (or tangent) sphere bundle in the homotopy category. Michael Spivak's thesis [Spi67] written under Milnor's direction develops this idea, which assumed great importance in the surgery theory of the 1960's and beyond.

Corollary *If Σ^n is a smooth homotopy n-sphere, then under the* J*-homomorphism* [WhdG42] $\mathrm{J}\colon \pi_{n-1}\,\mathrm{SO} \to \Pi_{n-1}$ *the classifying element of the stable tangent (or the stable normal) bundle of Σ maps to zero.*

For the moment it will suffice to accept that, for $n \geq 2$, there is a commutative diagram:

$$\begin{array}{ccccc}
\pi_{n-1}(\mathrm{SO}) & \xrightarrow{\ \mathrm{J}\ } & \pi_{n-1}(\mathrm{G}) & \cong & \Pi_{n-1} \\
\partial \uparrow \cong & & \cong \uparrow \partial & & \\
\pi_n(\mathrm{B_{SO}}) & \longrightarrow & \pi_n(\mathrm{B_G}) & &
\end{array}$$

where $\mathrm{B_G}$ classifies stable spherical fiber spaces, and the bottom arrow comes from viewing the complement of the zero section of a vector bundle as a spherical fiber space. Definitions of Π_{n-1} and of the J-homomorphism (due to H. Hopf, L.S. Pontrjagin, G. Whitehead and others) will appear presently.

Theorem 2 *Every smooth homotopy n-sphere Σ^n is stably parallelizable; in other words, its tangent bundle $\mathrm{T}\Sigma^n$ is stably trivial.*

Proof This is established by a sequence of miracles:

(a) For $n \equiv 3, 5, 6, 7$, modulo 8, Bott periodicity alone tells us that $\mathrm{T}\Sigma^n$ is stably trivial.
(b) For $n \equiv 0$, or 4 modulo 8, Σ^n is stably parallelizable. Indeed, the homotopy class in $\pi_{n-1}(\mathrm{SO})$ classifying the stable tangent bundle $\tau\Sigma^n$ of Σ^n is necessarily proportional to the Pontrjagin class $p_{n/4}(\tau\Sigma^n)$, which is 0 by the Thom–Hirzebruch signature formula, because vacuously $\sigma(\Sigma^n) = 0$.
(c) For $n \equiv 1$ or 2, modulo 8, J.F. Adams [Ad58], [Ad60] and [AdA66] proved that $\mathrm{J}: \mathbb{Z}_2 = \pi_{n-1}(\mathrm{SO}) \to \Pi_n$ is injective. The corollary above now implies that Σ^n is stably parallelizable. Incidentally, Adams' cited work makes good use of a striking 1958 theorem of Milnor [M58b] proving that the dual of the Steenrod algebra (for any prime p) is a polynomial algebra. □

I have mentioned the following corollary as a justification for Milnor's 'coboundary' approach to detecting exotic 7-spheres.

Corollary *Every smooth homotopy 7-sphere* Σ *is parallelizable, i.e., its tangent bundle* $\mathrm{T}\Sigma$ *with 7-dimensional fibers is itself trivial.*

Proof The stabilization map s for n-dimensional vector bundles over $\mathbb{S}^n$ appears in the following exact sequence arising from the fibration $\mathrm{SO}(n) \to \mathrm{SO}(n+1) \to \mathbb{S}^n$, which is the principle $\mathrm{SO}(n)$ bundle of the tangent bundle of $\mathbb{S}^n$:

$$\mathbb{Z} = \pi_n\big(\mathbb{S}^n\big) \xrightarrow{\partial} \pi_{n-1}\big(\mathrm{SO}(n)\big) \xrightarrow{s} \pi_{n-1}\big(\mathrm{SO}(n+1)\big) = \pi_{n-1}(\mathrm{SO}).$$

Here $\partial(1)$ is Feldbau's classifier for $\mathrm{T}\Sigma^n$; there is a similar sequence for any principal bundle over a sphere, (see [Strd51]).

Since $\mathrm{T}\mathbb{S}^7$ is trivial (as already observed) the exactness at $\pi_{n-1}(\mathrm{SO}(n))$ tells us that s is injective when $n = 7$. Hence $\mathrm{T}\Sigma^n$ is also (non-stably) trivial. □

Remark A more strenuous argument establishes the following generalization: *For any n, if* $\mathbb{S}^n \to \Sigma$ *is a homotopy equivalence from* $\mathbb{S}^n$ *to a smooth homotopy n-sphere* Σ*, then the vector bundle* $f^*(\mathrm{T}\Sigma^n)$ *is isomorphic to* $\mathrm{T}\mathbb{S}^n$. This implies that no tangent bundle invariant of $\mathrm{T}\Sigma^n$ can ever detect an exotic sphere Σ^n. For discussion and references, search on Internet for "Math Overflow" and "exotic spheres".

For tabulated values of the groups of exotic spheres involved see [KM63] noting the one correction on page 97 of [M07a] that doubled the order of bP_{19} to 261, 632. See also the Wikipedia entry for Exotic sphere. It seems that the tables of [KM63] have not yet been systematically extended!

We next discuss aspects of the spectacular 1963 Kervaire–Milnor classification [KM63] of smooth homotopy spheres, emphasizing the role of surgery in it.

3.4 An Exact Sequence Entrapping Θ_n, for $n \geq 5$

The Kervaire–Milnor calculation focused on a sequence

$$0 \to \mathrm{bP}_{n+1} \xrightarrow{i} \Theta_n \xrightarrow{j} \Pi_n/\,\mathrm{Image(J)} \qquad (*)$$

defined as follows:

- bP_{n+1} is the subgroup of Θ_n generated by homotopy n-spheres that occur as the boundary of a compact smooth $(n+1)$-manifold having trivial stable normal bundle. The map i is simply inclusion.
- Π_n is by definition the stable homotopy group $\pi_{n+N}(\mathbb{S}^N)$, $N \geq n+1$. By the famous construction of Pontrjagin–Thom (see Milnor's little 1965 primer [M65e]), it is naturally isomorphic to the cobordism group of closed oriented n-submanifolds of $\mathbb{R}^{n+N}$ each equipped with a trivialization of normal bundle (also called a *framing*). Π_n was proved to be finite for all n by J.P. Serre in [Ser51] of 1951. It is, in principle, calculable for each single n, but calculations are notoriously hard. For tabulations of these groups Π_n, consult the Wikipedia entry on Homotopy groups of spheres.[7]
- The classical J-homomorphism $\mathrm{J}\colon \pi_n(\mathrm{SO}) \to \Pi_n$ exploits a map $\mathbb{S}^n \to \mathrm{SO}$ to choose a framing of the standard sphere $\mathbb{S}^n$ in $\mathbb{R}^{n+N}$.
- j is defined on the class of a homotopy n-sphere M as follows. Embed M in $\mathbb{R}^{n+N}$ and (recalling that the stable tangent bundle τM is trivial) choose a trivialization of its normal bundle (also called a *normal framing*). This choice of framing clearly has no influence in $\Pi_n/\,\mathrm{Image(J)}$.

Observation *The sequence* $(*)$ *is exact.*

Proof This is immediate from the above definitions. □

Theorem *In the sequence* $(*)$ *the map* $j\colon \Theta_n \to \Pi_n/\,\mathrm{Image(J)}$ *is surjective for* $n \geq 5$, *except (perhaps) when* $n \equiv 2 \bmod 4$. *In other words, for these dimensions,* j *induces an isomorphism* $\Theta_n/\,\mathrm{Image(J)} \to \mathrm{Coker(J)}$.

Proof Assuming $n \not\equiv 2 \bmod 4$, we show that, given any normally framed n-manifold M in $\mathbb{R}^{n+N}$, it can be surgered to become a framed homotopy n-sphere. If $n \equiv 1$ or $3 \bmod 4$, the surgery obstruction group $\mathbb{L}_n$ is 0 so the surgery is possible. If $n = 4k \geq 8$, then the obstruction lies in $\mathbb{L}_{4k} = \mathbb{Z}$ and is the signature σM. But the Hirzebruch–Thom signature theorem tells us that M stably parallelizable implies that $\sigma M = 0$; thus, once again, the surgery is possible. □

Complement (For same data introduced for the theorem but $n = 4k+2$) *If* $n \equiv 2 \bmod 4$, *then the Arf-invariant obstruction to surgery on* M *lying in* $\mathbb{L}_{4k+2} = \mathbb{Z}_2$ *gives a well defined homomorphism* $\kappa\colon \Pi_{4k+2} \to \mathbb{Z}_2$.

[7] http://en.wikipedia.org/wiki/Homotopy_groups_of_spheres.

This complement follows straightforwardly from definitions. However the problem of determining when κ is surjective and when it is 0 remained open for 50 years and is still (slightly) open.

Corollary *For* $4k \geq 8$, $\Theta_{4k} = \Pi_{4k}$.

There is still no uniform way to calculate Π_{4k} for all k!

Proof Every element of Π_{4k} is represented by a framed homotopy sphere Σ. Since $\pi_{4k}(\mathrm{SO}) = 0$ by Bott periodicity, any other framing of Σ determines the same element of Π_{4k}. If Σ and Σ' represent the same element of Π_{4k}, then surgery (exploiting $\mathbb{L}_5 = 0$) shows they are h-cobordant. □

3.5 Analysis of the Subgroup bP *of* Θ_n

When this subgroup is nontrivial, it has a satisfyingly explicit generator obtained by Milnor's now famous *plumbing construction* introduced in [M59d] (see also [KM63] and [M64f]).

Theorem *In the sequence* (∗), *the subgroup* bP_{n+1} *of* Θ_n, $n \geq 5$, *is finite cyclic, and*:

(a) *For* $n \equiv 0$ *or* 2 *mod* 4, *the group* $\mathrm{bP}_{n+1} = 0$.
(b) *For* $n \equiv 3$ *mod* 4, *the order* $|\,\mathrm{bP}_{n+1}\,|$ *is* 1/8 *of the least positive signature of any connected closed* $4k$*-manifold that is* almost parallelizable—*which means parallelizable in the complement of a point. A canonical generator of* bP_{n+1} *is represented by any homotopy* $(4k-1)$*-sphere that is boundary of a smooth compact parallelizable with signature* 8, *for example the manifold* $P(\mathrm{E}_8, 4k)$ *obtained by plumbing together* 8 *copies of the compact unit tangent disk bundle of the standard* $2k$*-sphere according to the famous* E_8 *tree*:

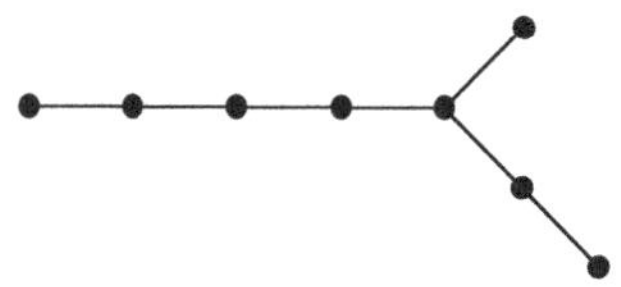

in which the vertices represent disk bundles and the edges represent plumbing operations.
(c) *For* $n \equiv 1$ *mod* 4, *the group* bP_{n+1} *is either* $\mathbb{Z}_2$ *or* 0. *More precisely, it is always generated by the boundary* ∂P *of the compact smooth* $(4k+2)$*-manifold* P *obtained by plumbing together two copies of the unit tangent disk bundle of* $\mathbb{S}^{2k+1}$ *according to the simple graph* •—•. *Furthermore the homotopy sphere* ∂P *is exotic if and only if the homomorphism* $\kappa : \Pi_{4k+2} \to \mathbb{Z}_2$ *(introduced above) is zero.*

Remark At the time of publication of [KM63], it had been proved that the groups bP_{10} (see Kervaire's [Kerv60] of 1960) and bP_{18} are $\mathbb{Z}_2$, and that bP_6 and bP_{14} are 0.

The proofs of (b) and (c) use the following lemma whose geometric proof is elementary.

Lemma *A closed connected manifold M of dimension m is almost parallelizable if and only if there exists a stable vector bundle γ over $\mathbb{S}^m$ and a map $f : M \to \mathbb{S}^m$ such that the induced bundle $f^*\gamma$ over M is bundle isomorphic to τM.*

Proofs in outline of the theorem By the definitions above, any element of bP_{n+1}, $n \geq 5$, can be represented by a smooth homotopy n-sphere Σ that is the boundary of a compact framed $(n+1)$-manifold W.

Proof of (a) For $n+1 \equiv 1$ or 3 mod 4, the surgery obstruction group $\mathbb{L}_{n+1}$ is 0; hence the $(n+1)$-manifold W can be surgered down to become a homotopy disk. Thus the class $[\Sigma]$ of Σ in bP_{n+1} is 0. □

Proof of (b) When $n+1 = 4k$, the obstruction group $\mathbb{L}_{n+1} = \mathbb{Z}$. But, if by luck σW is 0, one can proceed as in (a). We will reduce the matter to this lucky case. The signature σW of W is clearly defined since the PL manifold $\widehat{W}$ obtained by adding to W the cone on ∂W satisfies Poincaré duality. Furthermore, the quadratic form on $2k$-homology of W takes only even values since W is framed. Hence, the theory of even quadratic forms tells us that σW is a multiple of 8, say $8s$. We can assume without loss that s is negative. Write ∂P for $\partial P(E_8, 4k)$. From the disjoint sum of W with s copies of $P(E_8, 4k)$ we can readily construct a framed manifold W', whose boundary the $(1+s)$-fold connected sum $\partial W' = \partial W \# \partial P \# \partial P \# \cdots \# \partial P$, and whose signature is $-8s + 8s = 0$. The lucky case then applies to W'; thus the class $[\partial W']$ in bP_{4k} is 0. But $[\partial W'] = [\Sigma] + 8[\partial P]$ whence $[\Sigma] = -8[\partial P]$. This shows that $[\partial P] := [\partial P(E_8, 4k)]$ generates the group bP_{4k}.

It remains to show that the generator $[\partial P(E_8, 4k)]$ of bP_{4k} is of finite order, or equivalently that there exists a closed almost parallelizable manifold M of dimension $4k$ whose signature is non-zero. One such M can be derived from any nontrivial stable vector bundle γ over $\mathbb{S}^{4k}$, as follows.

Pause *If you seek information more precise than the mere finiteness of* bP_{4k}, *you can go from here directly to the complements stated below and the references given there.*

Such a γ is classified by a nonzero element $[\gamma]$ in $\pi_{4k}(B_{SO}) \cong \pi_{4k-1}(SO) \cong \mathbb{Z}$. The Pontrjagin class $p_k(\gamma) \in H^{4k}(\mathbb{S}^{4k}) \cong \mathbb{Z}$ is also nonzero [MSt74]. Both $[\gamma]$ and $p_k(\gamma)$ are additive for bundle sum and take values in an infinite cyclic group. The 'forgetting map' from vector bundles to spherical fiber spaces induces a map on stable homotopy groups $\varphi \colon \pi_{4k}(B_{SO}) \to \pi_{4k}(B_G) \cong \Pi_{4k-1}$, and the target group is finite by Serre [Ser51]. Consequently, for some finite positive integer s the s-fold vector bundle sum $s\gamma = \gamma \oplus \cdots \oplus \gamma$ is stably fiber homotopy trivial but has nonzero Pontrjagin class. Thus, if γ is sufficiently stabilized (say to dimension m), there exists a proper map from the total space $\mathrm{E}(s\gamma)$ onto $\mathbb{R}^{sm}$ that has degree $+1$ on each fiber of $s\gamma$. Then the preimage, M say, of a generic point in $\mathbb{R}^{sm}$ is a closed normally framed submanifold of $\mathrm{E}(s\gamma)$; its stable tangent bundle τM is therefore induced from the tangent bundle of $\mathrm{E}(s\gamma)$, which, in turn, is induced from the stable bundle $s\gamma$ over $\mathbb{S}^{4k}$; thus, by the preceding lemma, M is almost parallelizable. The signature formula [Hirz56], [MSt74] and the non-zero Pontrjagin class $\mathrm{p}_k(s\gamma)$ then show that the signature σM is non-zero. This completes the proof of case (b). □

Proof of (c) This is left aside since it is *easier* than the arguments above for case (b), see [KM63]. It is easier because the result falls far short of calculating bP_{4k+2}. □

3.6 Complements Concerning Boundaries of Parallelizable Manifolds

1st Complement Concerning bP in Dimensions $4k-1$ In [KM60] and [KM63], Kervaire and Milnor determine, using an integrality theorem for the Atiyah-Hirzebruch $\hat{A}$ genus of closed smooth 'spin' manifolds (see [AH59] and Theorem 26.3.2 in [Hirz66a]), that the order of bP_{4k}, $k \geq 2$, is exactly

$$2^{2k-2}\left(2^{2k-1}-1\right)\mathrm{Numerator}(4\,\mathrm{B}_k\,/k)$$

where B_k is the k-th Bernoulli number.[8] In Appendix B in [MSt74] of 1974, Milnor reveals, with massive support from the classical literature, that $\mathrm{Numerator}(\mathrm{B}_k\,/k)$ is ≥ 1 and grows monotonicly and exponentially with a certain asymptotically precise growth estimate. A fortiori, $|\,\mathrm{bP}_{4k}\,| > 1$ for $k \geq 2$. Homotopy theory uses closely related information: the image of the stable J-homomorphism $\mathrm{J} : \pi_{4k-1}(\mathrm{SO}) \to \Pi_{4k-1}$ is known to have order equal to the denominator of $\mathrm{B}_k\,/4k$; a 1989 reference concerning this is [DaM89].

2nd Complement Concerning bP in Dimensions $4k-1$ In his articles [Sm60] and [Sm61] on handlebody theory, Smale gives an easy proof of the non-smoothability of the closed topological 12-manifold $\hat{P}^{12}$ obtained from the plumbing manifold $P^{12} := P(\mathrm{E}_8, 12)$ defined above by adding the cone on its exotic 11-sphere boundary. As already mentioned, a few months before Smale [Sm60] was submitted, Kervaire, in [Kerv60], had established (nontrivially!) the stronger *homotopy non-smoothability* (see below) a similar 10-dimensional manifold derived from the graph •—•.

[8] Beware that number theorists denote this rational number $|B_{2k}|$.

Here is Smale's easy argument. Seeking a contradiction suppose that M^{12} is a smoothing of $\hat{P}^{12}$. Clearly the complement M_0^{12} of a point in M^{12} is homotopy equivalent to P^{12} or a wedge of 8 copies of $\mathbb{S}^6$. Since, by Bott periodicity, $\pi_5(\mathrm{SO}(N)) = 0$ for $N \geq 7$, it follows easily that the stable normal bundle of M_0^{12} is trivial. Thus M^{12} is almost parallelizable with signature 8 in contradiction to $|\,\mathrm{bP}_{4k}\,| > 1$ for $k \geq 2$.

This argument can be strengthened to show that $\hat{P}^{12}$ is *homotopy non-smoothable*: i.e., it is not even homotopy equivalent to any closed smooth manifold. Better, it can be enhanced as follows to prove the homotopy non-smoothability of half of the analogous PL manifolds $\hat{P}^{4k}$, $k \geq 2$, derived from the graph E_8. Namely those for which $4k \equiv 4 \bmod 8$, or equivalently $2k \equiv 2 \bmod 4$. In these dimensions, Bott periodicity tells us that there are at most 2 distinct oriented stable vector bundles over $\mathbb{S}^{2k}$. By Adams [Ad60] any one that is nontrivial is also fiber homotopically nontrivial. This lets one prove the following homotopy invariance.

Proposition *Consider a smooth compact parallelizable $4k$-manifold P, with $4k \equiv 4$ modulo 8, and $k \geq 2$, that is homotopy equivalent to a wedge of $2k$-spheres, and whose boundary is a homotopy sphere. Suppose that there exists a closed smooth $4k$-manifold M homotopy equivalent to the closed* PL *manifold $\widehat{P}$ obtained by adding to P the cone on ∂P. Then M is almost parallelizable and ∂P^{4k} is diffeomorphic to $\mathbb{S}^{4k}$.*

Proof Let M_0 be the complement $M - \operatorname{Int} D$ of the interior of a small compact smooth $4k$-disk D in M. A geometric lemma proves that $(M_0, \partial M_0)$ is homotopy equivalent to $(P^{4k}, \partial P^{4k})$. By the obvious generalization to manifolds with boundary, of the normal (or tangent) bundle fiber homotopy equivalence theorem (discussed in Sect. 3.3), one obtains a fiber homotopy equivalence between the (trivial!) stable normal bundle of P^{4k} and that of M_0. To now prove that the stable normal bundle υ of M_0 is trivial (not just fiber homotopically trivial), it suffices to transfer it by homotopy equivalence to P^{4k} and again by homotopy equivalence to the a homotopy equivalent wedge of $2k$-spheres. Over each $2k$-sphere, it is obviously fiber homotopy trivial; and hence trivial, whence applying the Theorem of Sect. 3.5, $\sigma(\hat{P}^{4k}) = \sigma(M)$ is divisible by $8|\,\mathrm{bP}_{4k}\,|$. This implies $\partial P^{4k} \approx \mathbb{S}^{4k}$. □

1st Complement Concerning bP in Dimensions $4k+1$: The Kervaire Conjecture Milnor and Kervaire conjectured cautiously in the last lines of [KM63] that $\mathrm{bP}_{4k+2} = \mathbb{Z}_2$ *except* for the dimensions 2, 6, and 14. Motivation came from the fact that parallelizability of $\mathbb{S}^3$, $\mathbb{S}^7$, and $\mathbb{S}^{15}$ was used to establish those exceptions. This is usually called the *Kervaire Conjecture* in honor of M. Kervaire's pioneering proof of the first non-exceptional case $n = 10$. In 1966, Brown and Peterson [BP66] confirmed the conjecture for all dimensions of the form $8k+2$. In 1969, W. Browder [Brow69], confirmed the conjecture for all dimensions *not* of the form $2^x - 2$. However the conjecture had to be revised when, in the 1970s and 1980s, it was proved wrong in dimensions $4k+2 = 30$ and 62, see [BaMT70] and [BaJM83]. Subsequent efforts to prove the conjecture wrong in all dimensions of the form $2^x - 2$ bore no fruit. Finally, in 2009 and 2010, M. Hill, M. Hopkins, and D. Ravenel posted at http://arXiv.org/ a highly sophisticated proof [HillHR09] for all dimensions *strictly* greater than $2^7 - 2 = 126$; see H. Miller [Miller10] for a very favorable summary. See Ravenel's Internet page [Ravenel] for many lectures including much history and folklore, notably a lecture [Rav12] from February 2012.

2nd Complement Concerning bP in Dimensions $4k + 1$: Brumfiel's Problem
In all dimensions n of the form $2^x - 2 > 126$ the article [HillHR09] establishes the Kervaire conjecture that $\mathrm{bP}_n \cong \mathbb{Z}_2$. For dimensions, $2^x - 2$, Greg Brumfiel has recently (re)posed the question whether this subgroup bP_n of Θ_{n-1} is also a retract of Θ_{n-1} thus making bP_n a direct summand of Θ_{n-1}. For all *other* dimensions $n \geq 6$ Brumfiel showed that bP_n is indeed a retract of Θ_{n-1}. See his article [Bru70].

4 A Metamorphosis

I conclude by briefly describing the metamorphosis of Milnor's surgery into a classification method applicable in principle to most compact manifolds of various species. One point I would like to make, and which may be surprising to newcomers, is that one can profitably perceive the classical surgery of [M59d] and [KM63] as a vital part of today's surgery.

4.1 Milnor's Microbundles

Before the possibility of exploiting surgery methods for quite general compact manifolds was publicly discussed, Milnor quietly laid the groundwork for parallel treatment of piecewise linear and topological manifolds along with smooth manifolds. He felt the lack of a well working notion of tangent and normal bundle in the world of piecewise linear or topological manifolds, and he responded by introducing microbundles for that role. They came in three variants: piecewise linear [M61f], topological [M63c], [M64b], and (trivially) smooth, and provided a uniform context for the development of manifold structure theories mediating between the notions of topological, piecewise linear, and smooth manifold structure, not to mention homotopy type. In [Thom60], René Thom had conjectured that the problem of introducing a compatible smooth structure on a piecewise linear manifold is equivalent to a bundle reduction problem. For Thom's conjecture, Milnor's microbundles provided a clear and plausible meaning and also a proof modulo a stabilization. Several other topologists gradually established the conjectured theories, see [Mu66], [Mu67], [Mu68], [HirM74], [Sieb71], [KirS77] (although smoothings of *topological* 4-manifolds are still mysterious). The obstruction groups to smoothing turned out to be, in high dimensions, Milnor's groups of homotopy spheres! When the time was ripe (1964–70) these theories, in turn, helped develop surgery theory on piecewise linear and then on topological manifolds of dimension ≥ 5; see [BrowH66], [Sieb71], [KirS77], [Mar77].

4.2 Surgery for Classical Smooth Manifolds

Perhaps the most abrupt and decisive leap forward for surgery theory in dimensions ≥ 5 was its extension from smooth closed oriented homotopy spheres to all smooth closed oriented and simply connected manifolds of dimension ≥ 5. That breakthrough came already in a 1962 article by Sergei Novikov [Nov62] (cf. [Brow62], [Nov00]). It intimately involved Thom complexes, the central artifacts of cobordism theory.

I can briefly describe Novikov's idea in a simplified but extremely useful form that avoids Thom complexes; I first encountered it in Dennis Sullivan's 1965 Princeton thesis [Sull66] (cf. [Sull67]) written under the direction of William Browder. (I believe Thom complexes are still required to decide whether there exists a closed smooth manifold in a given homotopy type. This existence problem was first treated by W. Browder in [Brow62] that followed shortly on [Nov62].)

Given a closed smooth simply connected manifold N of dimension $n \geq 5$, we seek homotopy equivalent manifolds M each equipped with a degree $+1$ homotopy equivalence $f: M \to N$. Such a pair M, f is called a *homotopy smoothing of* N.

Here is how to proceed. Consider ξ any oriented and fiber homotopicly trivial vector bundle over N, one with high fiber dimension k, together with a fiber-homotopy trivialization of ξ given by a proper map $t\colon \mathrm{E}(\xi) \to \mathbb{R}^k$ of the total space of $\mathrm{E}(\xi)$; it is proper, and of degree $+1$ on each fiber. By a deformation of t with compact support, we can arrange that t is transverse to the origin 0 of $\mathbb{R}^k$. Then the preimage $M := t^{-1}(0)$ is a closed and framed submanifold in $\mathrm{E}(\xi)$, and we orient M so that the bundle projection of ξ restricted to M has degree $+1$ as a map to N; call it f. In this situation, one has all the prerequisites for performing surgery on t, somewhat as in doing the surgery of [KM63] on sufficiently stable maps of spheres (see Sect. 3.4). Thus, a single obstruction is encountered in $\mathbb{L}_n$ to deforming t so that the pair (M, f) becomes a homotopy smoothing of N. It is not difficult to show that every homotopy smoothing arises in this way.

A quite similar surgery process with obstruction in $\mathbb{L}_{n+1}$ serves to decide whether two homotopy smoothings of N, say (M, f) and (M', f') are equivalent in the sense that there exists a degree $+1$ diffeomorphism $h\colon M \to M'$ such that hf is homotopic to f'.

Thus, with a minimum of modification, the surgery of [KM63] suddenly applied, in dimensions ≥ 5, to a horde of classical smooth manifolds!

4.3 Further Extensions of Surgery

Let me mention a few extensions that had great repercussions.

(a) W. Browder, J. Levine, and A. Haefliger applied surgery to manifold embeddings; [Brow68], [Haef62], [Haef68], [Lev65].

(b) C.T.C. Wall showed in [Wall66] that, for $n \geq 5$, surgery with target space having the homotopy type of a simply connected compact manifold or Poincaré space *with nonempty simply connected boundary* is unobstructed. Here, the surgery is understood to occur in certain natural ways that may alter both manifold boundary and manifold interior.
(c) In a massive technical 'tour de force' [Wall70], C.T.C. Wall extended surgery to all compact manifolds of dimension ≥ 5, not necessarily simply connected. For simplicity consider only connected closed manifolds with trivial Whitehead group. Formally speaking, the main novelty was the introduction of new surgery obstruction groups $\mathbb{L}_n(\pi, w)$ depending on n but only modulo 4. They depend functorially on the pair (π, w) consisting of π the fundamental group of the target of the surgery, and w its orientation map $\pi \to \mathbb{Z}_2$. Wall made good use of a generalization of (b) in thus founding non-simply connected surgery theory.
(d) D. Sullivan used surgery to initiate an analysis in depth of the homotopy type of the classifying space G/PL for stable PL bundles equipped with a fiber homotopy trivialization. See [Sull66], [Sull67], [Wall70], [MadM79]. This space G/PL is (in the homotopy category) the fiber of the forgetting map $B_{PL} \to B_G$. The analogous space G/O classifies the homotopy trivialized vector bundles encountered in Sect. 4.2. For G/TOP see [Sieb71], [KirS77], [MadM79]. This classifying space G/PL brings simply connected surgery into fruitful contact with Wall's often mysterious surgery obstruction groups $\mathbb{L}_n(\pi, w)$.
(e) Andrew Ranicki has gradually forged an autonomous and fruitful algebraic version of surgery theory. This was not easy. See [Ran92], [Ran01], [Ran02].

4.4 Conjectures

Here are three great conjectures that are currently guiding the development of surgery:

(i) The Borel conjecture.[9]
(ii) The Novikov conjecture [FRR95].
(iii) The Farrell–Jones conjecture.[10]

References

[Ad58] Adams J.F., On the nonexistence of elements of Hopf invariant one, Bull. Am. Math. Soc. 64 (1958) 279–282.

[9]See http://en.wikipedia.org/wiki/Borel_conjecture, http://www.maths.ed.ac.uk/~aar/surgery/borel.pdf, http://lcs98.free.fr/biblio/exposit/, http://egg.epfl.ch/~nmonod/geneve08/Slides_Lueck.pdf.

[10]See http://en.wikipedia.org/wiki/Farrell%E2%80%93Jones_conjecture.

[Ad60] Adams J.F., On the non-existence of elements of Hopf invariant one, Ann. of Math. (2) 72(1960) 20–104.
[AdA66] Adams J.F. and Atiyah M.F., K-theory and the Hopf invariant, Q. J. Math. Oxford, series II, 17(1966) 31–38.
[Alr30] Alexander J.W., The combinatorial theory of complexes, Ann. Math. 31(1930) 292–320.
[Alr32] Alexander J.W., Some problems in topology, Verhandlungen Kongress Zürich 1932, 1(1932) 249–257.
[Arf41] Arf C., Untersuchungen über quadratische Formen in Körpern der Charakteristik 2, J. Reine Angew. Math., 183(1941) 148–167. See [LorR11].
[AH59] Atiyah M.F. and Hirzebruch F., Riemann-Roch theorems for differentiable manifolds, Bull. Amer. Math. Soc. 65 (1959), 276–281.
[At61] Atiyah M.F., Thom complexes, Proc. Lond. Math. Soc., Ser. III., 11(1961) 291–310.
[BaJM83] Barratt M.G., Jones J.D.S., Mahowald M.E., The Kervaire invariant and the Hopf invariant, algebraic topology, Proc. Workshop, Seattle 1985, Lect. Notes Math. 1286(1987) 135–173.
[BaMT70] Barratt M.G., Mahowald M.E., and Tangora M.C., Some differentials in the Adams spectral sequence, II, Topology 9(1970) 309–316.
[Brie66] Brieskorn E., Beispiele zur Differentialtopologie von Singularitäten, Invent. Math. 2(1966) 1–14
[Brow62] W. Browder, Homotopy type of differential manifolds, Proc. Aarhus Topology Conference (1962); Vol 1, London Math. Soc. Lecture Notes 226(1995) 97–100.
[Brow68] Browder W., Embedding smooth manifolds, Proc. Internat. Congr. Math. (Moscow, 1966), Mir Moscow 1968, 712–719.
[Brow69] Browder W., The Kervaire invariant of framed manifolds and its generalization, Ann. Math. 90(1969) 157–186.
[Brow72] Browder W., Surgery on Simply-Connected Manifolds, Ergebnisse Band 65, Springer Berlin, 1972.
[BrowH66] Browder W. and Hirsch, M.W, Surgery on piecewise linear manifolds and applications, Bull. Am. Math. Soc. 72(1966) 959–964.
[BP66] Brown E.H. and Peterson F.P., The Kervaire invariant of $(8k+2)$-manifolds, Am. J. Math. 88(1966) 815–826.
[Bru70] Brumfiel G., The homotopy groups of BPL and PL/O III, Mich. Math. J. 17(1970) 217–224.
[Cai44] Cairns S., Introduction of a Riemannian geometry on a triangulable 4-manifold, Ann. of Math. (2) 45(1944) 218–219.
[Cap00] Cappell S. et al. (editors), Surveys on Surgery Theory. Vol. 1, Papers dedicated to C.T.C. Wall on the occasion to his 60th birthday, Princeton Univ. Press, Princeton, Ann. Math. Stud. 145, 2000.
[Cap01] Cappell S. et al.(editors), Surveys on Surgery Theory, Vol. 2: Papers dedicated to C.T.C. Wall on the occasion of his 60th birthday, Princeton Univ. Press, Princeton, Ann. Math. Stud. 149, 2001.
[CrE03] Crowley D. and Escher C., A classification of S^3-bundles over S^4, Differ. Geom. Appl. 18(2003), No. 3, 363–380.
[DaM89] Davis D. and Mahowald M., The image of the stable J-homomorphism, Topology 28(1989) 39–58.
[Don87] Donaldson S.K., Irrationality and the h-cobordism conjecture, J. Diff. Geom. 26 (1987) 141–168.
[EK61] Eells J., and Kuiper N., Closed manifolds which admit nondegenerate functions with three critical points, Indag. Math. 23(1961) 411–417.
[EK62] Eells J., and Kuiper N., An invariant for certain smooth manifolds, Ann. Mat. Pura Appl., ser. IV, v. 60, (1962) 93–110.

[FRR95] Ferry S., Ranicki A., and Rosenberg J. (editors), Novikov Conjectures, Index Theorems and Rigidity, Vol. 1 and Vol. 2, LMS Lecture Note Series 226 and 227, Cambridge Univ. Press, Cambridge, 1995.

[GrZ00] Grove K. and Ziller W., Curvature and symmetry of Milnor spheres, Ann. Math. 152(2000) 331–367.

[Haef61b] Haefliger A., Plongements différentiables de variétés dans variétés, Comment. Math. Helv. 36(1961) 47–82; see Theorem 5.1 on page 76.

[Haef62] Haefliger A., Knotted $(4k-1)$-spheres in $6k$-space, Ann. of Math.(2) (1962) 452–466.

[Haef68] Haefliger A., Knotted spheres and related geometric problems, Proc. Internat. Congr. Math. (Moscow, 1966), Mir Moscow, 1968, 437–445.

[Hau14] Hausdorff F., Grundzüge der Mengenlehre, Veit Leipzig 1914, 476 S; reprinted by Chelsea, USA, 1965.

[HillHR09] Hill M., Hopkins M. and Ravenel D., On the non-existence of elements of Kervaire invariant one, article arXiv:0908.3724, revision of November 2010.

[Hir63] Hirsch M., Obstruction theories for smoothing manifolds and maps, Bull. Amer. Math. Soc. 69(1963) 352–356.

[HirM74] Hirsch, M. W. and Mazur B., Smoothings of Piecewise Linear Manifolds, Ann. of Math. Study 80, Princeton Univ. Press, Princeton, 1974, ix+134 pages.

[Hirz53] Hirzebruch F., Über die quaternionalen projektiven Räume (On the quaternion projective spaces), Sitzungsber. Math.-Naturw. Kl., Bayer. Akad. Wiss. München, 1953, S. 301–312.

[Hirz54] Hirzebruch F., Some problems on differentiable and complex manifolds, Ann. of Math. (2) 60(1954) 213–236.

[Hirz56] Hirzebruch F., Neue topologische Methoden in der algebraischen Geometrie, Ergebnisse 9. Springer, Berlin, 1956, 165 S.

[Hirz66a] Hirzebruch F., Topological Methods in Algebraic Geometry, Springer Berlin, 1966, 232 pages. This is a translation and expansion of [Hirz56] with an appendix by R.L.E. Schwarzenberger.

[Hus66] Husemoller D., Fibre Bundles, McGraw-Hill New York, 1966; Springer-Verlag, GTM 20, 1975 and 1994.

[JW55] James I.M., and Whitehead J.H.C., The homotopy theory of sphere bundles over spheres. II, Proc. London Math. Soc. (3) 5, (1955) 148–166.

[JoW08] Joachim M. and Wraith D., Exotic spheres and curvature, Bull. Am. Math. Soc., 45(2008), No. 4, 595–616.

[John12] Johnson N., Visualizing seven-manifolds, an animation presented at the second Abel conference: A mathematical celebration of John Milnor, January 2012, see http://www.nilesjohnson.net/seven-manifolds.html.

[Kahn72] Kahn P.J., A note on topological Pontrjagin classes and the Hirzebruch index formula, Ill. J. Math. 16(1972) 243–256.

[Kerv60] Kervaire M.A., A manifold which does not admit any differentiable structure, Comment. Math. Helv. 34(1960) 257–270.

[KM60] (=[M60c]) Kervaire M.A., and Milnor, J.W., Bernoulli numbers, homotopy groups, and a theorem of Rohlin, Proc. Internat. Congress Math. Edinburgh 1958, pages 454–458. Cambridge Univ. Press, New York, 1960.

[KM63] (=[M63a]) Kervaire M.A., and Milnor J.W., Groups of homotopy spheres I, Ann. of Math. (2) 77(1963) 504–537.

[KirS71] Kirby R.C. and Siebenmann L.C., Some theorems on topological manifolds, in manifolds—Amsterdam 1970, Proc. NUFFIC Summer School Manifolds 1970, Lect. Notes Math. 197, 1–7, 1971.

[KirS77] Kirby R.C. and Siebenmann L.C., Foundational Essays on Topological Manifolds, Smoothing and Triangulations, Annals of Mathematics Study 88(1977).

[Kn25] Kneser H., Die Topologie der Mannigfaltigkeiten, Jahresber. Dtsch. Math.-Ver. 34(1925) 1–14.

[Lang62] Lang S., Introduction to Differentiable Manifolds, Interscience, John Wiley Sons, New York, 1962.

[Lev65] Levine J., A classification of differentiable knots, Ann. of Math. (2) 82(1965) 15–50.

[LorR11] Lorenz F. and Roquette P., Cahit arf and his invariant. Preprint, June 17, 2011. Current URL is http://www.rzuser.uni-heidelberg.de/%7Eci3/arf3-withpicture.pdf.

[MadM79] Madsen I. and Milgram R.J., The Classifying Spaces for Surgery and Cobordism of Manifolds, Annals of Mathematics Study 92(1979), xii+279 pages.

[Mar77] Marin A., La transversalité topologique, Ann. Math. (2) 106(1977) 269–293.

[Maz59] Mazur B., On embeddings of spheres, Bull. Amer. Math. Soc. 65(1959) 59–65.

[Maz61] Mazur B., Stable equivalence of differentiable manifolds, Bull. Am. Math. Soc. 67(1961) 377–384.

[Miller10] Miller H., Kervaire invariant one [after M.A. Hill, M.J. Hopkins, and D.C. Ravenel], Séminaire Bourbaki no 1029, Novembre 2010, 63ème année.

[Milnor] List of Publications for John Willard Milnor. There are currently three sources: (1) This volume, Chapter 22; (2) Any future volume of in the series: Collected Papers of John Milnor, published by the American Mathematical Society; (3) Internet: http://www.math.sunysb.edu/%7Ejack/milnor-pub.pdf.

[M56d] Milnor J.W., On manifolds homeomorphic to the 7-sphere, Ann. of Math. (2) 64(1956) 399–405.

[M56e] Milnor J.W., On the relationship between differentiable manifolds and combinatorial manifolds. First published in [M07a, pp. 19–28].

[M58b] Milnor J.W., The Steenrod algebra and its dual, Ann. of Math. (2) 67(1958) 150–171, see also [M09a, pp. 61–82].

[M59b] Milnor J.W., Differentiable structures on spheres, Am. J. Math., 81(1959) 962–972.

[M59c] Milnor J.W., Sommes de variétes différentiables et structures différentiables des sphères. Bull. Soc. Math. Fr., 87(1959) 439–444.

[M59d] Milnor J.W., Differentiable manifolds which are homotopy spheres, mimeographed at Princeton, dated January 23 1959, first published as pp. 65–88 of [M07a].

[M61a] Milnor J.W., A procedure for killing homotopy groups of differentiable manifolds. In Proc. Sympos. Pure Math., V. III, pages 39–55, Amer. Math. Soc., Providence, 1961.

[M61b] Milnor J.W., Two complexes which are homeomorphic but combinatorially distinct. Ann. of Math. (2) 74(1961) 575–590.

[M61f] Milnor J.W., Microbundles and differentiable structures, polycopied at Princeton U., 1961; first published as pages 173–190 in [M09a].

[M63c] Milnor J.W., Topological manifolds and smooth manifolds, in Proc. Internat. Congr. Math., Stockholm 1962, pages 132–138, Inst. Mittag-Leffler, Djursholm, Sweden 1963; also published as pages 191–197 in [M09a].

[M63d] Milnor J.W., Morse Theory, Based on Lecture Notes by M. Spivak and R. Wells, Annals of Math. Studies, no. 51. Princeton Univ. Press, Princeton, 1963. (Translated into Russian, Japanese, Korean).

[M64b] Milnor J.W., Microbundles I, Topology 3 (suppl. 1), 53–80.

[M64f] Milnor J.W., Differential Topology, pages 165–183 in Lectures on Modern Mathematics, vol. II, ed. T. Saaty, Wiley 1964 New York; also in Russian, Uspehi Mat. Nauk, 20(1965) no. 6, 41–54.

[M65c] Milnor J.W., Remarks concerning spin manifolds, In Differential and Combinatorial Topology, a Symposium in Honor of Marston Morse, S.S. Cairns, editor, pages 55–62. Princeton Univ. Press, Princeton, 1965 and [CP-3, 299–306].

[M65d] Milnor J.W., Lectures on the h-Cobordism Theorem. Notes by L. Siebenmann and J. Sondow. Princeton Univ. Press, Princeton, 1965. (Translated into Russian).

[M65e] Milnor J.W., Topology from the differentiable viewpoint, based on notes by David W. Weaver, University Press of Virginia, Charlottesville, Revised reprint, Princeton Landmarks in Mathematics, Princeton Univ. Press, Princeton, NJ, 1997. (Translated into Russian, Japanese).

[M00b] Milnor J.W., Classification of $(n-1)$-connected $2n$-dimensional manifolds and the discovery of exotic spheres. In Surveys on Surgery Theory, vol. 1, Annals of Math. Studies, no. 145, pages 25–30, Princeton Univ. Press, Princeton, 2000.

[M03a] Milnor J.W., Towards the Poincaré conjecture and the classification of 3-manifolds, Not. Am. Math. Soc., 50(10) (2003) 1226–1233. Also available in Gaz. Math. S.M.F. 99 (2004) 13–25 (in French).

[M06a] Milnor J.W., The Poincaré conjecture one hundred years later. In The Millennium Prize Problems, pages 71–83. Clay Math. Inst., Cambridge, 2006.

[M07a] Milnor J.W., Collected Papers of John Milnor III, Differential Topology, Amer. Math. Soc., Providence, 2007.

[M09a] Milnor J.W., Collected Papers of John Milnor IV, Homotopy, Homology and Manifolds, Edited by J. McCleary, Amer. Math. Soc., Providence, 2009.

[M09c] Milnor J.W., Fifty years ago: topology of manifolds in the 50s and 60s. In Low Dimensional Topology, volume 15 of IAS/Park City Math. Ser., T. Mrowka and P. Osváth, editors, pages 9–20, Amer. Math. Soc. Providence, 2009, see [M09a, p. 345–356].

[M11a] Milnor J.W., Differential topology forty-six years later, Notices Am. Math. Soc., 58(2011) 804–809.

[MSp60] (=[M60a]) Milnor J.W. and Spanier E., Two remarks on fiber homotopy type, Pac. J. Math. 10(1960) 585–590.

[MSt74] (=[M57a]) Milnor J.W. and Stasheff J.D., Characteristic Classes. Annals of Math. Studies, no. 76. Princeton University Press, Princeton, 1974. (Translated into Russian, Japanese).

[Moi52] Moise E.E., Affine structures in 3-manifolds. V., The triangulation theorem and hauptvermutung, Ann. of Math. (2) 56(1952) 96–114.

[Mors25] Morse M., Relation between the critical points of a real function of n independent variables, Trans. Am. Math. Soc. 27(1925) 345–396.

[Mu63] Munkres J.R., Elementary differential topology, Annals of Math Studies 54, Princeton U. Press, Princeton, 1963.

[Mu66] Munkres J.R., Concordance is equivalent to smoothability, Topology 5(1966) 371–389.

[Mu67] Munkres J.R., Concordance of differentiable structures—two approaches. Michigan Math. J. 14(1967) 183–191.

[Mu68] Munkres J.R., Compatibility of imposed differentiable structures. Illinois J. Math. 12(1968) 610–615.

[Nov62] Novikov S.P., Diffeomorphisms of simply connected manifolds Sov. Math. Dokl. 3(1962) 540–543; translation from Russian Dokl. Akad. Nauk SSSR 143(1962) 1046–1049.

[Nov65] Novikov S.P., Topological invariance of rational Pontrjagin classes (English), Sov. Math. Dokl. 6 (1965) 921–923; translation from Dokl. Akad. Nauk SSSR 163 (1965) 298–300 (Russian).

[Nov00] Novikov S.P., Surgery in the 1960's, pages 31–39 in [Cap00].

[Pal09] Palais R.S., Gleason's contribution to the solution of Hilbert's fifth problem, Not. Am. Math. Soc. 56(2009) 1243–1248.

[PW27] Peter F., and Weyl H., Die Vollständigkeit der primitiven Darstellungen einer geschlossenen kontinuierlichen Gruppe, Math. Ann. 97(1927) 737–755.

[Ran92] Ranicki A., Algebraic L-Theory and Topological Manifolds, Cambridge Tracts in Mathematics, 102(1992), Cambridge University Press, Cambridge, viii+358 pages.

[Ran01] Ranicki A., An introduction to algebraic surgery, Surveys on Surgery Theory, Vol. 2, 81–163, Ann. of Math. Study 149(2001), Princeton Univ. Press, Princeton.

[Ran02] Ranicki, A., The structure set of an arbitrary space, the algebraic surgery exact sequence and the total surgery obstruction, pages 515–538, Topology of High-Dimensional Manifolds, Nos. 1, 2 (Trieste, 2001), ICTP Lect. Notes, 9, Abdus Salam Int. Cent. Theoret. Phys., Trieste, 2002. (See http://lcs98.free.fr/biblio/exposit/.)

[RS11] Raussen M. and Skau C., Interview with John Milnor, Newsl. - Eur. Math. Soc., September 2011, (81), pages 31–40.

[Ravenel] Ravenel, D., See http://www.math.rochester.edu/people/faculty/doug

[Rav12] Ravenel D., A solution to the Arf–Kervaire invariant problem, illustrated lecture, Second Abel conference, A Mathematical Celebration of John Milnor, U. of Minn, Minneapolis, February 2012. (Preprint slides available at http://www.math.rochester.edu/people/faculty/doug/AKtalks.html and http://www.maths.ed.ac.uk/%7Eaar/hhrabel.pdf.)

[Reeb49] Reeb G., Stabilité des feuilles compactes à groupe de Poincaré fini, C.R. Acad. Sci., Paris 228(1949) 47–48.

[Reid38] Reidemeister K., Topologie der Polyeder und kombinatorische Topologie der Komplexe, Akademische Verlagsgesellschaft Geest und Portig, Leipzig, 1938; zweite (unveränderte) Auflage 1953.

[RoS72] Rourke C.P. and Sanderson B.J., Introduction to piecewise-linear topology, Ergebnisse, Bd. 69. Springer, Berlin, 1972.

[Ser51] Serre J.-P., Homologie singulière des espaces fibrés, Applications, Ann. of Math. (2) 54(1951) 425–505.

[Seif36] Seifert H., La théorie des noeuds, Enseign. Math. 35(1936) 201–212.

[Sieb71] Siebenmann L.C., Topological manifolds, Actes Congr. Internat. Math. 1970, 2, 133–163 (1971).

[Sieb80] Siebenmann L.C., Les Bisections expliquent le théorème de Reidemeister–Singer—un retour aux sources, dont la Section 5 (suite) l'Histoire des bisections, prépublication d'Orsay, 1980. (Printed version at http://lcs98.free.fr/biblio/prepub/.)

[Sm60] Smale S., The generalized Poincaré conjecture in higher dimensions, Bull. Am. Math. Soc. 66(1960) 373–375.

[Sm61] Smale S., Generalized Poincaré's conjecture in dimensions greater than four, Ann. of Math. (2) 74(1961) 391–406.

[Sm62] Smale S., On the structure of manifolds, Am. J. Math. 84(1962) 387–399.

[Spi67] Spivak M., Spaces satisfying Poincaré duality, Topology 6(1967) 77–101.

[Strd51] Steenrod N., The Topology of Fibre Bundles, Princeton Univ. Press, Princeton 1951.

[Stz08] Steinitz E., Beiträge zur Analysis Situs, Sitzungsber. Berl. Math. Ges. 7(1908) 29–49.

[Sull66] Sullivan D.P., Triangulating Homotopy Equivalences, Princeton Univ. Press, Princeton, 1966.

[Sull67] Sullivan D.P., Triangulating and Smoothing Homotopy Equivalences, Geometric Topology Seminar Notes, Princeton Univ. Press, Princeton, 1967.

[Thom54] Thom R., Quelques propriétés globales des variétés différentiables, Comment. Math. Helv. 28(1954) 17–86.

[Thom58] Thom R., Les classes caractéristiques de Pontrjagin des variétés triangulées, International Symposium on Algebraic Topology, Universidad Nacional Autonoma de Mexico and UNESCO, Mexico City, Mexico, 1958, 54–67.

[Thom60] Thom R., Des variétés triangulées aux variétés différentiables, Proc. Int. Congr. Math. 1958, 248–255 (1960).

[Ttz08] Tietze H., Über die topologischen Invarianten mehrdimensionaler Mannigfaltigkeiten, Monatshefte für Math. u. Physica 19(1908) 1–118.

[vonN27] von Neumann J., Zur Theorie der Darstellungen kontinuierlicher Gruppen, Sitzungsberichte Akad. Berlin 1927, 76–90.

[Wall66] Wall C.T.C., An extension of results of Novikov and Browder, Am. J. Math. 88(1966) 20–32.

[Wall70] Wall C.T.C., Surgery on Compact Manifolds, London Mathematical Society Monographs, No.1., Academic Press, New York, 1970, 280 pp.; 2nd edition (edited by A. Ranicki), Amer. Math. Soc. 302 pages (1999).

[Waa60] Wallace A.H., Modifications and cobounding manifolds, Can. J. Math. 12(1960) 503–528.

[Waa61] Wallace A.H., Modifications and cobounding manifolds II, Journal of Mathematics and Mechanics 10(1961), 773–809. This journal's current name is Indiana University Mathematics Journal.

[Weier85] Weierstrass K., Über die analytische Darstellbarkeit sogenannter willkürlicher Functionen einer reellen Veränderlichen, Berliner Berichte 1885, 633 – 640, 789–806.

[WhdG42] Whitehead G.W., On the homotopy groups of spheres and rotation groups, Ann. of Math. (2) 43(1942) 634–640.

[WhdJ40] Whitehead J.H.C., On C^1-complexes, Ann. of Math. (2) 41(1940) 809–824.

[WhdJ61] Whitehead J.H.C., Manifolds with transverse fields in Euclidean space, Ann. of Math. (2) 73(1961) 154–212.

[Why36] Whitney, H., Differentiable manifolds, Ann. of Math. (2) 37(1936) 645–680.

[Why44] Whitney H., The singularities of a smooth n-manifold in $(2n - 1)$-space, Ann. of Math. (2) 45(1944) 247–293.

List of Publications for John Willard Milnor

1950

[1] On the total curvature of knots. *Ann. of Math. (2)*, 52:248–257. [CP-I, pp. 3–14].[1]

[2] On a relationship between the Betti numbers of a hypersurface and an integral of its Gaussian curvature. First published in [CP-I, pp. 15–26].

1953

[3] Sums of positional games. In *Contributions to the theory of games, vol. 2*, Annals of Mathematics Studies, no. 28, pages 291–301. Princeton University Press, Princeton, NJ.

[4] The characteristics of a vector field on the two-sphere. *Ann. of Math. (2)*, 58:253–257. [CP-VI, pp. 15–19].

[5] On total curvatures of closed space curves. *Math. Scand.*, 1:289–296. [CP-I, pp. 27–36].

[6] (with I.N. Herstein). An axiomatic approach to measurable utility. *Econometrica*, 21:291–297.

1954

[7] Link groups. *Ann. of Math. (2)*, 59:177–195. [CP-II, pp. 7–25].

[8] Games against nature. In *Decision processes*, pages 49–59. Wiley, New York. (Reprinted in *Game Theory and Related Approaches to Social Behavior*, Wiley, New York, 1964.)

[9] (with G. Kalish, J. Nash, and E.D. Nering). Some experimental n-person games. In *Decision processes*, pages 301–327. Wiley, New York.

1956

[10] Construction of universal bundles. I. *Ann. of Math. (2)*, 63:272–284. [CP-IV, pp. 7–19].

[1] Here CP-n stands for "Collected Papers of John Milnor: Volume n". See [CP-I] [117], [CP-II] [120], [CP-III] [138], [CP-IV] [142], [CP-V] [145] and [CP-VI] [149].

DOI 10.1007/978-3-642-39449-2_22, © Springer-Verlag Berlin Heidelberg 2014

[11] Construction of universal bundles. II. *Ann. of Math. (2)*, 63:430–436. [CP-IV, pp. 21–27].

[12] On the immersion of n-manifolds in $(n + 1)$-space. *Comment. Math. Helv.*, 30:275–284. [CP-IV, pp. 141–150].

[13] On manifolds homeomorphic to the 7-sphere. *Ann. of Math. (2)*, 64:399–405. [CP-III, pp. 11–17].

[14] On the relationship between differentiable manifolds and combinatorial manifolds. First published in [CP-III, pp. 19–28].

1957

[15] The geometric realization of a semi-simplicial complex. *Ann. of Math. (2)*, 65:357–362. [CP-IV, pp. 29–34].

[16] Groups which act on S^n without fixed points. *Amer. J. Math.*, 79:623–630. [CP-II, pp. 97–104].

[17] Isotopy of links. In *Algebraic geometry and topology, a symposium in honor of S. Lefschetz*, pages 280–306. Princeton University Press, Princeton, NJ. [CP-II, pp. 27–53].

[18] (with L.S. Shapley). On games of survival. In *Contributions to the theory of games, vol. 3*, Annals of Mathematics Studies, no. 39, pages 15–45. Princeton University Press, Princeton, NJ.

1958

[19] On the existence of a connection with curvature zero. *Comment. Math. Helv.*, 32:215–223. [CP-I, pp. 37–47].

[20] The Steenrod algebra and its dual. *Ann. of Math. (2)*, 67:150–171. [CP-IV, pp. 61–82].

[21] (with R. Bott). On the parallelizability of the spheres. *Bull. Amer. Math. Soc.*, 64:87–89. [CP-III, pp. 229–231].

[22] Some consequences of a theorem of Bott. *Ann. of Math. (2)*, 68:444–449. [CP-III, pp. 233–238].

[23] On the Whitehead homomorphism *J*. *Bull. Amer. Math. Soc.*, 64:79–82. [CP-III, pp. 239–242].

[24] On simply connected 4-manifolds. In *Symposium internacional de topología algebraica. (International symposium on algebraic topology)*, pages 122–128. Universidad Nacional Autónoma de México and UNESCO, Mexico City. [CP-IV, pp. 151–157].

[25] Lectures on differential topology. First published in [CP-III, pp. 145–176].

[26] On the cobordism ring Ω^*. Research announcement 547-26. *Notices of the AMS*, 5:457. [CP-III, p. 255].

1959

[27] On spaces having the homotopy type of a CW-complex. *Trans. Amer. Math. Soc.*, 90:272–280. [CP-IV, pp. 35–43].

[28] Differentiable structures on spheres. *Amer. J. Math.*, 81:962–972. [CP-III, pp. 35–45].

[29] Sommes de variétes différentiables et structures différentiables des sphères. *Bull. Soc. Math. France*, 87:439–444. [CP-III, pp. 29–34].

[30] Differentiable manifolds which are homotopy spheres. First published in [CP-III, pp. 65–88].

1960

[31] (with E. Spanier). Two remarks on fiber homotopy type. *Pacific J. Math.*, 10:585–590. [CP-IV, pp. 167–172].

[32] On the cobordism ring Ω^* and a complex analogue. I. *Amer. J. Math.*, 82:505–521. [CP-III, pp. 257–273].

[33] (with M.A. Kervaire). Bernoulli numbers, homotopy groups, and a theorem of Rohlin. In *Proc. internat. congress math. 1958*, pages 454–458. Cambridge Univ. Press, New York. [CP-III, pp. 243–247].

1961

[34] A procedure for killing homotopy groups of differentiable manifolds. In *Proc. sympos. pure math., Vol. III*, pages 39–55. American Mathematical Society, Providence, RI. [CP-III, pp. 47–63].

[35] Two complexes which are homeomorphic but combinatorially distinct. *Ann. of Math. (2)*, 74:575–590. [CP-II, pp. 123–138].

[36] (with M.A. Kervaire). On 2-spheres in 4-manifolds. *Proc. Nat. Acad. Sci. USA*, 47:1651–1657. [CP-IV, pp. 159–165].

[37] Variedades diferenciables con frontera. *An. Inst. Mat. Univ. Nac. Autónoma México*, 1:82–116. (Revised and augmented translation: "Smooth manifolds with boundary") first published in [CP-III, pp. 191–222].

[38] Lectures on differentiable structures. First published in [CP-III, pp. 177–190].

[39] Microbundles and differentiable structures. First published in [CP-IV, pp. 173–190].

1962

[40] (with M.G. Barratt). An example of anomalous singular homology. *Proc. Amer. Math. Soc.*, 13:293–297. [CP-IV, pp. 107–111].

[41] A duality theorem for Reidemeister torsion. *Ann. of Math. (2)*, 76:137–147. [CP-II, pp. 139–149].

[42] A unique decomposition theorem for 3-manifolds. *Amer. J. Math.*, 84:1–7. [CP-II, pp. 237–243].

[43] A survey of cobordism theory. *Enseignement Math. (2)*, 8:16–23. [CP-III, pp. 283–290]. Erratum in [CP-III, pp. 291–292].

[44] On axiomatic homology theory. *Pacific J. Math.*, 12:337–341. [CP-IV, pp. 101–105].

[45] The work of J.H.C. Whitehead. In *The mathematical works of J.H.C. Whitehead*, volume 1, pages xxi–xxxiii. Pergamon Press, Oxford. [CP-IV, pp. 261–278].

1963

[46] (with M.A. Kervaire). Groups of homotopy spheres. I. *Ann. of Math. (2)*, 77:504–537. [CP-III, pp. 89–122].
[47] Spin structures on manifolds. *Enseignement Math. (2)*, 9:198–203. [CP-III, pp. 293–298].
[48] Topological manifolds and smooth manifolds. In *Proc. Internat. Congr. Mathematicians (Stockholm, 1962)*, pages 132–138. Inst. Mittag-Leffler, Djursholm. [CP-IV, pp. 191–197].
[49] *Morse theory*. Based on lecture notes by M. Spivak and R. Wells. Annals of Mathematics Studies, No. 51. Princeton University Press, Princeton, NJ. (Translated into Russian, Japanese and Korean).
[50] The representation rings of some classical groups. First published in [CP-V, pp. 143–154].

1964

[51] On the Betti numbers of real varieties. *Proc. Amer. Math. Soc.*, 15:275–280. [CP-I, pp. 133–140].
[52] Microbundles. I. *Topology*, 3(suppl. 1):53–80. [CP-IV, pp. 199–226].
[53] Eigenvalues of the Laplace operator on certain manifolds. *Proc. Nat. Acad. Sci. USA*, 51:542. [CP-I, pp. 49–51].
[54] Most knots are wild. *Fund. Math.*, 54:335–338. [CP-II, pp. 55–58].
[55] (with M.W. Hirsch). Some curious involutions of spheres. *Bull. Amer. Math. Soc.*, 70:372–377. [CP-II, pp. 105–110].
[56] Differential topology. In *Lectures on modern mathematics, Vol. II*, pages 165–183. Wiley, New York. Also available in *Uspehi Mat. Nauk*, 20(6(126)): 41–54, 1965 (in Russian). [CP-III, pp. 123–141].
[57] Some free actions of cyclic groups on spheres. In *Differential analysis, Bombay Colloq., 1964*, pages 37–42. Oxford Univ. Press, London. [CP-II, pp. 111–116].

1965

[58] (with J.C. Moore). On the structure of Hopf algebras. *Ann. of Math. (2)*, 81:211–264. [CP-V, pp. 37–90]. An earlier version from 1959 is published in [CP-V, pp. 7–36].
[59] On the Stiefel-Whitney numbers of complex manifolds and of spin manifolds. *Topology*, 3:223–230. [CP-III, pp. 307–314].
[60] Remarks concerning spin manifolds. In *Differential and combinatorial topology. A Symposium in Honor of Marston Morse*, pages 55–62. Princeton Univ. Press, Princeton, NJ. [CP-III, pp. 299–306].
[61] *Lectures on the h-cobordism theorem*. Notes by L. Siebenmann and J. Sondow. Princeton University Press, Princeton, NJ. (Translated into Russian).
[62] *Topology from the differentiable viewpoint*. Based on notes by David W. Weaver. The University Press of Virginia, Charlottesville, Va. (Translated into Russian). Revised reprint, Princeton Landmarks in Mathematics, Princeton University Press, Princeton, NJ, 1997.

[63] (with H. Bass). On unimodular groups over number fields. First published in [CP-V, pp. 165–179].

1966

[64] Whitehead torsion. *Bull. Amer. Math. Soc.*, 72:358–426. [CP-II, pp. 151–219].

[65] (with R.H. Fox). Singularities of 2-spheres in 4-space and cobordism of knots. *Osaka J. Math.*, 3:257–267. [CP-II, pp. 59–69].

1967

[66] (with H. Bass and J.-P. Serre). Solution of the congruence subgroup problem for SL_n $(n \geq 3)$ and Sp_{2n} $(n \geq 2)$. *Inst. Hautes Études Sci. Publ. Math.*, (33):59–137. Erratum: On a functorial property of power residue symbols, *ibid.* 44:241–244, 1974. [CP-V, pp. 181–259 and pp. 261–264].

1968

[67] On characteristic classes for spherical fibre spaces. *Comment. Math. Helv.*, 43:51–77. [CP-IV, pp. 227–253].

[68] A note on curvature and fundamental group. *J. Differential Geometry*, 2:1–7. [CP-I, pp. 53–61].

[69] *Singular points of complex hypersurfaces.* Annals of Mathematics Studies, No. 61. Princeton University Press, Princeton, NJ. (Translated into Russian and Japanese).

[70] Infinite cyclic coverings. In *Conference on the topology of manifolds (Michigan State Univ., E. Lansing, Mich., 1967)*, pages 115–133. Prindle, Weber & Schmidt, Boston, Mass. [CP-II, pp. 71–89].

[71] Growth of finitely generated solvable groups. *J. Differential Geometry*, 2:447–449. [CP-V, pp. 155–157].

[72] Uses of the fundamental group. First published in [CP-II, pp. 221–225].

1969

[73] (with G. Lusztig and F.P. Peterson). Semi-characteristics and cobordism. *Topology*, 8:357–359. [CP-IV, pp. 113–115].

[74] On isometries of inner product spaces. *Invent. Math.*, 8:83–97. [CP-V, pp. 331–345].

[75] A problem in cartography. *Amer. Math. Monthly*, 76:1101–1112. [CP-I, pp. 159–172].

[76] Algebraic K-theory and quadratic forms. *Invent. Math.*, 9:318–344. [CP-V, pp. 347–373].

1970

[77] (with O. Burlet). Torsion et type simple d'homotopie. In *Essays on topology and related topics (Mémoires dédiés à Georges de Rham)*, pages 12–17. Springer, New York. [CP-II, pp. 227–232].

[78] (with P. Orlik). Isolated singularities defined by weighted homogeneous polynomials. *Topology*, 9:385–393. [CP-I, pp. 141–151].

[79] Foliations and foliated vector bundles. First published in [CP-IV, pp. 279–320].

[80] Symmetric inner products over a Dedekind domain. First published in [CP-V, pp. 375–380].

1971

[81] Symmetric inner products in characteristic 2. In *Prospects in mathematics (Proc. Sympos., Princeton Univ., Princeton, N.J., 1970)*, pages 59–75. Ann. of Math. Studies, No. 70. Princeton Univ. Press, Princeton, NJ. [CP-5, pp. 381–397].

[82] *Introduction to algebraic K-theory*. Annals of Mathematics Studies, No. 72. Princeton University Press, Princeton, NJ. (Translated into Russian).

1972

[83] On the construction *FK*. In *Algebraic topology: a student's guide*, by J.F. Adams, London Mathematical Society Lecture Note Series, No. 4, pages 119–136. Cambridge University Press, Cambridge. [CP-IV, pp. 45–53].

1973

[84] (with D. Husemoller). *Symmetric bilinear forms*. Ergebnisse der Mathematik und ihrer Grenzgebiete, Band 73. Springer-Verlag, New York. (Translated into Russian).

1974

[85] (with J.D. Stasheff). *Characteristic classes*. Annals of Mathematics Studies, No. 76. Princeton University Press, Princeton, NJ. (Translated into Russian and Japanese).

1975

[86] Isolated critical points of complex functions. In *Differential geometry (Proc. Sympos. Pure Math., Vol. XXVII, Stanford Univ., Stanford, Calif., 1973), Part 1*, pages 381–382. Amer. Math. Soc., Providence, RI. [CP-I, pp. 153–156].

[87] On the 3-dimensional Brieskorn manifolds $M(p,q,r)$. In *Knots, groups, and 3-manifolds (Papers dedicated to the memory of R. H. Fox)*, pages 175–225. Ann. of Math. Studies, No. 84. Princeton Univ. Press, Princeton, NJ. [CP-II, pp. 245–295].

1976

[88] Curvatures of left invariant metrics on Lie groups. *Advances in Math.*, 21(3):293–329. [CP-I, pp. 73–111].

[89] Problems in differential geometry. In *Mathematical developments arising from Hilbert problems (Proc. Sympos. Pure Math., Northern Illinois Univ., De Kalb, Ill., 1974)*, pages 54–57. Proc. Sympos. Pure Math., Vol. XXVIII. Amer. Math. Soc., Providence, RI. [CP-I, pp. 63–71].

[90] Hilbert's problem 18: on crystallographic groups, fundamental domains, and on sphere packing. In *Mathematical developments arising from Hilbert problems (Proc. Sympos. Pure Math., Northern Illinois Univ., De Kalb, Ill., 1974)*, pages 491–506. Proc. Sympos. Pure Math., Vol. XXVIII. Amer. Math. Soc., Providence, RI. [CP-I, pp. 173–187].

1977

[91] On deciding whether a surface is parabolic or hyperbolic. *Amer. Math. Monthly*, 84(1):43–46. [CP-I, pp. 113–118].

[92] On fundamental groups of complete affinely flat manifolds. *Advances in Math.*, 25(2):178–187. [CP-I, pp. 119–130].

[93] (with W. Thurston). Characteristic numbers of 3-manifolds. *Enseignement Math. (2)*, 23(3-4):249–254. [CP-II, pp. 297–302].

1978

[94] (with L.S. Shapley). Values of large games. II. Oceanic games. *Math. Oper. Res.*, 3(4):290–307.

[95] Analytic proofs of the "hairy ball theorem" and the Brouwer fixed-point theorem. *Amer. Math. Monthly*, 85(7):521–524. [CP-I, pp. 235–242].

[96] How to compute volume in hyperbolic space. First published in [CP-I, pp. 189–212].

[97] Euler characteristic and finitely additive Steiner measures. First published in [CP-I, pp. 213–234].

1982

[98] Hyperbolic geometry: the first 150 years. *Bull. Amer. Math. Soc. (NS)*, 6(1):9–24. [CP-I, pp. 243–260].

1983

[99] On the homology of Lie groups made discrete. *Comment. Math. Helv.*, 58(1):72–85. [CP-IV, pp. 117–130].

[100] On polylogarithms, Hurwitz zeta functions, and the Kubert identities. *Enseign. Math. (2)*, 29(3-4):281–322. [CP-V, pp. 265–306].

[101] On the geometry of the Kepler problem. *Amer. Math. Monthly*, 90(6):353–365. [CP-I, pp. 261–279].

[102] The Schläfli differential inequality. First published in [CP-I, pp. 281–295].

1984

[103] Remarks on infinite-dimensional Lie groups. In *Relativity, groups and topology, II (Les Houches, 1983)*, pages 1007–1057. North-Holland, Amsterdam. [CP-V, pp. 91–141].

1985

[104] On the concept of attractor. *Comm. Math. Phys.*, 99(2):177–195. Correction and remarks, *Comm. Math. Phys.*, 102(3):517–519. [CP-VI, pp. 21–41 and pp. 43–45].

1986

[105] Directional entropies of cellular automaton-maps. In *Disordered systems and biological organization (Les Houches, 1985)*, volume 20 of *NATO Adv. Sci. Inst. Ser. F Comput. Systems Sci.*, pages 113–115. Springer, Berlin. [CP-VI, pp. 47–49].

1987

[106] The work of M.H. Freedman. In *Proceedings of the international congress of mathematicians, Vol. 1, 2 (Berkeley, Calif., 1986)*, pages 13–15, Providence, RI. Amer. Math. Soc. Also published in *Fields Medallists Lectures*, volume 5 of World Sci. Ser. 20th Century Math., pages 405–408. World Sci. Publ., River Edge, NJ, 1997. [CP-IV, pp. 321–323].

1988

[107] Nonexpansive Hénon maps. *Adv. in Math.*, 69(1):109–114. [CP-VI, pp. 81–86].

[108] On the entropy geometry of cellular automata. *Complex Systems*, 2(3):357–385. [CP-VI, pp. 51–79].

[109] (with W. Thurston). On iterated maps of the interval. In *Dynamical systems (College Park, MD, 1986–87)*, volume 1342 of *Lecture Notes in Math.*, pages 465–563. Springer, Berlin. [CP-VI, pp. 87–155].

1989

[110] Self-similarity and hairiness in the Mandelbrot set. In *Computers in geometry and topology (Chicago, IL, 1986)*, volume 114 of *Lecture Notes in Pure and Appl. Math.*, pages 211–257. Dekker, New York. [CP-VI, pp. 289–335].

[111] (with S. Friedland). Dynamical properties of plane polynomial automorphisms. *Ergodic Theory Dynam. Systems*, 9(1):67–99. [CP-VI, pp. 255–287].

1992

[112] Remarks on iterated cubic maps. *Experiment. Math.*, 1(1):5–24. [CP-VI, pp. 337–369].

[113] Hyperbolic components in spaces of polynomial maps. With an appendix by A. Poirer. Available as arXiv:math/9202210. See also [148].

1993

[114] (with M. Lyubich). The Fibonacci unimodal map. *J. Amer. Math. Soc.*, 6(2):425–457. [CP-VI, pp. 439–471].

[115] (with L.R. Goldberg). Fixed points of polynomial maps. II. Fixed point portraits. *Ann. Sci. École Norm. Sup. (4)*, 26(1):51–98. [CP-VI, pp. 473–520].

[116] Geometry and dynamics of quadratic rational maps. *Experiment. Math.*, 2(1):37–83. With an appendix by the author and Lei Tan. [CP-VI, pp. 371–438].

1994

[117] *Collected Papers of John Milnor: I. Geometry*. Publish or Perish Inc., Houston, TX. Available through the American Mathematical Society.

[118] Thurston's algorithm without critical finiteness. Linear and Complex Analysis Problem Book 3, Part 2, Havin and Nikolskii editors, Lecture Notes in Math no. 1474, pp. 434–436, Springer, Berlin.

[119] Problems on local connectivity. Linear and Complex Analysis Problem Book 3, Part 2, Havin and Nikolskii editors, Lecture Notes in Math no. 1474, pp. 443–446, Springer, Berlin.

1995

[120] *Collected Papers of John Milnor: II. The fundamental group*. Publish or Perish Inc., Houston, TX. Available through the American Mathematical Society.

[121] A Nobel Prize for John Nash. *Math. Intelligencer*, 17(3):11–17. Also available in *Pokroky Mat. Fyz. Astronom*, 41(4):169–179, 1996 (in Czech) and in *Gac. R. Soc. Mat. Esp.*, 5(3):560–577, 2002 (in Spanish).

[122] (with S.P. Dawson, R. Galeeva, and C. Tresser). A monotonicity conjecture for real cubic maps. In *Real and complex dynamical systems (Hillerød, 1993)*, volume 464 of *NATO Adv. Sci. Inst. Ser. C Math. Phys. Sci.*, pages 165–183. Kluwer Acad. Publ., Dordrecht. [CP-VI, pp. 157–175].

[123] On the Steenrod homology theory. In *Novikov conjectures, index theorems and rigidity, Vol. 1 (Oberwolfach, 1993)*, volume 226 of *London Math. Soc. Lecture Note Ser.*, pages 79–96. Cambridge Univ. Press, Cambridge. [CP-IV, pp. 83–100].

1997

[124] Fubini foiled: Katok's paradoxical example in measure theory. *Math. Intelligencer*, 19(2):30–32. [CP-VI, pp. 233–235].

1998

[125] John Nash, and "A beautiful mind". *Notices Amer. Math. Soc.*, 45(10):1329–1332.

1999

[126] The mathematical work of Curtis T. McMullen. In *The mathematical work of the 1998 Fields medalists*, volume 46 of *Notices Amer. Math. Soc.*, pages 23–26. [CP-VI, pp. 521–529].

[127] Growing up in the old Fine Hall. In *Prospects in mathematics (Princeton, NJ, 1996)*, pages 1–11. Amer. Math. Soc., Providence, RI.

[128] *Dynamics in one complex variable*. Friedr. Vieweg & Sohn, Braunschweig. Second edition, 2000. Third edition, volume 160 of *Annals of Mathematics Studies*, Princeton University Press, Princeton, NJ, 2006.

2000

[129] (with C. Tresser). On entropy and monotonicity for real cubic maps. *Comm. Math. Phys.*, 209(1):123–178. With an appendix by Adrien Douady and Pierrette Sentenac. [CP-VI, pp. 177–232].

[130] Classification of $(n-1)$-connected $2n$-dimensional manifolds and the discovery of exotic spheres. In *Surveys on surgery theory, Vol. 1*, volume 145 of *Ann. of Math. Stud.*, pages 25–30. Princeton Univ. Press, Princeton, NJ.

[131] Periodic orbits, externals rays and the Mandelbrot set: an expository account. *Astérisque*, (261):xiii, 277–333. Géométrie complexe et systèmes dynamiques (Orsay, 1995).

[132] Local connectivity of Julia sets: expository lectures. In *The Mandelbrot set, theme and variations*, volume 274 of *London Math. Soc. Lecture Note Ser.*, pages 67–116. Cambridge Univ. Press, Cambridge.

[133] On rational maps with two critical points. *Experiment. Math.*, 9(4):481–522.

2003

[134] Towards the Poincaré conjecture and the classification of 3-manifolds. *Notices Amer. Math. Soc.*, 50(10):1226–1233. Also available in *Gaz. Math.*, (99):13–25, 2004 (French). [CP-IV, pp. 325–336].

2004

[135] Pasting together Julia sets: a worked out example of mating. *Experiment. Math.*, 13(1):55–92.

2006

[136] The Poincaré conjecture. In *The millennium prize problems*, pages 71–83. Clay Math. Inst., Cambridge, MA. [CP-IV, pp. 337–344].

[137] On Lattès maps. In *Dynamics on the Riemann sphere*, pages 9–43. Eur. Math. Soc., Zürich.

2007

[138] *Collected Papers of John Milnor: III. Differential topology*. American Mathematical Society, Providence, RI.

[139] A concluding amusement: symmetry breaking. First published in [CP-III, pp. 315–317].

[140] (with A. Bonifant and M. Dabija). Elliptic curves as attractors in $\mathbb{P}^2$. I. Dynamics. *Experiment. Math.*, 16(4):385–420.

2008

[141] (with A. Bonifant). Schwarzian derivatives and cylinder maps. In *Holomorphic dynamics and renormalization*, volume 53 of *Fields Inst. Commun.*, pages 1–21. Amer. Math. Soc., Providence, RI.

2009

[142] *Collected Papers of John Milnor: IV. Homotopy, homology and manifolds.* American Mathematical Society, Providence, RI. Edited by John McCleary.

[143] Cubic polynomial maps with periodic critical orbit. I. In *Complex Dynamics, Families and Friends*, pages 333–411. A.K. Peters, Wellesley, MA.

[144] Fifty years ago: topology of manifolds in the 50's and 60's. In *Low dimensional topology*, volume 15 of *IAS/Park City Math. Ser.*, pages 9–20. Amer. Math. Soc., Providence, RI. [CP-IV, pp. 345–356].

2010

[145] *Collected Papers of John Milnor: V. Algebra.* American Mathematical Society, Providence, RI. Edited by H. Bass and T.-Y. Lam.

[146] (with A. Bonifant and J. Kiwi). Cubic polynomial maps with periodic critical orbit. II. Escape regions. *Conform. Geom. Dyn.*, 14:68–112. Errata, *Conform. Geom. Dyn.*, 14:190–193.

2011

[147] Differential topology forty-six years later. *Notices Amer. Math. Soc.*, 58(6): 804–809.

2012

[148] Hyperbolic components. In *Conformal dynamics and hyperbolic geometry*, volume 573 of *Contemp. Math.*, pages 183–232. Amer. Math. Soc., Providence, RI. With an appendix by A. Poirier.

[149] *Collected Papers of John Milnor: VI. Dynamics (1953–2000).* American Mathematical Society, Providence, RI. Edited by A. Bonifant.

2013

[150] *Collected Papers of John Milnor: VII. Dynamics (1984–2012).* American Mathematical Society, Providence, RI. Edited by A. Bonifant. In preparation.

[151] Arithmetic of unicritical polynomial maps. To appear in *Proceedings of the Conference "Frontiers in Complex Dynamics", Banff 2011.* See also arXiv:1203.5447.

Curriculum Vitae for John Willard Milnor

Born:	February 20, 1931 in Orange, USA
Degrees/education:	AB, Princeton University, 1951
	PhD, Princeton University, 1954
Positions:	Princeton University: Lecturer 1954–1955,
	Assistant Professor 1955–1957,
	Alfred P. Sloan Fellow 1955–1959,
	Associate Professor 1957–1960,
	Professor 1960–1967,
	Chair 1963–1965
	Massachusetts Institute of Technology: Professor 1968–1970
	Institute for Advanced Study: Long Term Member 1963–1970,
	Professor 1970–1990
	Institute for Mathematical Sciences, SUNY at Stony Brook:
	Professor and Director 1989–2007,
	Professor and Co-Director 2007–present
Visiting positions:	University of California at Los Angeles: Professor 1967–1968
Memberships:	National Academy of Sciences, 1963
	American Philosophy Society, 1965
	Russian Academy of Science (foreign member), 1994
	Norwegian Academy of Science and Letters, 2011
Awards and prizes:	Fields Medal, 1962
	National Medal of Science, US, 1967
	Wolf Prize, 1989
	Steele Prize, American Mathematical Society:
	for Research, 1982
	for Mathematical Exposition, 2004
	for Lifetime Achievement, 2011
	Abel Prize, 2011

H. Holden, R. Piene (eds.), *The Abel Prize 2008-2012*,
DOI 10.1007/978-3-642-39449-2_23, © Springer-Verlag Berlin Heidelberg 2014

Honorary degrees:	Syracuse University, 1965
	University of Chicago, 1967
Presidencies:	Vice President, American Mathematical Society 1975–1976

2012

Endre Szemerédi

"for his fundamental contributions to discrete mathematics and theoretical computer science, and in recognition of the profound and lasting impact of these contributions on additive number theory and ergodic theory"

ABEL
PRISEN

Autobiography

Endre Szemerédi

Unlike most mathematicians I did not start to study to be a mathematician. I was born in 1940 in Budapest, Hungary. I lost my mother when I was 8 years old. My two brothers and I were sent to different boarding schools for orphans. It was a few years after WWII.

I have always liked mathematics, and it actually helped me to survive in a way. I was very short and weak, and the strong tall guys would beat me up. I was kind of lucky, since the physically strongest guy was mathematically very weak. He could never solve his homework exercises. So I solved them for him even at the exams. He was an honest person, he always protected me. So mathematics and in return served my interest. All this happened in elementary school.

In high school I was good at mathematics, but never took part in competitions. For my father's request I was preparing to be a physician and I studied biology. At this time this was the most recognized profession in Hungary.

My education was not the usual education you get in Hungary if you want to be a mathematician. There are a few extremely good highschools concentrating in mathematics, both in Budapest and in the country side.

After a few months at medical school I realized that it was not for me and not knowing what exactly would suit me I went to work in a machine making factory. These two years were actually a good experience for me. We had to finish our work on time even if we felt that it was very monotonous. As a mathematician I can appreciate now that I can work on things that I like.

Electronic supplementary material Supplementary material is available in the online version of this chapter at http://dx.doi.org/10.1007/978-3-642-39449-2_24. Videos can also be accessed at http://www.springerimages.com/videos/978-3-642-39449-2.

E. Szemerédi (✉)
Rényi Institute of Mathematics, Budapest, Hungary
e-mail: szemered@cs.rutgers.edu

E. Szemerédi
Rutgers University, New Braunswick, USA

H. Holden, R. Piene (eds.), *The Abel Prize 2008-2012*,
DOI 10.1007/978-3-642-39449-2_24,

Once one of my best friends, Gabor Ellmann, who attended the university to become a mathematics and physics teacher, suggested that I try to go to the university to study the same thing. So I tried. At that time in Hungary you studied mathematics and physics for two years, in the third year if you decided to be a mathematician and were among the 15 best, you were able to specialize in mathematics. The others followed with math and physics plus pedagogy and psychology to become a mathematics and physics teacher in a high school.

After being admitted to the Eötvös Loránd University in 1960, and attending Professor Paul Turán's lecture series on number theory. I consider Professor Turán to be the one who actually helped me to decide to become a mathematician and he still is one of my icons. I never worked with him, I have only listened to his lectures and sometimes went to his seminars.

When I had finished at the university I was hired at the Mathematical Research Institute (later Rényi Alfréd Institute) of the Hungarian Academy of Sciences.

At that time Paul Erdős, the great Hungarian mathematician, often visited Hungary. His mother lived in Budapest. His main area was discrete mathematics. This included a lot of things. Among others elementary number theory, graph theory, random graph theory and so on. Paul Erdős and Alfred Renyi are the founders of random graph theory. Paul Erdős had many problems, conjectures. Some of them were not so hard, but the others were extremely difficult. Fortunately or unfortunately the solution to many of his problems required only elementary methods. Of course quite often the proofs using only elementary methods are not simple because one may have to put together basic ingredients in extremely complicated and sophisticated ways.

Knowing that I had a very limited knowledge, I was very happy that Paul Erdős was willing to work with me. With him and a fellow mathematician Andras Sarkozy we wrote a large number of papers on number theory. Then I decided that I would do something alone. My first try ended up in failure. Anyway I am going to tell it, because it is very closely related with my later works, and also it shows that a failure—even an embarrassing one—can be helpful later. The story is the following. I tried to prove that in a long arithmetic progression it is not possible that a positive percentage of the elements of the arithmetic progression are squares. In order to prove it, I took it for granted that if you have a positive percentage of the integers, then it contains an arithmetic progression of length 4. I proved that if you have an arithmetic progression of length 4, then not all of them can be squares. If you put these results together, you prove what you wanted. I was very proud of 'my result'. I showed the proof to Paul Erdős. Then he told me that there were some slight problems with the "proof". The first one was that I assumed something which had not been proved yet at that time, namely that any set of a positive percentage of the integers contains an arithmetic progression of length 4. But this was still OK. The second one was really shameful. Erdős told me that the other thing, stating that there are no four squares that form an arithmetic progression, was proved by Euler, already 250 years earlier. I felt that I must correct this mistake, because Erdős was my other icon and mentor.

Still at the Rényi Institute I started to work on the arithmetic progression problem. Van der Waerden proved his famous theorem, stating that if you divide the integers into finitely many classes, then some class contains arbitrarily long arithmetic progressions. Then Erdős and Turán conjectured in 1936 that the important thing is that the set is dense enough. If you have a positive proportion of the integers, then you have already long arithmetic progressions. More precisely speaking, for every $d > 0$ and every positive integer k there exists a number $n = n(d, n)$ such that every subset of the integers $\{1, 2, \ldots, n\}$, $n > n(d, k)$ of cardinality dn contains an arithmetic progression of length k. In 1953 Roth provided a beautiful proof, using harmonic analytic methods, that the conjecture is true for $k = 3$. He even proved that among at least $n/\log\log n$ integers there was always an arithmetic progression of length 3. Actually, one of my favorite mathematicians is K.F. Roth. When I first went abroad in 1967, I met him and read his proof. I knew very little about harmonic analysis, so I tried to use elementary methods. First I gave a very simple high school proof for $k = 3$. Then I proved it for $k = 4$. In 1967 Paul Erdős arranged for me an invitation to the University of Nottingham. There I was supposed to give a lecture about my proof. My English was practically non-existing. So I just drew some pictures, and Peter Elliot and Edward Wirsing, both number theorists, based on these pictures and my very bad English, wrote down the proof. I am very grateful for their great help.

In 1968 I went to Moscow to be a PhD student of Gelfond, a well known number theorist. By some unfortunate misspelling of the names I ended up with I.M. Gelfand who was one of the greatest mathematicians of the last century, but his area was very far from my expertise and soon I realized that I can not learn that kind of mathematics. I was lucky that Gelfond visited Budapest, and I, as a student studying in Moscow, was supposed to be his guide. He was a very nice, warm person. He agreed to arrange that I would be his student. Soon after his return to Moscow he tragically died.

András Hajnal, the well known logician and combinatorialist spent half a year in Moscow in 1969. We worked together and proved an important conjecture of Erdős. He was working at his desk while I worked walking in the woods. Considering that the Russian winter usually is pretty harsh, my working method may not have been an optimal one. I.M. Gelfand was very generous and agreed that this result was good enough for a PhD thesis. Later our result generated a lot of activities in graph packing.

In 1973 I proved the Erdős–Turán conjecture. My good friend András Hajnal helped me to write up the paper. Or better, say he listened to my explanations and then wrote it up.

After my proof many different proofs were found. Furstenberg gave an ergodic theoretical proof. His method is much deeper and much more powerful than my elementary method, and could be generalized into the multidimensional setting. He and Katznelson could prove in 1978 a multidimensional analogue, and they could finally prove in 1991 the density version of the Hales–Jewett theorem.

Timothy Gowers gave a much, much better bound than what I had. Even more importantly, he invented many fundamental methods which completely changed the landscape. We cannot overestimate the influence of his paper. Gowers used a higher order Fourier analysis and introduced his famous Gowers norm, which controls the randomness of the set in the question. For me the absolutely striking result was the

Endre Szemerédi, 1977

The Szemerédi family

Timothy Gowers (*left*) and Endre Szemerédi (*right*)

fundamental theorem of Ben Green and Terence Tao which states that among the primes there are arbitrarily long arithmetic progressions.

They and other mathematicians are those who really moved this field. Without them my theorem would be just a good theorem and nothing more. They strengthened it, invented many revolutionary new ideas, found connections, between different branches of mathematics and getting unbelievable results. They are doing unbelievable things.

My original proof contains a graph theoretical lemma called a week regularity lemma. The regularity lemma was needed for something else, namely for a graph theoretical problem. I listened to a lecture by Béla Bollobás, which was in 1974 in Calgary. He talked about the Erdős–Stone theorem. Bollobás and Erdős wanted to determine the right magnitude, and they had a very good bound, the order was $\log n$ and only the constant was missing a little bit. I decided to work on it, as I liked the problem and Béla presents things very nicely and extremely cleverly. Then I realized that maybe a lemma like the regularity lemma would help considerably. Actually, that was the real reason why the regularity lemma was invented. At least, this is my recollection. I would like to thank Vasek Chvatal for helping me to write up the regularity lemma. Later we together found the exact constant for the Erdős–Stone theorem. Although the regularity lemma was created for a particular problem, it later found many important applications. In discrete mathematics and theoretical computer science. Gowers, Rodle, Nagle, Skokan, Schacht found hypergraph regularity lemmas and a hypergraph removal lemma. These lemmas are extremely powerful and they have a wide range of application in arithmetic combinatorics and computer science.

I would like to mention briefly some other results.

1. My favorite work is the creation of the pseudo-random method together with Miklós Ajtai and János Komlós (it may be that the pseudo-random method existed in some other areas of mathematics). We used this method, when we tried to give a better estimate on density of an infinite Sidon sequence. Using this

method, we also disproved Heilbronn's conjecture, which was at that time already 40 years old.
2. Euclid's system of axioms state some of the basic facts about incidences between points and lines in the plane. In the 1940s, Paul Erdős started asking slightly more complicated questions about incidences that even Euclid would have understood. An interesting question is that at most how many incidences can occur among n points and n lines, where an incidence means that a line passes trough a point. My theorem with Trotter confirmed Erdős's rather surprising conjecture: The maximum number of incidences is much smaller in the real plane than in the projective one—much smaller than what we could deduce by simple combinatorial considerations.
3. Together with Paul Erdős, we discovered an interesting phenomenon and made the first nontrivial step in exploring it. We noticed, roughly speaking, that a set of numbers may have nice additive properties or nice multiplicative properties, but not both at the same time. This has meanwhile been generalized to finite fields and other structures by Bourgain, Katz, Tao, and others. Their results had far-reaching consequences in seemingly unrelated fields of mathematics.
4. We want to sort n numbers, that is to put them in increasing order by using comparisons of pairs of elements. Our algorithm is non-adaptive. The next comparison never depends on the outcome of the previous one. Moreover the algorithm can efficiently run simultaneously on cn processors such that every number is processed by only one of them at a time. Somewhat surprisingly our algorithm does not require more comparisons than any adaptive nonparallel-shorting algorithm.

After returning from Moscow I continued to work at the Rényi Alfréd Mathematical Institute. From time to time, either with the family or alone, I visited different universities, mainly in the US. Ron Graham suggested to Donald Knuth to invite me to Stanford University. There I started to work on theoretical computer science too. Donald Knuth and László Babai encouraged me to continue to work on that discipline. I thank them for this.

With the family we spent two years at the University of South Carolina, in Columbia. I am very thankful to Tom Trotter who, when I became seriously ill, helped enormously.

In 1990 I got a tenured position at Rutgers University. I am still a member of the Department of the Computer Science. I enjoy teaching both, undergraduate and graduate levels also supervising PhD students.

Being 72, I am a professor emeritus at the Rényi Institute. My plan is to start to study analytical number theory, although I am well aware that most probably I will never get any significant result. I am still working on my old problems.

I would like to say some words about the Abel Prize. As I said in my acceptance speech, I consider this award a recognition of discrete mathematics and theoretical computer science. Also this award could not have happened were it not for the fundamental work of many mathematicians who might have been influenced by some of my results and methods, but who have developed much stronger results and established deep connections between different branches of mathematics.

The Szemerédi family

I live with my wife Anna. Our five children, Andrea, Anita, Peter, Kati, Zsuzsi and five grandchildren, Krisztian, Tibi, Szandi, Matyi and Liza live in Budapest, London and Madrid. Nevertheless we see each other very often.

The Mathematics of Endre Szemerédi

W.T. Gowers

1 Introduction

Endre Szemerédi is famous for his work in combinatorics and theoretical computer science. He has published a very large number of papers, often involving extraordinarily intricate arguments, so it will not be possible in an article such as this to do justice to either the breadth or the depth of his work. Instead, therefore, I shall describe a representative sample of his best known theorems, and attempt to convey in an informal way some of the ideas that go into their proofs. The sample will consist of the following results. (The dates given below and throughout the paper are the dates of publication rather than the dates that the results were actually proved.)

- In 1975, he proved that every dense subset of the natural numbers contains arbitrarily long arithmetic progressions, solving a famous and decades-old problem of Erdős and Turán.
- As part of the proof of the Erdős-Turán conjecture he formulated and proved a lemma, now known as Szemerédi's regularity lemma, that became a central tool in extremal graph theory and an inspiration for many other results in graph theory and beyond.
- In 1978, with Imre Ruzsa, he proved, using the regularity lemma, that every graph with few triangles can be approximated by a graph with no triangles. This innocent-looking result has been at the heart of many developments in graph theory, hypergraph theory, additive combinatorics, and computer science.
- In 1980, with Miklós Ajtai and János Komlós, he showed that the Ramsey number $R(3,k)$ is at most $Ck^2/\log k$.

Electronic supplementary material Supplementary material is available in the online version of this chapter at http://dx.doi.org/10.1007/978-3-642-39449-2_25. Videos can also be accessed at http://www.springerimages.com/videos/978-3-642-39449-2.

W.T. Gowers (✉)
Royal Society 2010 Anniversary Research Professor, Centre for Mathematical Sciences, Wilberforce Road, Cambridge, CB3 0WB, UK
e-mail: wtg10@dpmms.cam.ac.uk

H. Holden, R. Piene (eds.), *The Abel Prize 2008-2012*,

DOI 10.1007/978-3-642-39449-2_25,

- In 1982, with János Komlós and János Pintz, he found a counterexample to an old conjecture of Heilbronn in combinatorial geometry.
- In 1983, with Miklós Ajtai and János Komlós, he constructed a parallel sorting network that sorts n objects in $O(\log n)$ rounds.
- Also in 1983, with William Trotter, he proved a theorem about point-line incidences that has become one of the central results in combinatorial geometry.
- In 1995, with Jeff Kahn and János Komlós, he obtained the first exponentially small upper bound for the probability that a random ± 1 matrix is singular.

2 Szemerédi's Theorem

As its name suggests, Szemerédi's most famous theorem is … Szemerédi's theorem [26], which states the following.

Theorem 1 *For every positive integer k and every $\delta > 0$ there exists N such that every subset $A \subset \{1, \ldots, N\}$ of size at least δN contains an arithmetic progression of length k.*

This result was first conjectured in 1936, by Erdős and Turán [12], so by the time Szemerédi proved it in 1975 it had been open for almost four decades. Since then, a number of other proofs have been discovered, but they have been discussed in several other places, and here it seems more appropriate to talk about Szemerédi's own proof. Unfortunately, his proof is long and extremely intricate, so it is out of the question to present it here, and difficult even to give an overview. However, it is significantly easier in the case $k = 3$, so I shall begin by describing the argument in that case.

2.0.1 Density Increment Strategies Common to many proofs of Szemerédi's theorem, and indeed of several other results in extremal combinatorics, is the so-called *density increment strategy*. If X is a combinatorial structure and A is a subset of X, then the *density* of A is $|A|/|X|$. Suppose now that X has many substructures that are very similar to X itself and we want to prove that if the density of A is at least δ then A has some property P. Suppose also that if any subset of A has property P then A itself has that property. (In particular, this is true when the property is of the form "contains a configuration of type T".) Then instead of aiming directly for the property P we can instead try to prove one of the following *two* statements.

1. A has property P.
2. There exists a substructure Y such that the density of $A \cap Y$ in Y is at least $\delta + c(\delta)$.

Here $c(\delta)$ is a positive constant that depends on δ and increases as δ increases. (One could get away with a weaker condition, but in practice this one always holds.) If the substructure Y is sufficiently similar to X, then it too will have plenty of substructures, so in the second case we can apply the argument again. The density cannot continue increasing for ever, so eventually, as long as the initial structure X is large enough, we obtain a subset of A with property P, which shows that A has property P.

Another way of looking at the density increment strategy is that its existence allows us to add an extra assumption about the set A: that it has density δ in X, and density at most $\delta + c(\delta)$ in every substructure Y of X. (The proof: if not, then we can pass to Y and we have a bigger density to work with; this cannot go on for ever.) This is extremely useful, because it allows us to assume that A has the kind of homogeneity that is usually associated with random sets, and random sets behave very nicely. This very rough idea is at the heart of all proofs of Szemerédi's theorem, but turning it into a rigorous proof is not at all easy.

2.1 Sketch Proof of Szemerédi's Theorem when $k = 3$

Our aim in this subsection is to present Szemerédi's proof of the following result, which is the first non-trivial case of Theorem 1.

Theorem 2 *For every $\delta > 0$ there exists N such that every subset $A \subset \{1, \dots, N\}$ of cardinality at least δN contains an arithmetic progression of length* 3.

This result was first proved by Klaus Roth [20] in 1953. Roth used Fourier analysis, while Szemerédi's argument is, as we shall see, purely combinatorial.

We shall apply the density increment strategy. In the context of this problem, there is a very natural collection of substructures of $\{1, 2, \dots, N\}$, namely the set of all arithmetic progressions that are subsets of $\{1, 2, \dots, N\}$. So by the discussion above, we are free to assume that A has density δ in $\{1, 2, \dots, N\}$ and density at most $\delta + c(\delta)$ in every arithmetic progression $Y \subset \{1, \dots, N\}$ of length at least m. What matters here is that m should tend to infinity with N, since then we can ensure that m is as large as we like by choosing N sufficiently large. We shall choose $c(\delta)$ and m later in the argument.

Let us now introduce a second idea that appears in many arguments that use a density increment strategy. Suppose you want to show that A intersects a substructure Y with density at least $\delta + c(\delta)$. Often it is not obvious how to find such a substructure in one step, but it is much clearer how to show that A intersects some kind of "nice" set W with density at least $\delta + 2c(\delta)$ (say). In principle, that allows one to obtain a density increment in two stages. In the first stage, one obtains the "nice" set W such that $|A \cap W| \geq (\delta + 2c(\delta))|W|$. In the second stage, one shows that W can be partitioned into large substructures $Y_1, \dots, Y_r$. By averaging, one then deduces that there exists i such that $|A \cap Y_i| \geq (\delta + 2c(\delta))|Y_i|$ and one has a density increment.

Sometimes, asking for a partition is too much, but one can get away with a slightly weaker assertion. This is where the factor 2 comes in. It is enough if *almost* all of W can be partitioned into large substructures $Y_1, \dots, Y_r$: if $|W \setminus (Y_1 \cup$

$\cdots \cup Y_r)| \le c(\delta)|W|$, then $|A \cap (Y_1 \cup \cdots \cup Y_r)| \ge (\delta + c(\delta))|Y_1 \cup \cdots \cup Y_r|$ and the averaging argument works again.

Another way of weakening the requirement that the substructures Y_i partition W is to ask instead that they form a *uniform covering* of W: that is, every element of W is contained in the same number of sets Y_i. It is easy to see that the averaging argument still works. And one can weaken that to an approximately uniform covering. However, approximate partitions will suffice here.

2.1.1 A Strategy for Obtaining a "Nice" Set

To see where a "nice" set W might come from in our case, we shall make a simple observation, but we need to set the scene first.

Let θ be a smallish positive absolute constant, and let us divide the interval $\{1, \ldots, N\}$ into three parts: the integers up to $(1/2 - \theta)N$, the integers between $(1/2 - \theta)N$ and $(1/2 + \theta)N$ and the integers between $(1/2 + \theta)N$ and N. That is, we split $\{1, \ldots, N\}$ up into a smallish interval around $N/2$ and the intervals on either side of it. Let us refer to these intervals as the left interval, the middle interval and the right interval, and write them as L, M and R. Let us also write A_L, A_M and A_R for $A \cap L$, $A \cap M$ and $A \cap R$.

The density-increment strategy allows us to assume that the density of A in each of L, M and R is at most $\delta + c(\delta)$, which implies, by an easy averaging argument, that it is approximately *equal* to δ in each of the three subintervals.

Now if we are given any subset B of A_M, then A has an empty intersection with $2.B - A_L$ (which is defined to be $\{2y - x : x \in A_L, y \in B\}$). That is because the numbers x and y belong to A and the triple $(x, y, 2y - x)$ is an arithmetic progression. (We are of course assuming that A does not contain an arithmetic progression of length 3.) Thus, A_R is disjoint from $2.B - A_L$.

The set $2.B - A_L$ is not just any old set: it is a sumset of two large and very homogeneous sets, and as such has a highly atypical structure. Could that structure allow us to partition its complement into long arithmetic progressions?

The answer is not immediately obvious, so let us try a simple-minded approach, picking a positive integer d and partitioning the complement of $2.B - A_L$ into maximal arithmetic progressions of common difference d. It turns out to be easy to characterize how many of these progressions there are, since if x is the minimal element of such an arithmetic progression, we know that $x \notin 2.B - A_L$ and $x - d \in 2.B - A_L$. That is, $x \in (2.B - A_L + d) \setminus (2.B - A_L)$.

This gives us a proof strategy. Suppose we can find a subset $B \subset A_M$ and a positive integer d such that $(2.B - A_L + d) \setminus (2.B - A_L)$ has cardinality $o(N)$. Then we can partition the complement of $2.B - A_L$ into $o(N)$ arithmetic progressions with common difference d. Since this complement contains A_R, which has density at least $\delta/4$, the average length of these progressions tends to infinity. By an easy averaging argument, we can throw away $o(N)$ points and partition the rest of the complement of $2.B - A_L$ into long arithmetic progressions. Since $2.B - A_L$ has density at least $\delta/4$ (because A_L does, and $2.B$ is non-empty) and A_R has density roughly δ in R, the average density of A inside these arithmetic progressions is at least $(1+c\delta)\delta = \delta + c\delta^2$ for some absolute constant c, and we have our density increment on a long arithmetic progression.

2.1.2 Implementing the Strategy

It remains to find a set B and a positive integer d such that $(2.B - A_L + d) \setminus (2.B - A_L)$ has cardinality $o(N)$. For this we use the simple observation that if $B \subset B'$ then $2.B - A_L \subset 2.B' - A_L$. This again leads to a proof strategy. Suppose we can find a sequence of subsets $B_0, B_1, \ldots, B_k$ of A_M such that each B_i is of the form $B_{i-1} \cup (B_{i-1} + d_i)$ for some positive integer d_i. Then the B_i are nested, so the sets $2.B_i - A_L$ are nested, from which it follows that there exists i such that $(2.B_i - A_L) \setminus (2.B_{i-1} - A_L)$ has cardinality at most N/k. But $2.B_i - A_L = (2.B_{i-1} - A_L + 2d_i) \cup (2.B_{i-1} - A_L)$, so

$$(2.B_i - A_L) \setminus (2.B_{i-1} - A_L) = (2.B_{i-1} - A_L + 2d_i) \setminus (2.B_{i-1} - A_L),$$

which is a set of the form that we would like to have cardinality $o(N)$. Therefore, our proof will be complete if we can get k to tend to infinity.

We have now reduced the problem to the task of finding a large *Hilbert cube* inside A_M: that is, a set of the form $\{x + \sum_{j=1}^{k} \varepsilon_j d_j : \varepsilon_j \in \{0,1\}\}$. If we can find that, then we can set $B_i = \{x + \sum_{j=1}^{i} \varepsilon_j d_j : \varepsilon_j \in \{0,1\}\}$ and we will have $B_i = B_{i-1} \cup (B_{i-1} + d_i)$ as desired.

Claim *Let I be an interval of integers of cardinality m and let E be a subset of I of density η. Then E contains a Hilbert cube of dimension at least $c \log\log m$.*

The proof is well known and very simple. There are $\eta m(\eta m - 1) \approx \eta^2 m^2$ pairs (x, y) of distinct elements of E and at most $2m$ possible differences, so at least one difference d_1 occurs at least $\eta^2 m/2$ times (up to a tiny error). Let $E_1 = E \cap (E - d_1)$. Then E_1 has cardinality at least $\eta^2 m/2$ and $E_1 \cup (E_1 + d_1) \subset E$. We repeat this observation for E_1 to find a subset E_2 and a positive integer d_2 such that $E_2 \cup (E_2 + d_2) \subset E_1$, and so on. At each stage of the iteration, we square the density of the set, and the iteration continues for as long as we still have a density of at least m^{-1}. From this we obtain a cube of dimension k provided that $(1/\eta)^{2^k} \le m$, which gives us the bound of $c \log\log m$.

With the claim established, the proof of the theorem is complete.

2.2 *What Happens when the Progressions Are Longer?*

The proof just sketched contains the germs of various ideas that appear in the proof of the general case of Szemerédi's theorem. However, the argument for longer progressions is *much* more difficult and complicated. There is a simple reason for this, which can be summarized in the form of a slogan: *it is very easy to find arithmetic progressions of length 2*. In the argument above, we made use of the fact that if a set A contains no arithmetic progressions of length 3, and if B and C are subsets of A, then A is disjoint from the set $2C - B$. That is, every pair $(b, c) \in B \times C$ gives us a number $2c - b$ that is not allowed to belong to A (as long as $b \neq c$).

If we want to try to do something similar for progressions of length 4, then we will find ourselves considering *three* sets, B, C and D. But now a triple $(b, c, d) \in B \times C \times D$ does not usually yield for us an element that cannot belong to A: it does so only if b, c and d lie in an arithmetic progression (in which case the next term in that progression is not allowed to belong to A). As just one example of the kind of difficulty this can cause, suppose we tried to imitate the proof for progressions of length 3 as follows: split the interval $\{1, \ldots, N\}$ into four subintervals, let $A_1, \ldots, A_4$ be the intersections of A with those intervals, find structured subsets S_2 and S_3 of A_2 and A_3 (to play the role of the Hilbert cube), and then use points in A_1 together with the sets S_2 and S_3 to find many points that cannot belong to A_4. Whereas in the argument for progressions of length 3, every point in A_L ruled out many points from A_R, now a point in A_1 would not rule out any points at all from A_4 unless it belonged to the set $2.S_2 - S_3$. If S_2 and S_3 are very small sets (as the Hilbert cube was in the argument for progressions of length 3), then there is no reason to suppose even that the set $A_1 \cap (2.S_2 - S_3)$ is non-empty. So for an approach like this to have any chance of working, the construction of the sets S_2 and S_3 would have to depend in an essential way on the sets A_1 and A_4—by contrast with the construction of the Hilbert cube in the earlier argument, which depended on A_M only.

We shall have a little more to say about the proof for longer progressions at the end of the next section.

3 Szemerédi's Regularity Lemma

In 1947, Erdős gave a remarkably simple proof [10] that the Ramsey number $R(k, k)$ is at least $2^{k/2}$. The proof can be summarized in a single sentence: if you take the complete graph on $2^{k/2}$ vertices and randomly colour its edges with two colours, then the expected number of monochromatic cliques of size k is less than 1. This proof gave birth to the subject of random graphs, and to the realization that random graphs are in many ways easy to understand. In particular, if G is a graph with n vertices and each pair of vertices xy forms an edge with probability p, with all these events being independent, then two things happen with high probability.

1. For every large set X of vertices the density of the induced subgraph with vertex set X is approximately p.
2. If v is a fixed constant and H is a graph with v vertices and e edges, then the number of copies of H in G is approximately $p^e(1 - p)^{\binom{v}{2} - e} n^v$.

To be clear, a *copy* of H means a function ϕ from the vertex set of H to the vertex set of G such that $\phi(x)\phi(y)$ is an edge of G if and only if xy is an edge of H.

3.1 Quasirandom Graphs and the Counting Lemma

Approximately 40 years later it was realized, as a result of work of Thomason [31] and Chung, Graham and Wilson [8], that the two properties above are *equivalent*. This leads to the extremely useful notion of a *quasirandom* graph, which is a graph with one, and hence both, of the above properties.

There are many further respects in which quasirandom graphs behave like random graphs. A particularly important one is an example of a *counting lemma*, for which we need the closely related notion of a quasirandom *bipartite* graph. A bipartite graph G of density p with vertex sets X and Y of sizes m and n is quasirandom if it has one of the following two properties, which again turn out to be equivalent.

1. For every large pair of subsets $X' \subset X$ and $Y' \subset Y$ the density of the induced bipartite subgraph with vertex sets X' and Y' is approximately p.
2. If v and w are fixed constants and H is a bipartite graph with vertex sets of sizes v and w vertices and e edges, then the number of copies of H in G with the vertex set of size v in X and the vertex set of size w in Y is approximately $p^e(1-p)^{vw-e}m^v n^w$.

The counting lemma is the following statement.

Lemma 1 *Let G be a k-partite graph with vertex sets $V_1, \dots, V_k$ such that V_i has cardinality n_i for each i and such that for each i, j the bipartite graph that joins V_i to V_j is quasirandom with density α_{ij}. Let H be a graph with vertex set $\{1, 2, \dots, k\}$. Then the number of copies $\phi : H \to G$ of H in G such that $\phi(i) \in V_i$ for each i is approximately $n_1 \dots n_k \prod_{ij \in E(H)} \alpha_{ij} \prod_{ij \notin E(H)}(1-\alpha_{ij})$.*

The counting lemma has a very intuitive interpretation. Imagine that the edges between V_i and V_j are put in with probability α_{ij} and that we randomly pick a vertex v_i from each V_i. Then the probability that the vertices $v_1, \dots, v_k$ span a copy of H, in the sense that $v_i v_j$ is an edge of G if and only if ij is an edge of H, will be $\prod_{ij \in E(H)} \alpha_{ij} \prod_{ij \notin E(H)}(1-\alpha_{ij})$. The counting lemma tells us that the same is true if the edges between each V_i and V_j form *quasirandom* bipartite graphs rather than random ones. Thus, when those graphs are quasirandom, the number of copies of H is "what one would expect".

3.2 Statement of the Regularity Lemma

The counting lemma concerns graphs with vertex sets that can be partitioned into a small number of sets in such a way that the edges between each pair form quasirandom bipartite graphs. This may seem like a rather artificially strong condition to impose on a graph. Remarkably, it is not strong at all: to oversimplify slightly, Szemerédi's regularity lemma tells us that *every* dense graph is of this apparently special form.

A more precise statement of the lemma is as follows. Given any two sets U, V of vertices in a graph, let $e(U, V)$ denote the number of pairs $(u, v) \in U \times V$ such that uv is an edge of G, and define the *density* $d(U, V)$ to be $e(U, V)/|U||V|$. We define a bipartite graph G with vertex sets X and Y and density p to be *ε-regular* if $|d(U, V) - p| \leq \varepsilon$ whenever $U \subset X$, $V \subset Y$, $|U| \geq \varepsilon|X|$ and $|V| \geq \varepsilon|Y|$.

Theorem 3 *For every $\varepsilon > 0$ there exists a positive integer K with the following property. For every finite graph G there is a partition of its vertex set into subsets $V_1, \ldots, V_k$ with sizes differing by at most 1, such that $k \leq K$ and such that for all but at most εk^2 pairs (i, j) the bipartite subgraph of G induced by V_i and V_j is ε-regular.*

The slight oversimplification alluded to earlier was that I implied that *all* the pairs were quasirandom, whereas the correct statement is that they are *almost* all quasirandom. However, this does not matter too much: for example, the counting lemma remains true if a few of the bipartite graphs are not quasirandom, since it is an approximate statement.

3.3 Sketch Proof of the Regularity Lemma

The regularity lemma has been intensively studied ever since it was originally formulated, and there are now several approaches to proving it. However, even Szemerédi's original approach is simple and conceptual, so that is the one I shall present here.

3.3.1 Energy Increment Strategies

The key idea is a cousin of the density increment strategy discussed earlier: it is what is nowadays often referred to as an *energy increment strategy*. Let G be a graph with vertex set V and let $V_1, \ldots, V_r$ be a partition of the vertex set of G. Let $|V_i| = \mu_i|V|$ for each i. Then the *mean square density* of the partition is $\sum_i \mu_i \mu_j d(V_i, V_j)^2$. This we can think of as a kind of "energy".

I have stated this definition without assuming that the V_i have approximately the same size. We shall need a further definition adapted to this context. Let B be the set of "bad" pairs: that is, pairs (i, j) such that the bipartite subgraph induced by V_i and V_j is not ε-regular. Let us say that the partition is *ε-regular* if $\sum\{\mu_i\mu_j : (i, j) \in B\} \leq \varepsilon$.

The energy increment strategy, like the density increment strategy, is a way of proving results without trying to do everything in one go. Here we shall prove that one of the following two statements is true.

- The partition $V_1 \cup \cdots \cup V_r$ is ε-regular.

- There is a refinement of the partition $V_1 \cup \dots \cup V_r$ into at most $m = m(r)$ sets $W_1, \dots, W_m$ such that the mean square density of the refined partition is greater than the mean square density of the original partition by at least $c(\varepsilon)$.

If we can prove something like that, then we are clearly done: we cannot keep increasing the mean square density for ever (as with the density increment strategy, we are assuming that $c(\varepsilon)$ is an increasing function of ε), so after a certain number of refinements we end up with an ε-regular partition. The size of this partition will be bounded above by a function obtained by iterating the function m some number of times that depends on ε only.

3.3.2 Implementing an Energy Increment Strategy

To get the energy increment strategy to work here, we need two simple lemmas.

Lemma 2 *Let G be a graph with vertex set V and let $\mathcal{P} = \{V_1, \dots, V_k\}$ be a partition of V. For each i let $V_{i1}, \dots, V_{ik_i}$ be a partition of V_i and let $\mathcal{Q}$ be the partition of V into the sets V_{ij}. Then the mean square density of G with respect to $\mathcal{Q}$ is at least as big as the mean square density of G with respect to $\mathcal{P}$.*

Proof This lemma can be proved by a direct calculation using the Cauchy-Schwarz inequality. To see in a more conceptual way why it is true, let H be the Hilbert space of real-valued functions defined on $V \times V$ with the norm $\|f\| = (\mathbb{E}_{x,y} f(x, y)^2)^{1/2}$. (Here, we write $\mathbb{E}_{x,y}$ as shorthand for $|V|^{-2} \sum_{x,y}$.) Let $P : H \to H$ be the averaging projection with respect to the partition of $V \times V$ into the sets $V_i \times V_j$. That is, if $(x, y) \in V_i \times V_j$, then $Pf(x, y) = \mathbb{E}_{u \in V_i, v \in V_j} f(u, v)$. Similarly, let Q be the averaging projection that averages over the sets $V_{ij} \times V_{rs}$. Then P and Q are orthogonal projections, and $PQ = P$. Also, if f is the characteristic function of the graph G, then $\|Pf\|^2$ and $\|Qf\|^2$ are the mean square densities of G with respect to the partitions $\mathcal{P}$ and $\mathcal{Q}$, respectively. But $\|Pf\|^2 = \|PQf\|^2 \le \|Qf\|^2$, so the lemma is proved. □

For the next lemma we need to adapt the notion of mean square density to bipartite graphs. If we have a bipartite graph G with vertex sets X and Y and we have partitions $X = X_1 \cup \dots \cup X_r$ and $Y = Y_1 \cup \dots \cup Y_s$, we say that the mean square density with respect to the two partitions is $\sum_{i,j} \mu_i \nu_j d(X_i, Y_j)^2$, where $\mu_i = |X_i|/|X|$ and $\nu_j = |Y_j|/|Y|$. For the next lemma it will be useful to interpret the mean square density probabilistically. Let D be the random variable that takes a random $(x, y) \in X \times Y$ to the density $d(X_i, X_j)$ for the unique pair (i, j) such that $x \in X_i$ and $y \in Y_j$. Then the mean square density is simply $\mathbb{E}D^2$.

Lemma 3 *Let G be a bipartite graph with vertex sets X and Y and density p. Suppose that there are subsets $X_0 \subset X$ and $Y_0 \subset Y$ such that $|X_0| \ge \varepsilon|X|$, $|Y_0| \ge \varepsilon|Y|$ and $|d(X_0, Y_0) - p| \ge \varepsilon$. Let $X_1 = X \setminus X_0$ and $Y_1 = Y \setminus Y_0$. Then the mean square density of G with respect to the partitions $X = X_0 \cup X_1$ and $Y = Y_0 \cup Y_1$ is at least $p^2 + \varepsilon^4$.*

Proof Again this lemma can be proved by direct calculation. However, it is nicer to use a probabilistic argument. As we have just commented, the mean square density is $\mathbb{E}D^2$, where D is the random variable that tells you the density of G in the pair $X_i \times X_j$ that your point lies in. Now $\mathbb{E}D = p$, so $\mathbb{E}D^2 = p^2 + \text{Var}(D)$. Since the probability that a random point $(x, y) \in X \times Y$ lies in $X_0 \times Y_0$ is at least ε^2, the hypotheses of the lemma imply that the variance of D is at least $\varepsilon^2.\varepsilon^2 = \varepsilon^4$, which proves the lemma. □

We shall now be a little more sketchy. Recall that we are supposing that we have a partition $V = V_1 \cup \cdots \cup V_r$ that is not ε-regular, and we would like to find a refinement that has a slightly larger mean square density. Again, let $\mu_i = |V_i|/|V|$ and let B be the set of all pairs (i, j) such that the pair (V_i, V_j) is not ε-regular. Let us write $G(V_i, V_j)$ for the bipartite graph we obtain if we restrict G to $V_i \times V_j$.

By Lemma 3, for each pair $(i, j) \in B$, we can find partitions $V_i = V_{ij}^0 \cup V_{ij}^1$ and $V_j = V_{ji}^0 \cup V_{ji}^1$ such that the mean square density of $G(V_i, V_j)$ with respect to these two partitions is at least $d(V_i, V_j)^2 + \varepsilon^4$.

For each i, we now pick a common refinement of all the partitions $V_i = V_{ji}^0 \cup V_{ji}^1$. We can do this with $s_i \leq 2^r$ sets $V_{i1}, \dots, V_{is_i}$. Then by Lemma 2, for each $(i, j) \in B$, the mean square density of $G(V_i, V_j)$ with respect to the partitions $V_i = V_{i1} \cup \cdots \cup V_{is_i}$ and $V_j = V_{j1} \cup \cdots \cup V_{js_j}$ is at least $d(V_i, V_j)^2 + \varepsilon^4$. For all other pairs (i, j), Lemma 2 implies that the mean square density is at least $d(V_i, V_j)^2$. Since a random pair $(x, y) \in V^2$ has a probability at least ε of belonging to $V_i \times V_j$ for some $(i, j) \in B$, this implies that the mean square density of G with respect to the partition into the sets V_{ij} exceeds the mean square desnity of G with respect to $V_1, \dots, V_r$ by at least ε^5.

This completes the energy increment strategy and shows that we can find an ε-regular partition into at most $K = K(\varepsilon)$ sets, where K is the function obtained by starting with 1 and iterating the function $r \mapsto r.2^r$ ε^{-5} times. In other words, K has a tower-type dependence on ε.

When we stated the regularity lemma earlier we included an extra condition that said that the sets V_i in the partition had roughly equal size. For most applications this is not necessary, but it can be quite convenient. To obtain it, one runs the above argument but at each iteration one approximates the partition one has obtained by one in which the sets all have roughly equal size. We omit the details.

3.4 The Regularity Lemma and Szemerédi's Theorem

For an excellent account of why the regularity lemma was useful to Szemerédi for proving his theorem on arithmetic progressions, I recommend a blog post on the topic by Terence Tao [28]. Here I shall attempt to convey the idea very briefly—the arguments are explained in more detail in the blog post.

Let A be a subset of $\{1, 2, \dots, N\}$. Recall from the discussion of Szemerédi's theorem that the density increment strategy allows us to assume that $|A \cap P| \leq (\delta +$

$c(\delta))|P|$ whenever P is an arithmetic progression of length m, provided only that m tends to infinity with N. Therefore, by averaging it follows that $|A \cap P| \approx \delta|P|$ for almost all such progressions P. The same is true of more general sets that are made out of arithmetic progressions, such as sets of the form $P_1 + \cdots + P_k$ where each P_i is an arithmetic progression. (A set of this kind is called a *k-dimensional* arithmetic progression.) Let us refer loosely to sets for which this kind of conclusion follows as "structured sets".

If we now take a structured set P and look at a set of translates $P, P + r, P + 2r, \dots, P + (M - 1)r$, we can define a sequence of subsets $A_0, A_1, \dots, A_{M-1}$ of P by setting $A_i = A \cap (P + ir) - ir = \{x \in P : x + ir \in A\}$. Not only do we expect almost all the A_i to have density roughly δ, but there are also interesting relationships *between* the A_i. To give a simple example, no element $x \in P$ can belong to more than $(\delta + c(\delta))M$ of the sets A_i, since otherwise more than $(\delta + c(\delta))M$ of the elements of the arithmetic progression $x, x + r, \dots, x + (M - 1)r$ would belong to A. It follows by averaging that if E is any subset of P, then there exists i such that $|A_i \cap E| \leq (\delta + c(\delta))|E|$.

With the help of van der Waerden's theorem, one can improve this result to one about several sets $E_1, \dots, E_m$. Suppose that we cannot find i such that $|A_i \cap E_j| \leq (\delta + c(\delta))|E_j|$ for every j. Then we can m-colour the set $\{0, 1, \dots, M - 1\}$ by taking the colour of i to be some j such that $|A_i \cap E_j| > (\delta + c(\delta))|E_j|$. But then van der Waerden's theorem gives us a long arithmetic progression of numbers i for which we can take the same j, and that, by a small modification of the remarks in the previous paragraph, cannot happen.

If we apply this result not just to the E_j but also to their complements, then we may conclude further that $|A_i \cap E_j| \approx \delta|E_j|$ for every j. (By "$\approx$" here, we mean that the difference between the two sides is at most a small multiple of $|P|$, so if E_j is a very small set, then it tells us nothing.)

Unfortunately, m is so small compared with M that this observation is not very helpful on its own. It is here that the regularity lemma comes in. Let $E_1, \dots, E_N$ be a collection of subsets of P, where N may be arbitrarily large. We shall use the regularity lemma to find i such that $|A_i \cap E_j| \approx \delta|E_j|$ for *almost* every j. Thus, we have made a small loss—having to change from "every" to "almost every"—but have also made a big gain—going from m sets, where m is much less than M, to N sets, where N can be as large as we like.

Whenever one has a collection of subsets of a finite set, one can think of it as a bipartite graph in which the subsets are neighbourhoods. Here we take the vertex sets as P and $\{1, \dots, N\}$, joining $x \in P$ to j if and only if $x \in E_j$. (Thus, $E_j \subset P$ is the neighbourhood of j.) Let us apply the regularity lemma to this graph, obtaining partitions of P and $\{1, \dots, N\}$ into a bounded number of sets. Let the partition of P be $P_1 \cup \cdots \cup P_m$. Then we can apply the earlier result to obtain i such that $|A_i \cap P_s| \approx \delta|P_s|$ for every s.

Now if (U, V) is a regular pair of density α and $A \subset U$, then the regularity condition implies that for almost every $v \in V$ the neighbourhood of v in U intersects A in a set of size approximately $\alpha|A|$. Since almost all pairs are regular after we have applied the regularity lemma, for most P_s we can conclude that $|A_i \cap P_s \cap E_j| \approx$

$\delta|E_j \cap P_s|$ for almost every j. Summing over all s, it follows that $|A_i \cap E_j| \approx \delta|E_j|$ for almost every j, as claimed.

Of course, it is far from obvious how these ideas lead to a proof of Szemerédi's theorem, but that is beyond the scope of this article. At least the above argument makes it plausible that the regularity lemma could be of use.

4 The Triangle Removal Lemma

In this section we present a beautiful result of Ruzsa and Szemerédi [22], which amongst other things gives us an alternative proof of Theorem 2 and makes full use of the regularity lemma.

Theorem 4 *For every $\varepsilon > 0$ there exists $\delta > 0$ such that if G is any graph with n vertices and at most δn^3 triangles, then one can remove a set of at most εn^2 edges from G and obtain a graph that is triangle free.*

In short: every graph with few triangles can be approximated by a graph with no triangles.

One might have thought that this result would either be false, or be true with a more or less trivial proof. However, it is neither: it is true with a non-trivial proof, and to determine even very roughly the correct dependence of δ on ε is still an important open problem. (The best known bound, due to Jacob Fox, is that δ can be taken to be $1/T(\log(1/\varepsilon))$, where T is a tower-type function. In the other direction, it is known that δ cannot be greater than $\exp(-\log(1/\varepsilon)^2)$, which is just a little bit worse than a power dependence.)

4.1 Sketch Proof of the Triangle Removal Lemma

The proof of the triangle removal lemma starts in a similar way to many applications of the regularity lemma. We carry out the following three steps.

1. Apply the regularity lemma to the graph G with a suitable parameter η, obtaining a partition $V(G) = V_1 \cup \dots \cup V_k$ into sets of approximately equal size.
2. Remove from G all edges that belong to bipartite subgraphs $G(V_i, V_j)$ that are not η-regular.
3. Remove from G all edges that belong to bipartite subgraphs $G(V_i, V_j)$ of density less than θ, for some suitable parameter θ.

The result is to create a graph G' such that every bipartite subgraph $G'(V_i, V_j)$ that is non-empty has density at least θ and is η-regular. This is very useful, because it means that we can apply the counting lemma.

To see how this works for the triangle removal lemma, observe first that we have removed at most $(\eta + \theta)n^2$ edges from G. Also, since G' is a subgraph of G, we know that G' contains at most δn^3 triangles.

But when δ is sufficiently small, this second observation actually implies that G' contains *no* triangles. To see why, suppose that xyz is a triangle and let r, s and t be such that $x \in V_r$, $y \in V_s$ and $z \in V_t$. Then the three bipartite graphs $G'(V_r, V_s)$, $G'(V_s, V_t)$ and $G'(V_r, V_t)$ all contain at least one edge, from which it follows that they are all η-regular with density at least θ. But then the counting lemma implies (provided η is sufficiently small in terms of θ) that those three bipartite graphs when put together contain at least $(\theta^3/2)|V_r||V_s||V_t|$ triangles. But each of V_r, V_s and V_t has size approximately n/k, where k depends on η only. This is a contradiction when δ is small enough.

Chasing the parameters, we need $\eta + \theta \leq \varepsilon$ with η at most some power of θ. And then we can set $\delta = \theta^3/4K^3$, where $K = K(\eta)$ is the upper bound on k that comes from the regularity lemma. This gives the tower-type bound for $1/\delta$ in terms of $1/\varepsilon$.

4.2 Applications of the Triangle Removal Lemma

Ruzsa and Szemerédi noticed that the triangle removal lemma gave another proof of Roth's theorem (that is, Szemerédi's theorem for progressions of length 3). In this section we present a slight modification of their argument, observed by Jozsef Solymosi [24], that yields a stronger result.

4.2.1 The Corners Theorem

Theorem 5 *For every $\delta > 0$ there exists N such that every subset $A \subset \{1, \dots, N\}^2$ of cardinality at least δN^2 contains a triple $\{(x, y), (x + d, y), (x, y + d)\}$ with $d \neq 0$.*

Configurations of the form $\{(x, y), (x + d, y), (x, y + d)\}$ with $d \neq 0$ are sometimes called *corners*, and this result is sometimes referred to as the corners theorem.

To deduce the corners theorem from the triangle removal lemma, we need to construct a graph. This is done as follows. Let $X = Y = \{1, \dots, N\}$ and let $Z = \{1, \dots, 2N\}$. We construct a tripartite graph G with vertex sets X, Y and Z (regarding X and Y as disjoint copies of $\{1, \dots, N\}$ rather than as the same set) as follows.

1. $x \in X$ is joined to $y \in Y$ if and only if $(x, y) \in A$.
2. $x \in X$ is joined to $z \in Z$ if and only if $(x, z - x) \in A$.
3. $y \in Y$ is joined to $z \in Z$ if and only if $(z - y, y) \in A$.

If xyz forms a triangle in G, then we can set $d = z - x - y$, and A contains the three points (x, y), $(x, y + d)$ and $(x + d, y)$. Thus, triangles in G correspond to corners in A. Or rather, they almost do, but if $x + y = z$ then they don't, since in that case $d = 0$ and the "corner" is just a single point.

We therefore can't quite deduce that G contains no triangles from the fact that A contains no corners. Surprisingly, it turns out that this is not a setback: it is of vital importance to the proof that there should be at least some triangles in G, as we shall see. What we *can* say is that there is a one-to-one correspondence between triangles in G and points in A. It follows that the number of triangles in G is at most N^2. Since the number of vertices in G is $4N$, the hypotheses of the triangle removal lemma are very strongly satisfied. It follows that we can remove $o(N^2)$ edges from G to form a triangle-free graph.

However, this is easily seen to be impossible. The triangles in G are edge disjoint, since if $x + y = z$ and $x' + y' = z'$ then any two of the equalities $x = x'$, $y = y'$ and $z = z'$ implies the third. Since there are at least δN^2 triangles, one must remove at least δN^2 edges from G to make it triangle free. This contradiction implies the corners theorem.

The above proof was not the first proof of the corners theorem: that was a result of Ajtai and Szemerédi [4] from 1975. Their proof naturally gave the slightly stronger result that we may take $d > 0$. However, as was observed by Ben Green, the two statements are equivalent, since one can begin by intersecting A with a random translate of $-A$ in order to obtain a dense subset B of A with the property that if it contains a corner with $d < 0$ then it must also contain a corner with $d > 0$. As with many of Szemerédi's proofs that are apparently superseded, the argument of Ajtai and Szemerédi has turned out to have unexpected importance, serving as a model for later arguments in situations where the regularity approach cannot easily be made to work.

4.2.2 Another Proof of Roth's Theorem

As suggested above, the corners theorem implies Roth's theorem. Here is the simple deduction. Let A be a subset of $\{1, \ldots, N\}$ of density δ and let $A' \subset \{1, \ldots, 2N\}^2$ consist of all points (x, y) such that $x - y \in A$. Then A' has density at least $\delta/4$, so by the corners theorem it contains a triple $\{(x, y), (x + d, y), (x, y + d)\}$. But then the three points $x - y - d, x - y$ and $x - y + d$ all lie in A and form an arithmetic progression.

4.2.3 Property Testing

There is considerable interest amongst theoretical computer scientists in algorithms that can test for properties of their input by making only a constant number of queries. Given that the input has size n, which tends to infinity, this might seem a hopeless task. However, one typically asks for approximate answers, and one wants them to be correct with high probability rather than total certainty.

A very simple example would be testing an input sequence of n 0s and 1s to see whether at least half of the bits are 1s. We cannot hope to do this with only constantly many queries, but what if we relax the requirement so that our aim is to output one of the following two statements and be *almost* certain that the statement we go for is true?

1. At least half the bits are 1s.
2. At most 51 % of the bits are 1s.

If we sample 10^8 bits at random, then the standard deviation of the number of 1s we get will be of order of magnitude 10^4, so with very high probability the proportion of 1s in our sample will differ from the true proportion by less than 0.5 %. Therefore, we can output the first statement if the proportion of 1s in our sample is at least 50.5 % and the second statement otherwise.

Now let us think about a more interesting problem. This time our input is a graph G with n vertices. What we would really like to determine is whether or not the graph contains a triangle, but we cannot hope to do that after looking at only a constant number of edges. However, what we can do is output, with confidence, one of the following two statements.

1. G contains a triangle.
2. G can be approximated by a triangle-free graph.

This follows directly from the triangle removal lemma. The algorithm is very simple indeed: we randomly sample a large but constant number of triples of vertices, seeing in each case whether we have the vertices of a triangle. If we ever do, then we output "G contains a triangle" and we are 100 % certain that is correct. If we never discover a triangle in our sample, then with high probability the proportion of triples in G that form triangles is very small. But then, by the triangle removal lemma, G can be approximated by a triangle-free graph.

This is typical of many property-testing results in that we either make one claim with complete certainty or the other one with near certainty.

A great deal is now known about properties that can be tested for in this way, and the regularity lemma is a central tool for proving such results.

5 A Sharp Upper Bound for the Ramsey Number $R(3, k)$

Ramsey's theorem states that for every pair of positive integers k and l, there exists a positive integer n such that every graph G with n vertices contains a clique of size k or an independent set of size l. (A *clique* is a set of vertices such that every pair of vertices in the set is joined by an edge. An *independent set* is the opposite: a set of vertices such that no two of them are joined by an edge.)

The Ramsey number $R(k, l)$ is the smallest n for which Ramsey's theorem is true. Unless k and l are small, it does not appear to be feasible to calculate Ramsey numbers exactly, so attention has turned to asymptotics. However, even these are difficult to obtain with any accuracy. For example, the best known upper and lower bounds for $R(k,k)$ are roughly 2^{2k} and $2^{k/2}$, respectively, so the gap between them is exponential.

A simple argument shows that $R(3,k)$ is at most $k(k+1)/2$. Indeed, let G be a graph with n vertices. We would like to show that G contains either a triangle or an independent set of size k. Let us assume that G does not contain a triangle. This tells us that the neighbourhood of each vertex x (that is, the set of vertices joined to x) contains no edges.

We shall use this observation repeatedly to create an independent set $x_1,\dots,x_k$. Let x_1 be an arbitrary vertex of G, throw away x_1 and all its neighbours, and let V_1 be the set of all remaining vertices. Since the neighbours of x_1 form an independent set, either we are done or there are at most $k-1$ of them. In the second case, let x_2 be an arbitrary vertex in V_1, throw away x_2 and all its neighbours from V_1 and let V_2 be the set of all remaining vertices. Since the neighbours of x_2 form an independent set, which remains an independent set when x_1 is included, either we are done or there are at most $k-2$ of them. Continuing in this way, we end up finding an independent set provided that $n \geq k + (k-1) + \cdots + 1 = k(k+1)/2$.

This bound was improved in 1968 by Graver and Yackel [13] to $Ck^2 \log\log k/\log k$. Then in a paper published in 1981 Ajtai, Komlós and Szemerédi [2] improved the bound to $Ck^2/\log k$. They subsequently found a simpler argument [1] that (slightly confusingly for the historian) was published in 1980. The 1981 paper remained important for two reasons: it made progress on another interesting problem, and it introduced the so-called *semirandom method* into combinatorics, which has become a major tool with many further applications. We shall say a little about semirandom methods in the next section, but here we give the simpler proof from the 1980 paper.

5.1 Choosing an Independent Set More Carefully

The basic strategy we presented above for proving the bound $R(3,k) \leq k(k+1)/2$ was to choose an independent set $\{x_1,\dots,x_k\}$ greedily, exploiting the fact that in a triangle-free graph with no independent set of size k, no vertex has degree more than $k-1$.

If we want to improve this argument, then a natural strategy is to be slightly less greedy. For example, perhaps we could try to find a vertex of degree less than $k-1$ so that we have fewer neighbours to worry about.

In general, that may not be possible, but if we look ahead a little further, then there is something else we can try to do, namely pick the vertices x_i in such a way that when we remove their neighbourhoods, we remove as many further edges as we can. That way, we can hope that as the selection proceeds, the average degree in the remaining graph goes down, which enables us to pick vertices with not too many neighbours.

The next lemma shows how to find a vertex whose removal will cause us to remove many edges. Let us define the *second degree* of a vertex x in a graph to be the sum of the degrees of all the neighbours of x, and denote it by $d_2(x)$. Equivalently, it is the number of paths of length 2 that start at x (counting "paths" that begin and end at x). We also write $d(x)$ for the degree of x.

Lemma 4 *Let G be a graph with average degree t. Then there exists a vertex x such that $d_2(x) \geq td(x)$.*

Proof We show first that $\sum_x d_2(x) = \sum_x d(x)^2$. To see this, let A be the adjacency matrix of G. Then

$$\sum_x d_2(x) = \sum_{x,y,z} A(x,y)A(y,z) = \sum_y \sum_{x,z} A(y,x)A(y,z) = \sum_y \Big(\sum_x A(y,x)\Big)^2$$
$$= \sum_y d(y)^2$$

which is of course equal to $\sum_x d(x)^2$. (All we have done here is count the set of paths of length 2 in two different ways.)

It follows that $\mathbb{E}_x d_2(x) = \mathbb{E}_x d(x)^2$. Since the variance of the degrees is non-negative, $\mathbb{E}_x d(x)^2 \geq (\mathbb{E}_x d(x))^2 = t\mathbb{E}_x d(x)$. Therefore, there exists x such that $d_2(x) \geq td(x)$, as claimed. □

Theorem 6 *Let G be a triangle-free graph with n vertices and average degree t. Then G contains an independent set of size at least $n \log t/8t$.*

Proof By Lemma 4 we can find a vertex x such that $d_2(x) \geq td(x)$. If $d(x) > 4t$ then let us remove x from the graph, and otherwise let us remove x and all its neighbours from the graph.

In the second case, we remove $d(x)+1$ vertices from the graph, and the sum of the degrees goes down by at least $2td(x)$. The latter bound follows from the fact that G contains no triangles, which means that no edge is joined to more than one neighbour of x.

In both cases, we can then choose the largest independent set in the remainder of the graph; in the second case we can add x to that independent set to get a larger independent set.

Now let us define a function $\phi : \mathbb{N}^2 \to \mathbb{N}$ as follows: $\phi(n,m)$ is the minimum size of the largest independent set that is contained in a triangle-free graph with n vertices and m edges. Our preliminary remarks have shown that

$$\phi(n,tn) \geq \min\Big\{\phi\big(n-1, t(n-4)\big), 1 + \min_{d \leq 4t} \phi\big(n-d-1, t(n-2d)\big)\Big\}.$$

Let us now prove by induction that $\phi(n,m) \geq n^2 m^{-1} \log(m/n)/8 = n \log(m/n)/8(m/n)$. Note that $2m/n$ is the average degree of the graph, so up to a constant a bound of $n/(m/n)$ is the simple-minded bound one would get if G was regular and one just removed an arbitrary vertex and its neighbours at each stage. The interest is in the logarithmic improvement.

A simple back-of-envelope calculation shows that

$$\frac{(n-1)^2}{t(n-4)}\log\left(t\frac{n-4}{n-1}\right)\geq\frac{n}{t}\log t$$

if $\log t\geq 6$.This proves the inductive step in the case that $\phi(n,tn)\geq\phi((n-1),t(n-4))$, provided that the average degree is not too small.

Another simple back-of-envelope calculation shows that

$$\frac{(n-d-1)^2}{8t(n-2d)}\log\left(t\frac{n-2d}{n-d-1}\right)+1\geq\frac{n}{8t}\log t$$

provided that $d\leq 4t$. Let us actually do this second calculation, since it is the important case—that is, the case that tells us what happens when we remove a vertex of roughly average degree and reasonably high second degree.

Since $(n-d-1)^2=n^2-2(d-1)n+(d-1)^2>n(n-2d)$, we can bound the first fraction on the left-hand side below by $n/8t$. Therefore, it remains to prove the inequality

$$\log t+\log\left(\frac{n-2d}{n-d-1}\right)+\frac{8t}{n}\geq\log t,$$

which is equivalent to the inequality

$$\log(1-2d/n)-\log\bigl(1-(d-1)/n\bigr)+8t/n\geq 0.$$

Using the approximation $\log(1+x)\approx x$ (and not being too careful about justifying it—let us assume that d/n is reasonably small and take on trust that the argument can be made completely rigorous) we need to show that $8t/n\geq(d+1)/n$, which is true since $d\leq 4t$. (The extra elbow room here compensates for the sloppiness before.)

We haven't quite finished, since it remains to discuss what happens if $\log t<6$. In this case, we use the fact that the logarithmic improvement is just a constant. To be efficient about it, we use Turán's theorem, which implies that the largest independent set in a graph of average degree t has size at least $1+(n-1)/t$. We need this to be at least $n\log t/8t$, which it is, since $\log t<6$. This completes the proof. □

Corollary 1 *The Ramsey number $R(3,k)$ is bounded above by $Ck^2/\log k$ for an absolute constant C.*

Proof The bound on $R(3,k)$ is equivalent to the assertion that a triangle-free graph G with n vertices contains an independent set of size at least $c\sqrt{n\log n}$ for an absolute constant c. This is certainly true if there is a vertex of degree at least $\sqrt{n\log n}$, since the neighbourhood of that vertex is an independent set. If not, then the average degree is at most $\sqrt{n\log n}$, and then Theorem 6 tells us that there is an independent set of size at least $c'n\log(\sqrt{n\log n})/\sqrt{n\log n}=c\sqrt{n\log n}$. □

In another famous result, Jeong Han Kim proved in 1995 a lower bound for $R(3,k)$ that matches the upper bound of Corollary 1 to within a constant [16]. Thus, the result of Ajtai, Komlós and Szemerédi was shown by Kim to be best possible.

6 A Counterexample to Heilbronn's Triangle Conjecture

Suppose that you take n points in a unit disc. Then any three of those points define a (possibly degenerate) triangle. How large can the area of the smallest of these triangles be? A trivial upper bound is Cn^{-1}: by the pigeonhole principle there must be three points that have x-coordinates equal to within $Cn^{-1}/2$, and the triangle defined by those three points then cannot have area greater than Cn^{-1}.

There is also a fairly simple lower bound of cn^{-2}, due to Erdős. For convenience let n be a prime p, and let X be the set of all points of the form $(x/p, y/p)$ such that $0 \le x, y \le p-1$ and $y \equiv x^2 \bmod p$. In other words, X is basically the graph of the function $x \mapsto x^2 \bmod p$. Now no three of these points lie in a line, since if they did, then they would also lie in a line mod p, and a quadratic function can equal a linear function in at most two places. Therefore, X contains no degenerate triangles. But the smallest possible area of a non-degenerate triangle with vertices in $\mathbb{Z}^2$ is $1/2$, so the smallest triangle with vertices in X has area at least $p^{-2}/2$.

Heilbronn's conjecture was that the lower bound of cn^{-2} was correct. The gap between n^{-2} and n^{-1} is embarrassingly large, and initial work of Roth and Schmidt brought it down only very slightly: Roth [19] obtained an upper bound of $Cn^{-1}(\log\log n)^{-1/2}$ in 1950, then Schmidt [23] reduced that to $Cn^{-1}(\log n)^{-1/2}$ in 1972. Also in 1972, Roth [21] eventually managed to obtain an improvement in the power of n, but to nowhere near n^{-2}.

In 1982 (the paper was received in 1980), Komlós, Pintz and Szemerédi disproved Heilbronn's conjecture by proving the following result [18].

Theorem 7 *It is possible to choose n points in the unit disc such that no three form a triangle of area less than* $cn^{-2}\log n$.

That is, they obtained a logarithmic improvement over Erdős's lower bound.

Of course, a logarithmic improvement is quite small, and one could respond by modifying the conjecture to say that the smallest triangle has area at most $n^{-2+\varepsilon}$. However, the proof was very interesting and influential.

One particularly interesting aspect of the argument was that it reduced a geometrical problem to a purely combinatorial one about hypergraphs. A *k-uniform hypergraph* is a set V of *vertices* and a set E of k-tuples of vertices. The k-tuples are called *hyperedges*, but they are often simply called *edges*. A 2-uniform hypergraph is just a graph in the normal sense.

An *independent set* in a k-uniform hypergraph is the obvious generalization of what it is for a graph: it is a set of vertices such that no k of them form an edge.

The basic strategy that Komlós, Pintz and Szemerédi used to obtain their lower bound for the Heilbronn problem was as follows.

- Begin by dropping $n^{1+\alpha}$ random points into the unit disc for some small constant $\alpha > 0$.
- Define a 3-uniform hypergraph H by taking the random points as its vertices and all triples of points that form triangles of area less than $cn^{-2}\log n$ as its edges.

- Show that with high probability H has certain combinatorial properties that show that it is "locally sparse".
- Deduce from these local sparseness properties that H contains an independent set of size n.

Since an independent set in H is a set of points in the unit disc such that no three form a triangle of area less than $cn^{-2}\log n$, this strategy, if it can be carried out, disproves the Heilbronn conjecture.

Before we discuss this strategy further, it is worth looking at an observation that Komlós, Pintz and Szemerédi make in their paper, which is that selecting points from a random set can be used to give a different proof of Erdős's lower bound. To see this, let us first consider the probability that three random points form a triangle of area less than a. If the distance between the first two points is r, then the third point needs to lie within a strip of width $4a/r$. The probability that the distance between the first two points lies between r and $r+\delta r$ is at most about $2\pi r\delta r$, so an upper bound for the probability that the three points form a triangle of area at most a is $\int_0^2 (4a/r)(2\pi r)\,dr = 16\pi a$.

Therefore, if we drop $2n$ random points into the unit disc, the expected number of triangles of area at most a is at most $16\pi a\binom{n}{3} \leq 10an^3$. If we choose a to be $n^{-2}/10$, then this is at most n. Therefore, we can remove n points from the set and obtain a set of n points with no triangles of area less than $n^{-2}/10$.

Why should it be possible to gain anything over this simple approach if we choose, and then discard, more points? Let me quote from an article by Imre Bárány [5].

> According to his coauthors, Szemerédi's philosophy, that random subgraphs of a graph behave very regularly, and his vision that such a proof should work, proved decisive. Since then, the method has been applied several times and with great success.

Bárány was in fact referring to the first proof of Theorem 6 above. However, the results are closely connected: in both cases, there is some kind of sparseness condition that allows one to find a slightly larger independent set than one might naively think is possible. To make the connection clearer, let us look at the simple argument above in a slightly different way. Suppose we have m random points in the unit disc, forming a set S, and we want to choose as many of them as we can while avoiding a triangle of area a. The expected number of triangles of area a is, as we have already seen, at most $16\pi am^3$. Therefore, each point belongs, on average, to at most $16\pi am^2$ such triangles. So if a is small enough for $16\pi am^2$ to be substantially less than m, then we could imagine an algorithm that simply picks a random point $x \in S$, then throws away it and all other points $y \in S$ such that there is a point $z \in S$ for which the triangle xyz has area at most a. Typically, we will throw away at most $16\pi am^2$ points at each stage, so we can hope to obtain a subset of S with at least $m^{-1}a^{-1}/16\pi$ points.

Forgetting about the absolute constants, if $m^{-1}a^{-1} = n$ and $m \geq n$, then $a^{-1} \geq n^2$, and the larger m is, the worse a becomes. So at first it looks as though the strategy outlined above is doomed to fail. However, the argument we have just given is quite clearly very inefficient: if xyz is a triangle of small area, there is no need to throw away both y and z: it is enough to throw away just one of them. To see why this is a big help when m is large, note that if x and y are very close, then the argument above requires us to throw away all points z in quite a wide strip about the line that joins x and y, when to avoid all those triangles of small area it would be enough just to throw away y.

Let us now see what the local sparseness properties are that enable one to choose a large independent set in a hypergraph. Once we have dropped $n^{1+\alpha}$ random points into the unit disc, the expected number of pairs of points within distance $n^{-2/3}$ is at most $n^{-4/3}n^{2+2\alpha} = n^{2/3+2\alpha}$. If α is small enough, this is substantially less than n, so we can discard a small fraction of the points and end up with no two of them closer than $n^{-2/3}$.

We have already seen how to estimate the number of edges in the hypergraph H. If $a = n^{-2}$, then it is at most $16\pi n^{1+3\alpha}$, so on average each vertex belongs to at most $16\pi n^{3\alpha}$ edges.

Define a 2-*cycle* in a 3-uniform hypergraph to be a pair of edges that intersect in a set of size 2, a *simple* 3-*cycle* to be a triple of edges of the form abx, bcy, acz, where all of a, b, c, x, y, z are distinct, and a *simple* 4-*cycle* to be a quadruple of edges of the form abx, bcy, cdz, adw, where again different letters stand for distinct vertices. In a similar way to the way we estimated the number of edges, one can show that if α is small enough, then the numbers of 2-cycles, simple 3-cycles and simple 4-cycles are all substantially less than n, so we can remove a small fraction of the vertices and obtain a hypergraph with no 2-cycles, simple 3-cycles or simple 4-cycles. Komlós, Pintz and Szemerédi called such a hypergraph *uncrowded* (though they say that the term was in fact invented by Joel Spencer—indeed, the phrasing in terms of hypergraphs was Spencer's idea as well).

The main result that Komlós, Pintz and Szemerédi proved was the following result about 3-uniform hypergraphs. Define the *degree* of a vertex to be the number of edges that contain that vertex.

Theorem 8 *Let G be an uncrowded 3-uniform hypergraph with n vertices and average degree d. Suppose that d is sufficiently large, and also at most $n^{1/20}$. Then G contains an independent set of size at least $c(n/d^{1/2})(\log d)^{1/2}$.*

We shall not prove this theorem here, but we can make a few remarks. First, note that there is an easy bound of $cn/d^{1/2}$, proved as follows. If you pick a triple at random, then the probability that it is an edge is proportional to d/n^2. Therefore, if you pick m vertices at random, then the expected number of edges that they span is Cm^3d/n^2. If this is less than $m/2$, then you can discard at most $m/2$ vertices and end up with an independent set. Solving for m we obtain the bound claimed. This is essentially the argument we used above to rederive the Erdős lower bound.

Easy examples show that this bound is best possible if we do not impose the uncrowdedness assumption. So what is that assumption doing for us?

To answer that question, note first that a very similar situation applied with Theorem 6. If G is a graph (that is, a 2-uniform hypergraph) with average degree t, then a random set of m vertices spans Ctm^2/n edges on average, and for this to be less than $m/2$ we need $m = cn/t$. Theorem 6 improves this bound by a logarithmic factor under the additional assumption that G is triangle free, which is equivalent to saying that the neighbourhood of each vertex forms an independent set.

This suggests that we should look for a condition on hypergraphs that could play a similar role. The uncrowdedness assumption implies the following. Suppose we pick a vertex in an uncrowded hypergraph and throw away all vertices that belong to edges that contain x. If y and z are two such vertices, then we will lose all edges that contain either y or z. With the uncrowdedness assumption, the edges that contain y are all disjoint from the edges that contain z. That is because an edge that contains both y and z would form a 2-cycle or a simple 3-cycle, and if an edge containing y overlaps an edge containing z, then we would have either a 2-cycle or a simple 4-cycle.

There is, however, an important respect in which Theorem 8 differs from Theorem 6. In the case of graphs, each time we pick a vertex to go into our independent set, we must throw away all its neighbours. But with a 3-uniform hypergraph, if we pick a vertex x, then what we must ensure is that for every edge xyz we do not pick both of y and z. If we do this in a crude way by discarding all vertices that belong to an edge that contains x, then on average we throw away d vertices each time, and even if we can make some kind of logarithmic gain, we will end up with the wrong power of d in our final answer. In other words, a greedy algorithm, even if the hypergraph is very regular, gives a much worse bound than the simple random selection described earlier.

Very roughly, the strategy of the proof is this. Instead of choosing a single point at a time, one chooses small random sets of points to add to the independent set. If C is the set of points that have already been chosen, then it is necessary to discard every point z such that there exist $x, y \in C$ such that xyz is an edge. So each time a few more random points are added to C, one discards the points that need to be discarded and then chooses the next small random set from the points that remain.

At each stage s, if C_s is the set of points chosen so far and R_s is the set of points that remain, there is an important graph with vertex set R_s, and also an important hypergraph. The hypergraph is just the restriction of the original hypergraph G to R_s. The graph is the set of all pairs yz in R_s such that xyz is an edge of G for some $x \in C_s$. For the proof to work, it is vital that when we add some randomly chosen points from R_s to C_s to create the set C_{s+1} and pass to a new set R_{s+1}, the set R_{s+1} should resemble a randomly chosen subset of R_s, in the sense that the degrees in the graph and hypergraph should go down in roughly the expected way.

This kind of technique has become known as the *semirandom method*, and has been used to solve many problems in extremal combinatorics that had previously appeared to be hopelessly difficult.

7 An Optimal Parallel Sorting Network

A well-known mathematical problem is to minimize the number of pairwise comparisons needed to sort n objects that are linearly ordered. A simple argument shows that $\log_2(n!) = cn\log n$ comparisons are necessary. Indeed, before we do any comparisons there are $n!$ possible orderings compatible with the information we have so far. But each time we do a comparison, there are two possible results, so in the worst case the number of compatible orderings is over half what it was before the comparison. This implies the bound stated.

It is also not very hard to match this lower bound with an upper bound of the same form, using a recursively defined algorithm known as Mergesort. Take your n objects and divide them into two groups of size $n/2$ (for convenience let us assume that n is a power of 2—it is easy to remove this condition afterwards). Apply Mergesort to each group (which we know how to do by induction). We now have two ordered groups A and B of $n/2$ points, which we "merge" into an ordering of all n points as follows. Let the elements of A be $a_1 < \cdots < a_m$ and let the elements of B be $b_1 < \cdots < b_m$. Then we compare a_1 with b_1, then b_2, and so on until we reach i such that $b_i < a_1 < b_{i+1}$. We then compare a_2 with b_{i+1}, b_{i+2} and so on until we find where we can slot in a_2. We keep going like this until the two sets have been fully merged. The number of comparisons we make when doing this process is at most $2m = n$, since the number $i + j$ increases each time we move to a new comparison between some a_i and some b_j. Therefore, if we define $f(k)$ to be the time that mergesort needs to sort 2^k objects, we have the recursion $f(k) \leq 2f(k-1) + 2^k$. We also know that $f(1) = 1$. It follows easily by induction that $f(k) \leq k.2^k$. Setting $n = 2^k$, we obtain a bound of $Cn\log n$. (If n is not a power of 2, we can add some dummy objects to bring the number up to the next power of 2.)

Given two bounds that are obtained by simple arguments and are equal up to a constant, one might think that there was little more to say. However, this is not the case. A general question of major importance in computer science is whether algorithms can be *parallelized*. That is, if you have a large number of processors (growing with the size of the problem), can you get the algorithm to run *much* faster?

Rather than discuss what parallel computation is in general, let us look at a simple model that is sufficient for understanding this problem. Imagine that we have n rocks that all look quite similar but that all have slightly different weights. Imagine also that we have a very accurate balance that will take at most one rock on each side. The sorting problem just discussed is equivalent to asking how many times we need to use the balance if we want to order the rocks by weight.

For the parallel sorting problem, we can have as many balances as we like. Let us assume that comparing two rocks takes some fixed time such as one minute. Then what we would like to minimize is the total time needed to determine the order of the rocks. Since we can do up to $n/2$ comparisons at the same time, and since $cn\log n$ comparisons are needed, we will need to take at least $c\log n$ minutes. But can we achieve a growth rate that is anything like as small as logarithmic?

Before I answer that question, I need to mention another word from the title of this section. I have been discussing the word "parallel" but have not yet paid any attention to the word "network", which is also a critical part of what Ajtai, Komlós and Szemerédi did. The idea here is that we decide in advance all the comparisons we are going to do.

More formally, a *comparator network* is a sequence of partitions of $\{1,\ldots,n\}$ into $n/2$ pairs. Given a comparator network, we define a sorting algorithm as follows. At the ith stage, we use the ith partition to decide which rocks to compare: if r and s are paired, then we take the rocks in the rth and sth places, compare them, and put them back in the two places they came from, but switching them round if necessary so that the heavier rock is to the right of the lighter one. The *depth* of a comparator network is just the length of the sequence of partitions. If the network correctly sorts every permutation of the rocks, then it is a *sorting network*.

It is initially somewhat counterintuitive that efficient sorting networks exist, since the comparisons that are made do not depend at all on the results of earlier comparisions (which is quite unlike the behaviour of Mergesort). However, in 1968, Batcher [6] constructed a relatively simple sorting network of depth $C(\log n)^2$. Here, briefly, is how it works.

First, he shows inductively that merging two increasing sequences of length 2^{k-1} can be done with a comparator network of depth k. The idea is straightforward. The odd terms of the sequences form two increasing sequences of length 2^{k-2}, so by induction they can be merged with a network of depth $k-1$. In parallel, one can merge the even terms. This now gives a sequence such that the odd terms are in the right order and the even terms are as well. But it is not hard to check that because the original sequence was increasing in both halves, the only way that the final sequence can be out of order is if the terms in places $2r$ and $2r+1$ are the wrong way round. This can be cured with one final round of comparisons, which makes a depth of k.

This enables Mergesort to be carried out on 2^k objects with a sorting network of depth $k(k+1)/2$, since if the depth needed is $d(k)$, then $d(k) \leq d(k-1)+k$: the $d(k-1)$ is needed to sort each half and the k is needed to do the merging. That gives the $C(\log n)^2$ claimed, with a good constant C. There are reasons to think that improving on a $(\log n)^2$ bound might be difficult, but the remarkable result of Ajtai, Komlós and Szemerédi [3] is that there is a sorting network with the trivially optimal depth of $C \log n$.

The full proof of this result is quite technical, though it has been simplified over the years. However, it is possible to give a flavour of the ideas. Let us begin with the concept of an *ε-approximate halver*. Let us say that a rock *is in the correct half* if it is one of the $n/2$ lightest rocks and is in one of the first $n/2$ places, or is one of the $n/2$ heaviest rocks and is in one of the last $n/2$ places. An ε-approximate halver is a comparator network such that for every initial permutation of the rocks, at most εn of them do not end up in the right half when we perform the corresponding sorting algorithm.

A natural way to build an ε-approximate halver of low depth is just to choose d partitions randomly. Suppose we do that and then perform the algorithm. Let us say that a place *has a rock of the right type* if the rock in that place is in the correct half. If at any stage, a place has a rock of the right type, then it will have a rock of the right type from that moment on. For instance, if one of the $n/2$ lightest rocks is in one of the first $n/2$ places, then it can only ever be replaced by a lighter rock, so the place will continue to have a rock of the right type.

But at each stage of the process, if there are θn rocks in the wrong half and hence $\theta n/2$ wrong rocks in each half, the probability that a rock in the wrong half gets compared with a wrong rock in the other half is at least θ, and if such a comparison takes place, then the two places are filled with rocks from the correct halves. So to argue very crudely, for as long as there are εn rocks in the wrong half, each place with a rock of the wrong type has a probability ε of being filled with a rock of the correct type. So after d rounds, it has a probability at least $1-(1-\varepsilon)^d \approx 1-\exp(-d\varepsilon)$ of being filled with a rock of the correct type. Therefore, we can take d to be around $\varepsilon^{-1}\log(\varepsilon^{-1})$.

If we could move *all* rocks to the correct half in a constant number of rounds, then we would almost be done: all we would have to do is repeat the procedure (in parallel) inside each half so that each rock was in the correct quarter, and so on all the way down. But it is easy to see that this is impossible. If there are fewer than $n/2$ rounds and we only ever compare rocks from different halves, then for each place in the first half there is some place in the second half that it never gets compared with. Pick an arbitrary place r in the first half and a place s in the second half that is never compared with r. Then put the lightest $n/2$ rocks in the first half and the heaviest $n/2$ rocks in the second half, except in places r and s. In those places put rocks of the wrong type. Then none of the comparisons will move any of the rocks.

That changes if one is allowed to make comparisons within each half, but even then a depth of $c\log n$ is necessary. The reason is that if the depth is d then there are at most 2^d places that can hold rocks that end up in any given place, so we can put a rock of the wrong type in the first place, say and also in all the 2^d places that can hold rocks that end up in the first place, then there will be a rock of the wrong type in the first place at the end of the process.

To get round this difficulty, Ajtai, Komlós and Szemerédi invented a complicated and extremely ingenious scheme for ensuring that rocks that get "left behind" are moved at a later stage. Thus, in a sense, their network was an approximation of the kind of network that we have just seen cannot exist.

There was one final ingredient of their argument, which turned the above ideas from a random sorting network into a deterministic one. That was to use bipartite expander graphs. A bipartite graph with vertex sets X and Y of the same size is called a (λ,α,d)*-expander* if for every subset $A\subset X$ of size at most $\alpha|X|$, the number of vertices in Y that are joined to at least one vertex in A is greater than $\lambda|A|$, and the same for subsets of Y.

It can be shown that whenever $\lambda\alpha<1$ and n is sufficiently large, there exists d depending on λ and α only, and a collection of d perfect matchings between two sets X and Y of size n, such that the union of these perfect matchings is a (λ,α,d)-expander.

To see how this might be useful, suppose we use such a collection of matchings to form a comparator network of depth d, taking $\alpha=\varepsilon$ and $\lambda=(1-\varepsilon)/\varepsilon$. It is easy to see that after applying the corresponding sorting algorithm, we cannot be left with $\varepsilon n/2$ heavy rocks in the light places and $\varepsilon n/2$ light rocks in the heavy places. To see this, suppose that we have a set A of $\varepsilon n/2$ places on the light side and a set

B of $\varepsilon n/2$ places on the heavy side. In each place on the light side, rocks only ever get lighter, and in each place on the heavy side, rocks only ever get heavier. The expansion property guarantees that there is an edge in the graph between A and B. Therefore, a comparison must have happened between a rock at place $a \in A$ and a rock at place $b \in B$, after which the rock in place a will always be lighter than the rock in place b. This shows that it cannot be the case that after the comparisons, all the rocks in A are in the heavy half and all the rocks in B are in the light half.

8 A Theorem on Point-Line Incidences

Suppose that you have n points $x_1, \ldots, x_n$ and m lines $L_1, \ldots, L_m$ in the plane. An *incidence* is simply a pair (i, j) such that $x_i \in L_j$. The following question sounds almost too simple to be interesting: how many incidences can there be? The answer, discovered by Szemerédi and Trotter, turned out to be very interesting indeed: their result is not simple at all, and its numerous consequences have made it a central result in combinatorial geometry.

The Szemerédi–Trotter theorem is the following statement [27].

Theorem 9 *Amongst any n points and m lines the number of incidences cannot be greater than $C(m + n + (mn)^{2/3})$.*

This bound looks a little strange at first, but a few observations make it seem more natural. To begin with, we could equally well write the bound as $C\max\{m, n, (mn)^{2/3}\}$. The form of the bound is telling us that there are essentially three competing examples, and which one is best depends on the relative sizes of m and n.

It is easy to see that we can have m incidences or n incidences: we just take m lines containing a point or n points along a line. To see how to obtain $(mn)^{2/3}$ incidences, consider the grid $\{1, 2, \ldots, r\} \times \{1, 2, \ldots, s\}$ in $\mathbb{Z}^2$. For each pair of points $(a, 1)$ and $(a + d, 2)$ such that $1 \le a \le r/2$ and $1 \le d \le r/2s$, the line joining $(a, 1)$ to $(b, 2)$ intersects this grid at all the s points $(a, 1), (a + d, 2), \ldots, (a + (s - 1)d, s)$. There are $r^2/4s$ such lines.

Therefore, we can find a set of rs points and $r^2/4s$ lines with $r^2/4$ incidences. So for given m and n we need to solve the equations $rs = n$ and $r^2/4s = m$. This requires n to be at most m^2 (up to a constant) and m to be at most n^2 (also up to a constant). But if these inequalities do not hold, then one of m and n is bigger than $(mn)^{2/3}$.

This shows that the bound obtained by Szemerédi and Trotter is best possible.

For almost all of this article, I have focused on Szemerédi's original arguments, or slightly cleaned up versions of the arguments that have been produced since. In this case, however, there is a beautiful short proof discovered by Székely [25] that can be presented in full, and it seems a pity not to give it. As with most of

Szemerédi's proofs, however, his original proof of Theorem 9 is still interesting and important: there are certain generalizations that can be proved with his method that do not appear to be provable using the technique I am about to describe.

8.1 Székely's Proof of the Szemerédi–Trotter Theorem

The observation on which Székely's proof crucially depends is that a set of points and lines can be used to define a graph, and that graph has many vertices and edges. The graph is a very obvious one: its vertices are the points, and two vertices are joined if they appear consecutively along one of the lines.

If there are n points, m lines and t incidences, then the graph has n vertices and $t-m$ edges. The reason for the last assertion is that a line with k points on it gives rise to $k-1$ edges. (I am assuming here, as I may, that each line contains at least one of the points.)

Something else that we know about this graph is that it can be drawn in the plane with at most $\binom{m}{2}$ *crossings*—that is, edges that are represented by intersecting curves (which happen in this case to be line segments). However, it turns out that we can also get a *lower* bound on the number of crossings, and that means that we are in business.

Lemma 5 *Let G be a graph with n vertices and m edges. Then any drawing of G in the plane (whether edges are represented by line segments or by more general curves) must have at least $m^3/72n^2$ crossings.*

Proof Euler's formula tells us that if G is a planar graph with V vertices, E edges and F faces, then $V-E+F=2$. Since every face is bounded by at least three edges (if $V\geq 3$), and every edge is contained in at most two faces, $2E\geq 3F$, so $V-E+2E/3\geq 2$, which implies that $E\leq 3V-6$.

To put this result a different way, if we have a drawing of a graph with n vertices and more than $3n-6$ edges, then there must be at least one crossing. It follows that a drawing of a graph with n vertices and m edges must have at least $m-3n$ crossings, since we can repeat the following process at least $m-3n$ times (in fact, at least $m-3n+6$ times): find an edge involved in a crossing and remove it, thereby destroying that crossing.

Now a simple averaging argument allows us to improve this bound for large m. Let G be a graph with n vertices and m edges, and choose a random subgraph H of G by picking each vertex independently with probability p. Suppose that G has been drawn with t crossings. Then the expected number of vertices in H is pn and the expected number of crossings is p^4t, since for a crossing to belong to the subgraph, all four vertices of the two crossing edges must survive.

But the expected number of edges is p^2m, so the expected number of crossings is also at least $pm-3pn$ by the bound above. It follows that $p^4t\geq p^2m-3pn$. Choosing p to be $6n/m$, we find that $t\geq p^{-2}m/2=m^3/72n^2$. □

Applying Lemma 5 to the graph described above, we deduce that $\binom{m}{2} \geq (t-m)^3/72n^2$. Therefore, if $t \geq 2m$, we may deduce that $m^2 \geq ct^3/n^2$, which gives us the upper bound $t \leq C(mn)^{2/3}$.

8.2 An Application of the Szemerédi–Trotter Theorem

A major conjecture in additive combinatorics, due to Erdős and Szemerédi [11], states that if A is a set of integers of size n, then one of $A + A = \{x + y : x, y \in A\}$ and $A.A = \{xy : x, y \in A\}$ must have size at least $n^{2-\varepsilon}$. Since the largest possible size of the sumset or product set is $n(n+1)/2$, this is saying that one or other of the two sets must have near-maximal size.

It is not easy to obtain any non-trivial lower bound, but with the help of the Szemerédi–Trotter theorem one can show that either $A + A$ or $A.A$ has size at least $n^{5/4}$. More precisely, we have the following result. The beautiful proof is due to Elekes [9].

Theorem 10 *Let A be a set of size n. Then $|A + A||A.A| \geq cn^{5/2}$.*

Proof As our set of points we take the Cartesian product $(A + A) \times (A.A)$ and suppose that this set has size t. As our lines we take every line of the form $\{(a + \lambda, \lambda b) : \lambda \in \mathbb{R}\}$ with $a, b \in A$. Each such line intersects $(A + A) \times (A.A)$ once for every $\lambda \in A$, and therefore in n points. Therefore, since there are n^2 lines, the number of incidences is n^3. By the Szemerédi–Trotter theorem it follows that $n^3 \leq C \max\{t, n^2, t^{2/3}n^{4/3}\}$. It follows that either $t \geq cn^3$, in which case we are trivially done, or $Ct^{2/3}n^{4/3} \geq n^3$, which translates into the stated bound $t \geq cn^{5/2}$. □

The Szemerédi–Trotter theorem and modifications of it have been used to obtain many partial results in combinatorial geometry. Some of these exploit the fact that we can replace the lines in the theorem by any collection of curves, provided that no two of those curves intersect in more than a bounded number of points. For example, the Erdős distance problem asks whether given any set of n points in the plane there must be at least $n^{1-\varepsilon}$ distinct distances between them. If there are very few distances, then there are many circles about points in the set that contain many other points in the set. This gives us a set of curves with many point-curve incidences. The argument is not as straightforward as that makes it sound, because two points can belong to several different circles, so the crossing lemma needs to be generalized to graphs with multiple edges and applied accordingly. But if two points x and y belong to many circles, then the centres of those circles all lie in a line. Therefore, if we have many examples of pairs of points that belong to many circles in the set, we have a system of lines that contain many points in the set and can apply the Szemerédi–Trotter theorem again.

Recently, in a major breakthrough, the Erdős distance problem was solved by Guth and Katz using different methods [14]. However, the Szemerédi–Trotter theorem continues to be a very important tool.

8.3 What Are the Extremal Sets in the Szemerédi–Trotter Theorem?

Let us end this section with a fascinating and somewhat open-ended question (which I learned from Jozsef Solymosi).

Question 1 Let P be a set of n points and let L be a set of n lines. Suppose that there are at least $10^{-10}n^{4/3}$ incidences between P and L. What can be said about the structures of P and L?

An answer to this question would fit very well a recurring theme in extremal combinatorics, which is to take an extremal result and to ask what happens in the near-extremal cases. For most such problems, we have an inequality and can say what happens when equality occurs. To give a simple example, if A is a set of n numbers, then $|A+A| \geq 2n-1$, and equality holds if and only if A is an arithmetic progression. However, with the Szemerédi–Trotter theorem, the exact best possible bound is not known, so obtaining a structural result for *any* bound seems to be challenging. In the case of sumsets, a beautiful theorem of Freiman completely characterizes, at least qualitatively, all sets A such that $|A+A| \leq C|A|$ for some fixed constant C: each such set has to be a large subset of a generalized arithmetic progression of low dimension. Here, one might be looking for some kind of grid-like structure. This would follow from known results (one of which is Freiman's theorem itself) if one could show that there had to be cn^3 quadruples of points in P that formed the vertices of (possibly degenerate) parallelograms.

9 The Probability that a Random ±1 Matrix is Singular

Let M be a random $n \times n$ matrix where each entry has a 50 % chance of being 1 and a 50 % chance of being -1, with all choices independent. What is the probability that M is singular? Equivalently, what is the probability that if you choose n random 01-sequences of length n, then one of them will be in the linear span of the others?

This very basic question is surprisingly difficult to answer. Even to show that the probability tends to zero was a non-trivial open problem, solved by Komlós in 1967 [17]. (In this case the discrepancy between publication date and the date of the actual proof is quite large: the result was obtained in 1963.) He proved that the probability is at most $C/\sqrt{n}$.

There is a natural conjecture for the correct bound, which is $(2+o(1))\binom{n}{2}2^{-(n-1)}$. The heuristic argument for this is that by far the easiest way to obtain a linear dependence amongst the rows of a random ± 1 matrix ought to be to have two rows or two columns that are equal up to a ± 1 multiple.

The truth of this conjecture is still an open problem, and one that appears to need a major new idea. Given that situation, the next strongest aim it was reasonable to have was to prove that the probability was exponentially small. This too seemed out of reach, so it was a big surprise when Kahn, Komlós and Szemerédi proved it in 1995 [15]. In the remainder of this section, let us look at some of the ideas that were involved in their proof.

9.1 The Need to Consider Dependences

One might attempt to prove the result in the following way, which works for many problems.

- Express the event E whose probability we are trying to estimate as a union of simple events.
- Give upper bounds for the probabilities of the simple events.
- Use the trivial "union bound" (that is, just add up the probabilities of the simple events) as an upper bound for the original event E.

In our case, E is the event that a random $n \times n$ ± 1 matrix M is singular. But M is singular if and only if $Ma = 0$ for some $a \in \mathbb{R}^n$, so an obvious candidate for the set of simple events is to take all events of the form "$Ma = 0$". Let us call this event E_a.

Obviously this won't do as it stands, since there are infinitely many possible a. However, we could try to identify a finite set A of vectors a such that if M is singular then there exists $a \in A$ such that $Ma = 0$. Such sets trivially exist: for each singular matrix M we pick a vector a such that $Ma = 0$ and then we put together these vectors to form our set A. However, they do not necessarily help us. For example, let F be the set of all ± 1 matrices that have two pairs of equal columns. For each matrix $M \in F$, let a be a vector with four non-zero coordinates that take the values $\pm\lambda$ and $\pm\mu$ in such a way that $Ma = 0$, and make sure that no one of these vectors is a multiple of another. The number of vectors we create is at least $2^{n(n-2)}$, and for each such vector a the probability that $Ma = 0$ is at least 2^{-2n}. Multiplying these numbers together we get $2^{n(n-4)}$, which is far bigger than 1 and therefore tells us nothing.

Of course, it was perverse of us to make sure that no two of the vectors were multiples of each other: if we had taken $\lambda = \mu = 1$ for every single vector, then the number of vectors would have dropped to cn^4. But the point is nevertheless made that for a union bound to work one would have to obtain a great deal of duplication of this kind, which is not obviously possible.

Kahn, Komlós and Szemerédi use a natural generalization of this approach. Instead of using the trivial fact that if $a = b$ then the events E_a and E_b are the same, so that only one of them needs to be considered in a union bound, they show that if several vectors a_i belong to a low-dimensional subspace S, then the events E_{a_i} are highly correlated, with the result that the event $\bigcup_i E_{a_i}$ has a much smaller probability than $\sum_i P(E_{a_i})$. In other words, linear dependencies lead to useful probabilistic dependencies.

9.1.1 The Probability that $Ma = 0$

Nevertheless, it is useful to think about the events E_a and in particular about their probabilities. Given a vector a, let $p(a)$ be the probability that $\sum_i \varepsilon_i a_i = 0$, where $(\varepsilon_1, \dots, \varepsilon_n)$ is a random ± 1 sequence. Then the probability that $Ma = 0$ is $p(a)^n$.

What sort of values can $p(a)$ take?

- If a is the vector $(1, 1, \dots, 1)$, then $p(a)$ is around $n^{-1/2}$.
- More generally, if a takes the value ± 1 d times and 0 otherwise, then $p(a)$ is around $d^{-1/2}$.
- The Littlewood-Offord inequality, or rather a slight improvement of it due to Erdős, implies a sort of converse to this observation: if the support of a has size d, then $p(a) \leq d^{-1/2}$.
- Sárközy and Szemerédi proved that if $a_1, \dots, a_n$ are distinct, then $p(a) \leq Cn^{-3/2}$.

Thus, for $p(a)$ to be large, we need a to have small support and many repeated entries.

9.1.2 Dealing with Vectors a for which $p(a)$ Is Very Small

A simple lemma shows that we can at least disregard all vectors a for which $p(a)$ is exponentially small.

Lemma 6 *For every $p \in [0, 1]$ the probability that there exists a such that $p(a) \leq p$ and $Ma = 0$ is at most np.*

Proof Let $E(p)$ be the event that such a vector a exists. Let us condition on the entire matrix M apart from the ith row, and bound from above the probability, given those $n(n-1)$ values, that $E(p)$ holds and the ith row is in the linear span of the other rows.

Now for $E(p)$ to hold conditional on these values, there must exist a with $p(a) \leq p$ that is orthogonal to all rows apart from the ith. Pick any such vector a. For the ith row to be a linear combination of the other rows, it is necessary that it too should be orthogonal to a, and this happens with probability at most p.

For $E(p)$ to hold in general, there must exist a row of M that is in the linear span of the other rows. The lemma follows. □

9.1.3 Applying Linear Dependence

A similar argument shows that we can improve the bound in Lemma 6 if we insist that a belongs to a k-codimensional subspace.

Lemma 7 *Let S be a k-codimensional subspace and let $0 \leq p \leq 1$. Then the probability that there exists $a \in S$ such that $p(a) \leq p$ and $Ma = 0$ is at most $\binom{n}{k+1} p^{k+1}$.*

Proof For such an a to exist, it is necessary that the kernel of M intersects S. Therefore, writing V_j for the orthogonal complement of the span of the first j rows of M, there can be at most $n-k-1$ values of j for which $V_j \cap S$ is a proper subset of $V_{j-1} \cap S$.

Now let us fix a set J of size $n-k-1$ and assume that $j \in J$ whenever $V_j \cap S \subset\neq V_{j-1} \cap S$. Let us condition on the values of M in the $n-k-1$ rows corresponding to J.

Now let $j \notin J$. By construction, $V_j \cap S = V_{j-1} \cap S$, so if $Ma = 0$ with $p(a) \leq p$, then $a \in V_{j-1}$, which implies that $a \in V_j$ and hence that a must be orthogonal to the jth row of M, which happens with probability at most p. We can apply this argument to each of the $k+1$ rows not corresponding to elements of J, and since those rows are independent, we obtain an upper bound of p^{k+1} for that choice of J. Applying the law of total probability and summing over all J gives the result. □

9.2 The Main Idea

Let us informally refer to a vector a as *bad* if $p(a)$ is large (meaning greater than $(1-\varepsilon)^n$ for some suitable ε). For Lemma 7 to be useful, we need to be able to show that we can cover the bad vectors efficiently with subspaces of fairly low dimension. To this end, Kahn, Komlós and Szemerédi prove a result that seems at first glance to be rather unlikely to be true. Let $a_1, \dots, a_n$ are integers and let $\mu > 0$ be a smallish absolute constant. Consider the following two random walks. At time t, the first walk chooses a random step of $\pm a_t$, each with probability $1/2$. The second walk chooses a random step of $\pm a_t$, each with probability μ, and a step of 0 with probability $1-2\mu$. Their result is that, no matter what the initial sequence $a_1, \dots, a_n$ was, as long as it has a reasonably large support, the probability that the first walk ends up at 0 is smaller by a factor of $O(\sqrt{\mu})$ than the probability that the second walk ends up at 0.

Because the binomial distribution is highly concentrated about its mean, the second walk is similar, but not identical, to a walk where we first randomly choose $d = 2\mu n$ of the a_i, replace all the others by 0, and then do a normal random walk with the new sequence. So it might seem that the result cannot be true if, for example, we take the sequence $1, 2, 4, \dots, 2^{n-2}, -(2^{n-1}-1)$, in which case the only way of getting back to the origin is to take all signs positive or all signs negative. However, in this case the probability of ending at 0 with the first walk is $2^{-(n-1)}$, while the probability with the second walk is $2\mu^n + (1-2\mu)^n$, which is much larger as long as μ is smaller than $1/4$. However, in this case the result is not really telling us very much: for it to be useful we need $p(a)$ not to be too small, which, roughly speaking, allows us to assume that the higher probability in the second walk arises for non-trivial reasons.

9.2.1 A Very Rough Sketch of the Main Argument

To see how all this helps, recall that our aim is to choose a collection of subspaces of not too large dimension that cover all the "bad" vectors a. A bad vector is one for which $p(a)$ is large, and the main step described above implies that if $p(a)$ is large, then for some d close to λn, the probability that a random ± 1 sum of d randomly chosen entries of a is 0 is larger than $p(a)$ by at least a factor $c\sqrt{\lambda}$.

Now let us choose a subspace S randomly as follows. Define a *d-vector* to be a vector in $\{1, -1, 0\}^n$ that takes non-zero values exactly d times. Also, given a vector a, define a *d-sum* of a to be a ± 1-sum of d terms of a. Equivalently, it is the inner product of a with a d-vector. For a suitable γ, choose $(1-\gamma)n$ d-vectors at random and let S be the orthogonal complement of the space spanned by these d-vectors. Thus, $a \in S$ if and only if every d-sum of a that corresponds to one of the d-vectors we have chosen is 0. If we know that the probability that a random d-sum of a is 0 is greater by a factor $c\sqrt{\lambda}$ than $p(a)$, then the probability that a belongs to the γn-dimensional subspace S is greater by a factor $(c\sqrt{\lambda})^{(1-\gamma)n}$ than the bound of roughly $p(a)^{(1-\gamma)n}$ that comes from Lemma 7.

We can now partition the interval $[(1-\varepsilon)^n, 1]$ into not too many subintervals of ps of approximately the same size. If we apply Lemma 7 to a particular value of p, then each subspace we apply it to contributes roughly $p^{(1-\gamma)n}$ (the binomial coefficient turns out not to make too much difference so I am ignoring it). But since the probability that a vector a with $p(a) \approx p$ belongs to a random such subspace is more like $(C\lambda^{-1/2})^n p^{(1-\gamma)n}$, the number of such subspaces that we need to cover all the a with $p(a) \approx p$ is roughly $(c\lambda^{1/2})^n p^{-(1-\gamma)n}$, and the total contribution is $(c\lambda^{1/2})^n$, which is exponentially small. Adding up the contributions of this kind, we find that they are dominated by the contribution of $n(1-\varepsilon)^n$ that came from the a for which $p(a)$ is very small, and the result is proved.

9.3 Subsequent Improvements

The bound obtained by Kahn, Komlós and Szemerédi was around $(0.999)^n$. In 2006 this was slightly improved, to $(0.953)^n$, by Tao and Vu [29]. The following year they obtained a bound of $(3/4+o(1))^n$ using methods from additive combinatorics [30]. The current record is $(1/\sqrt{2}+o(1))^n$. This is a very recent result (it appeared in 2013) of Bourgain, Vu and Wood [7].

10 Conclusion

Szemerédi's work has several qualities that make it stand out and that make him one of the great mathematicians of the second half of the twentieth century, not to mention the beginning of the twenty-first. An obvious one is the sheer difficulty of so many of his results. He has often solved open problems on which the mathematical community had become completely stuck, and his ingenious and delicate solutions have often left other mathematicians feeling that they were in a sense right to be stuck. Another quality that many of his results have had, and that the very best results in combinatorics have, is that the proofs have introduced techniques and ideas with applications that go far beyond the original problems that Szemerédi was solving. His influence permeates the whole of combinatorics and theoretical computer science, fully justifying the award of the Abel Prize.

References

1. Ajtai, M., Komlós, J., Szemerédi, E.: A note on Ramsey numbers. J. Comb. Theory, Ser. A **29**, 354–360 (1980)
2. Ajtai, M., Komlós, J., Szemerédi, E.: A dense infinite Sidon sequence. Eur. J. Comb. **2**, 1–11 (1981)
3. Ajtai, M., Komlós, J., Szemerédi, E.: An $O(n \log n)$ sorting network. Combinatorica **3**, 1–19 (1983)
4. Ajtai, M., Szemerédi, E.: Sets of lattice points that form no squares. Studia Sci. Math. Hung. **9**, 9–11 (1975)
5. Bárány, I.: Applications of graph and hypergraph theory in geometry. In: Goodman, J.E., Pach, J., Welzl, E. (eds.) Combinatorial and Computational Geometry. MSRI Publications, vol. 52, pp. 31–50 (2005)
6. Batcher, K.: Sorting networds and their applications. AFIPS Spring Joint Comput. Conf. **32**, 307–314 (1968)
7. Bourgain, J., Vu, V.H., Wood, P.M.: On the singularity probability of discrete random matrices. http://arxiv.org/abs/0905.0461
8. Chung, F., Graham, R.L., Wilson, R.M.: Quasi-random graphs. Combinatorica **9**, 345–362 (1989)
9. Elekes, Gy.: On the number of sums and products. Acta Arith. **81**, 365–367 (1997)
10. Erdős, P.: Some remarks on the theory of graphs. Bull. Am. Math. Soc. **53**, 292–294 (1947)
11. Erdős, P., Szemerédi, E.: On sums and products of integers. In: Erdős, P., Alpar, L., Halasz, G. (eds.) Studies in Pure Mathematics: To the Memory of Paul Turán, pp. 213–218. Birkhäuser, Basel (1983)
12. Erdős, P., Turán, P.: On some sequences of integers. J. Lond. Math. Soc. **11**, 261–264 (1936)
13. Graver, J.E., Yackel, J.: Some graph theoretic results associated with Ramsey's theorem. J. Comb. Theory **4**, 125–175 (1968)
14. Guth, L., Katz, N.H.: On the Erdős distinct distances problem in the plane. http://arxiv.org/abs/1011.4105
15. Kahn, J., Komlós, J., Szemerédi, E.: On the probability that a random ± 1-matrix is singular. J. Am. Math. Soc. **8**, 223–240 (1995)
16. Kim, J.H.: The Ramsey number $R(3,t)$ has order of magnitude $t^2/\log t$. Random Struct. Algorithms **7**, 173–207 (1995)
17. Komlós, J.: On the determinant of (0, 1) matrices. Studia Sci. Math. Hung. **2**, 387–399 (1968)
18. Komlós, J., Pintz, J., Szemerédi, E.: A lower bound for Heilbronn's problem. J. Lond. Math. Soc. **25**, 13–24 (1982)
19. Roth, K.F.: On a problem of Heilbronn. J. Lond. Math. Soc. **26**, 198–204 (1951)
20. Roth, K.F.: On certain sets of integers, I. J. Lond. Math. Soc. **28**, 104–109 (1953)
21. Roth, K.F.: On a problem of Heilbronn, II. Proc. Lond. Math. Soc. **25**, 193–212 (1972)
22. Ruzsa, I.Z., Szemerédi, E.: Triple systems with no six points carrying three triangles. In: Combinatorics, Proc. Fifth Hungarian Colloq., Keszthely, 1976. Coll. Math. Soc. J. Bolyai 18, Volume II, pp. 939–945. North Holland, Amsterdam (1978)

23. Schmidt, W.M.: On a problem of Heilbronn. J. Lond. Math. Soc. **4**, 545–550 (1972)
24. Solymosi, J.: Note on a generalization of Roth's theorem. In: Pach, J. (ed.) Discrete and Computational Geometry. Algorithms Combin., vol. 25, pp. 825–827. Springer, Berlin (2003)
25. Székely, L.A.: Crossing numbers and hard Erdős problems in discrete geometry. Comb. Probab. Comput. **6**, 353–358 (1997)
26. Szemerédi, E.: On sets of integers containing no k elements in arithmetic progression. Acta Arith. **27**, 199–245 (1975)
27. Szemerédi, E., Trotter, W.T.: Extremal problems in discrete geometry. Combinatorica **3**, 381–392 (1983)
28. Tao, T.: Some ingredients in Szemerédi's proof of Szemerédi's theorem. Blog post (2012). http://terrytao.wordpress.com/2012/03/23/some-ingredients-in-szemeredis-proof-of-szemeredis-theorem/
29. Tao, T., Vu, V.H.: On random ± 1 matrices: singularity and determinant. Random Struct. Algorithms **28**, 1–23 (2006)
30. Tao, T., Vu, V.H.: On the singularity probability of random Bernoulli matrices. J. Am. Math. Soc. **20**, 603–628 (2007)
31. Thomason, A.: Pseudo-random graphs. In: Karoński, M. (ed.) Proceedings of Random Graphs, Poznań, 1985, pp. 307–331. North Holland, Amsterdam (1987)

List of Publications for Endre Szemerédi

1964

[1] (with J. Komlós and A. Sárközi). On sums of powers of complex numbers. *Mat. Lapok*, 15:337–347.

1965

[2] (with A. Sárközi). Über ein Problem von Erdős und Moser. *Acta Arith.*, 11:205–208.
[3] (with A. Sárközy). On the sequence of squares. *Mat. Lapok*, 16:76–85.

1966

[4] (with P. Erdős and A. Sárközi). On divisibility properties of sequences of integers. *Studia Sci. Math. Hungar*, 1:431–435.
[5] (with P. Erdős and A. Sárközi). On the solvability of the equations $[a_i, a_j] = a_r$ and $(a'_i, a'_j) = a'_r$ in sequences of positive density. *J. Math. Anal. Appl.*, 15:60–64.
[6] (with P. Erdős and A. Sárközy). On the divisibility properties of sequences of integers. I. *Acta Arith*, 11:411–418.

1967

[7] (with P. Erdős and A. Šarkezi). On the solvability of certain equations in the dense sequences of integers. *Dokl. Akad. Nauk SSSR*, 176:541–544.
[8] (with P. Erdős and A. Sárközi). On an extremal problem concerning primitive sequences. *J. London Math. Soc.*, 42:484–488.
[9] (with P. Erdős and A. Sárközy). On a theorem of Behrend. *J. Austral. Math. Soc.*, 7:9–16.
[10] (with G. Katona). On a problem of graph theory. *Studia Sci. Math. Hungar*, 2:23–28.

1968

[11] (with P. Erdős and A. Sárközi). On the divisibility properties of sequences of integers. II. *Acta Arith.*, 14:1–12.

H. Holden, R. Piene (eds.), *The Abel Prize 2008-2012*,
DOI 10.1007/978-3-642-39449-2_26, © Springer-Verlag Berlin Heidelberg 2014

[12] (with P. Erdős and A. Sárközi). On the solvability of certain equations in sequences of positive upper logarithmic density. *J. London Math. Soc.*, 43:71–78.

[13] (with P. Erdős). On a problem of P. Erdős and S. Stein. *Acta Arith.*, 15:85–90.

1969

[14] (with P. Erdős and A. Sárközi). On some extremal properties of sequences of integers. *Ann. Univ. Sci. Budapest. Eötvös Sect. Math.*, 12:131–135.

[15] (with P. Erdős and A. Sárközi). Über Folgen ganzer Zahlen. In *Number Theory and Analysis (Papers in Honor of Edmund Landau)*, pages 77–86. Plenum, New York.

[16] On sets of integers containing no four elements in arithmetic progression. *Acta Math. Acad. Sci. Hungar.*, 20:89–104.

1970

[17] (with P. Erdős and A. Sárközi). On divisibility properties of sequences of integers. In *Number Theory (Colloq., János Bolyai Math. Soc., Debrecen, 1968)*, pages 35–49. North-Holland, Amsterdam.

[18] (with A. Hajnal). Proof of a conjecture of P. Erdős. In *Combinatorial theory and its applications, II (Proc. Colloq., Balatonfüred, 1969)*, pages 601–623. North-Holland, Amsterdam.

[19] (with A. Sárközy). On intersections of subsets of finite sets. *Mat. Lapok*, 21:269–278.

[20] On a conjecture of Erdős and Heilbronn. *Acta Arith.*, 17:227–229.

[21] On a problem of P. Erdős. In *Combinatorial theory and its applications, III (Proc. Colloq., Balatonfüred, 1969)*, pages 1051–1053. North-Holland, Amsterdam.

[22] On sets of integers containing no four elements in arithmetic progression. In *Number Theory (Colloq., János Bolyai Math. Soc., Debrecen, 1968)*, pages 197–204. North-Holland, Amsterdam.

1971

[23] On a problem of W. Schmidt. *Studia Sci. Math. Hungar.*, 6:287–288.

1972

[24] (with P. Erdős). On a Ramsey type theorem. *Period. Math. Hungar.*, 2:295–299. Collection of articles dedicated to the memory of Alfréd Rényi, I.

[25] (with A. Hajnal and E.C. Milner). A cure for the telephone disease. *Canad. Math. Bull.*, 15:447–450.

[26] (with J. Komlós and M. Sulyok). A lemma of combinatorial number theory. *Mat. Lapok*, 23:103–108.

[27] (with G. Petruska). On a combinatorial problem. I. *Studia Sci. Math. Hungar.*, 7:363–374.

[28] On graphs containing no complete subgraph with 4 vertices. *Mat. Lapok*, 23:113–116.

1973

[29] (with P. Erdős). On the number of solutions of $m = \sum_{i=1}^{k} x_i^k$. In *Analytic number theory (Proc. Sympos. Pure Math., Vol. XXIV, St. Louis Univ., St. Louis, Mo., 1972)*, pages 83–90. Amer. Math. Soc., Providence, RI.

[30] On a problem of P. Erdős. *Mathematika*, 20:226–228.

[31] On the difference of consecutive terms of sequences defined by divisibility properties. II. *Acta Arith.*, 23:359–361.

1974

[32] On a problem of Davenport and Schinzel. *Acta Arith.*, 25:213–224.

[33] (with M. Ajtai). Sets of lattice points that form no squares. *Stud. Sci. Math. Hungar.*, 9:9–11.

1975

[34] The number of squares in an arithmetic progression. *Studia Sci. Math. Hungar.*, 9(3-4):417.

[35] (with B. Bollobás and P. Erdős). On complete subgraphs of r-chromatic graphs. *Discrete Math.*, 13(2):97–107.

[36] (with S.L.G. Choi and P. Erdős). Some additive and multiplicative problems in number theory. *Acta Arith.*, 27:37–50. Collection of articles in memory of Juriĭ Vladimirovič Linnik.

[37] (with S.L.G. Choi and J. Komlós). On sum-free subsequences. *Trans. Amer. Math. Soc.*, 212:307–313.

[38] (with S. Józsa). The number of unit distance on the plane. In *Infinite and finite sets (Colloq., Keszthely, 1973; dedicated to P. Erdős on his 60th birthday), Vol. II*, pages 939–950. Colloq. Math. Soc. Janos Bolyai, Vol. 10. North-Holland, Amsterdam.

[39] (with J. Komlós and M. Sulyok). Linear problems in combinatorial number theory. *Acta Math. Acad. Sci. Hungar.*, 26:113–121.

[40] (with J. Komlós). Hamilton cycles in random graphs. In *Infinite and finite sets (Colloq., Keszthely, 1973; dedicated to P. Erdős on his 60th birthday), Vol. II*, pages 1003–1010. Colloq. Math. Soc. János Bolyai, Vol. 10. North-Holland, Amsterdam.

[41] On sets of integers containing no k elements in arithmetic progression. In *Proceedings of the International Congress of Mathematicians (Vancouver, B. C., 1974), Vol. 2*, pages 503–505. Canad. Math. Congress, Montreal, Que.

[42] On sets of integers containing no k elements in arithmetic progression. *Acta Arith.*, 27:199–245. Collection of articles in memory of Juriĭ Vladimirovič Linnik.

1976

[43] (with P. Erdős). On a problem of Graham. *Publ. Math. Debrecen*, 23(1-2):123–127.

[44] (with P. Erdős and R.L. Graham). On sparse graphs with dense long paths. In *Computers and mathematics with applications*, pages 365–369. Pergamon, Oxford.

[45] (with P. Erdős). On multiplicative representations of integers. *J. Austral. Math. Soc. Ser. A*, 21(4):418–427.

[46] (with L.J. Guibas). The analysis of double hashing. In *Eighth Annual ACM Symposium on Theory of Computing (Hershey, Pa., 1976)*, pages 187–191. Assoc. Comput. Mach., New York.

[47] On a problem of P. Erdős. *J. Number Theory*, 8(3):264–270.

1978

[48] (with M. Ajtai and J. Komlós). There is no fast single hashing algorithm. *Inform. Process. Lett.*, 7(6):270–273.

[49] (with B. Bollobás, P. Erdős, and M. Simonovits). Extremal graphs without large forbidden subgraphs. *Ann. Discrete Math.*, 3:29–41. Advances in graph theory (Cambridge Combinatorial Conf., Trinity Coll., Cambridge, 1977).

[50] (with P. Erdős). Combinatorial properties of systems of sets. *J. Combinatorial Theory Ser. A*, 24(3):308–313.

[51] (with R.L. Graham). On subgraph number independence in trees. *J. Combinatorial Theory Ser. B*, 24(2):213–222.

[52] (with L.J. Guibas). The analysis of double hashing. *J. Comput. System Sci.*, 16(2):226–274.

[53] (with I.Z. Ruzsa). Triple systems with no six points carrying three triangles. In *Combinatorics (Proc. Fifth Hungarian Colloq., Keszthely, 1976), Vol. II*, volume 18 of *Colloq. Math. Soc. János Bolyai*, pages 939–945. North-Holland, Amsterdam.

[54] Regular partitions of graphs. In *Problèmes combinatoires et théorie des graphes (Colloq. Internat. CNRS, Univ. Orsay, Orsay, 1976)*, volume 260 of *Colloq. Internat. CNRS*, pages 399–401. CNRS, Paris.

1979

[55] (with M. Ajtai and J. Komlós). Topological complete subgraphs in random graphs. *Studia Sci. Math. Hungar.*, 14(1-3):293–297.

1980

[56] (with M. Ajtai and J. Komlós). A note on Ramsey numbers. *J. Combin. Theory Ser. A*, 29(3):354–360.

[57] (with P. Erdős and A. Sárközy). On some extremal properties of sequences of integers. II. *Publ. Math. Debrecen*, 27(1-2):117–125.

[58] (with P. Erdős). Remarks on a problem of the *American Mathematical Monthly*. *Mat. Lapok*, 28(1-3):121–124.

[59] (with A. Gyárfás and Zs. Tuza). Induced subtrees in graphs of large chromatic number. *Discrete Math.*, 30(3):235–244.

[60] (with J. Komlós and M. Sulyok). Second largest component in a random graph. *Studia Sci. Math. Hungar.*, 15(4):391–395.

1981

[61] (with M. Ajtai and J. Komlós). The longest path in a random graph. *Combinatorica*, 1(1):1–12.

[62] (with M. Ajtai, P. Erdős, and J. Komlós). On Turán's theorem for sparse graphs. *Combinatorica*, 1(4):313–317.

[63] (with M. Ajtai and J. Komlós). A dense infinite Sidon sequence. *European J. Combin.*, 2(1):1–11.

[64] (with V. Chvátal). On the Erdős–Stone theorem. *J. London Math. Soc. (2)*, 23(2):207–214.

[65] (with J. Komlós and J. Pintz). On Heilbronn's triangle problem. *J. London Math. Soc. (2)*, 24(3):385–396.

1982

[66] (with M. Ajtai, V. Chvátal, and M.M. Newborn). Crossing-free subgraphs. In *Theory and practice of combinatorics*, volume 60 of *North-Holland Math. Stud.*, pages 9–12. North-Holland, Amsterdam.

[67] (with M. Ajtai, J. Komlós, J. Pintz, and J. Spencer). Extremal uncrowded hypergraphs. *J. Combin. Theory Ser. A*, 32(3):321–335.

[68] (with M. Ajtai, J. Komlós, and V. Rödl). On coverings of random graphs. *Comment. Math. Univ. Carolin.*, 23(1):193–198.

[69] (with M. Ajtai and J. Komlós). Largest random component of a k-cube. *Combinatorica*, 2(1):1–7.

[70] (with V. Chvátal). On an extremal problem concerning intervals. *European J. Combin.*, 3(3):215–217.

[71] (with P. Erdős and A. Hajnal). On almost bipartite large chromatic graphs. In *Theory and practice of combinatorics*, volume 60 of *North-Holland Math. Stud.*, pages 117–123. North-Holland, Amsterdam.

[72] (with M.L. Fredman and J. Komlós). Storing a sparse table with $O(1)$ worst case access time. In *23rd annual symposium on foundations of computer science (Chicago, Ill., 1982)*, pages 165–169. IEEE, New York.

[73] (with J. Komlós and J. Pintz). A lower bound for Heilbronn's problem. *J. London Math. Soc. (2)*, 25(1):13–24.

[74] (with Zs. Tuza). Upper bound for transversals of tripartite hypergraphs. *Period. Math. Hungar.*, 13(4):321–323.

1983

[75] (with M. Ajtai and J. Komlós). An $O(n \log n)$ sorting network. In *15th Annual ACM Symposium on Theory of Computing (Boston, MA, 1983)*, pages 1–9. IEEE, New York.

[76] (with W. Paul, N. Pippenger, and W.T. Trotter, Jr.). On determinism versus non-determinism and related problems. In *24th Annual Symposium on Foundations of Comupter Science (Tuscon, AZ, 1983)*, pages 429–438. IEEE, New York.

[77] (with M. Ajtai and J. Komlós). Sorting in $c \log n$ parallel steps. *Combinatorica*, 3(1):1–19.

[78] (with C. Chvatál, V. Rödl, and W.T. Trotter, Jr.). The Ramsey number of a graph with bounded maximum degree. *J. Combin. Theory Ser. B*, 34(3):239–243.

[79] (with V. Chvátal). Notes on the Erdős–Stone theorem. In *Combinatorial mathematics (Marseille-Luminy, 1981)*, volume 75 of *North-Holland Math. Stud.*, pages 183–190. North-Holland, Amsterdam.

[80] (with V. Chvátal). Short cycles in directed graphs. *J. Combin. Theory Ser. B*, 35(3):323–327.

[81] (with P. Erdős, A. Hajnal, and V.T. Sós). More results on Ramsey–Turán type problems. *Combinatorica*, 3(1):69–81.

[82] (with P. Erdős). On sums and products of integers. In *Studies in pure mathematics*, pages 213–218. Birkhäuser, Basel.

[83] (with J. Komlós). Limit distribution for the existence of Hamiltonian cycles in a random graph. *Discrete Math.*, 43(1):55–63.

[84] (with W.T. Trotter, Jr.). A combinatorial distinction between the Euclidean and projective planes. *European J. Combin.*, 4(4):385–394.

[85] (with W.T. Trotter, Jr.). Extremal problems in discrete geometry. *Combinatorica*, 3(3-4):381–392.

[86] (with W.T. Trotter, Jr.). Recent progress in extremal problems in discrete geometry. In *Graphs and other combinatorial topics (Prague, 1982)*, volume 59 of *Teubner-Texte Math.*, pages 316–319. Teubner, Leipzig.

1984

[87] (with M.L. Fredman and J. Komlós). Storing a sparse table with $O(1)$ worst case access time. *J. Assoc. Comput. Mach.*, 31(3):538–544.

[88] (with A. Gyárfás and J. Komlós). On the distribution of cycle lengths in graphs. *J. Graph Theory*, 8(4):441–462.

[89] (with H.A. Kierstead and W.T. Trotter, Jr.). On coloring graphs with locally small chromatic number. *Combinatorica*, 4(2-3):183–185.

[90] (with J. Komlós and J. Pintz). On a problem of Erdős and Straus. In *Topics in classical number theory, Vol. I, II (Budapest, 1981)*, volume 34 of *Colloq. Math. Soc. János Bolyai*, pages 927–960. North-Holland, Amsterdam.

[91] (with J. Spencer and W. Trotter, Jr.). Unit distances in the Euclidean plane. In *Graph theory and combinatorics (Cambridge, 1983)*, pages 293–303. Academic Press, London.

[92] (with L. Babei). On the complexity of matrix group problems. In *25th Annual Symposium on Foundations of Computer Science (Springer Island, FL, 1984)*, pages 229–240.

1985

[93] (with M. Ajtai and J. Komlós). First occurrence of Hamilton cycles in random graphs. In *Cycles in graphs (Burnaby, B.C., 1982)*, volume 115 of *North-Holland Math. Stud.*, pages 173–178. North-Holland, Amsterdam.

[94] (with A. Gyárfás, H.J. Prömel, and B. Voigt). On the sum of the reciprocals of cycle lengths in sparse graphs. *Combinatorica*, 5(1):41–52.

1986

[95] (with M. Ajtai, L. Babei, P. Hajnal, J. Komlós, P. Pudlák, V. Rödl, and G. Turán). Two lower bounds for branching problems. In *18th Annual ACM*

Symposium on Theory of Computing (Berkeley, CA, 1986), pages 30–38. IEEE, New York.

1987

[96] (with W. Maass and G. Schnitger). Two tapes are better than one for off-line Turing machines. In *19th Annual ACM Symposium on Theory of Computing (New York City NY, 1987)*, pages 94–100. IEEE, New York.

[97] (with M. Ajtai and J. Komlós). Deterministic Simulation in Logspace. In *19th Annual ACM Symposium on Theory of Computing (New York City NY, 1987)*, pages 132–140. IEEE, New York.

[98] (with L. Babai, P. Hajnal, and G. Turán). A lower bound for read-once-only branching programs. *J. Comput. System Sci.*, 35(2):153–162.

1988

[99] (with V. Chvátal). Many hard examples for resolution. *J. Assoc. Comput. Mach.*, 35(4):759–768.

[100] (with R. Cole, J. Salowe, and W.L. Steiger). Optimal slope selection. In *Automata, languages and programming (Tampere, 1988)*, volume 317 of *Lecture Notes in Comput. Sci.*, pages 133–146. Springer, Berlin.

[101] (with J. Pintz and W.L. Steiger). On sets of natural numbers whose difference set contains no squares. *J. London Math. Soc. (2)*, 37(2):219–231.

[102] (with P. Ragde, W. Steiger, and A. Wigderson). The parallel complexity of element distinctness is $\Omega\sqrt{\log n}$. *SIAM J. Discrete Math.*, 1(3):399–410.

1989

[103] (with J. Friedmann and J. Kahn). On the second eigenvalue in random regular graphs. In *21st Annual ACM Symposium on Theory of Computing (Seattle, WA, 1989)*, pages 587–598. IEEE, New York.

[104] (with M. Ajtai, J. Komlós, and W. Steiger). Almost Sorting in One Round. *Advances in Computing Research*, 5:117–125.

[105] (with M. Ajtai, D. Karabeg, and J. Komlós). Sorting in average time $o(\log n)$. *SIAM J. Discrete Math.*, 2(3):285–292.

[106] (with M. Ajtai, J. Komlós, and W.L. Steiger). Optimal parallel selection has complexity $O(\log\log n)$. *J. Comput. System Sci.*, 38(1):125–133. 18th Annual ACM Symposium on Theory of Computing (Berkeley, CA, 1986).

[107] (with R. Cole, J.S. Salowe, and W.L. Steiger). An optimal-time algorithm for slope selection. *SIAM J. Comput.*, 18(4):792–810.

[108] (with Z. Galil and R. Kannan). On 3-pushdown graphs with large separators. *Combinatorica*, 9(1):9–19.

[109] (with Z. Galil and R. Kannan). On nontrivial separators for k-page graphs and simulations by nondeterministic one-tape Turing machines. *J. Comput. System Sci.*, 38(1):134–149. 18th Annual ACM Symposium on Theory of Computing (Berkeley, CA, 1986).

[110] (with M. Geréb-Graus and R. Paturi). There are no p-complete families of symmetric Boolean functions. *Inform. Process. Lett.*, 30(1):47–49.

[111] (with J. Pintz and W.L. Steiger). Infinite sets of primes with fast primality tests and quick generation of large primes. *Math. Comp.*, 53(187):399–406.

1990

[112] (with M. Ajtai, H. Iwaniec, J. Komlós, and J. Pintz). Construction of a thin set with small Fourier coefficients. *Bull. London Math. Soc.*, 22(6):583–590.
[113] (with M. Ajtai and J. Komlós). Generating expanders from two permutations. In *A tribute to Paul Erdős*, pages 1–12. Cambridge Univ. Press, Cambridge.
[114] (with L. Babai, P. Pudlák, and V. Rödl). Lower bounds to the complexity of symmetric Boolean functions. *Theoret. Comput. Sci.*, 74(3):313–323.
[115] (with P. Hajnal). Brooks coloring in parallel. *SIAM J. Discrete Math.*, 3(1):74–80.
[116] Integer sets containing no arithmetic progressions. *Acta Math. Hungar.*, 56(1-2):155–158.

1992

[117] (with F.R.K. Chung and W.T. Trotter). The number of different distances determined by a set of points in the Euclidean plane. *Discrete Comput. Geom.*, 7(1):1–11.
[118] (with M. Ajtai and J. Komlós). On a conjecture of Loeb. In *Proc. 7th International Conference on the Theory and Applications of Graphs (Kalamazoo, MI, 1992)*, pages 1135–1146. Wiley, New York.
[119] (with N. Nissan and A. Wigderson). Undirected connectivity in $O(\log^{1.5} n)$ space. In *33rd Annual Symposium on Foundations of Computer Science (Pittsburgh, PA, 1992)*, pages 24–29. IEEE, New York.
[120] (with M. Ajtai and J. Komlós). Halvers and Expanders. In *33rd Annual Symposium on Foundations of Computer Science (Pittsburgh, PA, 1992)*, pages 686–692. IEEE, New York.
[121] (with M. Ajtai, N. Alon, J. Bruck, R. Cypher, C.T. Ho, and M. Noar). Fault tolerant graphs, perfect hash functions and disjoint paths. In *33rd Annual Symposium on Foundations of Computer Science (Pittsburgh, PA, 1992)*, pages 693–702. IEEE, New York.
[122] (with J. Pach and W. Steiger). An upper bound on the number of planar k-sets. *Discrete Comput. Geom.*, 7(2):109–123.

1993

[123] (with P. Erdős, A. Hajnal, M. Simonovits, and V.T. Sós). Turán–Ramsey theorems and simple asymptotically extremal structures. *Combinatorica*, 13(1):31–56.
[124] (with W. Maass, G. Schnitger, and G. Turán). Two tapes versus one for off-line Turing machines. *Comput. Complexity*, 3(4):392–401.
[125] (with A. Razborov and A. Wigderson). Constructing small sets that are uniform in arithmetic progressions. *Combin. Probab. Comput.*, 2(4):513–518.

1994

[126] (with A. Balog, J. Pelikán, and J. Pintz). Difference sets without κth powers. *Acta Math. Hungar.*, 65(2):165–187.

[127] (with A. Balog). A statistical theorem of set addition. *Combinatorica*, 14(3):263–268.

[128] (with P. Erdős, A. Hajnal, M. Simonovits, and V.T. Sós). Turán–Ramsey theorems and K_p-independence numbers. *Combin. Probab. Comput.*, 3(3):297–325. Also appeared in *Combinatorics, geometry and probability (Cambridge, 1993)*, pages 253–281, Cambridge Univ. Press, Cambridge, 1997.

[129] (with J. Komlós). Topological cliques in graphs. *Combin. Probab. Comput.*, 3(2):247–256. Also appeared in *Combinatorics, geometry and probability (Cambridge, 1993)*, pages 439–448, Cambridge Univ. Press, Cambridge, 1997.

[130] (with A. Sárközy). On a problem in additive number theory. *Acta Math. Hungar.*, 64(3):237–245.

1995

[131] (with M. Ajtai and J. Komlós). On a conjecture of Loebl. In *Graph theory, combinatorics, and algorithms, Vol. 1, 2 (Kalamazoo, MI, 1992)*, Wiley-Intersci. Publ., pages 1135–1146. Wiley, New York.

[132] (with N. Kahale, F.T. Leighton, Y. Ma, C.G. Plaxton, and T. Suel). Lower bounds for sorting networks. In *27th Annual ACM Symposium on Theory of Computing (Las Vegas, NV, 1995)*, pages 437–446. IEEE, New York.

[133] (with J. Kahn and J. Komlós). On the probability that a random ± 1-matrix is singular. *J. Amer. Math. Soc.*, 8(1):223–240.

[134] (with J. Komlós and G.N. Sárközy). Proof of a packing conjecture of Bollobás. *Combin. Probab. Comput.*, 4(3):241–255.

[135] (with L. Pyber and V. Rödl). Dense graphs without 3-regular subgraphs. *J. Combin. Theory Ser. B*, 63(1):41–54.

1996

[136] (with J. Komlós). Topological cliques in graphs. II. *Combin. Probab. Comput.*, 5(1):79–90.

[137] (with J. Komlós and G.N. Sárközy). On the square of a Hamiltonian cycle in dense graphs. In *Proceedings of the Seventh International Conference on Random Structures and Algorithms (Atlanta, GA, 1995)*, volume 9, pages 193–211.

1997

[138] (with J. Komlós and G.N. Sárközy). Blow-up lemma. *Combinatorica*, 17(1):109–123.

1998

[139] (with J. Komlós and Y. Ma). Matching nuts and bolts in $O(n \log n)$ time. *SIAM J. Discrete Math.*, 11(3):347–372.

[140] (with J. Komlós and G.N. Sarkozy). An algorithmic version of the blow-up lemma. *Random Structures Algorithms*, 12(3):297–312.

[141] (with J. Komlós and G.N. Sárközy). On the Pósa–Seymour conjecture. *J. Graph Theory*, 29(3):167–176.

[142] (with J. Komlós and G.N. Sárközy). Proof of the Seymour conjecture for large graphs. *Ann. Comb.*, 2(1):43–60.

[143] (with T. Łuczak and V. Rödl). Partitioning two-coloured complete graphs into two monochromatic cycles. *Combin. Probab. Comput.*, 7(4):423–436.

2000

[144] (with N. Alon, M. Capalbo, Y. Kohayakawa, V. Rödl, and A. Ruciński). Universality and tolerance (extended abstract). In *41st Annual Symposium on Foundations of Computer Science (Redondo Beach, CA, 2000)*, pages 14–21. IEEE Comput. Soc. Press, Los Alamitos, CA.

[145] (with J. Pelikán and J. Pintz). On the running time of the Adleman–Pomerance–Rumely primality test. *Publ. Math. Debrecen*, 56(3-4):523–534. Dedicated to Professor Kálmán Győry on the occasion of his 60th birthday.

[146] (with V. Rödl). On size Ramsey numbers of graphs with bounded degree. *Combinatorica*, 20(2):257–262.

2001

[147] (with N. Alon, M. Capalbo, Y. Kohayakawa, V. Rödl, and A. Ruciński). Near-optimum universal graphs for graphs with bounded degrees (extended abstract). In *Approximation, randomization, and combinatorial optimization (Berkeley, CA, 2001)*, volume 2129 of *Lecture Notes in Comput. Sci.*, pages 170–180. Springer, Berlin.

[148] (with J. Komlós and G.N. Sárközy). Proof of the Alon–Yuster conjecture. *Discrete Math.*, 235(1-3):255–269. Combinatorics (Prague, 1998).

[149] (with J. Komlós and G.N. Sárközy). Spanning trees in dense graphs. *Combin. Probab. Comput.*, 10(5):397–416.

2002

[150] (with B. Bollobás). Girth of sparse graphs. *J. Graph Theory*, 39(3):194–200.

[151] (with A. Khalfalah and S. Lodha). Tight bound for the density of sequence of integers the sum of no two of which is a perfect square. *Discrete Math.*, 256(1-2):243–255.

[152] (with J. Komlós, A. Shokoufandeh, and M. Simonovits). The regularity lemma and its applications in graph theory. In *Theoretical aspects of computer science (Tehran, 2000)*, volume 2292 of *Lecture Notes in Comput. Sci.*, pages 84–112. Springer, Berlin.

2003

[153] (with B. Csaba and A. Shokoufandeh). Proof of a conjecture of Bollobás and Eldridge for graphs of maximum degree three. *Combinatorica*, 23(1):35–72. Paul Erdős and his mathematics (Budapest, 1999).

[154] (with G.N. Sárközy and S.M. Selkow). On the number of Hamiltonian cycles in Dirac graphs. *Discrete Math.*, 265(1-3):237–250.

2005

[155] (with B. Sudakov and V.H. Vu). On a question of Erdős and Moser. *Duke Math. J.*, 129(1):129–155.

[156] (with V.H. Vu). Long arithmetic progressions in sum-sets and the number of x-sum-free sets. *Proc. London Math. Soc. (3)*, 90(2):273–296.

2006

[157] (with A. Gyárfás, M. Ruszinkó, and G.N. Sárközy). An improved bound for the monochromatic cycle partition number. *J. Combin. Theory Ser. B*, 96(6):855–873.

[158] (with A. Gyárfás, M. Ruszinkó, and G.N. Sárközy). One-sided coverings of colored complete bipartite graphs. In *Topics in discrete mathematics*, volume 26 of *Algorithms Combin.*, pages 133–144. Springer, Berlin.

[159] (with A. Khalfalah). On the number of monochromatic solutions of $x + y = z^2$. *Combin. Probab. Comput.*, 15(1-2):213–227.

[160] (with J. Polcyn, V. Rödl, and A. Ruciński). Short paths in quasi-random triple systems with sparse underlying graphs. *J. Combin. Theory Ser. B*, 96(4):584–607.

[161] (with V. Rödl and A. Ruciński). A Dirac-type theorem for 3-uniform hypergraphs. *Combin. Probab. Comput.*, 15(1-2):229–251.

[162] (with V. Rödl and A. Ruciński). Perfect matchings in uniform hypergraphs with large minimum degree. *European J. Combin.*, 27(8):1333–1349.

[163] (with V. Vu). Long arithmetic progressions in sumsets: thresholds and bounds. *J. Amer. Math. Soc.*, 19(1):119–169.

[164] (with V.H. Vu). Finite and infinite arithmetic progressions in sumsets. *Ann. of Math. (2)*, 163(1):1–35.

2007

[165] (with A. Gyárfás, M. Ruszinkó, and G.N. Sárközy). Three-color Ramsey numbers for paths. *Combinatorica*, 27(1):35–69. Corrigendum, *Combinatorica*, 28(4):499–502, 2008.

[166] (with A. Gyárfás, M. Ruszinkó, and G.N. Sárközy). Tripartite Ramsey numbers for paths. *J. Graph Theory*, 55(2):164–174.

[167] An old new proof of Roth's theorem. In *Additive combinatorics*, volume 43 of *CRM Proc. Lecture Notes*, pages 51–54. Amer. Math. Soc., Providence, RI.

2008

[168] (with A. Gyárfás and G.N. Sárközy). The Ramsey number of diamond-matchings and loose cycles in hypergraphs. *Electron. J. Combin.*, 15(1):Research Paper 126, 14.

[169] (with R. Martin). Quadripartite version of the Hajnal–Szemerédi theorem. *Discrete Math.*, 308(19):4337–4360.

[170] (with H.H. Nguyen and V.H. Vu). Subset sums modulo a prime. *Acta Arith.*, 131(4):303–316.

[171] (with V. Rödl, A. Ruciński, and M. Schacht). A note on perfect matchings in uniform hypergraphs with large minimum collective degree. *Comment. Math. Univ. Carolin.*, 49(4):633–636.

[172] (with V. Rödl and A. Ruciński). An approximate Dirac-type theorem for k-uniform hypergraphs. *Combinatorica*, 28(2):229–260.

2009

[173] (with A. Gyárfás and G.N. Sárközy). Stability of the path Ramsey number. *Discrete Math.*, 309(13):4590–4595.

[174] (with V. Rödl and A. Ruciński). Perfect matchings in large uniform hypergraphs with large minimum collective degree. *J. Combin. Theory Ser. A*, 116(3):613–636.

2010

[175] (with B. Csaba, I. Levitt, and J. Nagy-György). Tight bounds for embedding bounded degree trees. In *Fete of combinatorics and computer science*, volume 20 of *Bolyai Soc. Math. Stud.*, pages 95–137. János Bolyai Math. Soc., Budapest.

[176] (with I. Levitt and G.N. Sárközy). How to avoid using the regularity lemma: Pósa's conjecture revisited. *Discrete Math.*, 310(3):630–641.

[177] (with A. Gyárfás and G.N. Sárközy). Long monochromatic Berge cycles in colored 4-uniform hypergraphs. *Graphs Combin.*, 26(1):71–76.

[178] (with A. Gyárfás and G.N. Sárközy). Monochromatic Hamiltonian 3-tight Berge cycles in 2-colored 4-uniform hypergraphs. *J. Graph Theory*, 63(4):288–299.

[179] (with A. Gyárfás and G.N. Sárközy). Monochromatic matchings in the shadow graph of almost complete hypergraphs. *Ann. Comb.*, 14(2):245–249.

[180] (with H.A. Kierstead, A.V. Kostochka, and M. Mydlarz). A fast algorithm for equitable coloring. *Combinatorica*, 30(2):217–224.

2011

[181] (with A. Gyárfás, M. Ruszinkó, and G.N. Sárközy). Partitioning 3-colored complete graphs into three monochromatic cycles. *Electron. J. Combin.*, 18(1):Paper 53, 16.

[182] (with Y. Kohayakawa, V. Rödl, and M. Schacht). Sparse partition universal graphs for graphs of bounded degree. *Adv. Math.*, 226(6):5041–5065.

[183] (with V. Rödl and A. Ruciński). Dirac-type conditions for Hamiltonian paths and cycles in 3-uniform hypergraphs. *Adv. Math.*, 227(3):1225–1299.

2012

[184] (with G.N. Sárközy, F. Song, and S. Trivedi). A Practical Regularity Partitioning Algorithm and its Applications in Clustering. arXiv:1209.6540.

[185] (with J. Hladký, J. Komlós, D. Piguet, M. Simonovits, and M. Stein). The approximate Loebl–Komlós–Sós Conjecture. arXiv:1211.3050.

Curriculum Vitae for Endre Szemerédi

Born:	August 21, 1940 in Budapest, Hungary
Degrees/education:	Master of Science, Eötvös Loránd University, 1965
	PhD, Moscow State University, 1970
Positions:	Permanent Research Fellow, Rényi Institute of Mathematics, 1965–
	State of New Jersey Professor, Rutgers University, 1986–
Visiting positions:	Stanford University, 1974
	McGill University, 1980
	University of Southern California, 1981–1983
	University of Chicago, 1985–1986
	Fairchild Distinguished Scholar, California Institute of Technology, 1987–1988
	Aisenstadt Chair at CRM Montreal, 2003
	Institute for Advanced Study, 2007–2008 and 2009–2010
	Eisenbud Professor, MSRI, Berkeley, 2008
Memberships:	Corresponding Member Hungarian Academy of Sciences, 1982
	Hungarian Academy of Sciences, 1987
	National Academy of Sciences (USA), 2010
	Academia Europaea, 2012
	Norwegian Academy of Science and Letters, 2012
Awards and prizes:	Grünwald Prize, 1967
	Grünwald Prize, 1968
	Rényi Prize, 1973

H. Holden, R. Piene (eds.), *The Abel Prize 2008-2012*,
DOI 10.1007/978-3-642-39449-2_27, © Springer-Verlag Berlin Heidelberg 2014

	Pólya Prize for Achievement in Applied Mathematics (SIAM), 1975
	Prize of the Hungarian Academy of Sciences, 1979
	State of New Jersey Professorship, 1986
	Leroy P. Steele Prize for Seminal Contribution to Research, 2008
	Rolf Schock Prize in Mathematics, 2008
	DeLong Lecture Series University of Colorado, 2010
	Abel Prize, 2012
	George Washington Prize, USA, 2012
	Prima Prize, Hungary, 2012
	Szechenyi Prize, Hungary, 2012
Honorary degrees:	Charles University, Prague, 2010

A Letter from Niels Henrik Abel to August Leopold Crelle

Alf Bjørseth hands over the Abel manuscripts to Vigdis Moe Skarstein, National Librarian at The National Library of Norway, Oslo. (Photo: Heiko Junge/Scanpix)

The letter appeared as one of three handwritten Abel manuscripts acquired in 2007 from Institut Mittag-Leffler, Sweden, by Alf Bjørseth and his company Scatec and donated to the Norwegian Academy of Science and Letters on the occasion of its 150th anniversary. The originals are kept at the National Library of Norway. The letter is the last page of a manuscript of eight pages. The first seven pages contain the manuscript "Note sur quelques formules elliptiques" which appeared in Crelle's Journal, Band 4, pages 85–93 (1829). The article is contained in both editions of Abel's Collected Works. For further information regarding all the manuscripts acquired, please see A. Stubhaug: *I sporet etter Abel/Tracing Abel*, Scatec, n.d.

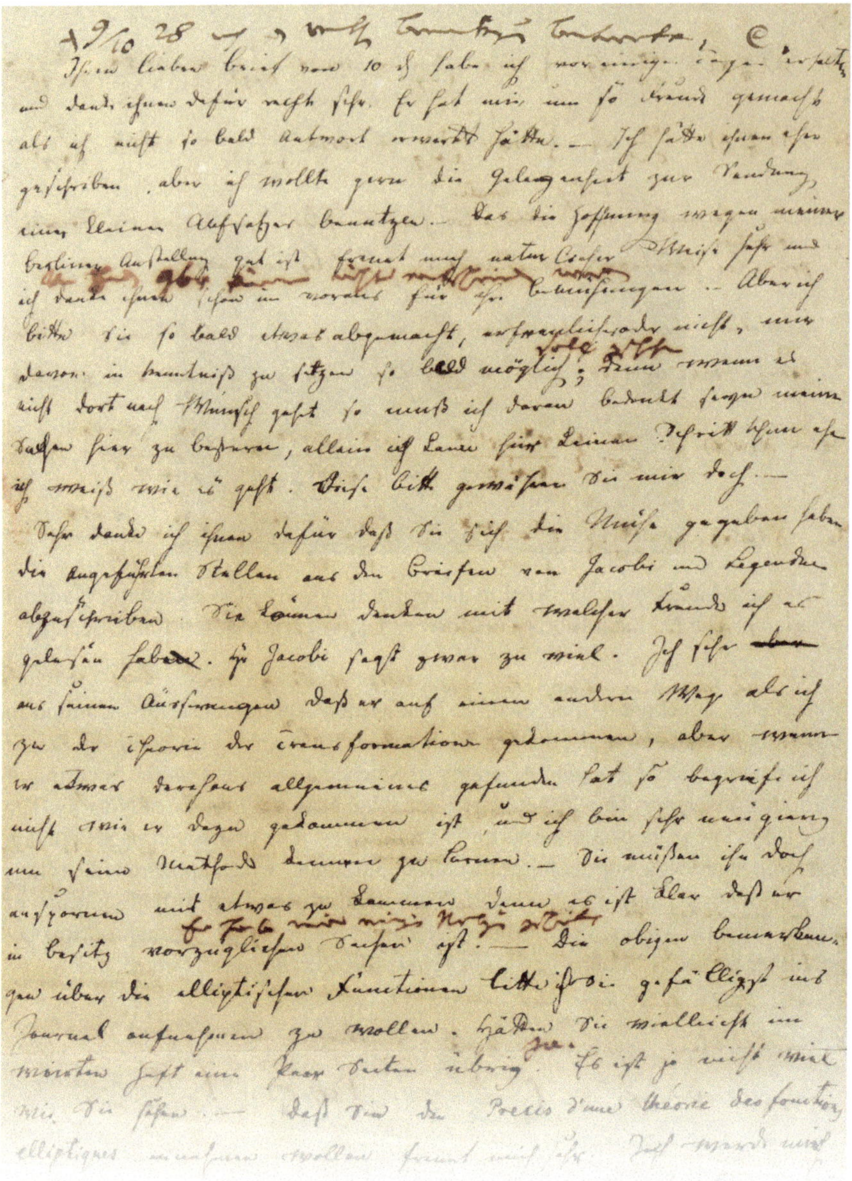

+ 9/10 28 ... beantwortet. C

Ihren lieben Brief vom 10 d. habe ich vor einigen Tagen erhalten und danke ihnen dafür recht sehr. Er hat mir um so Freude gemacht als ich nicht so bald Antwort erwartet hätte. — Ich hätte ihnen eher geschrieben, aber ich wollte gern die Gelegenheit zur Sendung einer kleinen Abhandlung benutzen. — Daß die Hoffnung wegen meiner Berliner Anstellung gut ist freuet mich natürlicher Weise sehr und ich danke ihnen schon im voraus für ihre Bemühungen. — Aber ich bitte Sie so bald etwas abgemacht, entscheidendes nicht, mir davon in Kenntniß zu setzen so bald möglich; denn wenn es nicht dort nach Wunsch geht so muß ich daran denken meine Sachen hier zu besorgen, allein ich kann hier keinen Schritt thun ehe ich weiß wie es geht. Diese Bitte gewähren Sie mir doch. —

Sehr danke ich ihnen dafür daß Sie sich die Mühe gegeben haben die angeführten Stellen aus den Briefen von Jacobi und Legendre abzuschreiben. Sie können denken mit welcher Freude ich es gelesen habe. Hr Jacobi sagt zwar zu viel. Ich sehe aus seinen Äußerungen daß er auf einem andern Weg als ich zu der Theorie der Transformation gekommen, aber wenn er etwas durchaus allgemeines gefunden hat so begreife ich nicht wie er dazu gekommen ist, und ich bin sehr neugierig um seine Methode kennen zu lernen. — Sie müssen ihn doch auffordern mit etwas zu kommen denn es ist klar daß er in Besitz vorzüglicher Sachen ist. — Die obigen Bemerkungen über die elliptischen Functionen bitte ich Sie gefälligst ins Journal aufnehmen zu wollen. Hätten Sie vielleicht im vierten Heft ein paar Seiten übrig. Es ist ja nicht viel wie Sie sehen. — Daß Sie den Precis d'une théorie des fonctions elliptiques aufnehmen wollen freut mich sehr. Ich werde mich in Bezug vorzüglichsten Sachen ist eine der obigen Abhandlungen über die elliptischen Functionen bitte ich Sie gefälligst im Journal aufnehmen zu wollen. Hätten Sie vielleicht im nächsten Heft ein paar Seiten übrig? Es ist zu nicht mehr wie Sie sehen. — Daß Sie den Precis d'une theorie des fonctions elliptiques annehmen wollen freut mich sehr. Ich werde mich anstrengen um diese Abhandlung so deutlich und gut als mir nur möglich ist zu machen, und ich hoffe daß Sie mir gelingen soll. Glauben Sie aber nicht es wäre das beste mit dieser Abhandlung statt mit der über die Gleichungen den Anfang zu machen? Dieses ist auch eine dringende Bitte an Sie. Denn erstens glaube ich daß die elliptischen Functionen mehr Interesse haben wird und zweitens so wird meine Gesundheit mir kaum erlauben mich mit der Gleichungen in einiger Zeit zu befassen. Ich bin nämlich längere Zeit krank gewesen so daß ich das Bett hüten mußte, und ob ich gleich jetzt reconvalescent bin so wird mir doch, wie der Arzt sagt, viel Anstrengung sehr nachtheilig seyn. Nun aber wird mir die Gleichungen unverhältnißmäßig mehr Anstrengung kosten als die elliptischen Functionen. Wenn Sie also nicht die Abhandlung über die Gleichungen verlangen (denn in diesem Falle sollen Sie sie haben) so werde ich mit der über die ell: Funct: anfangen. Übrigens soll die Gleichungen bald nachfolgen und wenn Sie nichts dagegen haben so denke ich die Abtheilungen so klein zu machen daß in jedem Heft etwas über elliptische Funct und etwas über die Gleichungen vorkommt. Nur für das erste Heft möchte ich gern bloß ellipt: F: haben. — Lassen Sie mir doch, ich bitte, ihre Meinung darüber wissen. —

Ihr ergebenster N. Abel

Christiania 25 Sept. 1828

Ihren lieben Brief vom 10 ds habe ich vor einigen Tagen erhalten und danke ihnen dafür recht sehr. Er hat mir um so Freude gemacht als ich nicht so bald Antwort erwartet hätte. – Ich hätte ihnen eher geschriben, aber ich wollte gern die Gelegenheit zur Sendung eines kleinen Aufsatzes benutzen. Das die Hoffnung wegen meiner Berliner Anstellung gut ist freuet mich naturlicher Weise sehr und ich danke ihnen schon im voraus für ihre Bemühungen. – Aber ich bitte Sie so bald etwas abgemacht, efreuliches oder nicht, mir davon in Kenntniss zu setzen so bald möglich; denn wenn es nicht dort nach Wunsch gehet so muss ich daran bedenkt seyn meine Sachen hier zu bessern, allein ich kann hier keinen Schritt thun ehe ich weiss wie es geht. Diese bitte gewähren Sie mir doch. –

Sehr danke ich ihnen dafür dass Sie sich die Mühe gegeben haben die angeführten Stellen aus den Briefen von Jacobi und Legendre abgeschrieben. Sie können denken mit welcher Freude ich es gelesen habe. Hr Jacobi sagt zwar zu viel. Ich sehe aus seinen Äusserungen dass er auf einen anderen Weg als ich zu der Theorie der Transformation gekommen, aber wenn er etwas durchaus allgemeines gefunden hat so begreife ich nicht wie er dazu gekommen ist, und ich bin sehr neugierig um seine Methode kennen zu lernen. – Sie müssten ihn doch ansporen mit etwas zu kommen denn es ist klar dass er in Besitz vorzüglichen Sachen ist. – De obigen Bemerkungen über die elliptischen Functionen bitte ich Sie gefälligst ins Journal aufnehmen zu wollen. Hätten Sie vielleicht im vierten Heft eine Paar Seiten übrig. Es ist ja nicht viel wie Sie sehen. – Dass Sie die "Précis d'une théorie des fonctions elliptiques" annehmen wollen freut mich sehr. Ich werde mich anstrengen um diese Abhandlung so deutlich und gut als mir nur möglich ist zu machen, und ich hoffe dass Sie mir gelingen solle. Glauben Sie aber nicht es wäre dass beste mit dieser Abhandlung statte mit der über die Gleichungen den Anfang zu machen? Dieses ist auch eine dringende Bitte an Sie, denn erstens glaube ich dass die elliptischen Functionen mehr Interesse haben wird und zweitens so wird meine Gesundheit mir kaum erlauben mich mit der Gleichungen in einiger Zeit zu befassen. Ich bin nämlich längere Zeit krank gewesen so dass ich das Bett hüten musste, und ob ich gleich jetzt restituert bin so wird mir doch, wie der Arzt sagt, viel Anstrengung sehr nachtheilig seyn. Nun aber wird mir die Gleichungen unverhältnismässig mehr Anstrengung kosten als die elliptischen Functionen. Wenn Sie also nicht die Abhandlung über die Gleichungen verlangen (denn in diesem Falle sollen Sie sie haben) so werde ich mit der über die ell: Funct: anfangen. Übrigens soll die Gleichungen bald nachfolgen und wenn Sie nichts dawieder haben so denke ich die Abtheilungen so klein zu machen dass in jedem Heft etwas über elliptische Funct und etwas über die Gleichungen vorkommt. Nur für das erste Heft mögte ich gern bloss ellipt: F: schicken. –Lassen Sie mir doch, ich bitte, ihre Meinung darüber wissen. –

Ihr ergebenster
N. Abel

Christiania September 25, 1828

I received your kind letter dated September 10 some days ago and I thank you greatly for it. It made me extra glad since I had not expected such a quick answer. I would have written you earlier, but I wanted to take the opportunity to also send you a short manuscript.

It pleases me greatly, of course, that the hope of an appointment in Berlin is good and I thank you in advance for your efforts. But I implore you, let me know as soon as possible when anything has been decided, be it favorable or not, for if it does not work out as I hope I must be prepared to improve on my conditions here.

I can take no steps until I know what happens. This request you will certainly grant me.

Thank you very much for taking the trouble to copy and send me excerpts from the letters you received from Jacobi and Legendre. You can just imagine with which pleasure I have read this. Herr Jacobi says indeed too much. I see from his remarks that he has arrived at the transformation theory (for elliptic functions (translator's note)) by taking a different path than I did, but if he has throughout found something general then I can not comprehend how he has arrived at this, and I am very curious to learn about his method. Please spur him on to reveal something, for it is clear that he is in possession of excellent things.

If you do not mind I would appreciate if you would publish in the journal the remarks on elliptic functions that are enclosed. Perhaps you have a few pages to spare in the upcoming fourth issue.

It pleases me greatly that you will print my "Précis d'une théorie des fonctions elliptiques". I shall exert myself to make it as clear and good as possible, and hope I shall succeed. But do you not think that it would be better to commence with this paper instead of the one on the equations? I ask you urgently.

Firstly, I believe that the elliptic functions will be of greater interest; secondly, my health will hardly permit me to occupy myself with the equations for a while. I have been ill for a considerable period of time, and compelled to stay in bed. Even if I am now recovered, the physician has warned me that any strong exertion can be very harmful.

Now the situation is this: the equations will require a disproportionately greater effort on my part than the elliptic functions. Therefore, I should prefer, if you do not absolutely insist on the article on equations – in that case you shall have it – to begin with the elliptic functions. The equations will follow soon afterward. If you have nothing against it, I would prefer to divide it into short sections, so that something about elliptic functions and something about equations could appear in each issue; for the first issue I should like to send only elliptic functions.

Please let me hear your opinion on this.

Yours sincerely,
N. Abel

Abel and the Theory of Algebraic Equations

(Reflections Stimulated by the Letter Abel Sent to Crelle on September 25, 1828)

Christian Skau

The letter exhibited is a copy of the original letter that Abel sent from Christiania (today's Oslo) September 25, 1828, to Crelle in Berlin in reply to a letter Crelle had sent to Abel on September 10. The annotations within the letter are in Crelle's handwriting. Crelle, who was the founder and editor of *Journal für die Reine und Angewandte Mathematik* (in the sequel referred to as Crelle's journal), was Abel's close friend who worked tirelessly to secure a professorship for him in Berlin. Abel published almost all his papers in Crelle's journal. For example, the first year the journal appeared, in 1826, six of the articles were written by Abel, among them the proof of the algebraic unsolvability of the general equation of degree higher than four. The Abel letter was discovered, and thus rescued from oblivion, by Mittag-Leffler, the founder and editor of *Acta Mathematica*. It became known to the mathematical community for the first time when a facsimile of the letter appeared in the 1903 edition of *Acta Mathematica*. The 1902, 1903 and 1904 editions of Acta, with the frontispiece: "Niels Henrik Abel—In Memoriam", were dedicated to Abel and his mathematical legacy on the occasion of the centennial of his birth. It contained articles by the most prominent mathematicians at the time, among them Hilbert, Frobenius, Minkowski, Picard and Poincaré, about theories and results that had their genesis in Abel's own discoveries.

1 Historical Context

The Abel letter to Crelle is of great interest because it throws considerable light on Abel's mathematical preoccupation the 14 weeks he had left before he was incapacitated and bedridden with tuberculosis and unable to work anymore—dying three months later on April 6, 1829, at the age of 26 years and 8 months. The letter raises

C. Skau (✉)
Department of Mathematical Sciences, NTNU, 7491 Trondheim, Norway
e-mail: csk@math.ntnu.no

H. Holden, R. Piene (eds.), *The Abel Prize 2008-2012*,
DOI 10.1007/978-3-642-39449-2_28,

the intriguing question of how the theory of equations would have evolved if Abel had not been impeded by poor health to write up his discoveries, as Crelle had urged him to do. In all likelihood the impact of Galois' work would have been less in a field which it came to dominate. According to Auguste Chevalier, Galois' close friend, it was in the year 1829 of Abel's death, that Galois (1811–1832), at the age of 17, made great advances in the theory of equations. At the end of 1831 Galois wrote a note on his relationship to Abel, probably provoked—if not angered—by the referee report by Poisson and Lacroix rejecting for publication his groundbreaking memoir *Sur les conditions de résolubilité des équations par radicaux* [6]. (Abel's work was referred to in the report.) Galois vehemently asserts his independence of Abel when he himself made his first important discoveries in the theory of equations. There is no reason to doubt that, but a familiar misquotation of a line spoken by the queen in Shakespeare's Hamlet, "methinks he doth protest too much", is highly pertinent, and we will get back to this issue. On the other hand, in the same note Galois also generously writes [15, pp. 238–239]:

> *Depuis une lettre particulière adressée par Abel à M. Legendre, annonçait qu'il avait eu le bonheur de découvrir une règle pour reconnaître si une équation était résoluble par radicaux; mais la mort anticipée de ce géomètre ayant privé la science des recherches * promises dans cette lettre, il n'en était pas moins nécessaire de donner la solution d'un problème qu'il m'est bien * pénible de posséder, puisque je dois cette possession à une des plus grandes pertes qu'aura faite la science.*
>
> [According to a certain letter addressed by Abel to Mr Legendre, he announced that he had the good fortune of finding a rule for recognizing if an equation is solvable by radicals, but the premature death of this geometer having deprived science of the research promised in this letter it became no less necessary to give to the world of learning the solution of a problem which is more painful to me to possess, because I owe this possession to one of the greatest losses science has suffered.]

Galois has a point here, as we will try to elucidate. However, none of this detracts from Galois' contribution. After all it is his admirable fundamental theorem, which goes far beyond the quest for a criterion for algebraic solvability of equations, that ranks as a watershed in mathematical history and from which emerged the fundamental group concept. We will give more details about what we have intimated at above, but first we will set this in a historical context and give some relevant background.

When Sylow in 1902 at the centennial celebration of Abel's birth wrote a lengthy survey [18] of Abel's discoveries, he was not aware of the letter to Crelle. At the 3rd Scandinavian Mathematical Congress in Oslo in 1913 Sylow gave the introductory talk [19]. The title of his talk was: *On Abel's work and his plans during the last phase of his life illuminated by documents that have come to light after the second edition of his collected works.* The second edition of Abel's Oeuvres Complètes was published in 1881 [2, 3] ([3] contains the posthumous articles). The editors were Lie and Sylow, and it contained detailed annotations, in contrast to the first edition that was published in 1839 [1], the editor being Holmboe, Abel's math teacher at the Cathedral School in Oslo. Sylow's main sources for his talk were the Abel letter together with a "lost" manuscript by Abel on the transformation theory of elliptic functions, including complex multiplication, and the various algebraic equations that are associated with this theory. Mittag-Leffler was able to trace this manuscript and published it in the aforementioned 1902-edition of Acta. Sylow had held the view that it was the competition between Abel and Jacobi in developing the theory of elliptic functions that was the reason Abel did not get around to write up his discoveries in the theory of equations. This he had to admit was only partially true. It was his poor health that held him back. Abel's letter shows that there had been an agreement between him and Crelle that he should first write about the theory of equations and then later about elliptic functions. (Abel had earlier published in Crelle's journal a long memoir on elliptic functions with the title *Recherches sur les fonctions elliptiques* [2, XVI].) Abel fears that the work on equations will be a much more arduous task than writing up a comprehensive survey article on elliptic functions demonstrating his methods. So following his doctor's advice about avoiding strenuous work he suggests to Crelle that he starts with the article on elliptic functions, but he writes that if Crelle insists he will instead start with the theory of equations. Crelle, of course, writes back and tells him to follow his doctor's advice. Abel's health was deteriorating during the autumn of 1828. Writing up the memoir *Précis d'une théorie des fonctions elliptiques* [2, XXVIII] (referred to as Précis in the sequel), took virtually all the time and strength he had left for work. Précis, or rather the part of it that he managed to complete, was published in Crelle's journal in June 1829 after his death. In the introduction Abel says that Précis will consist of excerpts from a monograph he had planned to write on the theory of elliptic functions. Its aim was to show his method and its breathtaking scope. It was to consist of two parts, but he only had time to finish the first part. Loosely speaking one can say that in Précis Abel treats the theory of elliptic functions essentially from an algebraic point of view.

2 Correspondence with Legendre

On November 25, 1828, Abel sent a letter to Legendre in reply to a letter he had received from him—sent from Paris on October 25. (The letter from Abel was published in Crelle's journal in 1830.) Besides answering questions that Legendre poses he gives a survey of his latest discoveries, some of which were to appear in Précis. It is almost touching to see the effort Abel makes to explain his theory to the aging Legendre, who had great difficulties understanding what Abel and Jacobi had done, the two of them having revolutionized the theory of elliptic integrals. Legendre had written a monograph on this subject, and he wanted to publish a supplement and include some of the new discoveries. Abel ended the letter with these words:

> *J'ai été assez heureux pour trouver une règle sure à l'aide de laquelle on pourra reconnaitre si une équation quelconque proposée est resoluble á l'aide*

de radicaux ou non. Un corollaire de ma théorie est qu'il est impossible de resoudre les équations générales superieures au quatrième degré.

[I have had the good fortune in finding a definite rule with whose help one can recognize whether any given equation can be solved by radicals or not. A corollary of my theory is that it is impossible to solve the general equation of degree greater than four.]

Legendre writes back on January 16, 1829, ending the letter by urging Abel to publish his discoveries in the theory of equations:

Vous m'annoncez, Monsieur, un très beau travail sur les équations algébriques, qui a pour objet de donner la résolution de toute équation numérique proposée, lorsqu'elle peut être développée en radicaux, et de déclarer insoluble sous ce rapport, toute équation qui ne satisferait pas aux conditions exigées; d'où résulte comme conséquence nécessaire que la résolution générale des équations au delà du quatrième degré, est impossible. Je vous invite à publier le plutôt que vouz pourrez, cette nouvelle théorie; elle vous fera beaucoup d'honneur, et sera généralement regardée comme la plus grande découverte qui restait à faire dans l'analyse.

[You announce to me, Sir, a very beautiful work on algebraic equations that has as object to resolve for each given numerical equation whether it can be solved by radicals, and to declare unsolvable those equations that do not satisfy the required conditions. A necessary consequence of this theory is that the general equation of degree higher than four can not be solved. I urge you to let this new theory appear in print as quickly as you are able. It will be of great honour to you, and will universally be considered the greatest discovery which remains to be made in mathematics.]

When the letter reached Abel the tuberculosis had incapacitated him and he was unable to work anymore. Legendre spread the rumour in the mathematical circles in Paris about Abel's claim that he had found a criterion to decide when a numerical equation could be solved algebraically ("by radicals"), and this we know also came to Galois' attention (cf. Galois' note referred to above).

3 The Addition Theorem

We will digress a little from our main story, but it is not irrelevant as it relates to the last few months of Abel's life. The correspondence between Abel and Legendre invites namely an obvious question: Why did Abel not remind and inquire Legendre about the great memoir (later given the sobriquet the *Paris Memoir*) that he had submitted to l'Académie des Sciences on October 30, 1826, during his soujourn in Paris? Abel knew that Legendre was one of the referees (the other was Cauchy), and even some of the results Legendre had asked about in his letter to Abel could be found in the memoir. Legendre had unquestionably forgotten the memoir as well as the fact that he had met Abel in Paris. Abel's manuscript sank deeper and deeper in the piles in Cauchy's study, and Legendre later confessed that he had never had it in his hands. Abel himself must have feared that his memoir had been lost. In a paper on hyperelliptic functions printed in Crelle's journal (appearing December 3, 1828) Abel mentioned its existence in a few words [2, XXI, footnote p. 445]. Jacobi, who immediately realized its importance, wrote to Legendre expressing his astonishment that Abel's masterpiece had not been published. In time this inquiry—after a convoluted process—led to the recovery and eventual publication of the memoir in 1841, twelve years after Abel's death [2, XII]. As a last effort, before tuberculosis had weakened him so much that he was not able to work anymore, he wrote a four-page letter to Crelle containing the statement and proof of the theorem that appears at the very outset of his Paris Memoir—later called Abel's addition theorem. This is the basic theorem upon which all the other results in the Paris Memoir depend—among these should be mentioned the first appearance ever in the mathematical literature of the genus concept for algebraic functions (or curves), which is included in a theorem, thus giving it life and significance. The date of the letter sent to Crelle is January 6, 1829, and it was published as a two-page article in Crelle's journal on March 28, 1829 [2, XXVII], nine days before Abel's death and eleven days before Crelle wrote triumphantly to Abel that he had secured a professorship for him in Berlin. The article contained no introduction, no superfluous remarks, no applications. It was a monument resplendent in its simple lines—the main theorem from his Paris Memoir, formulated in a few words. Abel fully understood the beauty of his result. In the last few lines he promised to give applications which would throw new light on analysis. Abel's addition theorem in its most general form can be stated, by a slight variant of how it is stated in [2, XXVII], like this:

Let y be an algebraic function of x given by an algebraic equation $f(x, y) = 0$. Let $\lambda(x, y) = 0$ be another algebraic equation, where the coefficients are certain parameters $a, a', a'', \ldots$, and let $(x_1, y_1), (x_2, y_2), \ldots, (x_\mu, y_\mu)$ be the intersection points of the two curves $f(x, y) = 0$ and $\lambda(x, y) = 0$. Let $R(x, y)$ be a rational function of x and y, and let $\psi(x) = \int R(x, y)dx$. ($\int R(x, y)dx$ is called an *Abelian integral.*) Then

$$\psi(x_1) + \psi(x_2) + \cdots + \psi(x_\mu) = u + k_1 \log v_1 + k_2 \log v_2 + \cdots + k_n \log v_n$$

where $u, v_1, v_2, \ldots, v_n$ are rational functions of $a, a', a'', \ldots$, and $k_1, k_2, \ldots, k_n$ are constants.

Abel's addition theorem was given the epithet *Monumentum aere perennius* [A monument more lasting than bronze] by Legendre, and Jacobi called it ... *die grösste mathematische Entdeckung unserer Zeit, obgleich erst eine künftige grosse Arbeit ihre ganze Bedeutung aufweise könne.* [... the greatest mathematical discovery of our time, even though only a great work in the future will reveal its full significance.] With the vantage point at the end of the 19th century—after Riemann's and Weierstrass' fundamental work on algebraic functions and curves—Picard gave the following assessment of Abel's addition theorem:

Le théorème parait tout à fait élémentaire, et il n'y a peut-être pas, dans l'histoire de la Science, de proposition aussi importante obtenue à l'aide de considérations aussi simples.

[The theorem appears as completely elementary, and perhaps there has never occurred in the history of science a proposition so important which is obtained by so simple considerations.]

In the proof of the addition theorem Abel uses the fundamental theorem of symmetric polynomials in an ingenious way. Recall that the fundamental theorem says that if $f(x_1,\dots,x_n)$ is a symmetric polynomial in the variables $x_1,\dots,x_n$ over the domain of rationality F (today we would say field), then $f(x_1,\dots,x_n)$ can be written as a polynomial in the elementary symmetric polynomials $s_1, s_2, \dots, s_n$, where

$$s_1 = \sum_{i=1}^{n} x_i, \qquad s_2 = \sum_{1\le i<j\le n} x_i x_j, \qquad \dots, \qquad s_n = x_1 x_2 \cdots x_n.$$

(Note that $s_i = (-1)^i a_i$, where $(x - x_1)(x - x_2)\cdots(x - x_n) = x^n + a_1 x^{n-1} + a_2 x^{n-2} + \cdots + a_n$.)

We shall see below effective and ingenious application of the fundamental theorem of symmetric polynomials in the context of algebraic equations.

4 Algebraic Equations—Primitive Elements

In a letter to Holmboe [3, XX, p. 260] in Oslo, sent from Paris and dated October 24, 1826, Abel says that he is working on his favorite topic, the theory of equations. [*Jeg arbeider nu paa Ligningernes Theorie, mit Yndlingsthema.*] He tells Homboe that he sees a way to solve the following general problem (which he formulates in French): *Déterminer la forme de toutes les équations algébriques qui peuvent être resolues algebriquement.* [Determine the form of all algebraic equations that can be solved algebraically.] He had earlier communicated to Crelle (in a letter dated March 14, 1826) the solution for equations of degree five (with rational coefficients, but it also holds for coefficients in any field of characteristic zero), saying that he is in possession of similar results for equations of degree 7, 11, 13, etc. The letter was printed in Crelle's journal in 1830 and reads in its entirety:

Wenn eine Gleichung des fünften Grades, deren Coëfficienten rationale Zahlen sind, algebraisch auflösbar ist, so kann man immer den Wurzeln folgende Gestalt geben:

$$\begin{aligned} x = c &+ A\,.\,a^{\frac{1}{5}}.\,a_1^{\frac{2}{5}}.\,a_2^{\frac{4}{5}}.\,a_3^{\frac{3}{5}} + A_1\,.\,a_1^{\frac{1}{5}}.\,a_2^{\frac{2}{5}}.\,a_3^{\frac{4}{5}}.\,a^{\frac{3}{5}} + A_2\,.\,a_2^{\frac{1}{5}}.\,a_3^{\frac{2}{5}}.\,a^{\frac{4}{5}}.\,a_1^{\frac{3}{5}} \\ &+ A_3\,.\,a_3^{\frac{1}{5}}.\,a^{\frac{2}{5}}.\,a_1^{\frac{4}{5}}.\,a_2^{\frac{3}{5}} \end{aligned} \tag{†}$$

wo

$$a = m + n\sqrt{1+e^2} + \sqrt{h\left(1+e^2+\sqrt{1+e^2}\right)}$$

$$a_1 = m - n\sqrt{1+e^2} + \sqrt{h\left(1+e^2-\sqrt{1+e^2}\right)}$$

$$a_2 = m + n\sqrt{1+e^2} - \sqrt{h\left(1+e^2+\sqrt{1+e^2}\right)}$$

$$a_3 = m - n\sqrt{1+e^2} + \sqrt{h\left(1+e^2-\sqrt{1+e^2}\right)}$$

$$A = K + K'a + K''a_2 + K'''aa_2, \qquad A_1 = K + K'a_1 + K''a_3 + K'''a_1a_3,$$

$$A_2 = K + K'a_2 + K''a + K'''aa_2, \qquad A_3 = K + K'a_3 + K''a_1 + K'''a_1a_3.$$

Die Grössen $c, h, e, m, n, K, K', K'', K'''$ *sind alle rationale Zahlen.*

Auf diese Weise lässt sich aber die Gleichung $x^5 + ax + b = 0$ *nicht auflösen, so lange a und b beliebige Grössen sind.*

Ich habe ähnliche Lehrsätze für Gleichungen vom 7ten, 11ten, 13ten etc. Grade.

Abel's mathematical diary from Paris contains 13 pages that gives evidence that he was thinking about algebraic equations. Here occurs an auxiliary quantity that Abel also introduced in Précis [2, XXVIII, p. 547], and which he uses as a powerful tool to prove some of his results. The same quantity occurs in his posthumously published and incomplete manuscript *Sur la résolution algébriques des équations* [3, XVIII]. In his memoir *Sur les conditions de résolubilité des équations par radicaux* [6], submitted to l'Académie des Sciences on January 16, 1831, Galois introduces exactly the same auxiliary quantity, which he denotes by V and he refers to the page in the Précis where it appears. Galois also refers to Abel about the same in his article *Sur la théorie des nombres*, published in Férussac's Bulletin des Sciences in June 1830. (Abel uses the same symbol V to denote the (unique) irreducible monic polynomial that has the quantity in question as a root, i.e., V denotes its minimal polynomial.) In the old literature V is called a *Galois resolvent*, and in modern terminology it is called a *primitive element*. It is a crucial tool used by Galois in introducing the group (later called the *Galois group*) associated to an equation—the group reflecting the essential character of the equation. Galois gives a proof of the existence of a primitive element [6, Lemma III], while Abel in Précis simply states:

C'est ce qui est facile à démontrer par la théorie des équations algébriques.

[It is easy to prove using the theory of algebraic equations.]

We want to sketch a proof of the existence of a primitive element because it gives a beautiful demonstration of the application of the fundamental theorem of symmetric polynomials—a proof technique that Abel also uses in his proof of his addition theorem. The proof idea is essentially due to Lagrange and it's a gem. (See

§100 of Lagrange's memoir from 1770: *Réflexions sur la résolution algébriques des équations* [13, Vol. 3].) Both Abel and Galois were familiar with Lagrange's work on the theory of equations. We do not know exactly how Abel would have proved his claim in Précis—the setting there is actually more general than the one Galois has. However, a pretty good idea of how he would proceed can be found in §2 of *Mémoire sur une classe particulière d'équations résolubles algébriquement* [2, XXV]. The proof technique that he displays in §2 is very close to Lagrange's proof alluded to above, and it can be easily adapted to prove the existence of a primitive element. We will instead state as a theorem the result Abel essentially states in Précis and we will give the proof that Weber presents in his *Lehrbuch der Algebra* [26, Vol. 1, §150]. That proof has vintage Abel flavour, if we may use such an expression. (We will use modern terminology and notation. All fields we are considering have characteristic zero.)

Theorem A *Let $f_1(x), f_2(x), f_3(x), \ldots, f_m(x)$ be a finite collection of polynomials over the field F such that each polynomial has distinct roots in $\bar{F}$, the algebraic closure of F. (However, the polynomials themselves can otherwise be arbitrary—they do not have to be distinct, for example.) Let $\alpha, \beta, \gamma, \ldots$ be roots of $f_1(x), f_2(x), f_3(x), \ldots,$ respectively, where the degrees of these polynomials are $m_1, m_2, m_3, \ldots,$ respectively. Then the field $E = F(\alpha, \beta, \gamma, \ldots)$ generated by $\alpha, \beta, \gamma, \ldots$ over F is simply generated, i.e. there exists a* primitive element *$\xi \in E$ such that $E = F(\xi)$. Furthermore ξ can be chosen to be of the form $\xi = a\alpha + b\beta + c\gamma + \cdots$, where $a, b, c, \ldots$ are integers.*

Proof Let $\xi = a\alpha + b\beta + c\gamma + \cdots$, where $a, b, c, \ldots$ are integers such that ξ and

$$\begin{aligned}\xi' &= a\alpha' + b\beta' + c\gamma' + \cdots\\ \xi'' &= a\alpha'' + b\beta'' + c\gamma'' + \cdots\\ &\vdots\end{aligned}$$

are distinct elements, where respectively $\alpha, \alpha', \alpha'', \ldots$ and $\beta, \beta', \beta'', \ldots$ and $\gamma, \gamma', \gamma'', \ldots,$ etc. are the roots of $f_1(x)$ and $f_2(x)$ and $f_3(x)$, etc. Here $\alpha, \beta, \gamma, \ldots$; $\alpha', \beta', \gamma', \ldots$; $\alpha'', \beta'', \gamma'', \ldots,$ are different combinations. The number of such combinations are obviously equal to $m = m_1 m_2 m_3 \cdots$. (Abel states in Précis that there exists integers $a, b, c, \ldots$ such that this can be achieved, something one easily sees by considering the pairwise differences of $\xi, \xi', \xi'', \ldots$.) Hence the polynomial $g(x)$ of degree $m = m_1 m_2 m_3 \cdots$ defined by

$$g(x) = (x - \xi)\big(x - \xi'\big)\big(x - \xi''\big)\cdots$$

is *symmetric* with respect to the α's, the β's, the γ's, …, respectively. By the fundamental theorem of symmetric polynomials, $g(x) \in F[x]$, i.e., $g(x)$ is a polynomial over F. Now any element $\theta \in E$ can be written as a polynomial in $\alpha, \beta, \gamma, \ldots$ (This was well known to Abel; it is proved by the Euclidean division algorithm for polynomials.) Let $\theta, \theta', \theta'', \ldots$ be the values that θ get by permuting the α's, the β's, the γ's, etc. as was done in obtaining $\xi, \xi', \xi'', \ldots$. Then

$$\psi(x) = g(x)\left(\frac{\theta}{x-\xi} + \frac{\theta'}{x-\xi'} + \frac{\theta''}{x-\xi''} + \cdots\right)$$

is a *symmetric* polynomial (of degree $m-1$) in the α's, the β's, the γ's, etc., and so again by the fundamental theorem of symmetric polynomials we have $\psi(x) \in F[x]$. Setting $x = \xi$, we get $\theta = \frac{\psi(\xi)}{g'(\xi)}$ which is what we wanted to prove. ($g'(x)$ denotes the derivative of $g(x)$.) □

5 Irreducibility Principle

Sylow [18] strongly emphasizes a novel feature that Abel brought to the theory of equations, namely the use that he made of an equation's (or rather, a polynomial's) irreducibility. Today this is completely elementary and "folklore" in algebra, and we tend to brush it aside as something obvious. However, by reading Abel's and Galois' original manuscripts one is struck by the ingenious way they use irreducibility, combined with the Euclidean division algorithm and the fundamental theorem of symmetric polynomials to overcome difficulties that apparently would require considerably more advanced and sophisticated methods. Sylow [18] says that he does not know of any examples where the concept of irreducibility appeared, at least as an indispensable tool, before Abel. To be sure, he points out that Gauss in Chap. VII of *Disquisitiones Arithmeticae* (1801) [7], where cyclotomic equations are studied, writes:

> *Omnique rigore demonstrare possumus, has equationes elevatas nullo modo nec evitari nec ad inferiores reduci posse.*
>
> [We can show with all rigour that these higher-degree equations can not be avoided in any way nor can they be reduced to lower-degree equations.]

By this Gauss undoubtedly means to say that the various cyclotomic polynomials that appear in his investigation are irreducible in the modern sense over their respective fields. But Gauss does not use irreducibility as a tool in his reasoning and arguments. This in contrast to both Abel and Galois who both define the concept and provide the proof—as if it was unknown—of the now well-known theorem about the irreducible equation. In modern language this is the theorem (referred to in the sequel as **Theorem B**) that says that if $p(x)$ and $f(x)$ are two polynomials over the same field with a common root, and $p(x)$ is irreducible, then $p(x)$ divides $f(x)$. In particular, all the roots of $p(x)$ will be roots of $f(x)$. (Cf. [2, XXV, Théorème I], [6, Lemme I].)

Abel (and Galois later) gives his theory the highest degree of generality by assuming a general domain of rationality (today we would say field), when he writes [2, XXV, p. 479]:

Une équation $\phi(x)=0$, dont les coefficiens sont des fonctions rationnelles d'un certain nombre de quantités connues $a, b, c, \ldots$, s'appelle irréductible, lorsqu'il est impossible d'exprimer aucune de ses racines par une équation moins élevée, dont les coefficiens soient également des fonctions rationnelles de $a, b, c, \ldots$

[An equation $\phi(x)=0$, where the coefficients are rational functions of a certain number of known quantities $a, b, c, \ldots$, is called irreducible if it is impossible to express any of its roots by an equation of lower degree whose coefficients are again rational functions of $a, b, c, \ldots$]

6 The Galois Group

We will describe how Galois constructed the group—later named after him—associated to an algebraic equation and give its characteristic property [6, Proposition I]. This will illustrate in a striking way the *irreducibility principle* that Abel introduced. Moreover, it will set the stage, so to say, for a comparison between Abel's and Galois' different approaches towards the theory of equations. (We will as before use modern terminology and notation.)

Let $f(x) \in F[x]$ be a polynomial of degree m over the field F with distinct roots $\alpha_1, \alpha_2, \ldots, \alpha_m$ (in the algebraic closure $\bar{F}$ of F). Let $E = F(\alpha_1, \alpha_2, \ldots, \alpha_m)$ be the splitting field of $f(x)$. Let $\xi \in E$ be a primitive element (i.e., $E = F(\xi)$) where ξ is of the form $\xi = a\alpha_1 + b\alpha_2 + c\alpha_3 + \cdots$, for some integers $a, b, c, \ldots$, as constructed in Theorem A. (In fact, choose the polynomials $f_1(x), f_2(x), f_3(x), \ldots, f_m(x)$ in Theorem A to be all equal to $f(x)$ and let $\alpha, \beta, \gamma, \ldots$ be $\alpha_1, \alpha_2, \alpha_3, \ldots$) Let $V(x) \in F[x]$ be the minimal polynomial of ξ, and let $V(x)$ be of degree n. Now by Theorem B, all the roots of $V(x)$ will also be roots of the polynomial $g(x)$ that appears in the proof of Theorem A. Hence these roots are all of the same form as ξ and so they lie in $F(\alpha_1, \alpha_2, \ldots, \alpha_m) = F(\xi) = E$. By the Euclidean division algorithm one shows that every element in E can be uniquely represented as a polynomial in ξ over F of degree $\leq n-1$. In particular, let $\xi_0 (= \xi), \xi_1, \xi_2, \ldots, \xi_{n-1}$ be the (distinct) roots of $V(x)$. There exist unique polynomials $\phi_0(x)(= x), \phi_1(x), \phi_2(x), \ldots, \phi_{n-1}(x)$ over F of degrees $\leq n-1$ such that

$$\xi_i = \phi_i(\xi); \quad i = 0, 1, 2, \ldots, n-1.$$

So $V(\phi_i(\xi)) = 0$ for all i, and by Theorem B we get that $V(x)$ divides $V(\phi_i(x)) \in F[x]$. Hence $V(\phi_i(\xi_j)) = 0$ for $0 \leq j \leq n-1$. We conclude that for all j the set $\{\xi_j (= \phi_0(\xi_j)), \phi_1(\xi_j), \phi_2(\xi_j), \ldots, \phi_{n-1}(\xi_j)\}$ is a subset of $\{\xi_0 (= \xi), \xi_1, \xi_2, \ldots, \xi_{n-1}\}$. We claim that the sets are equal. (This will, in particular, imply that for each j there exists some k such that $\xi = \phi_k(\xi_j)$, and so $E = F(\xi_j)$.) It is enough to show that $\phi_s(\xi_j) \neq \phi_t(\xi_j)$ if $s \neq t$. If $\phi_s(\xi_j) = \phi_t(\xi_j)$, then ξ_j is a root of the polynomial $\phi_s(x) - \phi_t(x) \in F[x]$. By Theorem B we get that $V(x)$ divides $\phi_s(x) - \phi_t(x)$, and so, in particular, $\phi_s(\xi) - \phi_t(\xi) = 0$. This implies that $\xi_s = \phi_s(\xi) = \phi_t(\xi) = \xi_t$, which is impossible, thus proving the claim. We define n distinct permutations $\sigma_0, \sigma_1, \sigma_2, \ldots, \sigma_{n-1}$ of the roots $\Sigma = \{\xi_0(=\xi), \xi_1, \xi_2, \ldots, \xi_{n-1}\}$ of $V(x)$ by letting (for $a \in \{0, 1, 2, \ldots, n-1\}$)

$$\sigma_a(\xi_k) = \phi_k(\xi_a); \quad k = 0, 1, 2, \ldots, n-1$$
$$\big(\text{In particular, } \sigma_a(\xi) = \phi_0(\xi_a) = \xi_a\big)$$

Notice that each of these permutations are uniquely determined by what $\xi = \xi_0$ is sent to. We claim that $G = \{\sigma_0, \sigma_1, \sigma_2, \ldots, \sigma_{n-1}\}$ is a group, where the group operation is composition of permutations, and σ_0 is the identity element. So let $\sigma_a, \sigma_b \in G$. We claim that $\sigma_c = \sigma_b \circ \sigma_a$, where $\xi_c = \sigma_b \circ \sigma_a(\xi)$. This will be a consequence of Theorem B. In fact, let $l \in \{0, 1, 2, \ldots, n-1\}$, and let $\xi_k = \phi_l(\xi_a)$. This implies that ξ is a root of the equation $\phi_k(x) = \phi_l(\phi_a(x))$, and so by Theorem B, $\xi_0(=\xi), \xi_1, \xi_2, \ldots, \xi_{n-1}$ are roots also. In particular, $\phi_k(\xi_b) = \phi_l(\phi_a(\xi_b))$. The left hand side of this equation is $\sigma_b \circ \sigma_a(\xi_l)$ while the right hand side is $\sigma_c(\xi_l)$, proving the claim. Now each σ_j induces a permutation of the roots $\Gamma = \{\alpha_1, \alpha_2, \ldots, \alpha_m\}$ of the original equation $f(x)$. In fact, for every $1 \le s \le m$ there exists a (unique) polynomial $\chi_s(x) \in F[x]$ of degree $\le n-1$ such that $\alpha_s = \chi_s(\xi)$. For $0 \le k \le n-1$ define $\sigma_k^*(\alpha_s) = \chi_s(\sigma_k(\xi)) = \chi_s(\xi_k)$. Then $\sigma_0^*, \sigma_1^*, \sigma_2^*, \ldots, \sigma_{n-1}^*$ will be n distinct permutations of Γ, and by again applying Theorem B one shows easily that the map $\sigma_j \to \sigma_j^*$ is a group isomorphism between $G = \{\sigma_0, \sigma_1, \sigma_2, \ldots, \sigma_{n-1}\}$ and $G^* = \{\sigma_0^*, \sigma_1^*, \sigma_2^*, \ldots, \sigma_{n-1}^*\}$, where the group operation of G^* is composition of permutations on Γ. G^* (or G) is called the Galois group of $f(x)$ over F, and we will henceforth identify G and G^*.

Remark The concept of automorphism of E (the splitting field of $f(x)$) over F came later with Dedekind. The influence of Dedekind in presenting Galois theory can be seen in the work of his de facto student and collaborator Heinrich Weber in the first volume of *Lehrbuch der Algebra* [26], which appeared in 1895. Weber's treatment has one foot in the 19th century and one in the 20th century. The modern presentation of Galois theory was heavily influenced by Emil Artin. This can be seen in van der Waerden's classical text *Moderne Algebra* from 1930 based upon Artin's lectures on Galois theory. However, according to van der Waerden [25] Artin was not happy with using a primitive element to prove the fundamental theorem of Galois theory. He eventually managed to avoid the primitive element. Although conceptually satisfying one pays a price: the proof is simpler, and certainly one applies more elementary means, if the primitive element is used.

Galois [6, Proposition I] gave the following characterization of the group G he associated to the polynomial $f(x) \in F[x]$ (with distinct roots $\alpha_1, \alpha_2, \ldots, \alpha_m$):

(i) Every equation $H(\alpha_1, \alpha_2, \ldots, \alpha_m) = 0$, where $H(x_1, x_2, \ldots, x_m)$ is a polynomial over F in m variables $x_1, x_2, \ldots, x_m$, is also valid if we permute the roots with an element $\sigma \in G$.
(ii) Every polynomial $H(\alpha_1, \alpha_2, \ldots, \alpha_m)$ over F in the roots $\alpha_1, \alpha_2, \ldots, \alpha_m$ which is invariant under all permutations $\sigma \in G$, is an element in F.

The proofs of (i) and (ii) are immediate consequences of Theorem A and Theorem B. In fact, let ξ as in Theorem A denote a primitive element of the form $\xi = a\alpha_1 + b\alpha_2 + c\alpha_3 + \cdots$, and let $\alpha_s = \chi_s(\xi)$ for $s = 1, 2, 3, \ldots$, where $\chi_s(x) \in F[x]$. Then the polynomial $K(x) = H(\chi_1(x), \chi_2(x), \ldots, \chi_m(x))$ has $x = \xi$ as a root. By Theorem B the minimal polynomial $V(x)$ of ξ divides $K(x)$, and consequently

$$0 = K(\xi_k) = H\big(\chi_1(\xi_k), \chi_2(\xi_k), \ldots, \chi_m(\xi_k)\big)$$

for $0 \leq k \leq n-1$, where $\xi_0(=\xi), \xi_1, \xi_2, \ldots, \xi_{n-1}$ are the roots of $V(x)$. Now $\sigma_k(\alpha_s) = \chi_s(\xi_k)$ for $s = 1, 2, \ldots, m$, and this proves (i).

As for the proof of (ii), the assumption implies that $L(\xi_j) = L(\xi)$ for $j = 0, 1, 2, \ldots, n-1$, where $L(x) = H(\chi_1(x), \chi_2(x), \ldots, \chi_m(x)) \in F[x]$ and $\chi_1(x)$, $\chi_2(x), \ldots$ are as above. Let $M(x)$ be the polynomial

$$M(x) = \big(x - L(\xi)\big)\big(x - L(\xi_1)\big)\big(x - L(\xi_2)\big) \cdots \big(x - L(\xi_{n-1})\big).$$

Since $L(x)$ is symmetric in the roots $\xi_0(=\xi), \xi_1, \xi_2, \ldots$ of $V(x)$, it follows that $M(x) \in F[x]$. Now

$$M(x) = \big(x - L(\xi)\big)^n,$$

and so $H(\alpha_1, \alpha_2, \ldots, \alpha_m) = H(\chi_1(\xi), \chi_2(\xi), \ldots, \chi_m(\xi)) = L(\xi)$ is an element in F, and this proves (ii).

Properties (i) and (ii) completely determine the Galois group G. Firstly (i) implies that G is the largest of all permutation groups G' on $\alpha_1, \alpha_2, \ldots, \alpha_m$ that satisfy property (i). One sees this by recalling that $V(\xi_j) = 0$ for $j = 0, 1, \ldots, n-1$, and ξ_j is a linear combination (over F) of $\alpha_1, \alpha_2, \alpha_3, \ldots$. Applying $\sigma \in G'$ one gets by (i) that $V(\xi') = 0$, where ξ_j is sent to ξ' by σ. Hence $\xi' = \xi_k$ for some k. Now $\xi_j = \phi_j(\xi)$, where we use the notation introduced above. The equality holds after applying σ on both sides, and so if $\sigma(\xi) = \xi_t$, we get that $\sigma = \sigma_t$, proving that $G' \subseteq G$.

Secondly, (ii) implies that G is the smallest of all permutation groups G'' on $\alpha_1, \alpha_2, \ldots, \alpha_m$ that satisfy property (ii). To see this let $\xi(=\xi_0)$ be mapped to $\xi, \xi', \xi'', \ldots$ under G''. Then the coefficients of the polynomial

$$g(x) = (x - \xi)\big(x - \xi'\big)\big(x - \xi''\big) \cdots$$

lie in F since G'' satisfies property (ii) and the coefficients are symmetric polynomials in $\xi, \xi', \xi'', \ldots$. By Theorem B we get that among $\xi, \xi', \xi'', \ldots$, we must find the roots $\xi_0(=\xi), \xi_1, \xi_2, \ldots, \xi_{n-1}$ of the (irreducible) polynomial $V(x)$ since $g(x)$ and $V(x)$ have the root ξ in common. Hence for every $t = 0, 1, 2, \ldots, n-1$, G'' contains a permutation σ such that $\sigma(\xi) = \xi_t$. This implies that $\sigma = \sigma_t$ by a similar argument as we gave above. So $G \subseteq G''$.

7 The Fundamental Theorem and Solvability Criterion

We summarize briefly the rest of the content in Galois' epoch-making memoir [6]. The setting is that $f(x)$ is a polynomial over F, E is the splitting field and $G = G(E/F)$ is the Galois group of $f(x)$. Based upon the foundation he has laid Galois describes what happens when one adjoins to F roots of another (polynomial) equation over F, thus getting an extension field K. He shows that the Galois groups of $f(x)$, now considered as a polynomial over K, is a subgroup H of the Galois group G. He also proves the crucially important result that if K is obtained by adjoining *all* the roots of an equation, then the group H becomes what we today call a *normal* subgroup of G. The correspondence between the subgroups H of the Galois group G and the fields K lying between F and E is a bijection, where $K = \{\alpha \in E | \sigma(\alpha) = \alpha \text{ for all } \sigma \in H\}$. This beautiful result is the *fundamental theorem of Galois theory*. Galois then applies his theory to give a criterion for when an equation can be solved by radicals—invoking the so-called Lagrange resolvent—and he applies that to irreducible equations of prime degree. The final theorem of his memoir is Proposition VIII:

> *Théorème. Pour qu'une équation irréducible de degré premier soit soluble par radicaux, il faut et il suffit que deux quelconques des racines étant connues, les autres s'en déduisent rationnellement.*
>
> [Theorem. In order that an irreducible equation of prime degree should be soluble by radicals, it is necessary and sufficient that any two of its roots being known, the others may be deduced from them rationally.]

Galois' proof of this, as presented in [6], is very elegant: he shows that the Galois group G is a subgroup of the (one-dimensional) affine linear group (mod p), where p is the degree of the polynomial, acting transitively on the p roots. It is easily shown that the only element of G that fixes two distinct roots is the identity element, and so they generate the splitting field.

In a letter to Crelle dated October 18, 1828 (published in Crelle's journal in 1830 [3, XXII, 5B]) Abel writes:

> *Si trois racines d'une équation quelconque irréductible dont le degré est un nombre premier, sont liées entre elles de sorte que l'une de ces racines puisse être exprimée rationnellement par les deux autres, l'équation en question sera toujours résoluble à l'aide de radicaux.*
>
> [If three roots of an arbitrary equation of which the degree is a prime number are related to each other in such a way that any one of these roots may be expressed rationally by means of the other two, the equation in question will always be soluble with the help of radicals.]

Note that Abel states (in a slightly imprecise way) only the sufficiency of the condition for solvability. However, he was fully aware of the necessity of the condition as can be seen from the previously mentioned (and posthumously published) manuscript *Sur la résolution algébriques des équations* [3, XVIII]. (Cf. also

[18, p. 21 (and pp. 22–23)].) So Abel was in possession of the result stated in Proposition VIII in Galois' memoir. However, Galois would not have known about this when he himself made this discovery, since he announced this particular result already in April 1830 in the paper *Analyse d'un mémoire sur la résolution algébrique des équations* (published in Férussac's Bulletin des Sciences). It is then quite understandable that he was irritated and angry that Lacroix and Poisson in their referee report (dated July 4, 1831)—rejecting his great memoir *Sur les conditions de résolubilité des equations par radicaux* [6]—use Abel as a truth witness, so to say, for the validity of Proposition VIII. The two referees admit they did not understand Galois' proof of Proposition VIII, but since Abel had stated (the sufficiency condition of) Proposition VIII in his letter to Crelle, they are inclined to accept that the result is true. Anyway, this might explain to a certain degree the vehemence with which Galois claims his independence of Abel in the note he wrote that we mentioned earlier.

8 Elliptic Functions and Algebraic Equations

Abel's investigation of algebraic equations and their solvability properties is inextricably linked to his study of elliptic functions. In fact, Abel approached the theory of elliptic functions mainly from an algebraic point of view, centered on the various algebraic equations that the transformation theory provided in such abundance. We can express this in the following way, and this applies to both Abel and Galois: The lifting of the veil concealing the secret of algebraic solvability of (numerical) equations did not occur in a vacuum. The rich source of (irreducible) algebraic equations, in particular, the "teilungsgleichungen" ("division equations") that the transformation theory of elliptic functions provided, was instrumental for Abel—as it in all likelihood was later for Galois—as suggestive examples that led to "lifting of the veil". Specifically, in his analysis of the teilungsgleichungen (more about that below), Abel saw how far Gauss' method for solving cyclotomic equations could be generalized. In Chap. VII: *Equations defining sections of a circle* of Disquisitiones Arithmeticae [7], Gauss had showed that the cyclotomic equations could be solved by radicals. Furthermore, he showed that if p is a prime of the form $p = 2^n - 1$ (this implies that p must be a Fermat prime, i.e., of the form $2^{2^m} - 1$), then the associated cyclotomic equation (of degree $p - 1$) could be solved by a succession of square roots—thus a regular p-gon could be constructed by ruler and compass. In *Mémoire sur une classe particulière d'équations résolubles algébriquement* [2, XXV], Abel proved the following theorem, stated in modern language:

> *Let E be the splitting field of a polynomial $f(x)$ over F. If the Galois group $G = G(E/F)$ of $f(x)$ is commutative, then $f(x) = 0$ can be solved algebraically ("by radicals").*

Leopold Kronecker and Camille Jordan would later call equations having the property stated in the theorem for *Abelian equations*. This is the reason why Abel's name became attached to commutative algebraic structures. Abel characterized his method as a generalization of the circle division theory, and he said that the associated equations could be solved by Gauss' method. This is also the way Galois would refer to this. The memoir *Mémoire sur une classe particulière d'équations résoluble algébriquement* [2, XXV] was sent to Crelle March 29, 1828, but was not published in Crelle's journal before March 28, 1829, a year later. The reason for this delay was that the memoir should contain more than the 5 paragraphs it consists of. It should contain a sixth paragraph that dealt with the division of the periods of an elliptic function that had complex multiplication. The seventh paragraph should have contained elliptic transformation formulas. Very likely even more paragraphs were intended to be included, one of which would treat those equations that determine the singular modules themselves (and so the associated elliptic functions will have complex multiplication). We know all this from looking at his mathematical diary. (Cf. [3, pp. 310–311] for more details.) Most of this would appear instead in his various publications on elliptic functions, including his Précis memoir. Crelle must have waited for these extra paragraphs which never arrived, and this explains the delay in the publication. The overarching explanation for Abel's change of mind regarding this was his competition with Jacobi in developing the theory of elliptic functions, and this also has some bearing on the Abel letter. In fact, Jacobi had published a short note in Crelle's journal, appearing March 25, 1828, containing an elegant—from an algebraic point of view—solution of a teilungsgleichung Abel had previously studied, and which was an alternative to the solution that Abel had given in his Recherches memoir on elliptic functions [2, XVI]. Abel decided to cast everything else aside and show that he was in possession of a much more comprehensive theory. This resulted in two remarkable papers. The first, sent from Oslo on May 27, 1828, was titled *Solution d'un problème général concernant la transformation des fonctions elliptiques* [2, XIX] (see also the related *Addition au mémoire précédent* [2, XX]), which was published in Astronomische Nachrichten—the same journal where Jacobi had earlier published a paper proving a theorem about transformations of elliptic functions. Jacobi, upon reading this paper, wrote to Legendre:

> *Elle est au-dessus de mes éloges comme elle est au-dessus de mes propre travanx.*
>
> [It stands as high above my praise as it surpasses my own works.]

It was this and similar sentiments expressing how Abel's work was appreciated that Crelle communicated in his letter to Abel, and which Abel—certainly pleased—reacted to by writing: "Herr Jacobi says indeed too much".

The other remarkable paper by Abel that we referred to above has the same title as his Recherches memoir, only that "Second Mémoir" is added to the title, and is dated August 27, 1828, so shortly before he became sick and compelled to stay in bed (cf. his letter to Crelle). This Second Memoir is the aforementioned "lost" manuscript that Mittag-Leffler found and printed in the 1902-memorial edition of Acta Mathematica, and that Sylow talked about at the 3rd Scandinavian Mathematical Congress in Oslo in 1913 [19]. Sylow speculates why Crelle only published the first paragraph of the Second Memoir in his journal, and under a different title,

Théorèmes sur les fonctions elliptiques [2, XXVI]. (It appeared March 28, 1829.) The main reason, Sylow surmises, for not publishing the next four paragraphs of the Second Mémoir is that Abel was not satisfied with the redaction of this part. However that may be, most of the content of the Second Mémoir was pieced together from fragments of manuscripts left by Abel in his *Nachlass* and published by Lie and Sylow in Abel's Oeuvres Complètes (1881) under the titles *Fragments sur les fonctions elliptiques* [3, XIX] and *Démonstration de quelques formules elliptique* [3, XV]. What strongly supports Sylow's supposition is that one of these fragments has the same title as the Second Mémoir, indicating that Abel was writing a new version of the manuscript he had sent Crelle. Furthermore, it is written on the same type of paper as what he had sent to Crelle. Anyway, Abel did not have time left to complete his revision of the Second Mémoir. Let us summarize briefly its content. It starts by looking at teilungsgleichungen and shows that Jacobi's solution formula follows from the treatment Abel gives of these. In §3 Abel's addition theorem is stated in the context of elliptic functions, and in §4 connections between the roots of the various teilungsgleichungen are given. In the last paragraph, §5, the so-called monodromy groups of the transformation equation is given with a sketch of the proof. Théorème XVI of the last paragraph, §5, is about complex multiplication. Mittag-Leffler, who was frustrated that Crelle had not published the Second Mémoir in its entirety, writes in his preface to the Acta publication of Abel's manuscript:

> *Si la publication de ce manuscrit n'apporte pas à la science actuelle des résultats nouveaux, elle semble pourtant d'une très grande valeur pour l'étude de l'enchaînement et du développement des idées d'Abel. On ne peut s'empêcher de penser que, si Crelle avait publié le mémoire en entier, les* Recherches sur les fonctions elliptiques *auraient constitué, dès le début, une doctrine plus complète et plus achevée, de nature à faire ressortir Abel aux yeux de ses contemporains comme le vrai et principal créateur de la théorie des fonctions elliptiques.*
>
> [Even if the publication of this manuscript does not bring any new results to science today, it is nevertheless of great value for the study of the interconnection and development of Abel's ideas. One cannot help thinking that if Crelle had published the memoir in its entirety, the *Recherches sur les fonctions elliptiques* would have established from the start a theory so complete and so comprehensive that Abel would have stood in the eyes of his contemporaries as the true and principal creator of the theory of elliptic functions.]

9 Transformation Theory and Teilingsgleichungen

For Abel the transformation theory of elliptic functions can be stated like this [2, XX, p. 429]: Find the conditions under which there exists an *algebraic* function $y = y(x)$ (i.e., $R(x, y) = 0$ for some rational function R in two variables over the complex numbers) which is a solution of the separable differential equation

$$\frac{dy}{\sqrt{(1-y^2)(1-l^2y^2)}} = a\frac{dx}{\sqrt{(1-x^2)(1-k^2x^2)}}. \qquad (*)$$

Here a is a constant (the *multiplier*), and l and k are two other constants (the *modules*). In Précis, where a much more general problem is posed, it is proved by algebraic means that one can reduce the situation to y being a *rational* function of x [2, XXVIII, Chapitre IV] (cf. also [5, p. 295])—the key tools being the irreducibility principle combined with the addition theorem applied to the special setting of elliptic integrals. The solution of (*) is by Abel transferred to finding all solutions of the equation

$$(1-y^2)(1-l^2y^2) = t^2(1-x^2)(1-k^2x^2)$$

where $y = U(x)/V(x)$, $t = t(x)$ are rational functions of x. Now the general case can be traced back to the special case where the degrees of the polynomials $U(x)$ and $V(x)$ are p and $p-1$, respectively, where p is a prime. (We say that the transformation is of degree p.) In this situation there exists (if $p \neq 2$) between l and k an algebraic equation of degree $p+1$ with integer coefficients, the so-called *modular equation*. In turn, the multiplier a is determined by this equation. Specifically, let l be given and let k be any particular root of the modular equation. Then the coefficients of $U(x)$ and $V(x)$ as well as a is uniquely determined by this choice of k. Abel showed that by composing two special transformations, both of degree p, one is led to teilungsgleichungen ("division equations") of degree p^2. This can be stated in the following way:

Let $\phi(u)$ be an elliptic function with (primitive) periods ω_1 and ω_2, say. Express $\phi(u/p)$ as an algebraic function of $\phi(u)$. In other words, find the algebraic equation (called a teilungsgleichung, plural: teilungsgleichungen) over some field F containing $\phi(u)$, which has $\phi(u/p)$ as a root and determine the solvability property of this equation. For example, $\phi(u)$ could be the inverse function of an elliptic integral of the first kind, to wit, $\phi(u)$ is defined by the formula

$$u = \int_0^{\phi(u)} \frac{dx}{\sqrt{(1-x^2)(1-k^2x^2)}}$$

the periods being determined by the module k.

The study of the algebraic character of these teilungsgleichungen as pertains to their solvability properties is what led Abel and, very likely, Galois to their discoveries. Let us be a little more specific. As background motivation let us start with the pure equation $x^p - a = 0$, where $a = e^u$, which has the root $x_0 = e^{u/p}$. (Note that the function $\phi(z) = e^z$ has a single period $2\pi i$.) A complete list of roots are $x_\nu = e^{\frac{u+\nu 2\pi i}{p}}$, where $\nu = 0, 1, \ldots, p-1$. Here one considers u as a (complex) variable and the domain of rationality, or field, F is assumed to be the rational numbers $\mathbb{Q}$ with e^u, as well as the primitive pth roots of unity $e^{\frac{s2\pi i}{p}}$, $s = 1, \ldots, p-1$, adjoined. The polynomial $x^p - a$ is irreducible over F, and the Galois group is a cyclic group of order p, generated by the permutation $x_s \to x_{s+1} \pmod p$. The primitive pth roots of unity are the roots of an irreducible polynomial over $\mathbb{Q}$ of degree $p-1$, called a cyclotomic polynomial. We will call the corresponding equation a *special* (or *period*) *teilungsgleichung*, while the equation $x^p - e^u = 0$ (with the pth roots of unity in the ground field) is called a *general teilungsgleichung*.

For elliptic functions we have an analogous situation. Let $\phi(u)$ be an elliptic function with (primitive) periods ω_1 and ω_2, and let p be a prime. Let the general teilungsgleichung, having $\phi(u/p)$ as a root, be $g(x) = 0$. The polynomial $g(x)$ is of degree p^2 (over the field F). In the coefficients appear, besides rational functions of $\phi(u)$, certain "invariants" that only depend upon the periods, notably the *module*. These should like the periods be considered to be variables and are rationally known, i.e., lie in the base field F. The p^2 roots of $g(x)$ are $x_{j,k} = \phi(\frac{u+j\omega_1+k\omega_2}{p})$; $j,k = 0,1,\ldots,p-1$. By setting $u=0$ one gets the roots of the special teilungsgleichung (removing $x_{0,0}$) which is irreducible (over an appropriate field) of degree p^2-1. If one assumes that the roots of the special teilungsgleichung lie in F, then $g(x)$ is irreducible and the roots of $g(x)=0$ can be algebraically solved ("by radicals"). This Abel showed [2, XVI, pp. 294–305], and from the viewpoint of Galois theory this can be explained as follows:

The Galois group G of $g(x)$ is generated by the two permutations—both of order p—namely $x_{j,k} \to x_{j+1,k}$ and $x_{j,k} \to x_{j,k+1} \pmod p$, cf. [9]. The group G is commutative, and so by Abel's memoir [2, XXV] the equation $g(x)=0$ is algebraically solvable.

The special teilungsgleichung (also called "periodenteilungsgleichung" in German)—as pertains to its solvability—is determined by $p+1$ equations each of degree $p-1$. These contain each in their coefficients a root of one and the same (irreducible) equation of degree $p+1$. If this (resolvent) equation can be solved by radicals, then so can each of the previous $p+1$ equations, and consequently the special teilungsgleichung. In fact, the Galois group after adjoining this root becomes commutative. All this is treated in Abel's memoir on elliptic functions [2, XVI, pp. 305–314], which was published before his memoir on equations [2, XXV], and so the results he obtained for teilungsgleichungen can be seen as models and inspirations for the latter. Now Abel also was aware of that the resolvent equation of degree $p+1$ can be algebraically solved when the elliptic function in question admits *complex multiplication*. To explain this concept let us return to the differential equation $(*)$, setting $l=k$:

$$\frac{dy}{\sqrt{(1-y^2)(1-k^2y^2)}} = a\frac{dx}{\sqrt{(1-x^2)(1-k^2x^2)}}. \qquad (**)$$

Abel showed that $(**)$ has an algebraic solution $y = y(x)$ if and only if a is a rational number or is of the form $a = m + i\sqrt{n}$, where m and $n(>0)$ are rational numbers [2, XVI, §X]. In the latter case the modules k (called *singular* modules) have to be very special: they are roots of specific types of algebraic equations, and these equations can be solved by radicals [2, XIX, p. 426]. In fact, these equations are essentially Abelian equations. So for elliptic functions associated to such modules the special teilungsgleichung can be solved by radicals. In particular, this applies to the teilungsgleichung associated to the division of the lemniscate arc. In this case, a and k in (**) are both equal to i, and so the elliptic integral is $\int \frac{dx}{\sqrt{1-x^4}}$. The inverse function, $\phi(x)$, which is an elliptic function, has the property that $\phi(iz) = i\phi(z)$. In Disquisitiones Arithmeticae [7, Chap. VII], Gauss made some cryptic remarks that the integral $\int \frac{dx}{\sqrt{1-x^4}}$, which computes the arc length of the lemniscate, has similar properties as the integral $\int \frac{dx}{\sqrt{1-x^2}}$, which computes the arc length of the circle. In his letter to Holmboe from Paris in December 1826 [3, XX, pp. 261–262], Abel writes:

> *I have discovered that one can divide the arc of the lemniscate by ruler and compass in 2^n+1 equal parts if this is a prime number. The division depend upon an equation of degree $(2^n+1)^2-1$, and I have shown that it can be solved by means of square roots. On the same occasion I have lifted the mystery which rested over Gauss' theory of the division of the circle; I see now clear as daylight how he has been led to it. All I have described about the lemniscate is the fruit of my efforts in the theory of equations. You will not believe how many delightful theorems I have discovered, for example the following: If an equation $P=0$, where the degree is $\mu\nu$, μ and ν being relatively prime, is solvable by radicals then P is decomposable in μ factors, each of degree ν, whose coefficients depend upon one single equation of degree μ, or, reciprocally, in ν factors, each of degree μ, whose coefficients depend upon one single equation of degree ν.*

The last sentence concerns what was later called imprimitivity, which is a key property for equations to be solvable if their degree is not a power of a prime. Galois would later make the same discovery. It is safe to say that the lemniscate problem brought Abel to consider a large class of equations with special properties which interested him greatly. With his characteristic ability to crystallize the essential of every problem, he created the theory of Abelian equations, which we already have encountered.

Let us get back to the special teilungsgleichungen. Abel conjectured that these equations could not be solved by radicals for primes $p>3$ if the module is non-singular. The proof of this was given by Galois in his letter to Chevalier (May 29, 1832) the night before his fatal duel. In fact, the resolvent equation of degree $p+1$ has a *simple* Galois group of order $(p-1)p(p+1)/2$ (after adjoining a specific square root to the ground field)—the group itself being $\mathrm{PSL}(2,\mathbb{Z}_p)$. It follows by Galois criterion that it can not be solved by radicals. As for the modular equation associated to transformations of (prime) degree p of elliptic functions, it leads to essentially the same problem as for the teilungsgleichung.

To put all this in some perspective we remark that the special teilungsgleichungen are closely related to the torsion points of an elliptic curve, where the elliptic curve is associated to an elliptic function via the Weierstrass parametrization. This is again intimately related to the so-called multiplication maps of an elliptic curve to itself. The importance of these maps for the study of the arithmetic of elliptic curves would be difficult to overestimate. For further information about this we refer to [21, Chap. III] and [22, Chap. VI].

Finally, we want to briefly describe an alternative—and more "modern"—way to look at the transformation theory of elliptic functions, a viewpoint that was also initiated by Abel. Instead of considering the differential equation $(*)$, we consider two elliptic functions $\phi(u) = \phi(u \mid \omega_1, \omega_2)$ and $\bar{\phi}(\bar{u}) = \bar{\phi}(\bar{u} \mid \bar{\omega}_1, \bar{\omega}_2)$ with (primitive) periods ω_1, ω_2 and $\bar{\omega}_1, \bar{\omega}_2$, respectively. The transformation theory, broadly speaking, is to investigate under which conditions there exists an algebraic relation between $\phi(u)$ and $\bar{\phi}(\bar{u})$ if $\bar{u} = mu$ for some constant m. Assuming that both ϕ and $\bar{\phi}$ are homogeneous functions in the three variables, one can assume that $m = 1$ (and hence $\bar{u} = u$), simply by writing $\bar{\omega}_i$ instead of $\frac{\bar{\omega}_i}{m}$, $i = 1, 2$. Without going into details, one can reduce the investigation into studying the relation between $\phi(u \mid \omega_1, \omega_2)$ and $\phi(u \mid \omega_1', \omega_2')$, where $\omega_1' = a\omega_1 + b\omega_2$, $\omega_2' = c\omega_1 + d\omega_2$, and $a, b, c, d \in \mathbb{Z}$, $ad - bc = n > 0$. One says that $\phi(u \mid \omega_1', \omega_2')$ arises from $\phi(u \mid \omega_1, \omega_2)$ by a transformation of degree n. All this is in principle treated in Précis (cf. [2, XXVIII, Introduction, Sect. 8]).

10 Posthumous Article

It remains to assess the promise Abel gives in his letter to Crelle to eventually write up his discoveries in the theory of equations; what would the content of such a work be? As we already have mentioned Abel had characterized a particular class of solvable equations, the so-called Abelian equations—being those with commutative Galois groups [2, XXV]. He had shown that a multitude of equations arising from the study of elliptic functions belonged to that class. So it is a safe bet that the memoir he intended to write would attack the general problem of when an algebraic equation can be solved by radicals. In his nachlass was found an unfinished manuscript, which was published in both the 1839 and the 1881 edition of his Oeuvres Complètes under the title *Sur la resolution algébrique des équations* [3, XVIII]. (In the 1881 edition Sylow made detailed annotation.) In the introduction Abel writes unequivocally:

> *Dans ce mémoire je vais traiter le problème de la résolution algébrique des équations, dans toute sa généralité.*
>
> [In this memoir I will treat the problem of the algebraic solution of equations in its full generality.]

We know fairly precisely when Abel wrote this manuscript by noticing where in his mathematical diary it is written. The letter of September 25, 1828, to Crelle tells us that it was written before that date. We know that as late as July 29 he was still working on the second article he sent to Astronomische Nachrichten [2, XX]. So the manuscript in question must in all likelihood have been written between those dates, and hence partly during the sickness period that he mentions in his letter to Crelle, and then being interrupted by his sickness.

This is an important piece of information in interpreting and forming an opinion of Abel's manuscript. For instance, it tells us that the reason he left the manuscript incomplete does not mean that he thought he would not be able to finish it and tie up the loose ends. Of course, this was not the first time Abel had thought about these things, as can be seen from the letters he had sent to Crelle and Holmboe earlier. In a lengthy and comprehensive introduction (in fact, there exists two versions!) which is fully redacted, Abel not only summarizes the content of the manuscript in question, but he also reflects upon mathematical methods and proofs in general. It is a "must" read for anyone interested in Abel's thinking about mathematics. Here is a small excerpt illustrating the flavour of his thinking:

On doit donner au problème une forme telle qu'il soit toujours possible de le résoudre, ce qu'en peut toujours faire d'un problème quelconque.... En présentant un problème de cette manière, l'énoncé même contient le germe de la solution, et montre la route qu'il faut prendre. Ce qui a fait que cette méthode ... a été peu usitée dans les mathématiques c'est l'extrème complication à laquelle elle paraît être assujettie dans la plupart des problèmes, surtout lorsqu'ils ont une certaine généralité; mais dans beaucoup de cas cette complication n'est qu'apparente et s'évanouira dès le premier abord. J'ai traité plusieurs branches de l'analyse de cette manière, et quoique je me sous souvent proposé des problèmes qui ont surpassé mes forces, je suis néanmoins parvenu à une grand nombre de résultats généraux qui jettent un grand jour sur la nature des quantités dont la connaissance est l'objet des mathématiques.

[One shall give the problem such a form that it is always possible to solve it, something that one can always do with any problem. ...In presenting a problem in this manner, the mere wording of it contains the germ to its solution and shows the route one should take. The reason why this method ... has been so little used in mathematics is the extreme complication to which it appears to be subject to in the plurality of problems, especially if these are of a certain general nature; but in many of these cases the complication is only seemingly and vanishes at first sight. I have treated several topics in analysis and algebra in this manner, and although I have often posed myself problems that surpasses my powers, I have nevertheless attained a great number of general results that have shed a broad light on the nature of these quantities, the knowledge of which is the object of mathematics.]

Abel states what is the main object of the memoir he is working on:

Trouver l'expression algébrique la plus générale qui puisse satisfaire à une équation (irréductible) d'un degré donné. On est conduit naturellement à considérer deux cas, selon que le degré de l'équation est un nombre premier ou non. Quoique nous n'ayons pas donné la solution complète de ce problème, néanmois la marche naturelle de la solution a conduit à plusieurs propositions générale, très remarquables en elles-même, et qui ont conduit à la solution du problème dont nous occupons.

[Find the most general algebraic expression that can satisfy an (irreducible) equation of a given degree. One is naturally led to consider two cases, whether the degree of the equation is a prime number or not. Although we have not given the complete solution to this problem, nevertheless the natural path to its solution has pointed the way to many general propositions—very remarkable in their own right—which in turn has led to the solution of the problem with which we are concerned.]

The above statement by Abel is significant in evaluating his approach to the theory of equations compared to that of Galois. In fact, the criterion that Galois gave for the solvability of an equation—to wit, that the Galois group is solvable—would for Abel only have been a stepping stone for a more ambitious goal: to find an explicit form of the solution of a solvable and irreducible equation of a given degree. As an illustrative example consider the formula (†) that he had sent to Crelle. The roots of an (irreducible) quintic over $\mathbb{Q}$ (actually, over any field of characteristic zero) have the form given in (†), where $c, e, h, m, n, K, K', K'', K'''$ are (rational) parameters. Conversely, any expression of this form is a root of a quintic equation. In particular, for any given quintic equation one can choose the parameters so that the formula in (†) gives a root. In *Sur la résolution algébriques des équations* [3, XVIII] (which we will refer to as **(A)** in the sequel), Abel attacks the general problem of finding the form of the roots of a solvable and irreducible equation. He reduces the problem to so-called primitive equations, i.e., those that can not be decomposed by *Gauss method*, as Galois would have formulated it. In today's language we would say that the equation can not be decomposed in a proper normal subfield of the splitting field. The primitive (solvable) equations are necessarily of prime power degree. Abel essentially gives the solution when the degree of the (irreducible) equation is a prime number. Admittedly, there are some loose ends in his proof, but as Sylow convincingly argues in his annotated comments [3, pp. 329–338], these can be fixed by giving an expression that occur in the proof a more careful reduction. While the manuscript is fully redacted in the beginning, it gradually gets more sketchy, giving formulas with little text. This is, as Sylow remarks, a peculiar feature of Abel's way of working, of which there are many examples: when he gets to a point where he sees some shortcomings in his argument, he will subsequently jot down some keywords and formulas, reminding himself when he later comes back to fix up the matter. An especially noteworthy example of this in (A) is on page 241, where one finds the irreducible equation $\phi\rho = 0$ (in Abel's notation) with roots $\rho, \rho_1, \rho_2, \ldots, \rho_{\nu-1}$. Sylow makes a convincing argument (cf. [3, pp. 336–337] and [18, p. 22 (and p. 23)]) that ρ must be interpreted as a primitive element for a special field extension. It is an intriguing question what Abel would have made of this in a final redaction. The similarity with how Galois introduced his group is striking.

The salient feature in (A) is the role played by the Lagrange resolvent

$$(\epsilon, x) = x_0 + \epsilon x_1 + \epsilon^2 x_2 + \cdots + \epsilon^{n-1} x_{n-1}.$$

(Here ϵ is a primitive nth root of unity and $x_0, x_1, x_2, \ldots, x_{n-1}$ are the roots of the polynomial in question.) How the Lagrange resolvent transforms under specific permutations of the roots is the key to the proof. Specifically, Abel shows that if $f(x)$ is a solvable irreducible polynomial of prime degree p with roots $x_0, x_1, x_2, \ldots, x_{p-1}$, then there exists a tower of fields from the ground field F to the splitting field $E = F(x_0, x_1, \ldots, x_{p-1})$ such that

$$F \subseteq \cdots \subseteq K\big(= F(s)\big) \subseteq K\big(s^{1/p}\big) = E$$

where $K = F(s)$ is a *normal* extension of F such that the Galois group $G(K/F)$ of K over F is Abelian (in fact, cyclic). (We have assumed that F contains appropriate roots of unity.) The polynomial $f(x)$ is irreducible over K, and splits in linear factors over $E = K(s^{1/p})$. Hence the problem of finding the form of the roots of $f(x)$ is reduced to the simpler problem: finding the form of the roots of an irreducible *Abelian* (in fact, cyclic) equation of degree a divisor of $p - 1$. This latter problem is not treated in (A). However, there is a page in his mathematical diary that deals with this, and as we already mentioned the solution for $p = 5$ (over $\mathbb{Q}$) was communicated to Crelle, cf. (†). It was Kronecker who finished the edifice of which Abel laid the foundation ([11] and [12]). In Weber's book [26, Vol. 1, Chap. 17], Kronecker's proof is presented in detail. (In the aforementioned 1903-edition of Acta there is an article by A. Wiman: *Über die metacyclischen Gleichungen von Primzahlgrad*, where another proof is given. For a "modern" proof using Galois theory as it is presented today, see [4].) What is noteworthy about (A) is that one can say with some justification that Abel decomposes the Galois group G in normal subgroups such that the consecutive quotients are Abelian groups, without actually "seeing" the group G itself. This correlates with him constructing a tower of *normal* fields over F, ending with the splitting field E:

$$F = L_0 \subseteq L_1 \subseteq L_2 \subseteq \cdots \subseteq L_n = E$$

such that each field is a *primary* extension of the previous one; in other words, $L_{i+1} = L_i(\alpha_1, \alpha_2, \ldots, \alpha_{m_i})$, where each α_k is a root of a pure equation $x^{p_i} - a_k = 0$; $a_k \in L_i$, p_i prime. From his analysis Abel is able to draw the conclusion that an irreducible equation of prime degree is solvable if and only the splitting field is generated by two arbitrary roots, which is Proposition VIII in Galois' memoir [6]. (We refer to [8] for a detailed analysis of (A).)

The following paragraphs at the end of the Introduction in (A) are rather telling:

> *Des théorèmes généraux auxquels on est ainsi parvenu, on déduit ensuite une règle générale pour reconnaître si une équation proposée est résoluble ou non. En effet, on est conduit à ce résultat remarquable, que si une équation irréductible est résoluble algébriquement, on pourra dans tous les cas trouver les racines à l'aide de la méthode de* Lagrange, *proposée pour la résolution des équations; savoir, en suivant la marche de* Lagrange *on doit parvenir à des équations qui aient au moins une racine qui puisse s'exprimer rationnellement par les coefficiens. Il y a plus,* Lagrange *a fait voir qu'on peut ramener la résolution d'une équation du degré ... à celle de ... équations respectivement des degrés ... à l'aide d'une équation du degré .... Nous démontrerons*

que c'est cette équation qui doit nécessairement avoir au moins une racine exprimable rationnellement par ses coefficiens pour que l'équation proposée soit résoluble algébriquement.

Donc, si cette condition n'est pas remplie, c'est une preuve incontestable que l'équation n'est pas résoluble; mais il est à remarquer qu'elle peut être remplie sans que l'équation soit en effet résoluble algébriquement. Pour le reconnaître, il faut encore soumettre les équations auxiliaires au même examen. Cependant dans le cas où le degré de la proposée est un nombre premier, la première condition suffira toujours, comme nous le montrerons. De ce qui précède, il a été facile ensuite de tirer comme corollaire qu'il est impossible de résoudre les équations générales.

[From the general theorems that have been attained, one deduces next a general criterion by which one can recognize whether a given equation is solvable or not. In fact, one is led to a remarkable result, namely if an irreducible equation is algebraically solvable one can in all cases find the roots by means of the method proposed by Lagrange for solving equations; specifically, by following Lagrange's approach one is bound to obtain equations that have at least one root that can be rationally expressed by the coefficients. (Cf. Lagrange: *Traité de la résolution des équations numériques de tous les degrés*, Note XIII (1806) [13, Vol. 8]. (Translator's note).) Furthermore, Lagrange has shown that one can reduce the solution of an equation of degree ... to ... equations, respectively of degrees ... by means of one equation of degree .... We will prove that it is this latter equation that must necessarily have at least one root that can be expressed rationally in terms of the coefficients, in order that the given equation can be solved algebraically.

Hence, if this condition is not satisfied it is an indisputable proof that the equation is not solvable; but we remark that the condition can be satisfied without the equation being algebraically solvable. To decide solvability, one must submit the auxiliary equations to the same test. However, in the case that the degree of the given equation is a prime number, the first condition is also necessary, as we will show. From the preceding it is easy to get as a corollary that it is impossible to solve general equations.]

It is not a far-fetched guess that it was exactly this analysis of solvability of equations by Lagrange that Abel refers to that led both him and Galois initially to discover the criterion for when an irreducible equation of prime degree can be solved by radicals. It is also noteworthy that Abel in the letter to Crelle where he communicated the formula (†) for the solution of the quintic, also writes the following:

Auf diese Weise lässt sich aber die Gleichung $x^5 + ax + b = 0$ nicht auflösen, so lange a and b beliebige Grössen sind.

[However, an equation of the form $x^5 + ax + b = 0$ can not be solved in this manner (i.e., by radicals) if a and b are arbitrary.]

Let us remark that by a so-called Tschirnhausen transformation

$$x = d_0 + d_1 y + d_2 y^2 + d_3 y^3 + d_4 y^4$$

one can transfer the quintic equation

$$y^5 + c_4 y^4 + c_3 y^3 + c_2 y^2 + c_1 y + c_0 = 0$$

to the form $x^5 + ax + b = 0$, where a, b and the d_i's can be expressed by adjoining successively square and cube roots starting from the ground field F (containing the c_i's). Hence if $x^5 + ax + b = 0$ can be solved by radicals then the original quintic can also be solved by radicals. Even though Abel did not pursue this, he knew full well that $x^5 + ax + b = 0$ can be solved by radicals if and only if the resolvent sextic equation that Lagrange had arrived at in his study of the general quintic equation (adapted in an obvious way to a numerical equation) has a root in the ground field. (Cf. Abel's remark just above. Cf. also [26, Vol. 1, §188, VI, p. 667].) For the equation $x^5 + ax + b = 0$ the resolvent sextic becomes

$$(z-a)^4(z^2 - 6az + 25a^2) = 5^5 b^4 z$$

This equation has a rational root (i.e., lying in F) if and only if a and b are of the form

$$a = \frac{5\mu^4(4\lambda+3)}{\lambda^2+1}, \qquad b = \frac{4\mu^5(2\lambda+1)(4\lambda+3)}{\lambda^2+1}$$

for some $\mu, \lambda \in F$, and so the equation $x^5 + ax + b = 0$ is solvable if and only if the coefficients are like this (cf. [17]); for the "only if" part we need that the equation is irreducible.) For the special case that $F = \mathbb{Q}$, we get by setting $\mu = -5$, $\lambda = -24/7$, the irreducible equation

$$x^5 - 2625x - 61500 = 0$$

whose solution was known to Euler:

$$x = \sqrt[5]{75(5+4\sqrt{10})} + \sqrt[5]{225(35+11\sqrt{10})} + \sqrt[5]{75(5-4\sqrt{10})} + \sqrt[5]{225(35-11\sqrt{10})}.$$

11 Kronecker's Reaction

Kronecker was one of the first, if not the first, mathematician that was intimately familiar with both Galois' and Abel's work on the theory of equations. It is of some interest to hear his thoughts on this new theory. (We refer to the interesting article: *From Abel to Kronecker: Episodes from 19th Century Algebra* by B. Petri and N. Schappacher in [14] for another—partially overlapping our—perspective on the Abel/Kronecker connection.) One should keep in mind that Abel's manuscript (A) first became known to the mathematical community in 1839, when the first edition of Abel's Oeuvres Complètes was published [1]. As for Galois' manuscripts, they were rescued from oblivion by Liouville, who published them in 1846 in the journal he had founded, *Journal de Mathématiques Pures et Appliquées*. In his article *Über die algebraisch auflösbaren Gleichungen* [11] from 1853, Kronecker writes:

> *Die bisherigen Untersuchungen über die Auflösbarkeit von Gleichungen, deren Grad eine Primzahl ist – namentlich die* Abel*schen und* Galois*schen, welche die Grundlage aller weiteren Forschungen in diesem Gebiete bilden – haben im Wesentlichen als Resultat zwei Kriterien ergeben, vermittelst deren man beurtheilen könnte, ob eine gegebene Gleichung auflösbar sei oder nicht. Indessen gaben diese Kriterien über die Natur der auflösbaren Gleichungen* selbst *eigentlich nicht das geringste Licht. Ja, man konnte eigentlich gar nicht wissen, ob (außer den von* Abel *im IV. Bande des Crelleschen Journals behandelten und den einfachsten mit den binomischen Gleichungen zusammenhängenden) überhaupt noch irgend welche Gleichungen existiren, welche die gegebenen Auflösbarkeits-Bedingungen erfüllen. Noch weniger konnte man solche Gleichungen* bilden, *und man ist auch durch sonstige mathematische Untersuchungen nirgends auf solche Gleichungen geführt worden. Dazu kommt noch, daß jene beiden erwähnten und wohl allgemein bekannten, von* Abel *und* Galois *gegebenen Eigenschaften der auflösbaren Gleichungen zufälliger Weise solche waren, die die wahre Natur dieser Gleichungen eher zu verdecken als aufzuklären geeignet sein dürften, wie ich das namentlich von dem einen jener beiden Kriterien späterhin zeigen werde. Und so blieben die auflösbaren Gleichungen* selbst *bisher in einem gewissen Dunkel, welches nur durch die übrigens, wie es scheint, wenig beachtete und ganz spezielle Notiz* Abel*s über die Wurzeln ganzzahliger Gleichungen fünften Grades ein wenig erhellt wurde. ...*
>
> Abel *hat in seiner fragmentarischen Abhandlung über die algebraische Auflösung der Gleichungen (No. XV. des zweiten Bandes der gesammelten Werke) unter andern Problemen wörtlich folgendes aufgestellt: "Den allgemeinsten algebraischen Ausdruck zu finden, welcher einer Gleichung von einem gegebenen Grade genügen könne." Fügt man diesem Probleme dasjenige hinzu, was erforderlich ist, um es zu einem bestimmten zu machen, so enthält es in der That alle Probleme in sich, die man in Bezug auf die Auflösbarkeit der Gleichungen stellen kann, und ist namentlich die wichtigste Verallgemeinerung des (als in gewissem Sinne zu speziell) unlösbaren Problems "die Wurzel einer Gleichung irgend eines Grades als algebraische Function ihrer Coëfficienten auszudrücken." Es ist nun aber, wie gesagt, bei obigem Probleme noch erforderlich, den Zusammenhang zwischen dem gesuchten algebraischen Ausdruck und den Coëfficienten der Gleichung zu bestimmen; deshalb ist die Aufgabe vielmehr dahin stellen:*
>
> *"Die allgemeinste algebraische Function irgend welcher Größen A, B, C, ... zu finden, welche einer Gleichung von einem gegebenen Grade genügt, deren Coëfficienten* rationale *Functionen jener Größen sind."*

[The study so far about solvability of equations of prime degree—above all due to Abel and Galois, which is the foundation of all further research in this area—has essentially resulted in two criteria by which one can decide whether a given equation is solvable or not. Nevertheless, these criteria throw absolutely no light on the equations themselves. Indeed, one knows in reality nothing at all which equations exist that satisfy the given solvability criteria (except for the special class of equations that Abel treated in the 4th volume of Crelle's journal *(Cf. [2, XXV])*, and which are closely related to the binomial equations). Even less can one construct such equations, and one is also by other mathematical considerations never led to such equations. In addition, both the aforementioned properties of solvable equations that Abel and Galois gave—which are probably commonly known—happen to be such that they rather conceal the true nature of these equations than throw light upon them, something which I will show later for one of these criteria in particular. Hence the solvable equations themselves remain so far in darkness, so to say, except for the otherwise scarcely noticed and quite special note by Abel about the roots of quintic equations with integer coefficients *(Cf. formula* (†) *above)*, which sheds a little light. ...

In his fragmentary memoir on algebraically solvable equations (No. XV. in the second volume of his Oeuvres Complètes *(Cf. [3, XVIII])*) Abel states the following problem: "Find the most general algebraic expression which can be root of an equation of a given degree". If one adds to this problem a necessary precision, then it comprises in fact all problems that one can state with respect to solvability of equations, and it is above all the most important generalization of the (in a certain sense too special) unsolvable problem: "express the root of an equation of any degree as an algebraic function of its coefficients." As we said, it is necessary with regard to the above problem to determine the connection between the sought after algebraic expression and the coefficients of the equation; therefore the problem should rather be stated as:

"Find the most general algebraic function of any quantities A, B, C, ... which is a root of an equation of a given degree, where the coefficients are *rational* functions of these quantities."]

In a letter to Dirichlet (dated January 31, 1853), commenting on (A), Kronecker is even more direct when he writes:

> *Überhaupt sieht man daraus die wahre Beschaffenheit der auflösbaren Gleichungen, die man aus den Galoisschen Untersuchungen durchaus nicht erkennen kann: denn Galois nimmt sich nur die eine Aufgabe vor "die Bedingung der Auflösbarkeit" zu finden, während Abel auch die andere berücksichtigt "alle auflösbaren Gleichungen zu finden."*
>
> [From this one sees the true nature of the solvable equations, which it is impossible to glean from Galois's investigations. For Galois only takes up the task to find "the condition of solvability", whereas Abel also takes into account the other one, "to find all solvable equations."]

There are several comments to be made to what Kronecker writes in [11] which are highly relevant for our topic. Firstly, Kronecker addresses straightforwardly the natural question one would ask if one is confronted with the problem of solvability of algebraic equations. In fact, this is exactly what Lacroix and Poisson wrote in their referee report on Galois' memoir: *The condition for solvability, if it exists, should be an external character which one might verify by inspection of the coefficients of a given equation, or at the worst, by solving other equations of degree lower than the one given.* This Abel realized was the wrong question to ask (cf. his comments in the Introduction to (A)), and Kronecker concurs. The reason is that it is "unsolvable" (note the double entendre). The right question to ask, which turns out to be "solvable", is the one Abel raises and which Kronecker makes precise: "*Find the most general algebraic function of any quantities $A, B, C, \ldots$ which is a root of an equation of a given degree, where the coefficients are* rational *functions of these quantities.*" The problem is reduced to so-called primitive equations and these are necessarily of prime power degrees $n = p^m$, p prime. Kronecker writes in [11]:

> *In diesen Fällen aber bietet das Problem, mit Ausnahme einiger bloßen Complicationen, auch keine größere Schwierigkeit dar, als wenn $n = p$ eine primzahl ist.*
>
> [In these cases, however—apart from a few complications—the problem does not present greater difficulty as when $n = p$ is a prime number.]

This is an understatement, though, as can be seen from Galois' fragment of a second memoir: *Des équations primitives qui sont solubles par radicaux* (see the reference to his collected works in [6]), and also by looking at Jordan's treatment of this in [10]. The complications are indeed considerable.

As an example, illustrating the general case, we mention that Abel's formula (†) is not "a solution of solvable quintics" in the sense that the quadratic formula is a solution of quadratics, because it does not give an algorithm for going from a given solvable quintic to an expression of its roots in terms of radicals. Instead, as we said earlier, given any solvable quintic, then it is possible to choose values for the parameters that appear in the expression so that it is a root of the given quintic. (For more specifics on this, cf. [4, Appendix 3].)

As we already remarked, Abel reduces the problem of finding the algebraic expression for the roots of a solvable and irreducible equation of prime degree to finding the form of the roots of an Abelian (actually, cyclic) equation. This is the problem Kronecker attacks in [11] and [12]. At the end of the article [11] he announces the following remarkable result:

> *Die Wurzel* jeder *Abelschen Gleichung mit ganzzahligen Coëfficienten kann als rationale Function von Wurzeln der Einheit dargestellt werden.*
>
> [The root of every Abelian equation with integer coefficients can be expressed as a rational function of roots of unity.]

This is the famous Kronecker–Weber theorem that was proved by Weber in 1886. (There is a small gap in his proof that was finally settled by Hilbert in 1896.) Another way to state the theorem is to say that the maximal Abelian extension $\mathbb{Q}^{ab}$ of $\mathbb{Q}$ is generated by roots of unity. (Remember that the roots of unity are of the form $e^{2\pi ir}$, where r is a rational number.) But Kronecker does not stop here. Being aware that Abel had studied Abelian extensions of the field $\mathbb{Q}(i)$ in connection with dividing the arc of the lemniscate, Kronecker in [11] raises the conjecture: *Every Abelian extension of* $\mathbb{Q}(i)$ *is contained in a field obtained from* $\mathbb{Q}(i)$ *by adjoining a certain value of the lemniscate elliptic function* $\phi = sl$. Specifically, he expected that every finite Abelian extension of $\mathbb{Q}(i)$ lies in some field $\mathbb{Q}(i, sl(\frac{\omega}{m}))$, where m is an integer and $\omega \approx 2.622$ is the lemniscate analogue of 2π. This is of course similar to the Kronecker–Weber theorem, with $sl(\omega/m)$ analogous to $e^{2\pi i/m}$, and it was rigorously proved by Takagi, who was a student of Hilbert, in his thesis from 1903. Extending Abel's work, Kronecker was able to generate Abelian extensions of any imaginary quadratic field using special values of elliptic and modular functions. In a letter to Dedekind dated March 15, 1880, he writes:

> *Es handelt sich um meinen liebsten Jugendtraum, nämlich um den Nachweis, dass die Abel'schen Gleichungen mit Quadratwarzeln rationaler Zahlen durch die Transformations – Gleichungen elliptischer Funktionen mit singulären Moduln grade so erschöpt werden, wie die ganzzahligen Abel'schen Gleichungen durch die Kreistheilungsgleichungen.*
>
> [It concerns the dearest dream of my youth, to wit, the proof that the Abelian equations with [*coefficients*] square roots of [*negative*] rational numbers are exhausted by the transformation equations of elliptic functions with singular moduli, exactly in the same way as the Abelian equations with integer [*coefficients*] are by the cyclotomic equations.]

To make a long story short, Kronecker's Jugendtraum would inspire Hilbert to formulate his 12th problem, one of the famous 23 problems he presented at the International Congress of Mathematicians (ICM) in Paris in 1900. Kronecker's Jugendtraum is intimately related to complex multiplication, and a fortiori to the elliptic functions that admit this. Hilbert held his 12th problem in high esteem as is documented by his comment about it at the 1932 ICM congress in Zürich: *The theory of complex multiplication* (of elliptic modular functions) *which forms a powerful link between number theory and analysis, is not only the most beautiful part of mathematics but also of science.*

This mathematical field, part of algebraic number theory, became known as Class Field Theory, and has been a very important research area in the 20th century. Broadly speaking, class field theory is the study of Abelian extensions of number fields K (i.e., K is a finite extension of the rationals $\mathbb{Q}$). The first two class field theories, the "classical" ones, were the very explicit cyclotomic and complex multiplication class field theories that we already have encountered. As we pointed out they use additional structures: in the case of the field of rational numbers they use roots of unity; in the case of imaginary quadratic extensions of the rational numbers they use elliptic curves (respectively, elliptic functions) with complex multiplication and their points of finite order (respectively, division of the periods). By 1930 the classical Kronecker Jugendtraum conjectures were proved. (We refer to a fascinating historical survey of this, with the tantalizing title: *On the History of Hilbert's Twelfth Problem. A Comedy of Errors* [20].)

Summing up the development we have just sketched it is safe to say that Class Field Theory has, at least partly, its genesis in Abel's work on the theory of equations. He put Abelian extensions at the forefront both in connection with the (general) teilungsgleichungen, and also with the special teilungsgleichungen in the complex multiplication case, thereby connecting it with torsion points on the associated elliptic curve (via the Weierstrass parametrization). Furthermore, he reduced the solution of solvable prime degree equations to Abelian (in fact, cyclic) equations. It certainly inspired Kronecker to his Jugendtraum conjecture. This is an example—of which there are many in mathematical history—of a problem, in this case solvability of equations, leading to theories that are much more important than the original problem; Fermat's theorem, concerning a special diophantine equation, is another example of this phenomenon. (However, truth be said, the problem of solvability of equations by radicals has enjoyed by itself—and still does—an incomparably higher fascination with mathematicians than the diophantine equations of Fermat, which are more of a curiosity!)

12 Galois' Legacy

Arguably the most important theory that came out of the quest for a solvability criterion for equations was Galois Theory—including the emergence of the group concept. It has all the hallmarks of a great mathematical theory:

- Solved a very old problem—the most important—about equations.
- Extremely comprehensive theory—goes far beyond the original question.
- Based on only a few principles of great elegance and simplicity which are formulated within a new framework with new concepts which demonstrate the greatest originality.
- The new viewpoints and concepts, especially the concept of a group, opened new paths and had a lasting influence on the whole of mathematics.

Galois deserves the highest accolades for his insistence on the conceptual character of mathematics, and his instinct for seeing problems according to the deep connections of their structure rather than their superficial appearance. In this respect he has much in common with Abel. While Galois' presentation of his theory seems so concise and clear to modern readers, this was not at all the case with many of his contemporaries. They found his theory almost incomprehensible, and it was only gradually that it gained the status it fully deserved. Galois was simply ahead of his times.

It was first with Camille Jordan's great treatise *Traité des substitutions et des équations algébriques* [10], which appeared in 1870, that Galois Theory became part of the mathematical canon, so to say. Jordan toned down the solvability aspect and stressed more the group theoretic aspect. His Traité can rightfully be considered to be the first text on group theory. (Incidentally, it is a little ironic that right after Traité was published, Sylow discovered and proved the three theorems about finite groups named after him; these fundamental theorems are therefore not in Traité.) Gradually Galois Theory was being severed, though not completely, from its connections with its past. No longer was it viewed as having any necessary connection with solvability of equations. Galois Theory became a theory about the structure of fields and their automorphisms.

13 Twists of Fate—Poetic Justice

If Abel had stayed healthy long enough to be able to complete and publish the manuscript (A), this would inevitably have had repercussions vis-à-vis the status of Galois in mathematics, since the latter should die so young. Abel would then almost certainly be given the credit for having solved the outstanding problem in the theory of equations: find a criterion for when a (numerical) algebraic equation can be solved by radicals. This notwithstanding that he may not have identified the group of the equation, as Galois did. But we think this should be seen in a different light: there is a poetic justice, if we may use the expression, that Abel did not publish (A), thereby reserving the glory for Galois. After all, Abel unwittingly had a similar encounter with fate that benefited him. In fact, in a letter sent from Paris to Holmboe at the beginning of December 1826 he writes that he plans to make a detour on his trip from Paris to Berlin to visit Göttingen and Gauss. However, he had changed his plans a few weeks later when he left Paris (the city he had wistfully call "the focus of all my mathematical desires", but which had been such a disappointment for him). Ostensibly he was almost broke and could not afford to visit Göttingen, but had to go straight to Berlin where he hoped money from Norway had arrived. (For more details, cf. [23].) So by a twist of fate Abel never met Gauss. Under ordinary circumstances it would have been marvellous if the two of them had met, but since Abel's life was so short a meeting with Gauss would probably have had repercussions, similarly as with Galois, with regard to his status in mathematics. In an imagined meeting between the two Abel would surely have told Gauss about his discoveries concerning elliptic functions, which he thought he alone was in possession of, only to be told by Gauss that he had made more or less the same discoveries much earlier. When Abel returned to Oslo in May 1827 after his trip abroad—a trip that had lasted altogether one year and eight months—he wrote up his discoveries in a long memoir that was published in two instalments in Crelle's journal.

After the first part of the memoir, *Recherches sur les fonctions elliptiques* [2, XVI], appeared in Crelle's journal on September 20, 1827 (the second part appeared May 26, 1828), Bessel, who was a close friend and confidant of Gauss, wrote a letter (dated November 30, 1827) to Gauss encouraging him to publish his discoveries on elliptic functions. In a letter dated March 30, 1828 (cf. Gauss, *Werke*, Band X, pp. 248–249) Gauss writes back:

> *Zur Ausarbeitung der seit vielen Jahren (1798) angestellten Untersuchungen über die transcendenten Functionen werde ich vorerst wol noch nicht kommen können, da erst noch mit manchen andern Dingen aufgeräumt werden muss. Hr. Abel ist mir, wie ich sehe, jetzt zuvorgekommen und überhebt*

mich in Beziehung auf etwa 1/3 dieser Sachen der Mühe, zumahl da er alle Entwickelungen mit Eleganz und Concision gemacht hat. Er hat gerade denselben Weg genommen, welchen ich 1798 einschlug, daher die grosse Übereinstimmung der Resultate nicht zu verwundern ist. Zu meiner Bewunderung erstreckt sich dies sogar auf die Form und zum Theil auf die Wahl der Zeichen, so dass manche seiner Formeln wie eine reine Abschrift der meinigen erscheinen. Jeder Misdeutung zuvorzukommen bemerke ich jedoch, dass ich mich nicht erinnere, von diesen Sachen irgend jemanden etwas mitgetheilt zu haben.

[I shall most likely not soon prepare my investigations on the transcendental functions which I have had for many years—since 1798—because I have many other matters which must be cleared up. Herr Abel has now, as I see, anticipated me and relieved me of the burden in regard to about one third of these matters, particularly since he has executed all developments with great stringency and elegance. He has followed exactly the same road which I traveled in 1798; it is no wonder that our results are so similar. To my surprise this extended also to the form and even, in part, to the choice of notations, so several of his formulas appeared as if they were copied from mine. But to avoid every misunderstanding, I must observe that I cannot recall ever having communicated any of these investigations to others.]

14 The Abel–Galois Linkage

As we have already mentioned both Abel and Galois had read and were familiar with the works of Lagrange and Gauss on the theory of equations. In Gauss' case this meant his investigation of cyclotomic equations [7, Chap. VII] and in Lagrange's case it meant his analysis of the solvability of the general equation, including his use of the resolvent named after him. The interesting question is what Galois learned from Abel? One word about access first: Abel wrote all his manuscripts in French except the very first ones which were written in Norwegian, but these were of minor importance. French was by far his best foreign language. It was only in more private letters to Crelle he would sometimes write in German, like the letter which is the background for this article, and also the note above containing the formula (†). However, in the first issue (in 1826) of Crelle's journal all the six papers by Abel were translated by Crelle into German. (One should keep in mind that Crelle's plan was to create a German language journal in mathematics.) In the ensuing years Crelle did not translate Abel's memoirs on elliptic functions or equations, presumably because of all the work it required, so these appeared in French. Furthermore, concerning the paper in the 1826 issue of Crelle's journal on the unsolvability of the general equation (of degree higher than four), Abel wrote a rather detailed survey article about this particular paper in French during his sojourn in Paris, and it was published in Férussac's Bulletin des Sciences in the autumn of 1826. We know that Galois, besides publishing papers himself in Férrusac's Bulletin (see above), did consult the current publications received in the office of that journal (including those from Crelle's journal). He enjoyed the support of the editor (and his friend) Charles Sturm. We know that Galois kept abreast with Abel's (and to a certain extent with Jacobi's) papers on the theory of equations and elliptic functions. For an excellent survey on this see René Taton's paper: *Évariste Galois and his contemporaries* [24]. Taton states unequivocally that it was the publications and the posthumous works (excepting (A), of course, which appeared in 1839) of Abel that influenced Galois the most and aroused in him the most lively and passionate reactions. Galois asserted, for example, that Abel's proof of the impossibility of solving the general equation of degree $n \geq 5$ depended on "des raisonnement relatifs au degré des équations auxiliaires" ["arguments concerning the degree of auxiliary equations"], and had no connection with his own theory. Referring to the memoir by Abel: *Sur une classe particulière d'équations résolvables algébriquement* [2, XXV], he observed that Abel had "rien laisse" ["not left anything"] relevant for the general solvability problem that he, Galois, treated. We have already encountered (see above) another outburst of Galois about his relation to Abel. What should one make of all this, with its passionate style? To make any judgement on this one should keep in mind that Galois had presented two articles on the theory of equations to the Académie des Sciences on May 25th and June 1st, 1829, by a very qualified judge, Cauchy. These articles are lost. Apparently on the advice of Cauchy he prepared a new memoir, *Sur les conditions pour qu'une équation soit soluble par radicaux*, which he submitted in February 1830 to be considered for the Grand Prix de Mathématiques of the Académie. At the end of June 1830 Galois learned that the prize had been awarded to Abel (posthumously) and Jacobi for their work on elliptic functions. He was disappointed and angry that the Académie preferred their work to his, and to top it off his manuscript was lost. Also, the shadow of Abel kept haunting him: after he submitted to the Académie his great memoir [6] on January 16, 1831, he did not anticipate being once more in competition with Abel, dead almost two years earlier. All this, together with his deep personal problems and disappointments—not to talk about his political ones, resulting in imprisonment—left an embittered young man who was fully aware of his own genius, but not getting the recognition he craved. With this as background it is not difficult to understand Galois' passionate outbursts. We think the careful and measured words of Sylow in [18, p. 24 (and p. 26)], trying to sum up in a few words the impact of Abel's work on Galois, are worth quoting (in its French translation of the original Norwegian):

> *Abel a donc certainement préparé d'une manière trés effective les découvertes de Galois, mais il n'y a rien dans ses mémoires ni dans ses papiers qui donne aucune indication certaine qu'il ait vu la théorie des équations d'un point de vue aussi général que Galois.*

> [Abel has therefore certainly effectively prepared the ground for Galois' discoveries, but there is nothing in his memoirs or his *Nachlass* that give any secure information that he has seen the theory of equations from such a general viewpoint as Galois.]

Just a few years separated Abel and Galois. The theory of equations made tremendous strides forward between 1824 (the year Abel proved the insolvability of the general equation of degree $n \geq 5$) and 1832, and came to an abrupt end with the death of the two protagonists, Abel and Galois. If they had a normal lifespan their contemporaries would have witnessed a competition that would have pushed the frontier of the theory of equations (read Galois Theory), and algebra in general, vastly forward.

The similarities between their two lives are striking in so many ways, even though they as personalities were entirely different. Here is a list:

- Both Abel and Galois were first and foremost algebraists.
- Their favorite topic was the theory of equations. (As a curious aside we mention that both initially thought they had solved the general quintic by radicals!)
- Their hopes for recognition and fame were dashed by l'Académie des Sciences; Abel because of neglect; Galois because of incomprehension.
- Both lost their fathers in tragic circumstances when they were at the tender age of 18. While Galois' father committed suicide, Abel's father made a scandal and was viciously ridiculed while he was an elected representative of the Storting (Parliament). Deeply depressed this hastened his death two years later, affecting the young Abel profoundly.
- They both had inspiring teachers in high-school that kindled their passion for mathematics when they were aged around 16 years; in Abel's case this was B.M. Holmboe; in Galois' case this was M. Vernier and, especially, L.P.E. Richard.
- Both learned mathematics at a deeper level by reading and studying the masters, especially Euler, Lagrange, Gauss, and to a lesser degree Legendre.

What can then be more fitting than end this article with a citation of something Abel wrote (in French) in the margin of the mathematical diary that he kept during his sojourn in Paris:

> *Au reste il me paraît que si l'on veut faire des progrès dans les mathématiques il faut étudier les maîtres et non pas les écoliers.*
>
> [It appears to me that if one wants to make progress in mathematics, one should study the masters, not the pupils.]

References

1. Abel, N.H.: Oeuvres Complètes (1839). Publiée par B.M Holmboe, Christiania
2. Abel, N.H.: Oeuvres Complètes. Nouvelle Édition, Tome 1 (1881). Publiée par L. Sylow et S. Lie, Christiania
3. Abel, N.H.: Oeuvres Complètes. Nouvelle Édition, Tome 2 (1881). Publiée par L. Sylow et S. Lie, Christiania
4. Edwards, H.M.: Roots of solvable polynomials of prime degree, preprint (2012), 12 p. (To appear in Exp. Math.) doi:10.1016/j.exmath.2013.09.005
5. Fricke, R.: Elliptische Funktionen. In: Enzyklopädie der Mathematischen Wissenschaften, IIB3. Analysis, pp. 177–348. Teubner, Leipzig (1901–1921)

6. Galois, É.: Sur les conditions de résolubilité des équations par radicaux, Oeuvres Mathématiques d'Évariste Galois (Préface par J. Liouville). J. Math. Pures Appl. **11**, 381–444 (1846) (Éditions Jacques Gabay (1989))
7. Gauss, C.F.: Disquisitiones Arithmeticae. Yale Univ. Press, Leipzig (1801), (English translation: Disquisitiones Arithmeticae (A.A. Clarke, translator) (1966))
8. Gårding, L., Skau, C.: Niels Henrik Abel and solvable equations. Arch. Hist. Exact Sci. **48**, 81–103 (1994)
9. Hölder, O.: Galois'che Theorie mit Anwendungen. In: Enzyklopädie der Mathematischen Wissenschaften, IB 3c,d. Arithmetik und Algebra, pp. 480–520. Teubner, Leipzig (1898–1904)
10. Jordan, C.: Traité des substitutions et des équations algébriques. Gauthier-Villars, Paris (1870)
11. Kronecker, L.: Über die algebraisch auflösbaren Gleichungen, I. Abhandlung, Monatsberichte Kgl. Preuss. Akad. Wiss. Berlin **IV**, 365–374 (1853), refers to Vol. 4 of Kronecker's Collected Works, 3–11
12. Kronecker, L.: Über die algebraisch auflösbaren Gleichungen, II. Abhandlung, Monatsberichte Kgl. Preuss. Akad. Wiss. Berlin **IV**, 203–215 (1856), refers to Vol. 4 of Kronecker's Collected Works, 27–37
13. Oeuvres de Lagrange, publiées par J.-A. Serret, Gauthier-Villars, Paris (1869)
14. Laudal, O.A., Piene, R. (eds.): The Legacy of Niels Henrik Abel, the Abel Bicentennial, Oslo, 2002. Springer, Berlin (2004)
15. Neumann, P.M.: The mathematical writings of Évariste Galois. Heritage of European Mathematics. European Math. Soc., Zürich (2011)
16. Ore, Ø.: Niels Henrik Abel. Mathematician Extraordinary. Chelsea, New York (1957)
17. Runge, C.: Über die auflösbaren Gleichungen von der Form $x^5 + ux + v = 0$. Acta Math. **7**, 173–186 (1885)
18. Sylow, L.: Abels studier og hans oppdagelser. Festskrift ved hundreaarsjubileet for Niels Henrik Abels fødsel (1902), 56 p. Kristiania (French translation: Les études d'Abel et ses découvertes)
19. Sylow, L.: Om Abels arbeider og planer i hans sidste tid belyst ved dokumenter, som er fremkomne efter den anden udgave av hans verker, 3rd edn. Scandinavian Mathematical Congress. Kristiania (1913). English translation in "I sporet etter Abel" ("Tracing Abel"), Scatec, 118–127 (2009)
20. Schappacher, N.: On the history of Hilbert's twelfth problem. A comedy of errors. In: Matériaux pour l'histoire des mathématiques au XXe siècle, Actes du colloque á la mémoire de Jean Dieudonné, Nice, 1996. Séminaires et Congrès, Société Mathématique de France, vol. 3, pp. 243–273 (1998)
21. Silverman, J.H.: The Arithmetic of Elliptic Curves. Graduate Texts in Mathematics, vol. 106. Springer, New York (1986)
22. Silverman, J.H., Tate, J.: Rational Points on Elliptic Curves. Undergraduate Texts in Mathematics. Springer, New York (1992)
23. Stubhaug, A.: Et foranskutt lyn. Niels Henrik Abel og hans tid. Aschehoug, Oslo (1996) (English translation: Called Too Soon by Flames Afar. Niels Henrik Abel and his Times, Springer, Berlin (2000))
24. Taton, R.: Évariste Galois and his contemporaries. Bull. Lond. Math. Soc. **15**, 107–118 (1983)
25. van der Waerden, B.L.: Die Galois-Theorie von Heinrich Weber bis Emil Artin. Arch. Hist. Exact Sci. **9**, 240–248 (1972)
26. Weber, H.: Lehrbuch der Algebra. Vieweg und Sohn, Braunschweig (1895) (Reprint of 2nd edition, AMS Chelsea Publishing, New York)

The Abel Committee

2008

Kristian Seip (Norwegian University of Science and Technology, Norway), chair
Hans Föllmer (Humboldt University, Germany)
Sir John Kingman (University of Bristol, UK)
Dusa McDuff (Columbia University, USA)
Efim Zelmanov (University of California at San Diego, USA)

2009

Kristian Seip (Norwegian University of Science and Technology, Norway), chair
Sir John Kingman (University of Bristol, UK)
Sergey Novikov (University of Maryland, USA)
Neil Trudinger (Australian National University, Australia)
Efim Zelmanov (University of California at San Diego, USA)

2010

Kristian Seip (Norwegian University of Science and Technology, Norway), chair
Björn Engquist (University of Texas at Austin, USA, and Royal Institute of Technology, Sweden)
Hendrik W. Lenstra (University of Leiden, The Netherlands)
Sergey Novikov (University of Maryland, USA)
Neil Trudinger (Australian National University, Australia)

2011

Ragni Piene (University of Oslo, Norway), chair
David Donoho (Stanford University, USA)
Björn Engquist (University of Texas at Austin, USA, and Royal Institute of Technology, Sweden)
Hendrik W. Lenstra (University of Leiden, The Netherlands)
M.S. Raghunathan (Tata Institute of Fundamental Research, India)

H. Holden, R. Piene (eds.), *The Abel Prize 2008-2012*,
DOI 10.1007/978-3-642-39449-2, © Springer-Verlag Berlin Heidelberg 2014

2012

Ragni Piene (University of Oslo, Norway), chair
Noga Alon (Tel Aviv University, Israel)
David Donoho (Stanford University, USA)
M.S. Raghunathan (Tata Institute of Fundamental Research, India)
Terence Tao (University of California at Los Angeles, USA)

The Niels Henrik Abel Board

2008

Ragnar Winther (chair)
Kari Gjetrang
Arne Bang Huseby
Idun Reiten
Leiv Storesletten
Reidun Sirevåg (observer)

2009

Ragnar Winther (chair)
Kari Gjetrang
Arne Bang Huseby
Idun Reiten
Leiv Storesletten
Reidun Sirevåg (observer)

2010

Ragnar Winther (chair)
Kari Gjetrang
Arne Bang Huseby
Idun Reiten
Leiv Storesletten
Øivind Andersen (observer)

H. Holden, R. Piene (eds.), *The Abel Prize 2008-2012*,
DOI 10.1007/978-3-642-39449-2,

2011

Helge Holden (chair)
Anne Borg
Kari Gjetrang
Arne Bang Huseby
Hans Munthe-Kaas
Øivind Andersen (observer)

2012

Helge Holden (chair)
Anne Borg
Kari Gjetrang
Arne Bang Huseby
Hans Munthe-Kaas
Øivind Andersen (observer)

The Abel Lectures 2003–2012[1]

2003

J.-P. Serre (Collège de France): *Finite subgroups of Lie groups*
T.A. Springer (University of Utrecht): *The compactification of a semi-simple group*
P. Sarnak (Princeton University): *L-functions and equidistributions*
B. Mazur (Harvard University): *Spectra and L-functions*

2004

Sir M.F. Atiyah (Edinburgh University): *Index theory in mathematics: a historical survey*
I.M. Singer (Massachusetts Institute of Technology): *Index theory in quantum physics*
J.-M. Bismut (Université Paris-Sud): *The Atiyah–Singer index theorem and the heat equation*
E. Witten (Institute for Advanced Study): *Some mathematical physics related to the work of Atiyah and Singer*

2005

P.D. Lax (New York University): *Abstract Phragmen–Lindelöf theorem & Saint Venant's principle*
S. Noelle (Rheinisch-Westfälische Technische Hochschule Aachen University): *Systems of conservation laws*
P. Sarnak (Princeton University): *Hyperbolic equations and spectral geometry*
S. Vanakides (Duke University): *Rigorous semiclassical asymptotics for integrable systems: The KdV and focusing NLS cases*

[1]Some of the lectures have been recorded, and they can be streamed from the Abel Prize web site www.abelprize.no or the Springer web site http://www.springerimages.com/videos/978-3-642-39449-2.

H. Holden, R. Piene (eds.), *The Abel Prize 2008-2012*,
DOI 10.1007/978-3-642-39449-2, © Springer-Verlag Berlin Heidelberg 2014

2006

L. Carleson (Royal Institute of Technology): *A Scandinavian chapter in analysis*
L.-S. Young (New York University): *A mathematical theory of strange attractors*
O. Schramm (Microsoft Research): *Conformally invariant random processes*
S.-Y.A. Chang (Princeton University): *Conformal invariants and differential equations*

2007

S.R.S. Varadhan (New York University): *A short history of large deviations*
G. Papanicolaou (Stanford University): *Stochastic analysis in finance*
O. Zeitouni (University of Minnesota): *Large deviations at work*
T. Lyons (University of Oxford): *Modelling diffusive systems*

2008

J.G. Thompson (University of Florida): *Dirichlet series and SL(2, Z)*
J. Tits (Collège de France): *Algebraic simple groups and buildings*
M. Broue (Université Paris Diderot): *Building cathedrals and breaking down reinforced concrete walls*
A. Lubotsky (Hebrew University): *Simple groups, buildings and applications*

2009

M.L. Gromov (Institut des Hautes Études Scientifiques and New York University): *Abel Lecture*
J. Cheeger (New York University): *How does he do it?*
M.R. Bridson (University of Oxford): *Geometry everywhere: Fiat lux!*
G. Sapiro (University of Minnesota): *One small step for Gromov, one giant leap for shape analysis: A window into the 2009 Abel Laureate's contribution in computer vision and computer graphics* [Science Lecture]

2010

J.T. Tate (University of Texas at Austin): *The arithmetic of elliptic curves*
R. Taylor (Harvard University): *The Tate conjecture*
A. Enge (INRIA Bordeaux-Sud-Oust): *The queen of mathematics in communication security* [Science Lecture]

2011

J. Milnor (Stony Brook University): *Spheres*
C.T. McMullen (Harvard University): *Manifolds, topology and dynamics*
M. Hopkins (Johns Hopkins University): *Bernoulli numbers, homotopy groups, and Milnor*
É. Ghys (École Normale Supérieure de Lyon): *A guided tour of the seventh dimension* [Science Lecture]

2012

E. Szemerédi (Hungarian Academy of Sciences and Rutgers University): *In every chaos there is order*

L. Lovász (Eötvös Loránd University): *The many facets of the Regularity Lemma*

T. Gowers (University of Cambridge): *The afterlife of Szemerédi's theorem*

A. Wigderson (Institute for Advanced Study): *Randomness and pseudorandomness* [Science Lecture]

The Abel Laureate Presenters 2003–2012

In March each year when the President of the Norwegian Academy of Science and Letters announces the Abel Laureate and the Chair of the Abel Committee states the reasons for the selection, a mathematician presents the work of the Laureate. Below we list the presenters for the period 2003–2012:

2003 (J.-P. Serre) Arne B. Sletsjøe, University of Oslo
2004 (M.F. Atiyah and I.M. Singer) John Rognes, University of Oslo
2005 (P.D. Lax) Helge Holden, Norwegian University of Science and Technology
2006 (L. Carleson) Marcus du Sautoy, University of Oxford
2007 (S.R.S. Varadhan) Tom Lindstrøm, University of Oslo
2008 (J.G. Thompson and J. Tits) Marcus du Sautoy, University of Oxford
2009 (M.L. Gromov) Vagn Lundsgaard Hansen, Technical University of Denmark
2010 (J.T. Tate) Marcus du Sautoy, University of Oxford
2011 (J.W. Milnor) Timothy Gowers, University of Cambridge
2012 (E. Szemerédi) Timothy Gowers, University of Cambridge

H. Holden, R. Piene (eds.), *The Abel Prize 2008-2012*,
DOI 10.1007/978-3-642-39449-2, © Springer-Verlag Berlin Heidelberg 2014

The Interviews with the Abel Laureates

Transcripts of parts of the interviews that Martin Raussen (Aalborg University) and Christian Skau (Norwegian University of Science and Technology) made with each laureate in connection with the Prize ceremonies, can be found in the following publications:

2008 John G. Thompson and Jacques Tits

EMS Newsletter, issue 69 (Sep. 2008) 31–38,
AMS Notices, **56** (2009) 471–478.

2009 Mikhail Leonidovich Gromov

EMS Newsletter, issue 73 (Sep. 2009) 19–30,
AMS Notices, **57** (2010) 391–403.

2010 John Torrence Tate

EMS Newsletter, issue 77 (Sep. 2010) 41–48,
AMS Notices, **58** (2011) 444–452.

2011 John W. Milnor

EMS Newsletter, issue 81 (Sep. 2011) 31–40,
AMS Notices, **59** (2012) 400–408.

2012 Endre Szemerédi

EMS Newsletter, issue 85 (Sep. 2012) 39–48,
AMS Notices, **60** (2013) 221–231.

The interviews can be viewed from the Abel Prize web site www.abelprize.no or the Springer web site http://www.springerimages.com/videos/978-3-642-39449-2.

H. Holden, R. Piene (eds.), *The Abel Prize 2008-2012*,
DOI 10.1007/978-3-642-39449-2, © Springer-Verlag Berlin Heidelberg 2014

Addenda, Errata, and Updates[1]

2003 Jean-Pierre Serre

Citation:

"for playing a key role in shaping the modern form of many parts of mathematics, including topology, algebraic geometry and number theory"

(i) Publications

2005

[276] arXiv:math/0503154v6, June 9, 2008.

2009

[286] La vie et l'œuvre scientifique de Henri Cartan. *Gaz. Math.* No. 121, 65–70.
[287] Un complément à la note de Lassina Dembélé: "A non-solvable Galois extension of **Q** ramified at 2 only" [C. R. Acad. Sci. Paris, Ser. I 347 (2009)]. *C. R. Acad. Sci. Paris* 347:117–118.
[288] Le groupe quaquaversal, vu comme groupe S-arithmétique, *Oberwolfach Reports* 6:1421–1422.

2010

[289] A tribute to Henri Cartan. *Notices Amer. Math. Soc.* 57(8): 946–949.
[290] Le groupe de Cremona et ses sous-groupes finis. *Séminaire Bourbaki. Volume 2008/2009. Exposés 997–1011. Astérisque* 332:Exp. No. 1000, vii, 75–100.
[291] Henri Cartan. *École normale supérieure, L'Archicube* 7:66–69.

2012

[292] *Lectures on $N_X(p)$*. Research Notes in Mathematics 11. CRC Press, Boca Raton, FL.

[1] H. Holden, R. Piene (eds.) *The Abel Prize 2003–2007. The First Five Years*, Springer, Heidelberg, 2010.

H. Holden, R. Piene (eds.), *The Abel Prize 2008-2012*,
DOI 10.1007/978-3-642-39449-2, © Springer-Verlag Berlin Heidelberg 2014

[293] (with J.-L. Nicolas). Formes modulaires modulo 2: l'ordre de nilpotence des opérateurs de Hecke. *C. R. Acad. Sci. Paris* 350: 343–348.
[294] (with J.-L. Nicolas). Formes modulaires modulo 2: structure de l'algèbre de Hecke. *C. R. Acad. Sci. Paris* 350: 449–454.

2013

[295] (with E. Bayer-Fluckiger and R. Parimala). Hasse principle for G-trace forms. *Izvestjia RAS/Ser. Math.* 77(3), 5–28.
[296] Un critère d'indépendance pour une famille de représentations ℓ-adiques. *Comm. Math. Helv.* 88(3):541–554

(ii) Addendum CV

Leroy P. Steele Prize 1995 Mathematical Exposition
Torino Academy of Sciences 2010
Academia Sinica, Taiwan 2010
Fellow, American Mathematical Society 2013
ICCM International Cooperation Award, Taipei, 2013

(iii) Articles in Connection with the Abel Prize

A. Chambert-Loir: Le prix Abel décerné à Jean-Pierre Serre. [French]. *Gaz. Math.* No. 99 (2004), 26–32.
G. Frei: Erstmalige Verleihung des Abel-Preises—Auszeichnung von Jean-Pierre Serre für sein Gesamtwerk. [German] *Mitt. Dtsch. Math.-Ver.* 2003, no. 2, 22–25.
N. Vila: Jean-Pierre Serre, primer Premi Abel. [Catalan] *SCM Not.* No. 19 (2003), 22–25.
R. Betti: A "Nobel" for Jean-Pierre Serre: the Abel Prize. [Italian] *Lett. Mat. Pristem* No. 47 (2003), 4.

2004 Sir Michael Atiyah and Isadore M. Singer

Citation:

"for their discovery and proof of the index theorem, bringing together topology, geometry and analysis, and their outstanding role in building new bridges between mathematics and theoretical physics"

(i) Publications by M. Atiyah

1996

[194a] Address of the President, Sir Michael Atiyah, O.M., given at the anniversary meeting on 30 November 1995. *Notes and Records Roy. Soc. London* 50(1) 101–113.

2007

[246] Published in *Biogr. Mem. Fellows R. Soc.* 53:63–76. Also published in *Bull. Lond. Math. Soc.* 42(1): 170–180 (2010).

2009

[253] From Probability to Geometry (I). Volume in Honor of the 60th Birthday of Jean-Michel Bismut. *Astérisque* No. 327 (2009), xvii (2010). Preface by Sir Michael Atiyah.

2010

[254] A tribute to Henri Cartan. *Notices Amer. Math. Soc.* 57(8): 946–949.
[255] (with R. Dijkgraaf and N. Hitchin). Geometry and physics. *Philos. Trans. R. Soc. Lond. Ser. A* 368:1914, 913–926.
[256] The art of mathematics. *Notices Amer. Math. Soc.* 57(1):8.
[257] Edinburgh Lectures on Geometry, Analysis and Physics. arXiv:1009.4827.
[258] Mathematical work of Nigel Hitchin. In *The Many Facets of Geometry. A Tribute to Nigel Hitchin*, Oxford Univ. Press, pages 11–16, Oxford.
[259] Working with Raoul Bott: from geometry to physics. In *A celebration of the mathematical legacy of Raoul Bott*, volume 50 of *CRM Proc. Lecture Notes*, Amer. Math. Soc., pages 51–61, Providence, RI.

2011

[260] (with G.W. Moore). A shifted view of fundamental physics. In: *Perspectives in Mathematics and Physics: Essays Dedicated to Isadore Singer's 85th Birthday* (T. Mrowka, ed.). International Press, Somerville, MA, pages 1–15.
[261] (with S.-T. Yau et al.). Shiing-Shen Chern (1911–2004). *Notices Amer. Math. Soc.* 58(9):1226–1249.
[262] (with V. Guillimin et al.). Remembering Johannes J. Duistermaat (1942–2010). *Notices Amer. Math. Soc.* 58(6):794–802.

2012

[263] (with N.S. Manton and B.J. Schroers). Geometric models of matter. *Proc. R. Soc. Lond. Ser. A Math. Phys. Eng. Sci.* 468, no. 2141, 1252–1279.
[264] (with M. Sanz-Solé, C. Bär, G.-M. Greuel, Y.I. Manin and J.-P. Bourguignon). Friedrich Hirzebruch memorial session at the 6th European Congress of Mathematics. Kraków, July 5th, 2012. Eur. Math. Soc. Newsl. No. 85, 12–20.
[265] (with C. LeBrun). Curvature, cones, and characteristic numbers. arXiv:1203.6389

(ii) Addendum CV

Grande Médaille of the French Academie des Sciences 2010
Grand Officier of the French Légion d'honneur 2011
Honorary degree, Hong Kong University of Science and Technology 2012
Fellow, American Mathematical Society 2013

Publications by I.M. Singer

2009

[107] (with V. Mathai, R.B. Melrose). The index of projective families of elliptic operators: the decomposable case. *Astérisque* No. 328: 255–296.

2011

[108] (with O. Alvarez). Beyond the string genus. *Nuclear Phys. B* 850(2): 349–386.

[109] (with S.-T. Yau et al.). Shiing-Shen Chern (1911–2004). *Notices Amer. Math. Soc.* 58(9):1226–1249.

(ii) Addendum CV

Fellow, American Mathematical Society 2013

(iii) Articles in Connection with the Abel Prize

K. Landsman: Abel Prize 2004: The Atiyah–Singer index theorem. [Dutch] *Nieuw Arch. Wiskd. (5) 5,* No. 3, 207–211 (2004).
S.-T. Yau (ed.): *The Founders of Index Theory: Reminiscences of and about Sir Michael Atiyah, Raoul Bott, Friedrich Hirzebruch, and I.M. Singer*. International Press, Somerville, MA, 2nd ed., 2009. lii+393 pp.
Pure and Applied Mathematics Quarterly, Volume 6, Number 2 (2010). Special Issue: In Honor of Michael Atiyah and Isadore Singer.

2005 Peter D. Lax

Citation:

"for his groundbreaking contributions to the theory and application of partial differential equations and to the computation of their solutions"

(i) Publications

1983

[132] Published in 1976, not in 1983. Second edition (with M. S. Terrell). *Calculus With Applications*, Springer-Verlag, New York, 2014.

2009

[228] Rethinking the Lebesgue integral. *Amer. Math. Monthly* 116(10):863–881.

2010

[229] (with F. Hirzebruch, B. Mazur, L. Conlon, E.B. Curtis, H.M. Edwards, J. Huebschmann and H. Shulman). Raoul Bott as we knew him. In: *A Celebration of the Mathematical Legacy of Raoul Bott.* (P.R. Kotiuga, ed.), CRM Proc. Lecture Notes, Vol. 50, American Mathematical Society, Providence, RI, pages 43–49.

2011

[230] (with E. Falbel, G. Francsics, J.R. Parker). Generators of a Picard modular group in two complex dimensions. *Proc. Amer. Math. Soc.* 139(7): 2439–2447.

2012

[231] (with L. Zalcman). *Complex Proofs of Real Theorems.* University Lecture Series, vol. 58, American Mathematical Society, Providence, RI

2013

[232] Stability of difference schemes. In: *The Courant–Friedrichs–Lewy (CFL) Condition. 80 Years After Its Discovery.* (C.A. de Moura, C.S. Kubrusly, eds.), Birkhäuser, pages 1–7.

(ii) Update CV

Fellow, Society for Industrial and Applied Mathematics 2009
Fellow, American Mathematical Society 2013

Errata for article "*A Survey of Peter D. Lax's Contributions to Mathematics*": pages 210–211: Some of the citations in the text are incorrect: [printed] ([should read]): [L116] ([L120]); [L195] ([L212]); [L199] ([L214]); [L200] ([L218]); [L187] ([L205]).

(iii) Articles in Connection with the Abel Prize

V. Mañosa: Peter Lax, 2005 Abel Prize. [Catalan] *SCM Not.* No. 21 (2005), 54–58.
S. Noelle: Abelpreis 2005 an Peter D. Lax. [German] *Mitt. Dtsch. Math.-Ver.* 13 (2005), no. 2, 84–89.
D. Serre: Le Prix Abel 2005 à Peter Lax. [French] *Matapli* No. 77 (2005), 8–14.
H. Holden: Peter D. Lax. Abelprisvinner 2005 [Norwegian] *Normat* 53 (2005) 145–154.

2006 Lennart Carleson

Citation:

"for his profound and seminal contributions to harmonic analysis and the theory of smooth dynamical systems"

(i) Publications

2007

[71] G.P. Curbera. *Mathematicians of the world, unite! The International Congress of Mathematicians—a human endeavor.* A.K. Peters, Ltd., Wellesley, MA. With a foreword by L. Carleson.

(ii) Addendum CV

Fellow, American Mathematical Society 2013

(iii) Articles in Connection with the Abel Prize

Lennart Carleson erhält den Abel-Preis. [German] *Mitt. Dtsch. Math.-Ver.* 14 (2006), no. 2, 87–88.
H. Duistermaat: Abel prize winner 2006; Lennart Carleson: achievements until now. *Nieuw Arch. Wiskd.* (5) 8 (2007), no. 3, 175–177.
J. Ortega-Cerdà, J.C. Tatjer: Lennart Carleson, 2006 Abel Prize. [Catalan] *Butl. Soc. Catalana Mat.* 22 (2007), no. 2, 153–164, 230 (2008).
A. Vargas: 2006 Abel Prize: Lennart Carleson. [Catalan] SCM Not. No. 23 (2007), 63–65.

2007 S.R. Srinivasa Varadhan

Citation:

"for his fundamental contributions to probability theory and in particular for creating a unified theory of large deviation"

(i) Publications

2008

[130] Published in 2009, not in 2008.

2009

[131] Scaling limits. In: *Perspectives in Mathematical Sciences. I,* Stat. Sci. Interdiscip. Res., vol. 7, World Sci. Publ., Hackensack, NJ, pages 247–262.
[132] Workshop on large deviations: lecture notes. *Bull. Kerala Math. Assoc. Special Issue*, pages 1–14.
[133] The role of weak convergence in probability theory. In: *Symmetry in Mathematics and Physics*, Contemp. Math., vol. 490, Amer. Math. Soc., Providence, RI, pages 3–10.

2010

[134] (with D. Stroock). Theory of diffusion processes. In: *Stochastic Differential Equations*, C.I.M.E. Summer School, 77, Springer, Heidelberg, pages 149–191.
[135] Large deviations. In: *Proceedings of the International Congress of Mathematicians. Volume I.* (R. Bhatia et al., eds.). Hindustan Book Agency, New Delhi, pages 622–639.
[136] (with Y. Kifer). Nonconventional limit theorems in discrete and continuous time via martingales. arXiv:1012.2223.

2011

[137] (with S. Chatterjee). The large deviation principle for the Erdős–Rényi random graph. *European J. Combin.* 32(7):1000–1017.

2012

[138] (with S. Chatterjee). Large deviations for random matrices. *Commun. Stoch. Anal.* 6, no. 1, 1–13.

[139] Large deviations for stochastic processes, *Bull. Amer. Math. Soc.* 49, no. 4, 597–601.

[140] (with Y. Kifer). Nonconventional large deviations theorems. arXiv:1206.0156.

[141] (with A. Dembo, M. Shkolnikov and O. Zeitouni). Large deviations for diffusions interacting through their ranks. arXiv:1211.5223.

2013

[142] *Prokhorov and contemporary probability theory.* In honor of Yuri V. Prokhorov. A. Shiryaev, S.R.S. Varadhan and E.L. Presman (editors). Springer Proceedings in Mathematics & Statistics 33.

[143] *Collected Papers of S.R.S. Varadhan.* Volume 1: Limit Theorems, Review Articles. Volume 2: PDE, SDE, Diffusions, Random Media. Volume 3: Large Deviations. Volume 4: Particle Systems and Their Large Deviations. (R. Bhatia, A. Bhatt, K.R. Parthasarathy, eds.) Springer Verlag and Hindustan Book Agency. 2690 pp.

(ii) Update CV

Leroy P. Steele Prize: Seminal Contribution to Research (joint with D. Stroock) 1996
Padma Bhushan, India, 2008
National Medal of Science (US) 2010
Fellow, Society for Industrial and Applied Mathematics 2009
Fellow, American Mathematical Society 2013

(iii) Articles in Connection with the Abel Prize

E. Bolthausen, A.-S. Sznitman: Zur Verleihung des Abel-Preises 2007 an S.R. Srinivasa Varadhan. [German] *Mitt. Dtsch. Math.-Ver.* 15 (2007), no. 3, 173–175.
F. den Hollander: Srinivasa Varadhan. [Dutch] *Nieuw Arch. Wiskd.* (5) 9 (2008), no. 3, 192–196.
S. Ramasubramanian: Large deviations: An introduction to 2007 Abel prize. [English] *Proc. Indian Acad. Sci., Math. Sci.* 118 (2008) No. 2, 161–182.
On the award of the Abel Prize to S.R.S. Varadhan. [Russian] Teor. Veroyatn. Primen. 52 (2007), no. 3, 417–418. (English translation in Theory Probab. Appl. 52 (2008), no. 3, 371.)

CPSIA information can be obtained at www.ICGtesting.com
Printed in the USA
LVOW01*0035290714

396455LV00002B/37/P

9 783642 394485